# Bifidobacteria: Microbiological Aspects

# Bifidobacteria: Microbiological Aspects

Edited by Melissa Fox

AMERICAN
MEDICAL PUBLISHERS
www.americanmedicalpublishers.com

American Medical Publishers,
41 Flatbush Avenue,
1st Floor, New York,
NY 11217, USA

Visit us on the World Wide Web at:
www.americanmedicalpublishers.com



ISBN: 978-1-63927-010-1



**Cataloging-in-Publication Data**

Bifidobacteria : microbiological aspects / edited by Melissa Fox.
p. cm.
Includes bibliographical references and index.
ISBN 978-1-63927-010-1
1. Bifidobacterium. 2. Actinomycetaceae. 3. Microbiology. 4. Bacteriology. I. Fox, Melissa.
QR82.A35 B54 2022
589.92--dc23

# Table of Contents

# Preface

Bifidobacteria is a genus of bacteria that dwells inside the intestines and stomach of mammals. They are classified as probiotic and assist the body in digestion and repelling harmful bacteria. They also helps in regulating the intestinal microbial homeostasis, the modulation of local and systemic immune responses, the inhibition of harmful bacteria and pathogens that infect the gut mucosa. They can be grown outside the body and then taken orally as medicine. Bifidobacteria are commonly used for the treatment of diarrhea, constipation, irritable bowel syndrome and the common cold. This book unravels the recent studies about Bifidobacteria. The topics included in this book on Bifidobacteria are of utmost significance and bound to provide incredible insights to readers. Those in search of information to further their knowledge will be greatly assisted by this book.

This book unites the global concepts and researches in an organized manner for a comprehensive understanding of the subject. It is a ripe text for all researchers, students, scientists or anyone else who is interested in acquiring a better knowledge of this dynamic field.

I extend my sincere thanks to the contributors for such eloquent research chapters. Finally, I thank my family for being a source of support and help.

**Editor**

# 1

# Bifidobacteria and their Role as Members of the Human Gut Microbiota

*Amy O'Callaghan and Douwe van Sinderen* *

*Alimentary Pharmabiotic Centre and School of Microbiology, University College Cork, Cork, Ireland*

Members of the genus *Bifidobacterium* are among the first microbes to colonize the human gastrointestinal tract and are believed to exert positive health benefits on their host. Due to their purported health-promoting properties, bifidobacteria have been incorporated into many functional foods as active ingredients. Bifidobacteria naturally occur in a range of ecological niches that are either directly or indirectly connected to the animal gastrointestinal tract, such as the human oral cavity, the insect gut and sewage. To be able to survive in these particular ecological niches, bifidobacteria must possess specific adaptations to be competitive. Determination of genome sequences has revealed genetic attributes that may explain bifidobacterial ecological fitness, such as metabolic abilities, evasion of the host adaptive immune system and colonization of the host through specific appendages. However, genetic modification is crucial toward fully elucidating the mechanisms by which bifidobacteria exert their adaptive abilities and beneficial properties. In this review we provide an up to date summary of the general features of bifidobacteria, whilst paying particular attention to the metabolic abilities of this species. We also describe methods that have allowed successful genetic manipulation of bifidobacteria.

**Keywords: *Bifidobacterium*, carbohydrate metabolism, genetic modification, probiotics, microbe-host interaction**

**Edited by:**
*Marco Ventura, University of Parma, Italy*

**Reviewed by:**
*Susana Delgado, Instituto de Productos Lácteos de Asturias-Consejo Superior Investigaciones Científicas, Spain*
*Christian Milani, Instituto de Productos Lácteos de Asturias - CSIC, Spain*

***Correspondence:**
*Douwe van Sinderen*
*d.vansinderen@ucc.ie*

## INTRODUCTION

The past 20 years has seen a research focus on those members of the gut microbiota that exhibit health-promoting or probiotic effects such as protection of the host against pathogens by competitive exclusion (Bernet et al., 1994; Hooper et al., 1999), modulation of the immune system (O'Hara and Shanahan, 2007), and provision of nutrients through the breakdown of non-digestible dietary carbohydrates (Roberfroid et al., 1995; Leahy et al., 2005). Furthermore, compositional alterations of the gastrointestinal tract (GIT) microbiota have been linked to certain gastrointestinal diseases such as inflammatory bowel disease (Ott et al., 2004) and necrotizing enterocolitis (De La Cochetiere et al., 2004). Particular interest has focused on members of the genus *Bifidobacterium*, some of which have been included as live components in a variety of so-called functional foods (Ventura et al., 2004). Bifidobacteria were first isolated from the feces of breast-fed infants in 1899 by Tissier and since then bifidobacteria have been isolated from a range of different ecological niches such as the oral cavity, sewage and the insect gut, the GIT of various mammals and more recently from water kefir (Klijn et al., 2005; Ventura et al., 2007; Laureys et al., 2016).

Although, it has been well established that bifidobacteria confer positive health benefits to the human host, there is a clear lack of knowledge concerning the molecular mechanisms that explain these probiotic traits of *Bifidobacterium* (Cronin et al., 2011). Deciphering whole genome sequences can shed light on the genetic basis of the probiotic action of bifidobacteria, or indeed the associated molecular adaptations that allow this gut commensal to take up residency in its highly competitive ecological niche (Ventura et al., 2014). Although, a significant sequencing effort of bifidobacterial genomes has generated a very extensive set of genomic data, yet this genomic information has hardly been explored at the functional level due to a lack of tools to make bifidobacteria genetically accessible (Serafini et al., 2012).

## GENERAL FEATURES OF BIFIDOBACTERIAL GENOMES

Since the publication of the first bifidobacterial genome in 2002, there has been a steady increase in the number of publicly available bifidobacterial genome sequences (Lee et al., 2008). The NCBI data base currently (April 2016) holds 254 publicly available bifidobacterial genome sequences, of which sixty one represent complete genome sequences (**Table** , source; http://www.ncbi.nlm.nih.gov/Taxonomy/Browser/wwwtax.cgi?id =1678NCBI, April 2016). Three or more complete genome sequences are available for certain bifidobacterial species, such as for *B. adolescentis, B. animalis, B. breve, B. bifidum, B. longum,* and *B. angulatum* (**Table** ).

The average size of a bifidobacterial genome is 2.2 Mb, although there is considerable size variation, for example *B. indicum* LMG11587 harbors a genome with a size of 1.73 Mb, wheras *B. scardovii* JCM12489 possesses a genome of 3.16 Mb in length. Bifidobacterial genomes typically encode 52–58 tRNA genes per genome, although there are exceptions, e.g., the genome of *B. longum* subsp. *infantis* ATCC15697 encompasses 79 tRNA-encoding genes. The number of rRNA operons within bifidobacterial genomes typically ranges from two to five, and it has been suggested that the number of rRNA operons present on a genome is correlated to the adaptation of a particular species to environmental conditions (Klappenbach et al., 2000). The G+C content of complete bifidobacterial genomes ranges from 59.2% (*B. adolescentis*) to 64.6% (*B. scardovii*), while the average gene number contained by a bifidobacterial genome is 1825 (**Table** ). The three species *B. indicum, B. coryneforme,* and *B. animalis* possess the lowest number of genes, consistent with their small genome size (Lee et al., 2008; Ventura et al., 2014).

## IMPACT ON HEALTH AND DISEASE

A diverse microbial community has evolved to adapt and survive in the human GIT and is commonly referred to as the gut microbiota (Guarner and Malagelada, 2003). The large intestine can contain up to $10^{12}$ bacterial cells/g of luminal content making this the most densely populated area of the gastrointestinal tract (Simon and Gorbach, 1984). Members of the gut microbiota interact with their (human) host in a variety of ways, thereby making them innocuous commensals, opportunistic pathogens or health-promoting or probiotic micro organisms (Guarner and Malagelada, 2003). Probiotics are defined as "live microorganisms that, when administered in adequate amounts, confer a health benefit on the host" (FAO/WHO, 2001; Hill et al., 2014), and research into the activities of purported health-promoting bacteria has increased substantially over the last 20 years (Leahy et al., 2005). Probiotic agents have been investigated in many clinical and animal model-based studies; however, we will summarize just a limited number of studies that specifically relate to bifidobacteria. Bifidobacteria have been commercially exploited as probiotic agents due to their associated health benefits and GRAS (Generally Recognised As Safe) status (Picard et al., 2005).

### Bifidobacteria and Colorectal Cancer

Several studies have investigated the potential of bifidobacteria to prevent and/or treat colorectal cancer. The majority of studies base their findings on murine models, and results suggest that a combination of prebiotics and bifidobacteria may reduce the occurrence of carcinogen-induced cancerous cells in mice (Sekine et al., 1985; Rowland et al., 1998; Rafter et al., 2007; Le Leu et al., 2010). For example, it was shown that *B. animalis* displays anti-mutagenic activity during growth in MRS broth thereby antagonizing the action of the carcinogen 2-amino-3-methylimidazo [4, 5-f] quinolone (Tavan et al., 2002). It has also been demonstrated under *in vivo* and *in vitro* conditions that a *B. longum* and a *B. breve* strain provide protection of DNA from induced damage by carcinogens, and inhibit the genotoxic effect of two different carcinogens when tested in a rat model (Pool-Zobel et al., 1996).

### Bifidobacteria and Diarrhoea

The use of bifidobacteria to treat various gastrointestinal disorders has also been reported. For example, successful treatment of diarrhea following administration of *B. longum* subsp. *infantis* CECT 7210 and *B. breve* K-110 was found to be due to inhibition of rotavirus, the predominant cause of sporadic diarrhea in infants (Bae et al., 2002; Chenoll et al., 2015). Another example involves a double-blind study investigating whether oral treatment with a commercial probiotic formula containing *B. bifidum* and *Streptococcus thermophiles* would reduce antibiotic-associated diarrhea in infants. This study found that there was a significant reduction in incidences of diarrhea for those infants fed the probiotic supplemented formula supplemented (Corrêa et al., 2005).

### Bifidobacteria and Necrotizing Entercolitis

A recent study reported lower incidences of necrotizing enterocolitis in preterm neonates following routine administration of *B. breve* M-16V (Patole et al., 2016). Administration of *B. breve* M-16V in association with breast-feeding was shown to be associated with a lower incidence of necrotising enterocolitis in neonates born before 34 weeks gestation, and, although not statistically significant, a lower incidence in this disease was reported for neonates born at a gestation age of less than 28 weeks (Patole et al., 2016).

**TABLE | Summary of all completely sequenced bifidobacterial genomes.**

| Microorganism | Genome Size (Mb) | Number of genes | G+C content (%) | tRNA | rRNA | GenBank |
|---|---|---|---|---|---|---|
| *B.actinocoloniiforme* DSM 22766 | 1.83 | 1502 | 62.7 | 47 | 6 | CP011786.1 |
| *B.adolescentis* ATCC 15703 | 2.09 | 1721 | 59.2 | 54 | 16 | AP009256.1 |
| *B.adolescentis* 22L | 2.2 | 1798 | 59.3 | 54 | 13 | CP007443.1 |
| *B.adolescentis* BBMN23 | 2.17 | 1812 | 59.3 | 55 | 13 | CP010437.1 |
| *B.angulatum* DSM20098 | 2.02 | 1615 | 59.4 | 53 | 12 | AP012322.1 |
| *B.angulatum* GT102 | 2.06 | 1651 | 59.3 | 53 | 3 | CP014241.1 |
| *B.animalis* subsp. *lactis* AD011 | 1.93 | 1615 | 60.5 | 52 | 7 | CP001213.1 |
| *B.animalis* subsp. *lactis* Bl-04 | 1.94 | 1608 | 60.5 | 52 | 12 | CP001515.1 |
| *B.animalis* subsp. *lactis* DSM10140 | 1.94 | 1607 | 60.5 | 51 | 12 | CP001606.1 |
| *B.animalis* subsp. *lactis* BB-12 | 1.94 | 1611 | 60.5 | 52 | 12 | CP001853.1 |
| *B.animalis* subsp. *lactis* V9 | 1.94 | 1610 | 60.5 | 52 | 12 | CP001892.1 |
| *B.animalis* subsp. *lactis* CNCM I-2494 | 1.94 | 1611 | 60.5 | 52 | 12 | CP002915.1 |
| *B.animalis* subsp. *lactis* BLC1 | 1.94 | 1608 | 60.5 | 52 | 12 | CP003039.2 |
| *B.animalis* subsp. *animalis* ATCC25527 | 1.93 | 1583 | 60.5 | 52 | 11 | CP002567.1 |
| *B.animalis* subsp. *lactis* B420 | 1.94 | 1610 | 60.5 | 52 | 12 | CP003497.1 |
| *B.animalis* subsp. *lactis* Bi-07 | 1.94 | 1608 | 60.5 | 52 | 12 | CP003498.1 |
| *B.animalis* subsp. *lactis* Bl12 | 1.94 | 1608 | 60.5 | 52 | 12 | CP004053.1 |
| *B.animalis* subsp. *lactis* ATCC27673 | 1.96 | 1624 | 60.6 | 52 | 12 | CP003941.1 |
| *B.animalis* RH | 1.93 | 1606 | 60.5 | 52 | 8 | CP007755.1 |
| *B.animalis* subsp. *lactis* KLDS2.0603 | 1.95 | 1610 | 60.5 | 52 | 15 | CP007522.1 |
| *B.animalis* A6 | 1.96 | 1623 | 60.5 | 52 | 16 | CP010433.1 |
| *B.animalis* subsp. *lactis* BF052 | 1.94 | 1608 | 60.5 | 52 | 12 | CP009045.1 |
| *B.*asteroides PRL2011 | 2.17 | 1727 | 60.1 | 45 | 6 | CP003325.1 |
| *B.bifidum* PRL2010 | 2.21 | 1791 | 62.7 | 52 | 9 | CP001840.1 |
| *B.bifidum* S17 | 2.19 | 1819 | 62.8 | 53 | 9 | CP002220.1 |
| *B.bifidum* BGN4 | 2.22 | 1832 | 62.6 | 52 | 9 | CP001361.1 |
| *B.bifidum* ATCC29521 | 2.21 | 1838 | 62.7 | 54 | 6 | AP012323.1 |
| *B.bifidum* BF3 | 2.21 | 1813 | 62.6 | 52 | 9 | CP010412.1 |
| *B.breve* UCC2003 | 2.42 | 2049 | 58.7 | 54 | 6 | CP000303.1 |
| *B.breve* ACS-071-V-Sch8b | 2.33 | 1956 | 58.7 | 53 | 9 | CP002743.1 |
| *B.breve* 12L | 2.24 | 1883 | 58.9 | 52 | 6 | CP006711.1 |
| *B.breve* JCM7017 | 2.29 | 1916 | 58.7 | 54 | 6 | CP006712.1 |
| *B.breve* JCM7019 | 2.36 | 2045 | 58.6 | 56 | 6 | CP006713.1 |
| *B.breve* NCFB2258 | 2.32 | 1946 | 58.7 | 53 | 6 | CP006714.1 |
| *B.breve* 689b | 2.33 | 1970 | 58.7 | 53 | 6 | CP006715.1 |
| *B.breve* S27 | 2.29 | 1926 | 58.7 | 53 | 9 | CP006716.1 |
| *B.breve* DSM20213 | 2.27 | 1973 | 58.9 | 53 | 6 | AP012324.1 |
| *B.breve* BR3 | 2.42 | 2232 | 59.1 | 54 | 9 | CP010413.1 |
| *B.catenulatim* DSM16992 | 2.08 | 1717 | 56.2 | 55 | 16 | AP012325.1 |
| *B.coryneforme* LMG18911 | 1.76 | 1423 | 60.5 | 46 | 9 | CP007287.1 |
| *B.dentium* Bd1 | 2.64 | 2177 | 58.5 | 56 | 13 | CP001750.1 |
| *B.dentium* JCM1195 | 2.64 | 2177 | 58.5 | 56 | 13 | AP012326.1 |
| *B.indicum* LMG11587 | 1.73 | 1403 | 60.5 | 47 | 9 | CP006018.1 |
| *B.kashiwanohense* PV20-2 | 2.37 | 2007 | 56.1 | 58 | 16 | CP007456.1 |
| *B.kashiwanohense* JCM15439 | 2.34 | 1965 | 56.3 | 54 | 16 | AP012327.1 |
| *B.longum* NCC2705 | 2.26 | 1797 | 60.1 | 57 | 12 | AE014295.3 |
| *B.longum* DJO10A | 2.38 | 1998 | 60.1 | 58 | 12 | CP000605.1 |
| *B.longum* subsp. infantis ATCC15697 | 2.83 | 2594 | 59.9 | 79 | 12 | CP001095.1 |
| *B.longum* subsp. *longum* JDM301 | 2.48 | 2062 | 59.8 | 55 | 9 | CP002010.1 |
| *B.longum* subsp. *longum* BBMN68 | 2.27 | 1873 | 59.9 | 54 | 9 | CP002286.1 |
| *B.longum* subsp. *longum* JCM1217 | 2.39 | 2001 | 60.3 | 73 | 12 | AP010888.1 |

*(Continued)*

**TABLE | Summary of all completely sequenced bifidobacterial genomes.**

| Microorganism | Genome Size (Mb) | Number of genes | G+C content (%) | tRNA | rRNA | GenBank |
|---|---|---|---|---|---|---|
| *B.longum* subsp. infantis 157F | 2.4 | 2044 | 60.1 | 59 | 12 | AP010890.1 |
| *B.longum* subsp. *longum* KACC91563 | 2.39 | 1979 | 59.8 | 56 | 9 | CP002794.1 |
| *B.longum* BXY01 | 2.48 | 2065 | 59.8 | 55 | 9 | CP008885.1 |
| *B.longum* subsp. *longum* GT15 | 2.34 | 1947 | 60 | 56 | 14 | CP006741.1 |
| *B.longum* 105-A | 2.29 | 1874 | 60.1 | 56 | 12 | AP014658.1 |
| *B.longum* subsp. infantis BT1 | 2.58 | 2399 | 59.4 | 56 | 9 | CP010411.1 |
| *B.longum* BG7 | 2.45 | 2116 | 60 | 57 | 9 | CP010453.1 |
| *B.longum* subsp. *longum* NCIMB8809 | 2.34 | 1959 | 60.1 | 56 | 9 | CP011964.1 |
| *B.longum* subsp. *longum* CCUG30698 | 2.45 | 2106 | 60.2 | 57 | 6 | CP011965.1 |
| *B.pseudocatenulatum* DSM20438 | 2.31 | 1864 | 56.4 | 54 | 19 | AP014658.1 |
| *B.pseudolongum* PV8-2 | 2.03 | 1704 | 63.3 | 53 | 12 | CP007457.1 |
| *B.scardovii* JCM12489 | 3.16 | 2418 | 64.6 | 56 | 9 | AP012331.1 |

### Bifidobacteria and Inflammatory Bowel Disease

Although, the exact mechanism of action is not understood, reduction in the symptoms of inflammatory bowel disease following treatment by probiotic strains has been reported (Venturi et al., 1999). Patients suffering from ulcerative colitis were given a probiotic preparation that includes three *Bifidobacterium* strains, four *Lactobacillus* strains and one *S. thermophilus* strain. Fifteen out of the 20 patients remained in remission throughout the trial, suggesting that administration of this bacterial cocktail is beneficial in maintaining remission from ulcerative colitis (Venturi et al., 1999; Gionchetti et al., 2000).

### Bifidobacteria and Colon Regularity

A number of studies have reported improvements in colon regularity following ingestion of fermented milk products that contain *B. animalis* (Marteau et al., 2002; Guyonnet et al., 2007; Meance et al., 2011). Two studies have associated the administration of certain bifidobacterial strains with the alleviation of constipation (Kumemura et al., 1992; Kleessen et al., 1997). However, further investigation is needed in order to identify the precise mechanism(s) of action elicited by bifidobacteria in the prevention and treatment of constipation (Leahy et al., 2005).

### Bifidobacteria and Competitive Exclusion

Bifidobacteria have also been reported to prevent gastrointestinal infections by competitive exclusion of pathogens based on common binding sites on epithelial cells (Duffy et al., 1994a,b; Perdigon et al., 1995; Picard et al., 2005; Gueimonde et al., 2007). Administration of high levels of bifidobacteria was shown to decrease the viable counts of *Clostridium perfringens*, a known producer of undesirable toxins (Tanaka et al., 1983).

## BIFIDOBACTERIA AND FUNCTIONAL FOODS

The inclusion of micro-organisms in the human diet has been on-going for thousands of years (Leahy et al., 2005). Throughout history the most common form of administration of microorganisms was through fermented dairy products and this is still the case today (Leahy et al., 2005). Certain lactic acid bacteria, in particular certain members of the genus *Lactobacillus*, and members of the *Bifidobacterium* genus make up the vast majority of the functional ingredients present in currently commercialized probiotic food products (Salminen and Wright, 1998; Ouwehand et al., 2002). Prebiotics have been defined as "selectively fermented ingredients that allow for specific changes, both in the composition and/or activity of the gastrointestinal microflora that confer benefits upon host well-being and health" (Hijova et al., 2009). This definition has been revisited several times since it was first introduced in 1995, although these alternative definitions are in agreement that prebiotics need to be "specific" or "selective" (Gibson and Roberfroid, 1995; Roberfroid et al., 2010; Rastall and Gibson, 2015). In a recent review the definition of prebiotics was revisited and proposed as follows: "a prebiotic is a non-digestible compound that, through its metabolisation by microorganisms in the gut, modulates composition and/or activity of the gut microbiota, thus conferring a beneficial physiological effect on the host" (Bindels et al., 2015).

The newly proposed definition moves away from the requirement of "specific effect" and puts forward the arguments that: (i) our knowledge does not allow for a reliable differentiation between beneficial and detrimental members of the microbiota, (ii) a diverse community is essential for intestinal homeostasis and host physiology, (iii) the metabolic benefits assigned to prebiotics do not require a "selective" fermentation, and (iv) community-wide molecular approaches have revealed that established prebiotics are not as specific as previously assumed (Bindels et al., 2015).

One outcome from the fermentation of prebiotics by the gut microbiota is the production of short chain fatty acids (SCFAs), such as acetate, butyrate and propionate (Broekaert et al., 2011). SCFA production in the GIT results in a lower pH, improved availability of calcium and magnesium, and inhibition of potentially pathogenic bacteria (Teitelbaum and Walker, 2002; Wong et al., 2006). Both bifidobacteria and lactobacilli produce acetate (and lactate), thus contributing to the SCFA-mediated health effects of prebiotics, although these two microorganisms do not produce butyrate and/or propionate (Fukuda et al., 2011; Bindels et al., 2015). The latter SCFAs are produced by members of the *Bacteroides* phylum and the *Clostridium* clusters XIVa and IV (Louis et al., 2010; Reichardt et al., 2014; Bindels et al., 2015). Furthermore, a recent study has demonstrated that acetate produced by *B. longum* NCC2705 acts as an essential co-substrate for butyrate production and growth by *Eubacterium rectale* ATCC 33656 (Rivière et al., 2015).

Non-digestible oligosaccharides (NDOs), typically obtained from complex carbohydrates or enzymatically produced from disaccharides, represent a group of glycans that include various prebiotics (Grootaert et al., 2007). Examples of this are fructo-oligosaccharides (FOS) and galacto-oligosaccharides (GOS), which are among the best documented and most commonly used prebiotics on the European and Japanese markets (Grootaert et al., 2007). The prebiotic effects of FOS, GOS, inulin and lactulose have been thoroughly assessed in human trials and many studies suggest that these carbohydrates are selective by increasing bifidobacterial numbers and decreasing the numbers of *E. coli* and enterococci (Menne et al., 2000; Kolida et al., 2002; Bosscher et al., 2006; Kapiki et al., 2007; Davis et al., 2010; Veereman-Wauters et al., 2011; Walton et al., 2012).

Due to the professed prebiotic effects of arabinoxylan (AX) and its derivatives arabinoxylo-oligosaccharides (AXOS) and xylo-oligosaccharides (XOS), these carbohydrates have in recent times enjoyed increasing scientific interest (Broekaert et al., 2011). The bifidogenic effect of AX has been confirmed in a number of *in vitro* studies (Van Laere et al., 2000; Crittenden et al., 2002), while the ability of bifidobacteria to metabolize XOS and AXOS in pure culture has also been demonstrated (Jaskari et al., 1998; Van Laere et al., 2000; Crittenden et al., 2002; Palframan et al., 2003; Moura et al., 2007). AXOS consumption amongst members representing eleven different bifidobacterial species suggests that AXOS metabolism is strain dependent and rather complex (Riviere et al., 2014). In this study, five different AXOS utilization clusters were identified based on principal component analysis of the different arabinose substituent and/or xylose backbone consumption patterns. The first and largest cluster (Cluster I) was composed of 15 different strains representing seven different species (*B. adolescentis, B. angulatum, B. bifidum, B. breve, B. dentium, B. longum,* and *B. thermophilum*). Strains within this cluster were unable to utilize the substitutions or xylan backbone of AXOS, although some strains were able to utilize the monosaccharides xylose and arabinose. Cluster II was composed of eight *B. longum* strains that were unable to utilize the xylan backbone, yet were able to utilize the arabinose substitutions on AXOS (both mono- and di-substituted), as well as the arabinose and xylose monosaccharides. Members of the third cluster (Cluster III), encompassing 10 strains representing six different species (*B. adolescentis, B. angulatum, B. longum, B. animalis, B. gallicum,* and *B. pseudolongum*), were shown to metabolize the xylan backbone of AXOS, albeit only up to xylotetraose, while eliciting no or limited utilization of the AXOS substitutions. Cluster IV contains two *B. longum* strains that share the ability to completely utilize AXOS, whereas the only member of Cluster V, *B. catenulatum* LMG 11043, was shown to display non-preferential degradation of XOS and a broad degradation pattern of arabinose substitutions (Riviere et al., 2014). A study investigating the *in vitro* fermentation of wheat-derived AX (AX-W) by human fecal microbiota reported that fermentation of AX-W was associated with the proliferation of bifidobacteria, lactobacilli and eubacteria (Hughes et al., 2007).

Several *in vivo* studies have also confirmed the bifidogenic effect of AX. An *in vivo* study in humanized rats demonstrated that long chain AX specifically stimulates the abundance of several different bacterial species in the cecum (relative bifidobacterial abundance in the cecum of the control group was 0.03 ± 0.01%, compared to 2.81 ± 1.46% in the group that were fed long chain AX; Van Den Abbeele et al., 2011). The findings of this latter study were validated by a recent study which detected the presence of two different *B. longum* species during the fermentation of long chain AX in an *in vitro* model of the proximal colon (Truchado et al., 2015). Another *in vivo* study found that when (high-fat) diet-induced obese mice were fed an AX-supplemented diet, a significant increase in caecal bifidobacterial numbers was observed (Neyrinck et al., 2011). Along with this increase in caecal bifidobacteria, AX supplementation restored (some of) the high-fat diet-induced changes to the microbial community.

Synbiotics are mixtures of one or more probiotics combined with one or more prebiotics (Patel and Dupont, 2015). Numerous *in vivo* studies have been conducted aimed at investigating the efficacy of bifidobacteria-based synbiotics in the treatment of gastrointestinal diseases and conditions. One such study investigated the synbiotic effect of *B. animalis* subsp. *lactis* B94 in combination with inulin on acute infectious diarrhea in children. Patients were administered the synbiotic agent once a day for five days and stool was examined for infectious agents such as rotavirus, *Salmonella, Shigella, Campylobacter, Cryptosporidium, Adenovirus, Entamoeba histolytica,* and *Clostridium difficile.* A marked decrease in the number of diarrhea stools was reported after 3 days of administration for the synbiotic group as compared to the control group, particularly for patients with rotavirus infection (Islek et al., 2014). A clinical trial investigated the effects of consumption of a synbiotic on the symptoms of Crohn's disease (Steed et al., 2010). The synbiotic, comprised of *B. longum,* inulin, and oligofructose, was consumed by patients twice daily over a 6 month period, and significant improvements in clinical outcomes were reported including a reduction in some activity indices of Crohn's disease (Steed et al., 2010). As a third example, the beneficial effect of a *B. breve* strain plus GOS synbiotic was investigated with regards to ulcerative colitis. The bifidobacterial strain was ingested three times a day whereas GOS was consumed once a day for 1 year. The clinical status of the treatment group significantly improved such as a marked improvement in colonoscopy scores and significant decreases in inflammatory markers. Furthermore, although no significant change in bifidobacterial numbers for those consuming the symbiotic was noticed, reduced fecal counts of *Bacteriodaceae* and reduced fecal pH was noted (Ishikawa et al., 2011).

## BIFIDOBACTERIAL CARBOHYDRATE METABOLISM

The human genome is predicted to encode just eight glycosyl hydrolases (GHs) that are directly linked to carbohydrate digestion. Therefore, many complex dietary carbohydrates remain un-digested and end up in the colon where they may be degraded by members of the microbiota (El Kaoutari et al., 2013).

The human GIT is home to complex microbial community that encompasses approximately 100-fold more genes than the number of genes present in the host genome (Backhed et al., 2005). Colonization of the human GIT, which is believed to occur immediately after birth, is influenced by various factors such as the method of delivery (i.e., vaginal or cesarean), type of feeding (breast-fed or formula-fed), exposure to antibiotics, frequency, and nature of diseases and hygiene conditions (Fanaro et al., 2003). Bifidobacteria dominate the total gut bacterial population in healthy breast-fed infants (Harmsen et al., 2000; Favier et al., 2002; Leahy et al., 2005), although this dominance decreases following weaning (Ventura et al., 2004). During adult life the bifidobacterial population stabilizes to represent 3–6% of the total gut microbial population, whereas in elderly (>65 years) the bifidobacterial numbers usually decline with age (Hopkins et al., 2001; Satokari et al., 2003).

The abundance and make-up of the gut microbiota is (among others) dependent on the diet of its host, and members of the microbiota have evolved effective mechanisms to utilize available nutrients (Vaughan et al., 2005; Ju-Hoon and O'Sullivan, 2010). Digestible and simple sugars such as lactose and sucrose are metabolized in the upper gut by the host and bacteria such as lactobacilli, a prevalent inhabitant of the upper GIT (Ganong, 2005; Vaughan et al., 2005). A diverse set of non-digestible carbohydrates are metabolized in the lower gut, including complex plant-derived polysaccharides (e.g., pectin, gums, hemicellulose, and xylans), host-derived carbohydrates (such as mucin and glycosphingolipids), and extracellular polysaccharides that are produced by members of the gut microbiota (Hooper et al., 2002; Korakli et al., 2002; Pokusaeva et al., 2011a). It is therefore not surprising that on average more than 12% of the annotated open reading frames within bifidobacterial genomes is predicted to encode enzymes involved in carbohydrate metabolism (Milani et al., 2014). In fact a recent study performed on the genome sequences from the type strains of each of the 47 *Bifidobacterium* (sub)species found that 5.5% of the core bifidobacterial genomic coding sequences (BifCOGs) is associated with carbohydrate metabolism (Milani et al., 2015).

Bifidobacteria present in the infant gut are presumed to metabolize human milk oligosaccharides (HMOs), and the genomes of *B. bifidum* and *B. longum* subsp. *infantis* are indeed tailored toward HMO metabolism (Sela et al., 2008; Duranti et al., 2015). However, other bifidobacterial species such as *B. breve* and *B. longum* subsp. *longum* are also commonly present in the infant gut. Although, they do not encode the same HMO catabolic arsenal found in *B. bifidum* and *B. longum* subsp. *infantis*, they can degrade certain HMOs and may also scavenge on carbohydrates that are released by other (bifido)bacteria (Egan et al., 2014a; Chaplin et al., 2015). After weaning the composition of the bifidobacterial population changes toward species capable of adapting to the metabolism of plant-derived sugars. For example, *B. longum* subsp. *longum* and *B. adolescentis* can utilize such diet-derived carbohydrates, while *B. bifidum* may shift its HMO-metabolic abilities toward mucin degradation (Schell et al., 2002; Turroni et al., 2010; Sela, 2011; Duranti et al., 2014; Egan et al., 2014a).

Prediction of the number of complete pathways used by bifidobacteria to degrade simple and complex sugars has been performed. The species *B. biavatti* specifies the largest number of pathways (14 complete pathways), whereas members of the species *B. bombi, B. crudilactis, B. longum* subsp. *infantis, B. minimum,* and *B. ruminantium* specifying just nine complete pathways (Milani et al., 2015). Bifidobacteria lack a number of key enzymes involved in the Emden-Meyerhof Parnas (EMP) pathway, instead, bifidobacteria metabolize hexose sugars through a metabolic pathway named the "bifid shunt" which is centered around the key enzyme, fructose-6-phosphoketolase (EC 4.1.2.2) (**Figure** ; De Vries and Stouthamer, 1967; De Vuyst et al., 2014). Furthermore, the action of additional enzymes allows for a variety of carbon sources (including pentose sugars) to be channeled through this pathway (Pokusaeva et al., 2011a). Fermentation through the bifid shunt is quite advantageous for bifidobacteria as this pathway allows for the production of more energy from carbohydrates compared to that produced by the EMP fermentative pathway (Salminen and Wright, 1998; Palframan et al., 2003). The bifid shunt theoretically yields 2.5 ATP moles from every 1 mole of glucose fermented, as well as 1.5 mole of acetate and 1 mole of lactate (Palframan et al., 2003). The ratios of acetate to lactate can be influenced, however, by the particular carbohydrate being fermented as well as the growth phase and bifidobacterial species being examined (Palframan et al., 2003). Furthermore, rapid consumption of an energy source was shown to result in the production of large amounts of lactate and low amounts of acetate, ethanol and formate, whereas less lactate is produced along with an increase in production of acetate, formate and ethanol when the energy source is consumed at a slow(er) rate (**Figure** ; Van Der Meulen et al., 2004, 2006a,b; Falony et al., 2009).

### Carbohydrate Uptake Strategies by Bifidobacteria

Bifidobacteria internalize carbohydrates by ATP-dependent ABC transporters, proton symporters or phosphoenolpyruvate-phosphotransferase systems (PEP-PTS) (Turroni et al., 2012). ABC transporters couple ATP hydrolysis to efficient internalization of sugars and appear to represent the primary carbohydrate transport systems for bifidobacteria (Ventura et al., 2007; Davidson et al., 2008; Jojima et al., 2010). PEP-PTS systems allow the concomittant transport and phosphorylation of carbohydrates, while they may also be involved in the regulation of various metabolic pathways (Postma et al., 1993). The PTS component of the system is involved in the internalization and concomitant phosphorylation of carbohydrates, while PEP acts as the (indirect) phosphate donor to the recipient carbohydrate (Ventura et al., 2007). These systems are found in many bacteria

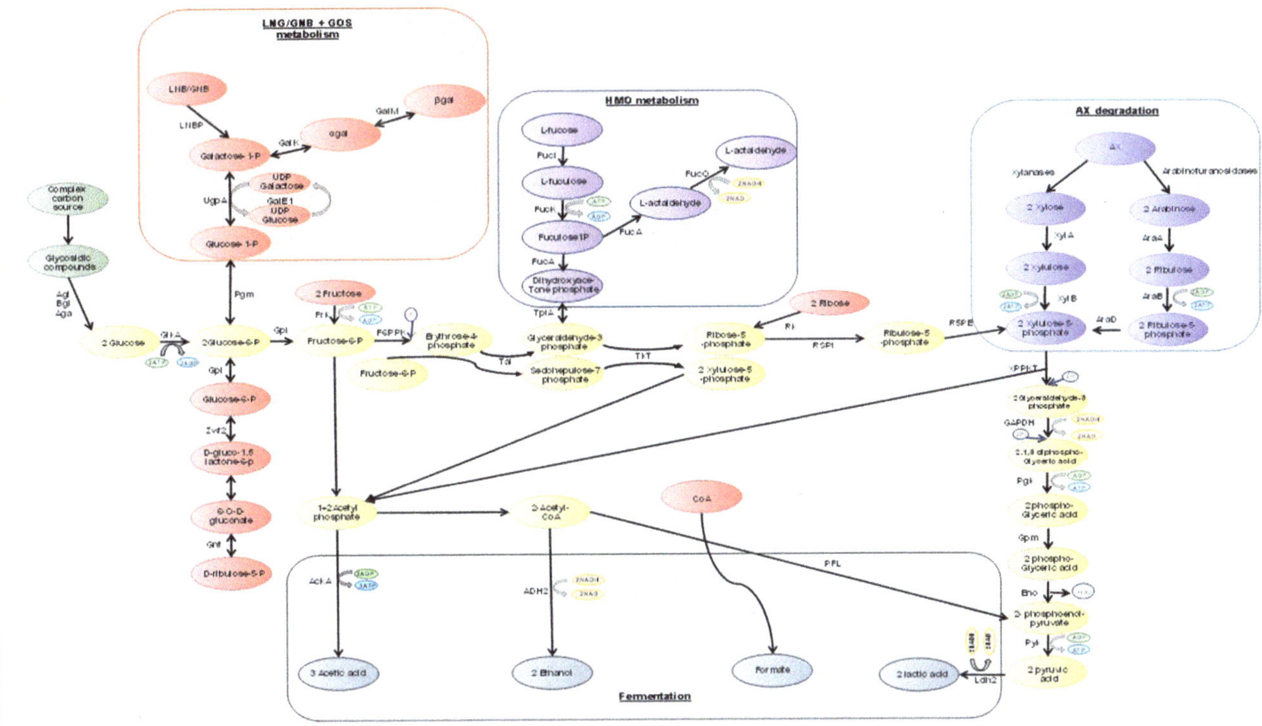


**FIGURE | A schematic representation of carbohydrate degradation through the "bifid shunt" in bifidobacteria.** Abbreviations: AckA, acetate kinase; Adh2, Aldehyde-alcohol dehydrogenase 2; Aga, α-galactosidase; Agl, α-glucosidase; AraA, L-arabinose isomerase; AraB, Ribulokinase; AraD, L-ribulose-5-phosphate 4-epimerase; Bgl, β-glucosidsae; Eno, enolase; GalE1, UDP-glucose 4-epimerase; GalK, galactokinase; GalM, glactose mutarotase; GAPDH, glyceraldehyde-3-phosphate dehydrogenase C; GlkA, glucokinase; Gnt, 6-phosphogluconate dehydrogenase; Gpi, glucose 6-phosphate isomerase; Gpm, phosphoglycerate mutase; FrK, frucktokinase; F6PPK, fructose-6-phodphoketolase; FucI, L-fucose isomerase; FucK, L-fuculose kinase; FucA, L-fuculose-1-phosphate aldose; FucO, lactaldehyde reductase; Ldh2, lactate dehydrogenase; LNBP, lacto-N-biose phosphorylase; Pgk, phosphoglyceric kinase; Pgm, phosphoglucomutase; Pfl, formate acetyltransferase; Pyk, pyruvate kinase; Rk, ribokinase; R5PI, ribose-5-phosphate isomerase; R5PE, ribulose-5-phosphate epimerase; Tal, transaldase; Tkt, transketolase; TpiA, trisephosphate isomerase; UgpA, UTP-glucose-1-phosphate uridylyltransferase; XPPKT, xylulose-5-phosphate/fructose-6-phosphate phosphoketolase; XylA, xylose isomerase; XylB, xylulose kinase; Zwf2, glucose-6-phosphate 1-dehydrogenase; Pi, phosphate (based on a figure from a previous review Pokusaeva et al., 2011a).

and have also been identified in most, but not all, bifidobacterial genomes (Postma et al., 1993; Pokusaeva et al., 2011a). The action of the PEP-PTS system was demonstrated experimentally in bifidobacteria whereby a PEP-PTS system from *B. breve* UCC2003 was found to be involved in the internalization of glucose (Degnan and Macfarlane, 1993). Since then a number of PEP-PTS systems have been identified and investigated in other bifidobacterial species (Lorca et al., 2007; Parche et al., 2007; Barrangou et al., 2009; Turroni et al., 2012; O'Connell Motherway et al., 2013).

The *B. longum* subsp. *longum* DJO10A and NCC2705 genomes are predicted to represent 10 and 13 different ABC transporters, respectively, responsible for the uptake of carbohydrates, while they each encode just a single glucose-specific PEP-PTS system (Lorca et al., 2007; Parche et al., 2007). Analysis of the *B. bifidum* PRL2010 genome sequence revealed that this strain encodes two ABC transporters, four PEP-PTS systems and four secondary transporters that are expected to transport mono- and disaccharides (Turroni et al., 2012). Transcriptional analysis revealed that one ABC transporter and two of the PEP-PTS systems are associated with the internalization of degradation products of host-derived glycans, in particular those that are found in mucin. The ABC transporters identified in *B. bifidum* PRL2010 were found to be linked to the uptake of monosaccharides such as glucose, ribose, fructose and galactose or disaccharides such as turanose (Turroni et al., 2012). Like *B. bifidum* PRL2010, *B. breve* UCC2003 is also predicted to encode four PEP-PTS systems and it has been shown experimentally that one system in *B. breve* UCC2003 is a fructose-inducible fructose/glucose uptake system (O'Connell Motherway et al., 2013). However, *B. breve* UCC2003 typically employs ABC-type transporters for the uptake of carbohydrates (Pokusaeva et al., 2011a, O'Connell Motherway et al., 2013; Egan et al., 2014b). In contrast, the strain *B. animalis* subsp. *lactis* B1-04 does not possess any PEP-PTS system and only encodes two copies of an ATP-binding protein linked to carbohydrate internalization (Barrangou et al., 2009). This low number of carbohydrate uptake systems may be a reflection of genome decay due to the commercial exploitation of *B. animalis* subsp. *lactis*, for which purpose it is extensively cultivated in nutritionally rich media (Lee et al., 2008).

## Bifidobacterial Glycosyl Hydrolases

Carbohydrates can be modified by a range of different enzymes including hexosyl- and phosphotransferases, hydrolases and isomerizes (Pokusaeva et al., 2011a). In the presence of water, glycosyl hydrolases (GHs) hydrolyse the glycosidic bond between two or more sugars or alternatively between a carbohydrate and a non-carbohydrate moiety (Pokusaeva et al., 2011a). GHs are assigned the enzyme commission or EC number EC 3.2.1x whereby the first three numbers indicate that these particular enzymes cleave glycosyl linkages, with the assignment of the final number being based on the target substrate or mechanism of action displayed by the enzyme. Members of one GH family may not only exhibit different substrate specificity, but may also exhibit a different mode of action (Van Den Broek et al., 2008). Classification of GHs can be found at http://www.cazy.org/Glycoside-Hydrolases.html (Lombard et al., 2014). For bifidobacteria GHs are the most prevalent group of carbohydrate-modifying enzymes and since the publication of the first bifidobacterial genome more information on these GHs has become available (Van Den Broek et al., 2008).

None of the carbohydrate active GHs encoded by the human genome appear to be involved in the breakdown of FOS, GOS, XOS, inulin, or arabinoxylan (Guarner and Malagelada, 2003; El Kaoutari et al., 2013). A recent study investigated the distribution of different carbohydrate-active enzymes among 177 bacterial genomes of the human microbiome, including genomes of 12 members of the *Actinobacteria* phylum, half of which were bifidobacteria (El Kaoutari et al., 2013). Polysaccharide lyases and GHs accounted for 59% of all carbohydrate-active enzymes identified. From these observations it can be said that the microbiota endows metabolic activities that make up for the paucity of GHs encoded by the human genome (El Kaoutari et al., 2013).

According to the current GH classifications, *B. scardovii* and *B. indicum* LMG11587 are predicted to encode the highest (126 GHs) and lowest number of GHs (25 GHs), respectively, among the currently sequenced bifidobacterial genomes (**Table** ). Classification according to the Carbohydrate Active Enzymes (CAZy) system revealed that 3385 genes belonging to the bifidobacterial pan-genome are predicted to represent carbohydrate active enzymes, including members of 57 GH families, 13 GT families and 7 carbohydrate esterases (Milani et al., 2015). Those enzymes belonging family GH13 are the most commonly found in bifidobacterial genomes and are known to be active against a wide range of carbohydrates including the plant-derived carbohydrates starch and related substrates, trehalose, stachyose, raffinose, and melibiose (Pokusaeva et al., 2011a; Milani et al., 2015). The bifidobacterial glycobiome contains a large number of enzymes belonging to the families GH29, GH95, GH20, GH112, GH38, GH125, GH101, and GH129 which are involved in the degradation of host-derived glycans. Members of the *B. scardovii, B. longum* subsp. *infantis,* and *B. bifidum* species in particular encode the most extensive set of enzymes active against host-derived glycans (Milani et al., 2015). Unlike other species, those bifidobacterial species isolated from honey/bumblebees encode a very limited set of GH13 representatives. However, these species specify a larger set of GH43 and GH3 enzymes. These families are active against plant-derived polysaccharides, such are arabino(xylan), (arabino)galactan, and cellodextran (Milani et al., 2015). Furthermore, a substantial number of genes encoding putative arabino(xylan)-degrading enzymes are present in certain bifidobacterial genomes, for example *B. longum* subsp. *longum* NCC2705, hinting at the importance of these enzymes toward the colonization of these microorganisms in the GIT (Schell et al., 2002; Van Den Broek et al., 2008).

Enzymes active against such arabinose- and xylose-containing carbohydrates have been characterized from *B. longum, B. adolescentis, B. animalis* subsp. *lactis,* and *B. breve*, and were first described in *B. adolescentis* (Van Laere et al., 1997; Lagaert et al., 2011). Seven bifidobacterial arabinofuranosidases have currently been characterized belonging to the families GH43 and GH51, five of which are produced by *B. adolescentis*, while two originate from *B. longum* (Van Laere et al., 1999; Margolles and De Los Reyes-Gavilan, 2003; Lagaert et al., 2010; Lee et al., 2011; Suzuki et al., 2013). Five intracellular xylanases have been characterized in bifidobacteria to date for both *B. breve* and *B. adolescentis* (Lagaert et al., 2007, 2011; Hyun et al., 2012; Amaretti et al., 2013). It is worth noting that all of the 14 currently characterized bifidobacterial arabino(xylan)-degrading enzymes are (predicted to be) intracellular. Due to their sizes, arabino(xylan)s cannot be transported inside the cell and the lack of extracellular enzymes specified by these bifidobacterial strains hints that they may rely on the extracellular hydrolytic activity of other members of the gut microbiota. Furthermore, the degradation of arabino(xylan) by extracellular enzymes has been observed previously for some bifidobacterial strains, and such extracellular enzymes may therefore be of significant interest as they are expected to provide an ecological advantage in the GIT (Riviere et al., 2014).

Investigations have been performed in order to determine those bifidobacterial GHs that are expected to be either secreted into the environment or associated with the cell surface. The majority of GHs located within the bifidobacterial pan-genome are predicted to be intracellular with 10.9% of GHs predicted to be secreted. The members of the family GH13 represent the largest proportion of such (predicted) secreted GHs, followed by members of the GH43 and GH51 families (Milani et al., 2015).

# CARBOHYDRATE CROSS-FEEDING BY BIFIDOBACTERIA

Several recent studies have investigated the impact on the gut microbiome by bifidobacterial cross-feeding of carbohydrates. Various studies have demonstrated that some members of the bifidobacterial community can co-operate in order to degrade large and complex polysaccharides into more simple sugars which are in turn then available to other members of the gut microbiota (De Vuyst and Leroy, 2011). This has been demonstrated for plant derived polysaccharides and also for

**TABLE** | **Bifidobacterial glycosyl hydrolases.**

| CAZy family (GH) | Substrate specificities | *B. adolescentis* 15703 | *B. adolescentis* 22L | *B. adolescentis* BBMN23 | *B. angulatum* DSM20098 | *B. animalis* A6 | *B. animalis* RH | *B. animalis* subsp. *animalis* ATCC25527 | *B. animalis* subsp. *lactis* ADO11 | *B. animalis* subsp. *lactis* ATCC27673 | *B. animalis* subsp. *lactis* B420 | *B. animalis* subsp. *lactis* Bb12 | *B. animalis* subsp. *lactis* BF052 | *B. animalis* subsp. *lactis* Bi-07 | *B. animalis* subsp. *lactis* Bl-04 | *B. animalis* subsp. *lactis* Bl12 | *B. animalis* subsp. *lactis* BLC1 | *B. animalis* subsp. *lactis* CNCM I-2494 | *B. animalis* subsp. *lactis* DSM 10140 | *B. animalis* subsp. *lactis* KLDS2.0603 | *B. animalis* subsp. *lactis* V9 |
|---|---|---|---|---|---|---|---|---|---|---|---|---|---|---|---|---|---|---|---|---|---|
| 1 | β-glucosidase, β-galactosidase | 2 | 2 | 0 | 1 | 1 | 1 | 1 | 1 | 1 | 1 | 1 | 1 | 1 | 1 | 1 | 1 | 1 | 1 | 1 | 1 |
| 2 | β-galactosidase | 3 | 4 | 2 | 2 | 2 | 2 | 2 | 2 | 2 | 2 | 2 | 2 | 2 | 2 | 2 | 2 | 2 | 2 | 2 | 2 |
| 3 | β-glucosidase, β-hexosaminidase, β-D-glucosideglucohydrolase, | 6 | 5 | 6 | 3 | 3 | 3 | 3 | 3 | 2 | 3 | 3 | 3 | 3 | 3 | 3 | 3 | 3 | 3 | 3 | 3 |
| 5 | β-mannosidase, β-glucosidase, β-exoglucanase | 2 | 2 | 2 | 0 | 1 | 1 | 1 | 1 | 2 | 1 | 1 | 1 | 1 | 1 | 1 | 1 | 1 | 1 | 1 | 1 |
| 8 | Xylanase | 1 | 1 | 1 | 1 | 0 | 0 | 0 | 0 | 0 | 0 | 0 | 0 | 0 | 0 | 0 | 0 | 0 | 0 | 0 | 0 |
| 13 | α-1, 4-glucosidase, amylopullulanase, sucrose Phosphorylase, α-amylase | 13 | 16 | 17 | 13 | 13 | 13 | 13 | 11 | 10 | 13 | 14 | 13 | 13 | 11 | 10 | 11 | 13 | 11 | 13 | 11 |
| 23 | Transglycosylase | 3 | 2 | 2 | 2 | 2 | 1 | 2 | 2 | 2 | 2 | 2 | 2 | 2 | 2 | 2 | 2 | 2 | 2 | 2 | 2 |
| 25 | β-N-acetylmuramidase | 1 | 2 | 3 | 1 | 2 | 2 | 1 | 2 | 1 | 2 | 2 | 2 | 2 | 2 | 2 | 2 | 2 | 2 | 2 | 2 |
| 26 | Endo-1, 4-β-mannosidase | 1 | 1 | 0 | 0 | 0 | 0 | 0 | 0 | 0 | 0 | 0 | 0 | 0 | 0 | 0 | 0 | 0 | 0 | 0 | 0 |
| 27 | α-galactosidase | 1 | 0 | 0 | 0 | 0 | 0 | 0 | 0 | 0 | 0 | 0 | 0 | 0 | 0 | 0 | 0 | 0 | 0 | 0 | 0 |
| 30 | β-D-xylosidase, endo-1, 6-β-glucosidase, Glucosylceramidase | 1 | 1 | 1 | 0 | 2 | 2 | 1 | 2 | 1 | 2 | 2 | 2 | 2 | 2 | 2 | 2 | 2 | 2 | 2 | 2 |
| 31 | α-xylosidase | 1 | 1 | 2 | 2 | 0 | 0 | 0 | 0 | 0 | 0 | 0 | 0 | 0 | 0 | 0 | 0 | 0 | 0 | 0 | 0 |
| 32 | β-fructofuranosidase, sucrose-6-phosphate hydrolase | 2 | 2 | 2 | 2 | 1 | 1 | 1 | 1 | 1 | 1 | 1 | 1 | 1 | 1 | 1 | 1 | 1 | 1 | 1 | 1 |
| 35 | β-galactosidase | 1 | 0 | 0 | 0 | 0 | 0 | 0 | 0 | 0 | 0 | 0 | 0 | 0 | 0 | 0 | 0 | 0 | 0 | 0 | 0 |
| 36 | α-galactosidase, raffinose synthase, | 2 | 3 | 2 | 2 | 3 | 3 | 3 | 3 | 3 | 3 | 3 | 3 | 3 | 3 | 3 | 3 | 3 | 3 | 3 | 3 |
| 38 | α-mannosidase | 1 | 1 | 1 | 0 | 0 | 0 | 0 | 0 | 0 | 0 | 0 | 0 | 0 | 0 | 0 | 0 | 0 | 0 | 0 | 0 |
| 39 | β-xylosidase | 0 | 1 | 0 | 0 | 0 | 0 | 0 | 0 | 0 | 0 | 0 | 0 | 0 | 0 | 0 | 0 | 0 | 0 | 0 | 0 |
| 42 | β-galactosidase | 4 | 3 | 3 | 2 | 2 | 2 | 0 | 2 | 1 | 2 | 2 | 2 | 2 | 2 | 2 | 2 | 2 | 2 | 2 | 2 |
| 43 | Endo-1, 5-α-L-arabinosidase, α-L-arabinofuranosidase, Endo-1,4-β-xylanase, β-1, 4-xylosidase | 7 | 7 | 8 | 5 | 3 | 3 | 3 | 3 | 3 | 3 | 3 | 3 | 3 | 3 | 3 | 3 | 3 | 3 | 3 | 3 |
| 51 | α-L-arabinofuranosidase | 2 | 1 | 1 | 1 | 1 | 1 | 1 | 1 | 1 | 1 | 2 | 1 | 1 | 1 | 1 | 1 | 1 | 1 | 1 | 1 |
| 77 | 4-α-glucanotransferase | 2 | 2 | 2 | 2 | 2 | 2 | 2 | 2 | 2 | 2 | 2 | 2 | 2 | 2 | 2 | 2 | 2 | 2 | 2 | 2 |
| 94 | Cellobiose-phosphorylase | 0 | 1 | 1 | 0 | 1 | 1 | 0 | 1 | 1 | 1 | 1 | 1 | 1 | 1 | 0 | 1 | 1 | 1 | 1 | 1 |
| 120 | β-xylosidase | 1 | 1 | 1 | 1 | 0 | 0 | 0 | 0 | 0 | 0 | 0 | 0 | 0 | 0 | 0 | 0 | 0 | 0 | 0 | 0 |
| 121 | β-galactosidase | 0 | 0 | 1 | 0 | 0 | 0 | 0 | 0 | 0 | 0 | 0 | 0 | 0 | 0 | 0 | 0 | 0 | 0 | 0 | 0 |
| 127 | β-L-arabinofuranosidase | 1 | 2 | 2 | 0 | 1 | 1 | 0 | 1 | 1 | 1 | 1 | 1 | 1 | 1 | 1 | 1 | 1 | 1 | 1 | 1 |
| Total | | 58 | 61 | 60 | 40 | 40 | 39 | 34 | 38 | 35 | 40 | 42 | 40 | 40 | 38 | 36 | 38 | 40 | 38 | 40 | 38 |

*(Continued)*

TABLE | Continued

| CAZy family (GH) | Substrate specificities | *B. asteroides* PRL2011 | *B. bifidum* ATCC29521 | *B. bifidum* BGN4 | *B. bifidum* PRL2010 | *B. bifidum* S17 | *B. breve* 12L | *B. breve* 6589b | *B. breve* ACS-071-V-Sch8b | *B. breve* DSM20213 | *B. breve* JCM7017 | *B. breve* JCM7019 | *B. breve* NCFB2258 | *B. breve* S27 | *B. breve* UCC2003 | *B. catenulatum* DSM 16992 | *B. coryneforme* LMG 18911 | *B. dentium* Bd1 | *B. dentium* JCM1195 | *B. indicum* LMG 11587 |
|---|---|---|---|---|---|---|---|---|---|---|---|---|---|---|---|---|---|---|---|---|
| 1 | β-glucosidase, β-galactosidase | 0 | 2 | 1 | 1 | 1 | 1 | 1 | 2 | 1 | 2 | 3 | 1 | 2 | 2 | 0 | 1 | 2 | 2 | 1 |
| 2 | β-galactosidase | 2 | 3 | 3 | 3 | 3 | 2 | 3 | 4 | 2 | 2 | 4 | 3 | 3 | 5 | 4 | 0 | 6 | 6 | 0 |
| 3 | β-glucosidase, β-hexosaminidase, β-D-glucosideglucohydrolase, | 4 | 1 | 1 | 1 | 1 | 3 | 6 | 5 | 6 | 4 | 4 | 5 | 4 | 3 | 7 | 4 | 11 | 12 | 4 |
| 4 | 6-phospho-β-glucosidase | 1 | 0 | 0 | 0 | 0 | 0 | 0 | 0 | 0 | 0 | 0 | 0 | 0 | 0 | 0 | 0 | 0 | 0 | 0 |
| 5 | β-mannosidase, β-glucosidase, β-exoglucanase | 1 | 0 | 0 | 0 | 0 | 2 | 2 | 2 | 2 | 2 | 2 | 2 | 2 | 2 | 2 | 0 | 4 | 4 | 0 |
| 13 | α-1,4-glucosidase, amylopullulanase, sucrose phosphorylase, α-amylase | 4 | 7 | 7 | 9 | 7 | 14 | 14 | 13 | 12 | 13 | 12 | 13 | 12 | 14 | 9 | 2 | 14 | 14 | 2 |
| 16 | Adenine glycosylase | 0 | 1 | 0 | 0 | 0 | 0 | 0 | 0 | 0 | 0 | 0 | 0 | 0 | 0 | 0 | 0 | 0 | 0 | 0 |
| 18 | Chitinase | 0 | 0 | 0 | 0 | 0 | 1 | 2 | 1 | 0 | 1 | 1 | 1 | 2 | 1 | 0 | 0 | 0 | 0 | 0 |
| 20 | β-hexosaminidase | 0 | 4 | 4 | 4 | 4 | 1 | 1 | 1 | 1 | 1 | 1 | 1 | 1 | 1 | 0 | 0 | 0 | 0 | 0 |
| 23 | Transglycosylase | 2 | 2 | 3 | 2 | 2 | 4 | 3 | 2 | 3 | 3 | 2 | 3 | 3 | 2 | 3 | 2 | 3 | 3 | 2 |
| 25 | β-N-acetylmuramidase | 0 | 3 | 1 | 2 | 1 | 1 | 1 | 1 | 1 | 2 | 1 | 1 | 1 | 1 | 1 | 1 | 2 | 2 | 1 |
| 26 | Endo-1, 4-β-mannosidase | 0 | 0 | 0 | 0 | 0 | 0 | 0 | 0 | 0 | 0 | 0 | 0 | 0 | 0 | 0 | 0 | 2 | 2 | 0 |
| 27 | α-galactosidase | 1 | 0 | 0 | 0 | 0 | 0 | 0 | 0 | 0 | 0 | 0 | 0 | 0 | 0 | 0 | 0 | 1 | 1 | 0 |
| 28 | Galacturan 1, 4-α-galacturonidase, pectinesterase | 2 | 0 | 0 | 0 | 0 | 0 | 0 | 0 | 0 | 0 | 0 | 0 | 0 | 0 | 0 | 0 | 0 | 0 | 0 |
| 29 | α-L-fucosidase | 1 | 1 | 1 | 1 | 1 | 0 | 0 | 0 | 0 | 0 | 0 | 0 | 0 | 0 | 0 | 0 | 1 | 1 | 0 |
| 30 | β-D-xylosidase, endo-1, 6-β-glucosidase, Glucosylceramidase | 3 | 0 | 0 | 0 | 0 | 0 | 1 | 0 | 0 | 0 | 1 | 0 | 0 | 1 | 1 | 3 | 1 | 1 | 3 |
| 31 | α-xylosidase | 3 | 0 | 0 | 0 | 0 | 0 | 0 | 0 | 0 | 1 | 1 | 1 | 1 | 2 | 1 | 1 | 4 | 4 | 1 |
| 32 | β-fructofuranosidase, sucrose-6-phosphate hydrolase | 1 | 2 | 0 | 0 | 0 | 1 | 1 | 1 | 1 | 2 | 1 | 1 | 2 | 2 | 2 | 2 | 2 | 2 | 2 |
| 33 | Sialidase | 0 | 4 | 3 | 3 | 3 | 1 | 1 | 1 | 1 | 1 | 1 | 1 | 1 | 1 | 0 | 0 | 0 | 0 | 0 |
| 35 | β-galactosidase | 0 | 0 | 0 | 0 | 0 | 0 | 0 | 0 | 0 | 0 | 1 | 0 | 0 | 1 | 0 | 0 | 0 | 0 | 0 |
| 36 | α-galactosidase, raffinose synthase, | 2 | 1 | 1 | 1 | 1 | 3 | 2 | 2 | 2 | 2 | 3 | 2 | 2 | 2 | 4 | 2 | 5 | 5 | 2 |
| 38 | α-mannosidase | 2 | 0 | 0 | 0 | 0 | 2 | 2 | 2 | 2 | 2 | 3 | 3 | 2 | 3 | 1 | 0 | 0 | 0 | 0 |
| 39 | β-xylosidase | 0 | 0 | 0 | 0 | 0 | 0 | 0 | 0 | 0 | 0 | 0 | 0 | 0 | 0 | 1 | 0 | 0 | 0 | 0 |
| 42 | β-galactosidase | 2 | 2 | 2 | 2 | 1 | 2 | 2 | 3 | 2 | 3 | 2 | 2 | 2 | 2 | 3 | 2 | 4 | 4 | 2 |
| 43 | Endo-1, 5-α-L-arabinosidase, α-L-arabinofuranosidase, Endo-1,4-β-xylanase, β-1, 4-xylosidase | 9 | 3 | 2 | 2 | 2 | 0 | 1 | 1 | 1 | 1 | 0 | 0 | 0 | 0 | 7 | 3 | 12 | 12 | 3 |
| 51 | α-L-arabinofuranosidase | 2 | 1 | 1 | 1 | 1 | 0 | 0 | 0 | 0 | 0 | 0 | 0 | 0 | 0 | 3 | 1 | 2 | 2 | 0 |
| 53 | Endogalactanase | 0 | 0 | 0 | 0 | 0 | 1 | 1 | 1 | 0 | 1 | 0 | 1 | 1 | 1 | 0 | 0 | 2 | 2 | 0 |

*(Continued)*

**TABLE | Continued**

| CAZy family (GH) | Substrate specificities | *B. asteroides* PRL2011 | *B. bifidum* ATCC29521 | *B. bifidum* BGN4 | *B. bifidum* PRL2010 | *B. bifidum* S17 | *B. breve* 12L | *B. breve* 6589b | *B. breve* ACS-071-V-Sch8b | *B. breve* DSM20213 | *B. breve* JCM7017 | *B. breve* JCM7019 | *B. breve* NCFB2258 | *B. breve* S27 | *B. breve* UCC2003 | *B. catenulatum* DSM 16992 | *B. coryneforme* LMG 18911 | *B. dentium* Bd1 | *B. dentium* JCM1195 | *B. indicum* LMG 11587 |
|---|---|---|---|---|---|---|---|---|---|---|---|---|---|---|---|---|---|---|---|---|
| 59 | β-galactosidase/galactocerebrosidase | 0 | 0 | 0 | 0 | 0 | 0 | 1 | 0 | 0 | 0 | 0 | 0 | 0 | 0 | 0 | 0 | 0 | 0 | 0 |
| 65 | Kojibiose phosphorylase | 1 | 0 | 0 | 0 | 0 | 0 | 0 | 1 | 0 | 0 | 0 | 1 | 1 | 1 | 0 | 1 | 0 | 0 | 1 |
| 67 | α-glucuronidase | 1 | 0 | 0 | 0 | 0 | 0 | 0 | 0 | 0 | 0 | 0 | 0 | 0 | 0 | 0 | 0 | 0 | 0 | 0 |
| 77 | 4-α-glucanotransferase | 0 | 1 | 1 | 1 | 1 | 2 | 2 | 2 | 2 | 2 | 2 | 2 | 2 | 2 | 2 | 0 | 2 | 2 | 0 |
| 78 | α-L-rhamnosidase | 1 | 0 | 0 | 0 | 0 | 0 | 1 | 0 | 0 | 0 | 1 | 0 | 0 | 0 | 0 | 1 | 1 | 1 | 1 |
| 84 | α-L-rhamnosidase | 0 | 2 | 2 | 2 | 2 | 0 | 0 | 0 | 0 | 0 | 0 | 0 | 0 | 0 | 0 | 0 | 0 | 0 | 0 |
| 85 | Endo-β-N-acetylglucosaminidase D | 0 | 0 | 0 | 0 | 0 | 0 | 1 | 0 | 0 | 1 | 1 | 1 | 0 | 1 | 0 | 0 | 0 | 0 | 0 |
| 89 | α-N-acetylglucosaminidase, β-N-hexosaminidase | 0 | 1 | 1 | 1 | 1 | 0 | 0 | 0 | 0 | 0 | 0 | 0 | 0 | 0 | 0 | 0 | 0 | 0 | 0 |
| 94 | Cellobiose-phosphorylase | 0 | 0 | 0 | 0 | 0 | 0 | 0 | 0 | 0 | 0 | 0 | 0 | 0 | 0 | 0 | 0 | 1 | 1 | 0 |
| 95 | α-L-fucosidase | 0 | 1 | 1 | 1 | 1 | 1 | 1 | 1 | 1 | 1 | 1 | 1 | 1 | 1 | 0 | 0 | 0 | 0 | 0 |
| 101 | endo-α-N-acetylgalactosaminidase | 0 | 1 | 1 | 1 | 1 | 0 | 0 | 0 | 0 | 0 | 0 | 0 | 0 | 0 | 0 | 0 | 0 | 0 | 0 |
| 105 | rhamnogalacturonyl hydrolase, d-4,5-unsaturated β-glucuronyl hydrolase | 1 | 0 | 0 | 0 | 0 | 0 | 0 | 0 | 0 | 0 | 0 | 0 | 0 | 0 | 0 | 0 | 0 | 0 | 0 |
| 110 | Exo-α-galactosidase | 0 | 1 | 1 | 1 | 1 | 0 | 0 | 0 | 0 | 0 | 0 | 0 | 0 | 0 | 0 | 0 | 0 | 0 | 0 |
| 112 | Lacto-N-biose phosphorylase | 0 | 2 | 2 | 2 | 2 | 1 | 1 | 1 | 1 | 1 | 1 | 1 | 1 | 1 | 0 | 0 | 0 | 0 | 0 |
| 115 | xylan α-1, 2-glucuronidase, α-(4-O-methyl)-glucuronidase | 2 | 0 | 0 | 0 | 0 | 0 | 0 | 0 | 0 | 0 | 0 | 0 | 0 | 0 | 0 | 0 | 0 | 0 | 0 |
| 120 | β-xylosidase | 0 | 0 | 0 | 0 | 0 | 0 | 0 | 0 | 0 | 0 | 0 | 0 | 0 | 0 | 1 | 0 | 0 | 0 | 0 |
| 121 | β-galactosidase | 0 | 0 | 0 | 0 | 0 | 0 | 0 | 0 | 0 | 0 | 0 | 0 | 0 | 0 | 1 | 0 | 1 | 1 | 0 |
| 123 | Glycosphingolipid β-N-acetylgalactosaminidase | 0 | 1 | 1 | 1 | 1 | 0 | 0 | 0 | 0 | 0 | 0 | 0 | 0 | 0 | 0 | 0 | 0 | 0 | 0 |
| 125 | Exo-α-1, 6-mannosidase | 0 | 0 | 0 | 0 | 0 | 1 | 1 | 1 | 1 | 1 | 1 | 1 | 1 | 1 | 0 | 0 | 1 | 1 | 0 |
| 127 | β-L-arabinofuranosidase | 0 | 0 | 0 | 0 | 0 | 1 | 1 | 0 | 1 | 0 | 2 | 1 | 1 | 0 | 3 | 0 | 2 | 2 | 0 |
| 129 | α-N-acetylgalactosaminidase | 0 | 1 | 1 | 1 | 1 | 1 | 1 | 0 | 1 | 1 | 1 | 1 | 1 | 1 | 0 | 0 | 0 | 0 | 0 |
| NC | Unknown | 0 | 0 | 0 | 0 | 0 | 1 | 1 | 1 | 1 | 1 | 1 | 1 | 1 | 1 | 0 | 0 | 1 | 1 | 0 |
| Total | | 48 | 48 | 41 | 43 | 39 | 47 | 55 | 49 | 45 | 51 | 54 | 51 | 50 | 55 | 57 | 26 | 87 | 88 | 25 |

*(Continued)*

**TABLE | Continued**

| CAZy family (GH) | Substrate specificities | *B. kashiwanohense* JCM 15439 | *B. kashiwanohense* PV20-2 | *B. longum* 105-A | *B. longum* BXY01 | *B.* lognum DJO10A | *B. longum* NCC2705 | *B. longum* subsp. *infantis* 157F | *B. longum* subsp. *infantis* ATCC15697 | *B. longum* subsp. *infantis* JCM1222 | *B. longum* subsp. *longum* BBMN68 | *B. longum* subsp. *longum* F8 | *B. longum* subsp. *longum* GT15 | *B. longum* subsp. *longum* JCM1217 | *B. longum* subsp. *longum* JDM301 | *B. longum* subsp. *longum* KACC91563 | *B. pseudocatenulatum* D2CA | *B. pseudocatenulatum* DSM20438 | *B. pseudolongum* PV8-2 | *B. scardovii* JCM12489 | *B. thermophilum* RBL67 |
|---|---|---|---|---|---|---|---|---|---|---|---|---|---|---|---|---|---|---|---|---|---|
| 1 | β-glucosidase, β-galactosidase | 2 | 1 | 1 | 0 | 0 | 0 | 0 | 1 | 1 | 0 | 0 | 0 | 0 | 0 | 0 | 1 | 0 | 1 | 1 | 0 |
| 2 | β-galactosidase | 4 | 2 | 2 | 3 | 2 | 2 | 2 | 2 | 3 | 1 | 2 | 2 | 2 | 2 | 2 | 2 | 4 | 3 | 7 | 0 |
| 3 | β-glucosidase, β-hexosaminidase, β-D-glucosideglucohydrolase, | 5 | 6 | 6 | 3 | 3 | 3 | 5 | 2 | 2 | 4 | 3 | 5 | 4 | 3 | 4 | 5 | 5 | 2 | 10 | 1 |
| 5 | β-mannosidase, β-glucosidase, β-exoglucanase | 1 | 2 | 2 | 2 | 2 | 2 | 2 | 2 | 2 | 2 | 2 | 2 | 1 | 2 | 1 | 2 | 2 | 0 | 2 | 3 |
| 8 | Xylanase | 2 | 0 | 0 | 1 | 0 | 0 | 0 | 0 | 0 | 0 | 0 | 0 | 0 | 1 | 0 | 1 | 2 | 0 | 0 | 0 |
| 10 | Endoxylanase | 1 | 0 | 0 | 0 | 0 | 0 | 0 | 0 | 0 | 0 | 0 | 0 | 0 | 0 | 0 | 0 | 0 | 0 | 0 | 0 |
| 13 | α-1, 4-glucosidase, amylopullulanase, sucrose phosphorylase, α-amylase | 9 | 8 | 8 | 12 | 13 | 8 | 11 | 8 | 9 | 13 | 10 | 12 | 13 | 12 | 13 | 15 | 17 | 13 | 12 | 14 |
| 16 | Adenine glycosylase | 0 | 0 | 0 | 0 | 0 | 0 | 0 | 0 | 0 | 0 | 0 | 0 | 0 | 0 | 0 | 0 | 0 | 0 | 1 | 0 |
| 18 | Chitinase | 1 | 1 | 1 | 0 | 0 | 0 | 1 | 1 | 1 | 0 | 0 | 0 | 0 | 0 | 0 | 0 | 1 | 0 | 1 | 1 |
| 20 | β-hexosaminidase | 1 | 1 | 1 | 1 | 1 | 1 | 1 | 3 | 3 | 1 | 1 | 1 | 1 | 1 | 1 | 1 | 1 | 0 | 3 | 0 |
| 23 | Transglycosylase | 4 | 1 | 1 | 2 | 4 | 3 | 3 | 2 | 2 | 2 | 3 | 3 | 2 | 2 | 3 | 1 | 3 | 2 | 3 | 2 |
| 25 | β-N-acetylmuramidase | 1 | 1 | 1 | 0 | 1 | 0 | 1 | 4 | 4 | 1 | 1 | 0 | 1 | 0 | 0 | 1 | 1 | 1 | 4 | 1 |
| 26 | Endo-1, 4-β-mannosidase | 0 | 0 | 0 | 0 | 0 | 0 | 0 | 0 | 0 | 0 | 0 | 0 | 0 | 0 | 0 | 0 | 1 | 0 | 2 | 0 |
| 27 | α-galactosidase | 1 | 0 | 0 | 1 | 1 | 1 | 1 | 0 | 0 | 1 | 0 | 1 | 1 | 1 | 1 | 1 | 1 | 0 | 1 | 0 |
| 28 | Galacturan 1,4-α-galacturonidase, pectinesterase | 0 | 0 | 0 | 0 | 0 | 0 | 0 | 0 | 0 | 0 | 0 | 0 | 0 | 0 | 0 | 0 | 0 | 0 | 2 | 0 |
| 29 | α-L-fucosidase | 2 | 0 | 0 | 1 | 0 | 0 | 0 | 3 | 3 | 0 | 0 | 0 | 0 | 1 | 0 | 0 | 0 | 0 | 2 | 0 |
| 30 | β-D-xylosidase, endo-1, 6-β-glucosidase, Glucosylceramidase | 1 | 0 | 0 | 1 | 1 | 1 | 1 | 0 | 0 | 2 | 0 | 1 | 2 | 1 | 1 | 2 | 1 | 1 | 2 | 0 |
| 31 | α-xylosidase | 3 | 2 | 2 | 1 | 3 | 3 | 2 | 0 | 0 | 2 | 3 | 1 | 2 | 1 | 2 | 3 | 2 | 1 | 6 | 1 |
| 32 | β-fructofuranosidase, sucrose-6-phosphate hydrolase | 2 | 2 | 2 | 2 | 1 | 1 | 1 | 2 | 2 | 1 | 1 | 2 | 1 | 2 | 1 | 1 | 1 | 1 | 2 | 2 |
| 33 | Sialidase | 0 | 0 | 0 | 0 | 0 | 0 | 0 | 2 | 2 | 0 | 0 | 0 | 0 | 0 | 0 | 0 | 0 | 0 | 1 | 0 |
| 35 | β-galactosidase | 1 | 0 | 0 | 1 | 0 | 0 | 0 | 0 | 0 | 0 | 0 | 0 | 0 | 1 | 0 | 0 | 0 | 0 | 1 | 0 |
| 36 | α-galactosidase, raffinose synthase, | 2 | 2 | 2 | 2 | 2 | 1 | 1 | 1 | 1 | 2 | 2 | 1 | 2 | 2 | 2 | 3 | 3 | 4 | 12 | 2 |
| 38 | α-mannosidase | 1 | 1 | 1 | 3 | 3 | 3 | 2 | 2 | 2 | 2 | 2 | 2 | 0 | 3 | 0 | 0 | 0 | 0 | 2 | 0 |
| 39 | β-xylosidase | 0 | 0 | 0 | 0 | 0 | 0 | 0 | 0 | 0 | 0 | 0 | 0 | 0 | 0 | 0 | 0 | 0 | 0 | 3 | 0 |

*(Continued)*

**TABLE** | Continued

| CAZy family (GH) | Substrate specificities | *B. kashiwanohense* JCM 15439 | *B. kashiwanohense* PV20-2 | *B. longum* 105-A | *B. longum* BXY01 | B. lognum DJO10A | *B. longum* NCC2705 | *B. longum* subsp. *infantis* 157F | *B. longum* subsp. *infantis* ATCC15697 | *B. longum* subsp. *infantis* JCM1222 | *B. longum* subsp. *longum* BBMN68 | *B. longum* subsp. *longum* F8 | *B. longum* subsp. *longum* GT15 | *B. longum* subsp. *longum* JCM1217 | *B. longum* subsp. *longum* JDM301 | *B. longum* subsp. *longum* KACC91563 | *B. pseudocatenulatum* D2CA | *B. pseudocatenulatum* DSM20438 | *B. pseudolongum* PV8-2 | *B. scardovii* JCM12489 | *B. thermophilum* RBL67 |
|---|---|---|---|---|---|---|---|---|---|---|---|---|---|---|---|---|---|---|---|---|---|
| 42 | β-galactosidase | 3 | 3 | 3 | 4 | 3 | 2 | 3 | 3 | 3 | 2 | 2 | 3 | 2 | 4 | 2 | 3 | 4 | 2 | 6 | 1 |
| 43 | Endo-1, 5-α-L-arabinosidase, α-L-arabinofuranosidase, Endo-1,4-β-xylanase, β-1, 4-xylosidase | 9 | 6 | 6 | 10 | 11 | 9 | 7 | 1 | 1 | 11 | 7 | 7 | 11 | 10 | 8 | 12 | 15 | 7 | 17 | 0 |
| 50 | β-agarase | 0 | 0 | 0 | 1 | 0 | 0 | 0 | 0 | 0 | 0 | 0 | 0 | 0 | 1 | 0 | 0 | 0 | 0 | 0 | 0 |
| 51 | α-L-arabinofuranosidase | 2 | 1 | 1 | 4 | 5 | 5 | 4 | 1 | 1 | 3 | 4 | 4 | 4 | 4 | 4 | 2 | 3 | 1 | 6 | 1 |
| 53 | Endogalactanase | 0 | 0 | 0 | 1 | 1 | 1 | 1 | 0 | 1 | 1 | 1 | 1 | 1 | 1 | 1 | 0 | 0 | 0 | 0 | 0 |
| 59 | β-galactosidase/galactocerebrosidase | 0 | 0 | 0 | 0 | 0 | 0 | 0 | 0 | 0 | 0 | 0 | 0 | 0 | 0 | 0 | 1 | 0 | 0 | 1 | 0 |
| 65 | Kojibiose phosphorylase | 0 | 0 | 0 | 0 | 0 | 0 | 1 | 0 | 0 | 0 | 0 | 0 | 0 | 0 | 0 | 0 | 0 | 0 | 0 | 0 |
| 76 | α-1, 6-mannanase | 0 | 0 | 0 | 0 | 0 | 0 | 0 | 0 | 0 | 0 | 0 | 0 | 0 | 0 | 0 | 0 | 0 | 0 | 1 | 0 |
| 77 | 4-α-glucanotransferase | 2 | 2 | 2 | 2 | 2 | 2 | 2 | 2 | 2 | 2 | 2 | 2 | 2 | 2 | 2 | 2 | 2 | 2 | 2 | 2 |
| 78 | α-L-rhamnosidase | 0 | 1 | 1 | 0 | 0 | 0 | 0 | 0 | 0 | 0 | 0 | 0 | 0 | 0 | 0 | 0 | 0 | 0 | 2 | 0 |
| 85 | Endo-β-N-acetylglucosaminidase D | 0 | 0 | 0 | 1 | 1 | 1 | 1 | 0 | 0 | 1 | 1 | 1 | 0 | 1 | 0 | 0 | 0 | 0 | 0 | 0 |
| 94 | Cellobiose-phosphorylase | 0 | 0 | 0 | 0 | 0 | 0 | 0 | 0 | 0 | 0 | 0 | 0 | 0 | 0 | 0 | 0 | 0 | 1 | 1 | 0 |
| 95 | α-L-fucosidase | 1 | 1 | 1 | 1 | 0 | 0 | 0 | 1 | 1 | 0 | 0 | 0 | 0 | 1 | 0 | 0 | 1 | 0 | 0 | 0 |
| 101 | endo-α-N-acetylgalactosaminidase | 0 | 0 | 0 | 0 | 1 | 1 | 1 | 0 | 0 | 1 | 0 | 1 | 1 | 0 | 1 | 0 | 0 | 0 | 0 | 0 |
| 112 | Lacto-N-biose phosphorylase | 0 | 0 | 0 | 1 | 1 | 1 | 1 | 1 | 1 | 1 | 1 | 1 | 1 | 1 | 1 | 0 | 0 | 0 | 1 | 0 |
| 115 | xylan α-1, 2-glucuronidase, α-(4-O-methyl)-glucuronidase | 0 | 0 | 0 | 0 | 0 | 0 | 0 | 0 | 0 | 0 | 0 | 0 | 0 | 0 | 0 | 0 | 0 | 0 | 1 | 0 |
| 120 | β-xylosidase | 1 | 0 | 0 | 1 | 2 | 0 | 1 | 0 | 0 | 0 | 1 | 1 | 0 | 1 | 1 | 1 | 2 | 0 | 1 | 0 |
| 121 | β-galactosidase | 0 | 0 | 0 | 0 | 0 | 1 | 1 | 0 | 0 | 1 | 1 | 1 | 1 | 0 | 1 | 0 | 1 | 0 | 0 | 0 |
| 125 | Exo-α-1, 6-mannosidase | 0 | 0 | 0 | 1 | 1 | 1 | 1 | 0 | 1 | 1 | 1 | 1 | 0 | 1 | 0 | 0 | 0 | 0 | 1 | 0 |
| 127 | β-L-arabinofuranosidase | 1 | 0 | 0 | 1 | 1 | 2 | 2 | 0 | 0 | 2 | 2 | 2 | 2 | 1 | 2 | 0 | 1 | 0 | 2 | 0 |
| 129 | α-N-acetylgalactosaminidase | 0 | 0 | 0 | 1 | 1 | 1 | 1 | 1 | 1 | 0 | 1 | 1 | 1 | 1 | 1 | 0 | 0 | 0 | 1 | 0 |
| 130 | β-mannose phosphorylase | 0 | 0 | 0 | 0 | 0 | 0 | 0 | 0 | 0 | 0 | 0 | 0 | 0 | 0 | 0 | 0 | 0 | 0 | 1 | 0 |
| NC | Unknown | 0 | 0 | 0 | 0 | 0 | 0 | 0 | 1 | 0 | 0 | 0 | 0 | 1 | 0 | 0 | 0 | 0 | 0 | 2 | 0 |
| Total | | 63 | 44 | 44 | 65 | 67 | 56 | 61 | 46 | 49 | 60 | 54 | 59 | 59 | 64 | 55 | 60 | 74 | 42 | 126 | 31 |

host-derived carbohydrates such as mucin (Milani et al., 2014, 2015; Egan et al., 2014a,b; Turroni et al., 2016).

Bifidobacteria were found to shape the microbiome of the murine gut either through direct action or by cross-feeding activities (Turroni et al., 2016). This study demonstrated that the addition of two or more bifidobacterial strains resulted in the enhanced levels of persistence of such strains within the murine cecum. Furthermore, members of this genus were capable of modulating gene expression toward an increase in glycan metabolism encompassing both host glycans and diet derived carbohydrates. The bifidobacterial strains had a further influence over the production of SCFAs (Turroni et al., 2016).

## CONTROL OF BIFIDOBACTERIAL CARBOHYDRATE METABOLISM

Carbon catabolite repression (CCR) is a regulatory system present in many bacteria in which the expression or activity of proteins involved in the utilization or uptake of available carbon sources is inhibited by the presence of a preferred carbon source (Postma et al., 1993; Saier and Ramseier, 1996). There is evidence that a CCR mechanism is operating in bifidobacteria, although as of yet a CCR-type regulatory system has not been described for any member of this genus. The first report on CCR-related metabolism in bifidobacteria was in *B. animalis* subsp. *lactis* (Trindade et al., 2003). This study reported an induction in sucrose metabolizing activities when this strain was grown on sucrose, raffinose or oligofructose, whereas a repression in the same metabolic activities was reported for growth on glucose (Trindade et al., 2003). An apparent reverse CCR was reported for *B. longum* NCC2705 in that glucose transport was repressed when lactose was present in the growth medium (Parche et al., 2006). There are two separate reports of CCR-like regulation in *B. breve* UCC2003 (Ryan et al., 2005; Pokusaeva et al., 2010). The first study reported that transcription of the *rbs* operon, responsible for ribose metabolism, is induced when *B. breve* UCC2003 is grown on ribose, whereas transcription of this operon is not induced (or repressed) when the strain is grown on a combination of ribose and glucose (Pokusaeva et al., 2010). The second study on *B. breve* UCC2003 demonstrated that when grown on sucrose or Actilight (a commercial source of short chain FOS), transcription of the *fos* operon, which is involved in the metabolism of FOS, is induced (Ryan et al., 2005). However, this operon was not induced (or repressed) when *B. breve* UCC2003 is grown on a combination of sucrose and glucose, or sucrose and fructose (Ryan et al., 2005).

Transcriptional repressors, such as LacI-type transcriptional regulators, are DNA binding proteins that physically interact with a specific DNA sequence, called the operator, in the vicinity of a regulated promoter, thereby preventing the binding of an RNA polymerase and the initiation of transcription (Ravcheev et al., 2014). Transcriptional repressors may also cause a "road block" for the DNA polymerase thus preventing the progression of transcription (Ravcheev et al., 2014). A substantial number of these repressor proteins have been identified in bifidobacterial genomes, e.g., *B. longum* NCC2705 is predicted to encode 22 LacI-type transcriptional repressors. The presence of a sugar-binding motif in each of these predicted 22 LacI-type proteins indicates that they are predicted to be involved in the regulation of carbohydrate metabolism (Schell et al., 2002). Six different LacI-type regulators have been characterized in *B. breve* UCC2003 and include $LacI_{fos}$, GalR, CldR, and RbsR, which regulate transcription of the *fos* operon, the galactan utilization cluster, the cellodextrin utilization cluster, and the ribose metabolism cluster, respectively (Ryan et al., 2005; Pokusaeva et al., 2010, 2011b; O'Connell Motherway et al., 2011a). More recently, two *B. breve* UCC2003-encoded LacI-type transcriptional repressors, named MelR1 and MelR2, were shown to control the melezitose utilization cluster (O'Connell et al., 2014).

Other carbohydrate-dependent regulatory systems exist, for example, the RafR regulator is a transcriptional activator of the raffinose metabolism gene cluster in *B. breve* UCC2003 (O'Connell et al., 2014), while in the same bacterium a GntR-type transcriptional repressor was shown to control transcription of the large *nag/nan* gene cluster, thereby regulating metabolism of sialic acid (Egan et al., 2015).

## BIOTECHNOLOGY OF BIFIDOBACTERIA

Research into bifidobacterial metabolism has led to the prebiotic concept which in turn confers health benefits to the human host by stimulating the metabolism and activity of bifidobacteria in the GIT (Hijova et al., 2009). However, the actual mechanisms of action that are responsible for such probiotic effects are far from fully understood (Sun et al., 2012). The gold standard approach to investigate the role of a single gene and its products is by site-directed mutagenesis and subsequent phenotypic analysis of the generated mutant(s) (Sun et al., 2012). Unfortunately, bifidobacteria are rather recalcitrant to artificial DNA acquisition methods which means that genetic modification of bifidobacteria has up until recently proven to be impossible. This part of the review will focus on genetic approaches and techniques aimed at improving the genetic accessibility of bifidobacteria.

## BIFIDOBACTERIAL MUTAGENESIS STRATEGIES

There are a small number of reports on targeted gene inactivation in bifidobacteria, although in recent years a number of techniques have been developed resulting the successful inactivation of genes in bifidobacteria. In this part of the review, we describe the fundamental concept of each strategy and discuss some of benefits and pitfalls associated with each.

### Single-Crossover Plasmid Insertion

This method involves the use of a non-replicating plasmid to select for homologous recombination events (**Figure** ). The first successful mutation created in a bifidobacterial strain was in the gene *apuB* using this single-crossover plasmid insertion approach whereby a combination of plasmids facilitated the conditional replication of the non-replicative plasmid pORI19

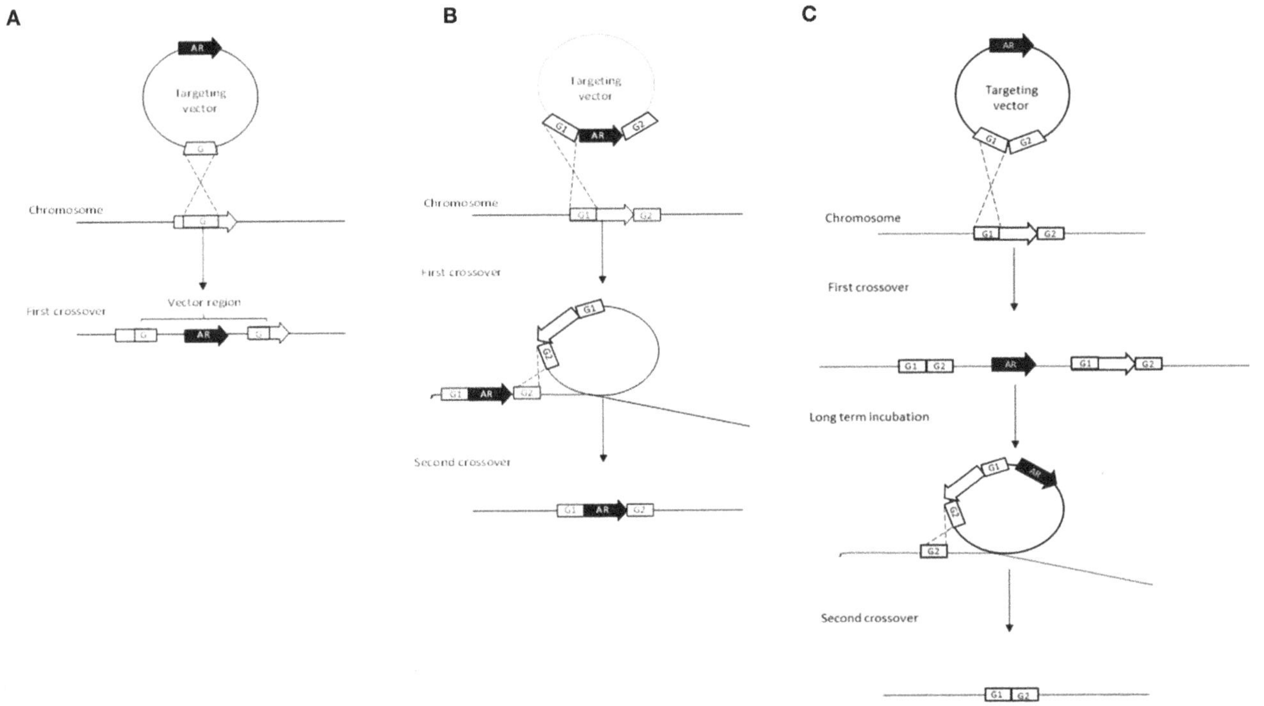


**FIGURE | Schematic representation of targeted gene inactivation approaches.** The targeted gene is indicated by an open arrow and homologous regions are indicated as open squares and given the names G1 and G2. The antibiotic resistance gene is indicated as a black arrow with the letters AR. **(A)** represents the single-crossover plasmid insertion, **(B)** illustrates the double-crossover gene disruption and finally **(C)** illustrates the double-crossover markerless gene deletion approach. (Figure was adapted from Fukiya et al., 2012).

(**Table** ; O'Connell Motherway et al., 2008). Although, it successfully led to the inactivation of the *apuB* gene, this method was quite time consuming and laborious (Fukiya et al., 2012).

As already discussed above, by-pass of native R-M systems and the single-crossover plasmid insertion approach using a non-replicating plasmid has been successfully and repeatedly employed for mutagenesis (O'Connell Motherway et al., 2009, 2011b, 2013; Fouhy et al., 2013; Christiaen et al., 2014; Egan et al., 2014a,b, 2015; O'Connell et al., 2014).

## Double-Crossover and Double-Crossover markerless Gene Deletion

Double-crossover gene disruption was first demonstrated in *B. longum* NCC2705 (**Table** ; Fukuda et al., 2011). This method makes use of a non-replicating plasmid and a double-cross over event in order to create a gene deletion (Fukiya et al., 2012). The non-replicative vector harbors two homologous regions of the target gene between which an antibiotic resistance gene is inserted (**Figure** ). During the first crossover event, homologous recombination occurs between one of the two homologous regions on the vector and on the chromosome. The second (desired) crossover occurs between the second homologous region on the vector and the chromosome, and results in the replacement of the wild-type allele with the mutated one and an antibiotic marker (**Figure** ). In the case of *B. longum* NCC2705, this approach allows for the successful deletion of (part of) the gene *BL0033*, a solute binding protein of an ABC transporter that is highly induced when *B. longum* NCC2705 is grown on fructose (Fukuda et al., 2011).

The double-crossover marker-less strategy has been patented and draws some similarities to the double-crossover gene disruption approach (**Table** , **Figure** C; Arigoni and Delley, 2013). Like the previous approach, a non-replicative plasmid is used for the first crossover, however, in this case the antibiotic marker is located beside the mutated targeted gene (**Figure** ). The second crossover event occurs during long-term sub-culturing of the first crossover integrants. As illustrated in **Figures 1**, **2C** and as described previously for the double-crossover gene disruption, a markerless gene deletion can be generated following a second crossover. An amended version of this method was used to successfully create a marker-less gene deletion in *B. longum* 105-A which involved the conditional replication of a vector plasmid (**Table** ; Hirayama et al., 2012).

The plasmid used for the homologous recombination events is a conditionally replicating plasmid due to a deletion of the gene encoding the replication protein RepA. This plasmid can only replicate in bifidobacteria when the RepA protein is provided *in trans*. The regions flanking the target gene to be deleted are cloned into the conditional replication vector (Hirayama et al., 2012). Once the plasmid has integrated into the target gene via the first homologous recombination or cross-over

**TABLE | Summary of bifidobacterial mutagenesis strategies.**

| Concept | Benefits | Pitfalls | Example |
|---|---|---|---|
| **SINGLE-CROSSOVER PLASMID INSERTION** | | | |
| • Non-replicative plasmid insertion | • Reproducible | • Prior knowledge of the strain is essential | • Disruption of *galA* and *apuB* genes in *B. breve* UCC2003 (O'Connell Motherway et al., 2009) |
| • Single cross-over homologous recombination | • Requires only a single transformation round | • Successful knock-outs left with antibiotic marker | |
| • Internal region of target gene and marker | | • Requires high transformation efficiencies<br>• Unstable mutations | |
| **DOUBLE-CROSSOVER GENE DELETION** | | | |
| • Non-replicative plasmid insertion | • Stable mutation | • Successful knock-outs left with antibiotic marker | • Disruption of *BL0033* in *B. longum* NCC2705 (Fukuda et al., 2011) |
| • Double crossover homologous recombination | • Gene target is deleted | • Time-consuming and laborious- multiple transformations and extensive screening of transformants | |
| • 5′ and 3′ regions of target gene separated by marker | | • Requires high transformation efficiencies | |
| **DOUBLE CROSS-OVER MARKER-LESS GENE DELETION** | | | |
| • Non-replicative plasmid insertion | • Stable mutation | • Time-consuming and laborious- multiple transformations and extensive screening of transformants | • Disruption of *aga* in *B. longum* 105-A (Hirayama et al., 2012) |
| • Double cross-over homologous recombination | • Gene deletion | | |
| • 5′ and 3′ regions of target gene adjacent to marker | • Successful knock-outs left without antibiotic marker | | |
| • Marker-less gene disruption | | | |
| **TEMPERATURE SENSITIVE (TS) PLASMID FOR GENE DISRUPTION** | | | |
| • Ts plasmid unable to replicate at high temperatures | • Does not require high transformation efficiencies | • Time-consuming and laborious- multiple transformations and extensive screening of transformants | • Disruption of *apuB* in *B. breve* UCC2003 (O'Connell Motherway et al., 2008) |
| • Ts plasmid contains homologous sequence and marker | | | |
| • Homologous recombination | | | |
| **TRANSPOSON MUTAGENESIS FOR GENE DISRUPTION** | | | |
| • Random mutagenesis with Transposome complex | • Generation of a large mutant bank | • Successful knock-outs left with antibiotic marker | • Creation of a mutant library in *B. breve* UCC2003 (Ruiz et al., 2013) |
| • Large number of gene disruptions | • High throughput screening of bank | • Requires high transformation efficiencies | |
| • Antibiotic marker | | | |

event, a second plasmid that encodes RepA is introduced which induces the conditional replication of the integrated plasmid (**Figure C**). During the second cross-over and excision, the integrated plasmid is expected to be excised along with the target gene. Second cross-over recombinants that lost the first plasmid can be selected for using appropriate antibiotic markers as the second plasmid is incompatible with the conditional replicating vector (Hirayama et al., 2012).

The double cross-over approaches are more time consuming due to multiple transformations and extensive screening for positive mutants, although the improved double crossover gene deletion system aims to address this (**Table** ; Fukiya et al., 2012). For single cross-over mutants it is theoretically possible for additional cross-over events to occur between the homologous regions left in the target gene thus resulting in the excision of the integrated plasmid (Fukiya et al., 2012). It is also worth noting that the marker-less method is a superior approach as it does not leave an antibiotic resistance gene, the presence of which may cause polar effects affecting genes down-stream of the mutated gene (**Table** ; Fukiya et al., 2012).

## Homologous Recombination Mediated by a Temperature Sensitive Plasmid

The temperature sensitive (Ts) plasmid strategy has been applied successfully to various microorganisms in order to create gene knock-outs (Hamilton et al., 1989; Maguin et al., 1992; Biswas et al., 1993; Takamatsu et al., 2001; Fuchs et al., 2006; Chen et al., 2011). The advantage of this approach over some of the other strategies described above, is that Ts plasmids do not require high transformation efficiencies. Therefore, this would be an ideal and widely applicable approach for (transformation recalcitrant) bifidobacterial strains (**Table** ; Sakaguchi et al., 2012).

The successful creation of insertion mutants using a Ts plasmid has been achieved in *B. breve* UCC2003, *B. longum* 105-A and *B. longum* NCC2705 (O'Connell Motherway et al., 2008; Sakaguchi et al., 2012). A plasmid harboring *repA* was first introduced into *B. breve* UCC2003 followed by the introduction of a non-replicating plasmid that harbored an internal fragment of the target gene, *apuB* (**Table** , **Figure** ; O'Connell Motherway et al., 2008). Once both plasmids were successfully introduced, growth of *B. breve* UCC2003 was shifted to 42°C, thus preventing further replication of the *repA*-containing plasmid. As a consequence this also blocks the replication of the pORI19-derivative. Selection on antibiotic results in the integration of the pORI19-derivative into the *B. breve* UCC2003 genome at the site of the *apuB* gene (**Figure** ; O'Connell Motherway et al., 2008). This temperature-sensitive plasmid approach was also used to create insertion mutants in *B. longum* 105-A and *B. longum* NCC2705 (Sakaguchi et al., 2012).This method was applied to create an insertion mutant in the strain *B. longum* 105-A using *pyrR* as the gene target and was further validated by re-creating a gene deletion in the gene *BL0033* in the strain *B. longum* NCC2705 (**Table** ; Sakaguchi et al., 2012).

## Transposons for Mutagenesis in Bifidobacteria

A recent study has described the creation of a mutant library in *B. breve* UCC2003 using a Tn5-based transposon mutagenesis system (Ruiz et al., 2013). This study is the first to demonstrate the application of random-mutagenesis technology in bifidobacteria (**Table** ). This approach involved the construction of a tetracycline-resistant Tn5 transposon and preparation of a transposome complex that was subsequently introduced into *B. breve* UCC2003 cells by electroporation (Ruiz et al., 2013). Screening of the mutant bank led to the identification of mutants that display defective growth on selected carbohydrates, such as *Bbr_0010* which encodes a β-galactosidase involved in the metabolism of lactose (Ruiz et al., 2013).

The benefit of this method is that it can be used for high-throughput screening in order to identify genes that are fundamental for a given phenotype (Judson and Mekalanos, 2000). However, similarly to targeted insertion mutant systems, high transformation efficiencies are crucial for this system to be effective (Ruiz et al., 2013).

## CONJUGATION IN BIFIDOBACTERIA

Conjugation can also be used by bacteria to transfer genetic material, as was first described in *E. coli* and involving cell-to-cell contact (Lederberg and Tatum, 1946). Conjugation-based techniques have been widely described for Gram-negative bacteria, though less so for Gram-positive bacteria (Schroder and Lanka, 2005).

Currently, transformation by electroporation is the most popular method used to genetically manipulate bifidobacteria. Shortfalls maybe overcome by the use of a

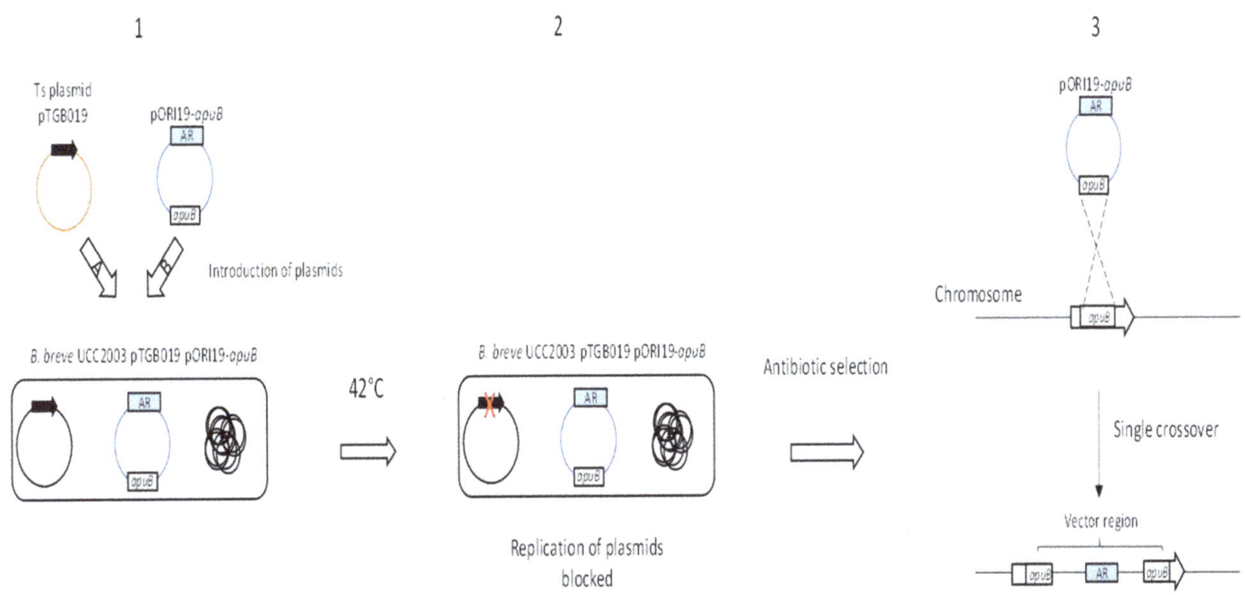


**FIGURE | Schematic representation of insertion mutagenesis using a temperature sensitive plasmid.** The targeted gene is indicated by the text *apuB* in the open gray box and the antibiotic resistance gene is indicated as a blue box with the letters AR. The black arrow represents the replication gene *repA* and the red-cross indicates a gene that is non-functional. Step **1** illustrates the introduction of (a) the plasmid pTGB019 into *B. breve* UCC2003 followed by (b) the introduction of the non-replicative plasmids pORI19-apuB. Step **2** illustrates a shift in temperature resulting in the blocked replication of pTGB019 and pORI19-apuB as a consequence of a non-functional *repA* gene. Finally in Step **3**, presence of antibiotic selects for integration of pORI19-derivatives into the desired site on the *B. breve* UCC2003 genome.

conjugation-based approach as it holds a number of advantages over transformation (Dominguez and O'Sullivan, 2013). Firstly, unlike electroporation, the size of the vector does not affect conjugation efficacy (Szostková and Horáková, 1998; Isaacs et al., 2011). Secondly, the presence of R-M systems is not an issue for conjugative methods as during conjugation DNA is transferred as a single strand from donor to recipient, in which form it is insensitive to most R-M systems (Dominguez and O'Sullivan, 2013).

The first, though unsuccessful, attempt at conjugative transfer in bifidobacteria was reported in 1998 (Shkoporov et al., 2008). A recent publication describes a conjugative gene transfer method and the first successful transfer of DNA between *E. coli* and different *Bifidobacterium* species (Dominguez and O'Sullivan, 2013). Based on the RP4 conjugative machinery of *E. coli* WM3064 (pBB109), the *E. coli-Bifidobacterium* shuttle vector pDOJHR-WD2 was constructed. This plasmid was transformed into *E. coli* WM3064 harboring pBB109. This latter plasmid encodes the relaxase function to catalyze nicking at the *oriT* site on pDOJHR-WD2, while RP4, which is integrated into the *E. coli* WM3064 chromosome, specifies Tra functions (Dominguez and O'Sullivan, 2013). Successful transfer of pDOJHR-WD2 from *E. coli* to members of four different bifidobacterial species was reported with an efficiency ranging from $1.8 \times 10^{-4}$ to $7.5 \times 10^{-6}$ transconjugants per recipient (Dominguez and O'Sullivan, 2013).

The recent sequencing of the *B. breve* JCM 7017 genome led to the discovery of a 190 kb megaplasmid designated as pMP7017 (Bottacini et al., 2015). Analysis of the plasmid sequence led to the identification of genes that are predicted to encode conjugative machinery. Sequence analysis of pMP7017 led to the observation that, although the overall G+C content is akin to that of its host, it appears that this megaplasmid was formed from the co-integration of two separate modules (Bottacini et al., 2015). The functionality of the conjugative mechanism was demonstrated by the successful transfer of pMP7017 from *B. breve* JCM7017 to *B. breve* and *B. longum* representatives (Bottacini et al., 2015).

The recent development of conjugation-based gene transfer systems and the discovery of the first native conjugative bifidobacterial plasmid is quite significant (Dominguez and O'Sullivan, 2013; Bottacini et al., 2015), and may assist in future tool development for the genetic modification of bifidobacteria.

## MICROBE-HOST INTERACTIONS

Insertion mutagenesis has proven to be a fundamental tool in the development of functional genomics in bifidobacteria. Two studies in particular demonstrate how the use of insertion mutagenesis can reveal how candidates genes are involved in the beneficial effects conferred to the host by bifidobacteria (Fukiya et al., 2012).

In the first example, Fukuda et al. combined comparative genomics, gene expression profiling and gene knock-outs to demonstrate that acetate production by the strain *B. longum* subsp. *longum* JCM 1217 is linked to the protection of host epithelial cells from lethal doses of the shiga toxin, Stx2 (Fukuda et al., 2011). Early observations highlighted that germ-free mice pre-colonized with *B. longum* subsp. *longum* JCM 1217 and subsequently infected with enterohaemorrhagic *E. coli* O157:H7 (EHEC O157:H7), survived better than germ-free mice that were infected with EHEC O157:H7. In contrast, the same survival was not observed when, prior to infection with EHEC O157:H7, mice were colonized with *B. adolescentis* JCM 1275. Comparative genomic analysis performed between protective *B. longum* subsp. *longum* and non-protective *B. adolescentis* implicated an ABC-type transporter system being involved in protection, after which a knock-out mutation was created in the gene encoding the corresponding solute binding protein using a double crossover gene disruption approach (Fukuda et al., 2011). Due to the consequently lower levels of acetate produced, the mutant strain was unable to protect mice from EHEC O157:H7 infection, whereas when the non-protective *B. adolescentis* strain expressed the ABC transporter cluster this resulted in moderate increases in mouse survival following EHEC O157:H7 infection as well as increased acetate production.

The second example of a functional microbe-host interaction study is the identification of candidate genes directly responsible for the colonization of *B. breve* UCC2003 in a murine colonization model (O'Connell Motherway et al., 2011b). Genome and *in vivo* transcriptome analyses of *B. breve* UCC2003 revealed that a gene cluster, responsible for the production of so-called type IVb or tight adherence (Tad) pili,is required for host colonization (O'Connell Motherway et al., 2011b).

As well as investigating candidate genes directly involved in host-microbe interactions, insertion mutagenesis has also been exploited in functional genomic studies investigating the physiological characteristics of bifidobacteria, such as carbohydrate metabolism (O'Connell Motherway et al., 2011a; O'Connell et al., 2013, 2014; Egan et al., 2014b, 2015).

## CONCLUSION

It is well established that bifidobacteria confer positive health benefits to their host via their metabolic activities. The availability of complete bifidobacterial genomes and corresponding comparative analysis allows for the identification of mechanisms underlying bifidobacterial metabolic activity. Carbohydrate utilization studies and identification of metabolic pathways also provides fundamental information allowing for the identification of novel and effective prebiotic compounds.

Plant-derived and host derived carbohydrates have been shown to stimulate the growth of some bifidobacterial species. To identify and obtain full knowledge of the genes implicated in carbohydrate degradation and utilization, characterization and mutagenesis of candidate genes is required. However, bifidobacteria are notoriously recalcitrant to genetic modification.

It is therefore essential that future studies continue to address the shortage of effective molecular tools. The development of these tools is essential to unravel the underlying molecular mechanisms that explain how bifidobacteria interact with their human host.

## AUTHOR CONTRIBUTIONS

All authors listed, have made substantial, direct and intellectual contribution to the work, and approved it for publication.

## FUNDING

This publication has emanated from research conducted with the financial support of Science Foundation Ireland (SFI) under Grant Number SFI/12/RC/2273. AO was supported by an enterprise partnership scheme of the Irish Research Council.

## REFERENCES

Amaretti, A., Bernardi, T., Leonardi, A., Raimondi, S., Zanoni, S., and Rossi, M. (2013). Fermentation of xylo-oligosaccharides by *Bifidobacterium* adolescentis DSMZ 18350: kinetics, metabolism, and beta-xylosidase activities. *Appl. Microbiol. Biotechnol.* 97, 3109–3117. doi: 10.1007/s00253-012-4509-y

Arigoni, F., and Delley, M. (2013). *Genetic Remodeling in Bifidobacterium.* Vevey: Google Patents.

Backhed, F., Ley, R. E., Sonnenburg, J. L., Peterson, D. A., and Gordon, J. I. (2005). Host-bacterial mutualism in the human intestine. *Science,* 307, 1915–1920. doi: 10.1126/science.1104816

Bae, E.-A., Han, M. J., Song, M.-J., and Kim, D.-H. (2002). *Purification of Rotavirus Infection-Inhibitory Protein from Bifidobacterium Breve K-110.* Seoul: COREE, REPUBLIQUE DE, Korean Society for Applied Microbiology.

Barrangou, R., Briczinski, E. P., Traeger, L. L., Loquasto, J. R., Richards, M., Horvath, P., et al. (2009). Comparison of the complete genome sequences of Bifidobacterium animalis subsp. lactis DSM 10140 and Bl-04. *J. Bacteriol.* 191, 4144–4151. doi: 10.1128/JB.00155-09

Bernet, M. F., Brassart, D., Neeser, J. R., and Servin, A. L. (1994). Lactobacillus acidophilus LA 1 binds to cultured human intestinal cell lines and inhibits cell attachment and cell invasion by enterovirulent bacteria. *Gut,* 35, 483–489. doi: 10.1136/gut.35.4.483

Bindels, L. B., Delzenne, N. M., Cani, P. D., and Walter, J. (2015). Towards a more comprehensive concept for prebiotics. *Nat. Rev. Gastroenterol. Hepatol.* 12, 303–310. doi: 10.1038/nrgastro.2015.47

Biswas, I., Gruss, A., Ehrlich, S. D., and Maguin, E. (1993). High-efficiency gene inactivation and replacement system for gram-positive bacteria. *J. Bacteriol.* 175, 3628–3635.

Bosscher, D., Van Loo, J., and Franck, A. (2006). Inulin and oligofructose as prebiotics in the prevention of intestinal infections and diseases. *Nutr. Res. Rev.* 19, 216–226. doi: 10.1017/S0954422407249686

Bottacini, F., O'Connell Motherway, M., Casey, E., McDonnell, B., Mahony, J., Ventura, M., et al. (2015). Discovery of a conjugative megaplasmid in bifidobacterium breve. *Appl. Environ. Microbiol.* 81, 166–176. doi: 10.1128/AEM.02871-14

Broekaert, W. F., Courtin, C. M., Verbeke, K., Van De Wiele, T., Verstraete, W., and Delcour, J. A. (2011). Prebiotic and other health-related effects of cereal-derived arabinoxylans, arabinoxylan-oligosaccharides, and xylooligosaccharides. *Crit. Rev. Food Sci. Nutr.* 51, 178–194. doi: 10.1080/10408390903044768

Chaplin, A. V., Efimov, B. A., Smeianov, V. V., Kafarskaia, L. I., Pikina, A. P., and Shkoporov, A. N. (2015). Intraspecies genomic diversity and long-term persistence of *Bifidobacterium longum. PLoS ONE* 10:e0135658. doi: 10.1371/journal.pone.0135658

Chen, X., Dego, O. K., Almeida, R. A., Fuller, T. E., Luther, D. A., and Oliver, S. P. (2011). Deletion of sua gene reduces the ability of Streptococcus uberis to adhere to and internalize into bovine mammary epithelial cells. *Vet. Microbiol.* 147, 426–434. doi: 10.1016/j.vetmic.2010.07.006

Chenoll, E., Rivero, M., Codoñer, F. M., Martinez-Blanch, J. F., Ramón, D., Genovés, S., et al. (2015). Complete genome sequence of *Bifidobacterium longum* subsp. *infantis* Strain CECT 7210, a probiotic strain active against rotavirus infections. *Genome Announcements* 3:e00105-15. doi: 10.1128/genomea.00105-15

Christiaen, S. E., O'Connell Motherway, M., Bottacini, F., Lanigan, N., Casey, P. G., Huys, G., et al. (2014). Autoinducer-2 plays a crucial role in gut colonization and probiotic functionality of *Bifidobacterium breve* UCC2003. *PLoS ONE* 9:e98111. doi: 10.1371/journal.pone.0098111

Corrêa, N. B. O., Péret Filho, L. A., Penna, F. J., Lima, F. M. L. S., and Nicoli, J. R. (2005). A randomized formula controlled trial of *bifidobacterium* lactis and streptococcus thermophilus for prevention of antibiotic-associated diarrhea in infants. *J. Clin. Gastroenterol.* 39, 385–389. doi: 10.1097/01.mcg.0000159217.47419.5b

Crittenden, R., Karppinen, S., Ojanen, S., Tenkanen, M., Fagerström, R., Mättö, J., et al. (2002). *In vitro* fermentation of cereal dietary fibre carbohydrates by probiotic and intestinal bacteria. *J. Sci. Food Agric.* 82, 781–789. doi: 10.1002/jsfa.1095

Cronin, M., Ventura, M., Fitzgerald, G. F., and Van Sinderen, D. (2011). Progress in genomics, metabolism and biotechnology of bifidobacteria. *Int. J. Food Microbiol.* 149, 4–18. doi: 10.1016/j.ijfoodmicro.2011.01.019

Davidson, A. L., Dassa, E., Orelle, C., and Chen, J. (2008). Structure, function, and evolution of bacterial ATP-binding cassette systems. *Microbiol Mol. Biol. Rev.* 72, 317–364. doi: 10.1128/MMBR.00031-07

Davis, L. M., Martinez, I., Walter, J., and Hutkins, R. (2010). A dose dependent impact of prebiotic galactooligosaccharides on the intestinal microbiota of healthy adults. *Int. J. Food Microbiol.* 144, 285–292. doi: 10.1016/j.ijfoodmicro.2010.10.007

Degnan, B. A., and Macfarlane, G. T. (1993). Transport and metabolism of glucose and arabinose in *Bifidobacterium* breve. *Arch. Microbiol.* 160, 144–151. doi: 10.1007/BF00288717

De La Cochetiere, M. F., Piloquet, H., Des Robert, C., Darmaun, D., Galmiche, J. P., and Roze, J. C. (2004). Early intestinal bacterial colonization and necrotizing enterocolitis in premature infants: the putative role of *Clostridium. Pediatr. Res.* 56, 366–370. doi: 10.1203/01.PDR.0000134251.45878.D5

De Vries, W., and Stouthamer, A. H. (1967). Pathway of glucose fermentation in relation to the taxonomy of bifidobacteria. *J. Bacteriol.* 93, 574–576.

De Vuyst, L., and Leroy, F. (2011). Cross-feeding between bifidobacteria and butyrate-producing colon bacteria explains bifdobacterial competitiveness, butyrate production, and gas production. *Int. J. Food Microbiol.* 149, 73–80. doi: 10.1016/j.ijfoodmicro.2011.03.003

De Vuyst, L., Moens, F., Selak, M., Riviere, A., and Leroy, F. (2014). Summer Meeting 2013: growth and physiology of bifidobacteria. *J. Appl. Microbiol.* 116, 477–491. doi: 10.1111/jam.12415

Dominguez, W., and O'Sullivan, D. J. (2013). Developing an efficient and reproducible conjugation-based gene transfer system for bifidobacteria. *Microbiology* 159, 328–338. doi: 10.1099/mic.0.061408-0

Duffy, L. C., Zielezny, M. A., Riepenhoff-Talty, M., Dryja, D., Sayahtaheri-Altaie, S., Griffiths, E., et al. (1994a). Reduction of virus shedding by *B. bifidum* in experimentally induced MRV infection. Statistical application for ELISA. *Dig. Dis. Sci.* 39, 2334–2340. doi: 10.1007/BF02087647

Duffy, L. C., Zielezny, M. A., Riepenhoff-Talty, M., Dryja, D., Sayahtaheri-Altaie, S., Griffiths, E., et al. (1994b). Effectiveness of *Bifidobacterium bifidum* in mediating the clinical course of murine rotavirus diarrhea. *Pediatr. Res.* 35, 690–695. doi: 10.1203/00006450-199406000-00014

Duranti, S., Milani, C., Lugli, G. A., Turroni, F., Mancabelli, L., Sanchez, B., et al. (2015). Insights from genomes of representatives of the human gut commensal *Bifidobacterium bifidum. Environ. Microbiol.* 17, 2515–2531. doi: 10.1111/1462-2920.12743

Duranti, S., Turroni, F., Lugli, G. A., Milani, C., Viappiani, A., Mangifesta, M., et al. (2014). Genomic characterization and transcriptional studies of the starch-utilizing strain *bifidobacterium* adolescentis 22L. *Appl. Environ. Microbiol.* 80, 6080–6090. doi: 10.1128/AEM.01993-14

Egan, M., Motherway, M. M., Kilcoyne, M., Kane, M., Joshi, L., Ventura, M., et al. (2014a). Cross-feeding by *Bifidobacterium breve* UCC2003 during co-cultivation with Bifidobacterium bifidum PRL2010 in a mucin-based medium. *BMC Microbiol.* 14:282. doi: 10.1186/s12866-014-0282-7

Egan, M., O'Connell Motherway, M., and Van Sinderen, D. (2015). A GntR-type transcriptional repressor controls sialic acid utilization in *Bifidobacterium breve* UCC2003. *FEMS Microbiol. Lett.* 362, 1–9. doi: 10.1093/femsle/fnu056

Egan, M., O'Connell Motherway, M., Ventura, M., and Van Sinderen, D. (2014b). Metabolism of sialic acid by *Bifidobacterium* breve UCC2003. *Appl. Environ. Microbiol.* 80, 4414–4426. doi: 10.1128/AEM.01114-14

El Kaoutari, A., Armougom, F., Gordon, J. I., Raoult, D., and Henrissat, B. (2013). The abundance and variety of carbohydrate-active enzymes in the human gut microbiota. *Nat. Rev. Microbiol.* 11, 497–504. doi: 10.1038/nrmicro3050

Falony, G., Lazidou, K., Verschaeren, A., Weckx, S., Maes, D., and De Vuyst, L. (2009). *In vitro* kinetic analysis of fermentation of prebiotic inulin-type fructans by Bifidobacterium species reveals four different phenotypes. *Appl. Environ. Microbiol.* 75, 454–461. doi: 10.1128/AEM.01488-08

Fanaro, S., Chierici, R., Guerrini, P., and Vigi, V. (2003). Intestinal microflora in early infancy: composition and development. *Acta Paediatr Suppl.* 91, 48–55. doi: 10.1111/j.1651-2227.2003.tb00646.x

FAO/WHO (2001). *Health and Nutritional Properties of Probiotics in Food Including Powder Milk with Live Lactic Acid Bacteria.* Cordoba: FAO/WHO.

Favier, C. F., Vaughan, E. E., De Vos, W. M., and Akkermans, A. D. (2002). Molecular monitoring of succession of bacterial communities in human neonates. *Appl. Environ. Microbiol.* 68, 219–226. doi: 10.1128/AEM.68.1.219-226.2002

Fouhy, F., O'Connell Motherway, M., Fitzgerald, G. F., Ross, R. P., Stanton, C., Van Sinderen, D., et al. (2013). *In Silico* assigned resistance genes confer *Bifidobacterium* with partial resistance to aminoglycosides but not to ÎŠ-lactams. *PLoS ONE* 8:e82653. doi: 10.1371/journal.pone.0082653

Fuchs, T. M., Klumpp, J., and Przybilla, K. (2006). Insertion-duplication mutagenesis of *Salmonella enterica* and related species using a novel thermosensitive vector. *Plasmid* 55, 39–49. doi: 10.1016/j.plasmid.2005.06.003

Fukiya, S., Hirayama, Y., Sakanaka, M., Kano, Y., and Yokota, A. (2012). Technological Advances in *Bifidobacterial* molecular genetics: application to functional genomics and medical treatments. *Biosci. Microbiota Food Health* 31, 15–25. doi: 10.12938/bmfh.31.15

Fukuda, S., Toh, H., Hase, K., Oshima, K., Nakanishi, Y., Yoshimura, K., et al. (2011). Bifidobacteria can protect from enteropathogenic infection through production of acetate. *Nature* 469, 543–547. doi: 10.1038/nature09646

Ganong, W. (2005). *Ganong's Review of Medical Physiology, 24th Edn,* Europe, McGraw-Hill Education.

Gibson, G. R., and Roberfroid, M. B. (1995). Dietary modulation of the human colonic microbiota: introducing the concept of prebiotics. *J. Nutr.* 125, 1401–1412.

Gionchetti, P., Rizzello, F., Venturi, A., and Campieri, M. (2000). Probiotics in infective diarrhoea and inflammatory bowel diseases. *J. Gastroenterol. Hepatol.* 15, 489–493. doi: 10.1046/j.1440-1746.2000.02162.x

Grootaert, C., Delcour, J. A., Courtin, C. M., Broekaert, W. F., Verstraete, W., and Van De Wiele, T. (2007). Microbial metabolism and prebiotic potency of arabinoxylan oligosaccharides in the human intestine. *Trends Food Sci. Technol.* 18, 64–71. doi: 10.1016/j.tifs.2006.08.004

Guarner, F., and Malagelada, J. R. (2003). Gut flora in health and disease. *Lancet* 361, 512–519. doi: 10.1016/S0140-6736(03)12489-0

Gueimonde, M., Margolles, A., G., De Los Reyes-Gavilán, C., and Salminen, S. (2007). Competitive exclusion of enteropathogens from human intestinal mucus by *Bifidobacterium* strains with acquired resistance to bile — A preliminary study. *Int. J. Food Microbiol.* 113, 228–232. doi: 10.1016/j.ijfoodmicro.2006.05.017

Guyonnet, D., Chassany, O., Ducrotte, P., Picard, C., Mouret, M., Mercier, C. H., et al. (2007). Effect of a fermented milk containing *Bifidobacterium animalis* DN-173 010 on the health-related quality of life and symptoms in irritable bowel syndrome in adults in primary care: a multicentre, randomized, double-blind, controlled trial. *Aliment. Pharmacol. Ther.* 26, 475–486. doi: 10.1111/j.1365-2036.2007.03362.x

Hamilton, C. M., Aldea, M., Washburn, B. K., Babitzke, P., and Kushner, S. R. (1989). New method for generating deletions and gene replacements in *Escherichia coli. J. Bacteriol.* 171, 4617–4622.

Harmsen, H. J., Wildeboer-Veloo, A. C., Raangs, G. C., Wagendorp, A. A., Klijn, N., Bindels, J. G., et al. (2000). Analysis of intestinal flora development in breast-fed and formula-fed infants by using molecular identification and detection methods. *J. Pediatr. Gastroenterol. Nutr.* 30, 61–67. doi: 10.1097/00005176-200001000-00019

Hijova, E., Chmelarova, A., Bomba, A., and Zitnan, R. (2009). Prebiotic foodstuffs and their health benefits in experiment. *Bratisl. Lek. Listy* 110, 523–525.

Hill, C., Guarner, F., Reid, G., Gibson, G. R., Merenstein, D. J., Pot, B., et al. (2014). Expert consensus document: the International Scientific Association for probiotics and prebiotics consensus statement on the scope and appropriate use of the term probiotic. *Nat. Rev. Gastroenterol. Hepatol.* 11, 506–514. doi: 10.1038/nrgastro.2014.66

Hirayama, Y., Sakanaka, M., Fukuma, H., Murayama, H., Kano, Y., Fukiya, S., et al. (2012). Development of a double-crossover markerless gene deletion system in *Bifidobacterium* longum: functional analysis of the α-galactosidase gene for raffinose assimilation. *Appl. Environ. Microbiol.* 78, 4984–4994. doi: 10.1128/AEM.00588-12

Hooper, L. V., Midtvedt, T., and Gordon, J. I. (2002). How host-microbial interactions shape the nutrient environment of the mammalian intestine. *Annu. Rev. Nutr.* 22, 283–307. doi: 10.1146/annurev.nutr.22.011602.092259

Hooper, L. V., Xu, J., Falk, P. G., Midtvedt, T., and Gordon, J. I. (1999). A molecular sensor that allows a gut commensal to control its nutrient foundation in a competitive ecosystem. *Proc. Natl. Acad. Sci. U.S.A.* 96, 9833–9838. doi: 10.1073/pnas.96.17.9833

Hopkins, M., Sharp, R., and Macfarlane, G. (2001). Age and disease related changes in intestinal bacterial populations assessed by cell culture, 16S rRNA abundance, and community cellular fatty acid profiles. *Gut* 48, 198–205. doi: 10.1136/gut.48.2.198

Hughes, S. A., Shewry, P. R., Li, L., Gibson, G. R., Sanz, M. L., and Rastall, R. A. (2007). *In Vitro* Fermentation by Human Fecal Microflora of Wheat Arabinoxylans. *J. Agric. Food Chem.* 55, 4589–4595. doi: 10.1021/jf070293g

Hyun, Y. J., Kim, B., and Kim, D. H. (2012). Cloning and characterization of ginsenoside Ra1-hydrolyzing beta-D-xylosidase from Bifidobacterium breve K-110. *J. Microbiol. Biotechnol.* 22, 535–540. doi: 10.4014/jmb.1110.10001

Isaacs, F. J., Carr, P. A., Wang, H. H., Lajoie, M. J., Sterling, B., Kraal, L., et al. (2011). Precise manipulation of chromosomes *in vivo* enables genome-wide codon replacement. *Science* 333, 348–353. doi: 10.1126/science.1205822

Ishikawa, H., Matsumoto, S., Ohashi, Y., Imaoka, A., Setoyama, H., Umesaki, Y., et al. (2011). Beneficial effects of probiotic bifidobacterium and galacto-oligosaccharide in patients with ulcerative colitis: a randomized controlled study. *Digestion* 84, 128–133. doi: 10.1159/000322977

Islek, A., Sayar, E., Yilmaz, A., Baysan, B. O., Mutlu, D., and Artan, R. (2014). The role of Bifidobacterium lactis B94 plus inulin in the treatment of acute infectious diarrhea in children. *Turk. J. Gastroenterol.* 25, 628–633. doi: 10.5152/tjg.2014.14022

Jaskari, J., Kontula, P., Siitonen, A., Jousimies-Somer, H., Mattila-Sandholm, T., and Poutanen, K. (1998). Oat beta-glucan and xylan hydrolysates as selective substrates for *Bifidobacterium* and *Lactobacillus* strains. *Appl. Microbiol. Biotechnol.* 49, 175–181. doi: 10.1007/s002530051155

Jojima, T., Omumasaba, C. A., Inui, M., and Yukawa, H. (2010). Sugar transporters in efficient utilization of mixed sugar substrates: current knowledge and outlook. *Appl. Microbiol. Biotechnol.* 85, 471–480. doi: 10.1007/s00253-009-2292-1

Judson, N., and Mekalanos, J. J. (2000). Transposon-based approaches to identify essential bacterial genes. *Trends Microbiol.* 8, 521–526. doi: 10.1016/S0966-842X(00)01865-5

Ju-Hoon, L., and O'Sullivan, D. J. (2010). Genomic Insights into Bifidobacteria. *Microbiology and Molecular Biology Reviews,* 74, 38. doi: 10.1128/MMBR.00004-10

Kapiki, A., Costalos, C., Oikonomidou, C., Triantafyllidou, A., Loukatou, E., and Pertrohilou, V. (2007). The effect of a fructo-oligosaccharide supplemented formula on gut flora of preterm infants. *Early Hum. Dev.* 83, 335–339. doi: 10.1016/j.earlhumdev.2006.07.003

Klappenbach, J. A., Dunbar, J. M., and Schmidt, T. M. (2000). rRNA operon copy number reflects ecological strategies of bacteria. *Appl. Environ. Microbiol.* 66, 1328–1333. doi: 10.1128/AEM.66.4.1328-1333.2000

Kleessen, B., Sykura, B., Zunft, H. J., and Blaut, M. (1997). Effects of inulin and lactose on fecal microflora, microbial activity, and bowel habit in elderly constipated persons. *Am. J. Clin. Nutr.* 65, 1397–1402.

Klijn, A., Mercenier, A., and Arigoni, F. (2005). Lessons from the genomes of *Bifidobacteria*. *FEMS Microbiol. Rev.* 29, 491–509. doi: 10.1016/j.fmrre.2005.04.010

Kolida, S., Tuohy, K., and Gibson, G. R. (2002). Prebiotic effects of inulin and oligofructose. *Br. J. Nutr.* 87 (Suppl. 2), S193–S197. doi: 10.1079/BJN/2002537

Korakli, M., Ganzle, M. G., and Vogel, R. F. (2002). Metabolism by bifidobacteria and lactic acid bacteria of polysaccharides from wheat and rye, and exopolysaccharides produced by *Lactobacillus* sanfranciscensis. *J. Appl. Microbiol.* 92, 958–965. doi: 10.1046/j.1365-2672.2002.01607.x

Kumemura, M., Hashimoto, F., Fujii, C., Matsuo, K., Kimura, H., Miyazoe, R., et al. (1992). Effects of administration of 4g-beta-d-galactosylsucrose on fecal microflora, putrefactive products, short-chain fatty-acids, weight, moisture and ph, and subjective sensation of defecation in the elderly with constipation. *J. Clin. Biochem. Nutr.* 13, 199–210. doi: 10.3164/jcbn.13.199

Lagaert, S., Pollet, A., Delcour, J. A., Lavigne, R., Courtin, C. M., and Volckaert, G. (2010). Substrate specificity of three recombinant alpha-L-arabinofuranosidases from *Bifidobacterium* adolescentis and their divergent action on arabinoxylan and arabinoxylan oligosaccharides. *Biochem. Biophys. Res. Commun.* 402, 644–650. doi: 10.1016/j.bbrc.2010.10.075

Lagaert, S., Pollet, A., Delcour, J. A., Lavigne, R., Courtin, C. M., and Volckaert, G. (2011). Characterization of two beta-xylosidases from Bifidobacterium adolescentis and their contribution to the hydrolysis of prebiotic xylooligosaccharides. *Appl. Microbiol. Biotechnol.* 92, 1179–1185. doi: 10.1007/s00253-011-3396-y

Lagaert, S., Van Campenhout, S., Pollet, A., Bourgois, T. M., Delcour, J. A., Courtin, C. M., et al. (2007). Recombinant expression and characterization of a reducing-end xylose-releasing exo-oligoxylanase from *Bifidobacterium* adolescentis. *Appl. Environ. Microbiol.* 73, 5374–5377. doi: 10.1128/AEM.00722-07

Laureys, D., Cnockaert, M., De Vuyst, L., and Vandamme, P. (2016). Bifidobacterium aquikefiri sp. nov., isolated from water kefir. *Int. J. Syst. Evol. Microbiol.* 66, 1281–1286. doi: 10.1099/ijsem.0.000877

Leahy, S. C., Higgins, D. G., Fitzgerald, G. F., and Van Sinderen, D. (2005). Getting better with bifidobacteria. *J. Appl. Microbiol.* 98, 1303–1315. doi: 10.1111/j.1365-2672.2005.02600.x

Lederberg, J., and Tatum, E. L. (1946). Gene recombination in *Escherichia coli*. *Nature* 158, 558. doi: 10.1038/158558a0

Lee, J. H., Hyun, Y. J., and Kim, D. H. (2011). Cloning and characterization of alpha-L-arabinofuranosidase and bifunctional alpha-L-arabinopyranosidase/beta-D-galactopyranosidase from *Bifidobacterium* longum H-1. *J. Appl. Microbiol.* 111, 1097–1107. doi: 10.1111/j.1365-2672.2011.05128.x

Lee, J. H., Karamychev, V. N., Kozyavkin, S. A., Mills, D., Pavlov, A. R., Pavlova, N. V., et al. (2008). Comparative genomic analysis of the gut bacterium Bifidobacterium longum reveals loci susceptible to deletion during pure culture growth. *BMC Genomics* 9:247. doi: 10.1186/1471-2164-9-247

Le Leu, R. K., Hu, Y., Brown, I. L., Woodman, R. J., and Young, G. P. (2010). Synbiotic intervention of *Bifidobacterium lactis* and resistant starch protects against colorectal cancer development in rats. *Carcinogenesis* 31, 246–251. doi: 10.1093/carcin/bgp197

Lombard, V., Golaconda Ramulu, H., Drula, E., Coutinho, P. M., and Henrissat, B. (2014). The carbohydrate-active enzymes database (CAZy) in (2013). *Nucleic Acids Res.* 42, D490–D495. doi: 10.1093/nar/gkt1178

Lorca, G. L., Barabote, R. D., Zlotopolski, V., Tran, C., Winnen, B., Hvorup, R. N., et al. JR. (2007). Transport capabilities of eleven gram-positive bacteria: comparative genomic analyses. *Biochim. Biophys. Acta* 1768, 1342–1366. doi: 10.1016/j.bbamem.2007.02.007

Louis, P., Young, P., Holtrop, G., and Flint, H. J. (2010). Diversity of human colonic butyrate-producing bacteria revealed by analysis of the butyryl-CoA:acetate CoA-transferase gene. *Environ. Microbiol.* 12, 304–314. doi: 10.1111/j.1462-2920.2009.02066.x

Maguin, E., Duwat, P., Hege, T., Ehrlich, D., and Gruss, A. (1992). New thermosensitive plasmid for gram-positive bacteria. *J. Bacteriol.* 174, 5633–5638.

Margolles, A., and De Los Reyes-Gavilan, C. G. (2003). Purification and functional characterization of a novel alpha-L-arabinofuranosidase from Bifidobacterium longum B667. *Appl. Environ. Microbiol.* 69, 5096–5103. doi: 10.1128/AEM.69.9.5096-5103.2003

Marteau, P., Cuillerier, E., Meance, S., Gerhardt, M. F., Myara, A., Bouvier, M., et al. (2002). *Bifidobacterium animalis* strain DN-173 010 shortens the colonic transit time in healthy women: a double-blind, randomized, controlled study. *Aliment. Pharmacol. Ther.* 16, 587–593. doi: 10.1046/j.1365-2036.2002.01188.x

Meance, S., Cayuela, C., Turchet, P., Raimondi, A., Lucas, C., and Antoine, J.-M. (2011). A fermented milk with a *Bifidobacterium* probiotic strain DN-173 010 shortened oro-fecal gut transit time in elderly. *Microb. Ecol. Health Dis.* 13, 217–222.

Menne, E., Guggenbuhl, N., and Roberfroid, M. (2000). Fn-type chicory inulin hydrolysate has a prebiotic effect in humans. *J. Nutr.* 130, 1197–1199.

Milani, C., Andrea Lugli, G., Duranti, S., Turroni, F., Mancabelli, L., Ferrario, C., et al. (2015). Bifidobacteria exhibit social behavior through carbohydrate resource sharing in the gut. *Sci. Rep.* 5, 15782. doi: 10.1038/srep15782

Milani, C., Lugli, G. A., Duranti, S., Turroni, F., Bottacini, F., Mangifesta, M., et al. (2014). Genomic encyclopedia of type strains of the genus bifidobacterium. *Appl. Environ. Microbiol.* 80, 6290–6302. doi: 10.1128/AEM.02308-14

Moura, P., Barata, R., Carvalheiro, F., Girio, F., Loureiro-Dias, M. C., and Esteves, M. P. (2007). *In vitro* fermentation of xylo-oligosaccharides from corn cobs autohydrolysis by Bifidobacterium and Lactobacillus strains. *Lwt-Food Sci. Technol.* 40, 963–972. doi: 10.1016/j.lwt.2006.07.013

Neyrinck, A. M., Possemiers, S., Druart, C., Van De Wiele, T., De Backer, F., Cani, P. D., et al. (2011). Prebiotic effects of wheat arabinoxylan related to the increase in bifidobacteria, *Roseburia* and *Bacteroides/Prevotella* in diet-induced obese mice. *PLoS ONE* 6:e20944. doi: 10.1371/journal.pone.0020944

O'Connell, K. J., O'Connell Motherway, M., Liedtke, A., Fitzgerald, G. F., Ross, R. P., Stanton, C., et al. (2014). Transcription of two adjacent carbohydrate utilization gene clusters in *Bifidobacterium* breve UCC2003 Is controlled by laci- and Repressor Open Reading Frame Kinase (ROK)-Type regulators. *Appl. Environ. Microbiol.* 80, 3604–3614. doi: 10.1128/AEM.00130-14

O'Connell, K. J., O'Connell Motherway, M., O'Callaghan, J., Fitzgerald, G. F., Ross, R. P., Ventura, M., et al. (2013). Metabolism of four alpha-glycosidic linkage-containing oligosaccharides by *Bifidobacterium* breve UCC2003. *Appl. Environ. Microbiol.* 79, 6280–6292. doi: 10.1128/AEM.01775-13

O'Connell Motherway, M., Fitzgerald, G. F., Neirynck, S., Ryan, S., Steidler, L., and Van Sinderen, D. (2008). Characterization of ApuB, an extracellular type II amylopullulanase from *Bifidobacterium breve* UCC2003. *Appl. Environ. Microbiol.* 74, 6271–6279. doi: 10.1128/AEM.01169-08

O'Connell Motherway, M., Fitzgerald, G. F., and Van Sinderen, D. (2011a). Metabolism of a plant derived galactose-containing polysaccharide by Bifidobacterium breve UCC2003. *Microb. Biotechnol.* 4, 403–416. doi: 10.1111/j.1751-7915.2010.00218.x

O'Connell Motherway, M., Kinsella, M., Fitzgerald, G. F., and Sinderen, D. (2013). Transcriptional and functional characterization of genetic elements involved in galacto-oligosaccharide utilization by *Bifidobacterium breve* UCC2003. *Microb. Biotechnol.* 6, 67–79. doi: 10.1111/1751-7915.12011

O'Connell Motherway, M., O'Driscoll, J., Fitzgerald, G. F., and Van Sinderen, D. (2009). Overcoming the restriction barrier to plasmid transformation and targeted mutagenesis in *Bifidobacterium* breve UCC2003. *Microb. Biotechnol.* 2, 321–332. doi: 10.1111/j.1751-7915.2008.00071.x

O'Connell Motherway, M., Zomer, A., Leahy, S. C., Reunanen, J., Bottacini, F., Claesson, M. J., et al. (2011b). Functional genome analysis of Bifidobacterium breve UCC2003 reveals type IVb tight adherence (Tad) pili as an essential and conserved host-colonization factor. *Proc. Natl. Acad. Sci. U.S.A.* 108, 11217–11222. doi: 10.1073/pnas.1105380108

O'Hara, A. M., and Shanahan, F. (2007). Mechanisms of action of probiotics in intestinal diseases. *Sci. World J.* 7, 31–46. doi: 10.1100/tsw.2007.26

Ott, S. J., Musfeldt, M., Wenderoth, D. F., Hampe, J., Brant, O., Folsch, U. R., et al. (2004). Reduction in diversity of the colonic mucosa associated bacterial microflora in patients with active inflammatory bowel disease. *Gut* 53, 685–693. doi: 10.1136/gut.2003.025403

Ouwehand, A. C., Salminen, S., and Isolauri, E. (2002). Probiotics: an overview of beneficial effects. *Antonie Van Leeuwenhoek* 82, 279–289. doi: 10.1023/A:1020620607611

Palframan, R. J., Gibson, G. R., and Rastall, R. A. (2003). Carbohydrate preferences of Bifidobacterium species isolated from the human gut. *Curr. Issues Intest. Microbiol.* 4, 71–75.

Parche, S., Amon, J., Jankovic, I., Rezzonico, E., Beleut, M., Barutcu, H., et al. (2007). Sugar transport systems of *Bifidobacterium* longum NCC2705. *J. Mol. Microbiol. Biotechnol.* 12, 9–19. doi: 10.1159/000096455

Parche, S., Beleut, M., Rezzonico, E., Jacobs, D., Arigoni, F., Titgemeyer, F., et al. (2006). Lactose-over-glucose preference in *Bifidobacterium* longum NCC2705: glcP, encoding a glucose transporter, is subject to lactose repression. *J. Bacteriol.* 188, 1260–1265. doi: 10.1128/JB.188.4.1260-1265.2006

Patel, R., and Dupont, H. L. (2015). New Approaches for Bacteriotherapy: prebiotics, new-generation probiotics, and synbiotics. *Clin. Infect. Dis.* 60, S108–S121. doi: 10.1093/cid/civ177

Patole, S. K., Rao, S. C., Keil, A. D., Nathan, E. A., Doherty, D. A., and Simmer, K. N. (2016). Benefits of *Bifidobacterium* breve M-16V supplementation in preterm neonates - a retrospective cohort study. *PLoS ONE* 11:e0150775. doi: 10.1371/journal.pone.0150775

Perdigon, G., Alvarez, S., Rachid, M., Aguero, G., and Gobbato, N. (1995). Immune system stimulation by probiotics. *J. Dairy Sci.* 78, 1597–1606. doi: 10.3168/jds.S0022-0302(95)76784-4

Picard, C., Fioramonti, J., Francois, A., Robinson, T., Neant, F., and Matuchansky, C. (2005). Review article: bifidobacteria as probiotic agents – physiological effects and clinical benefits. *Aliment. Pharmacol. Ther.* 22, 495–512. doi: 10.1111/j.1365-2036.2005.02615.x

Pokusaeva, K., Fitzgerald, G. F., and Van Sinderen, D. (2011a). Carbohydrate metabolism in *Bifidobacteria*. *Genes Nutr.* 6, 285–306. doi: 10.1007/s12263-010-0206-6

Pokusaeva, K., Neves, A. R., Zomer, A., O'Connell-Motherway, M., Macsharry, J., Curley, P., et al. (2010). Ribose utilization by the human commensal Bifidobacterium breve UCC2003. *Microb. Biotechnol.* 3, 311–323. doi: 10.1111/j.1751-7915.2009.00152.x

Pokusaeva, K., O'Connell-Motherway, M., Zomer, A., Macsharry, J., Fitzgerald, G. F., and Van Sinderen, D. (2011b). Cellodextrin utilization by *bifidobacterium* breve UCC2003. *Appl. Environ. Microbiol.* 77, 1681–1690. doi: 10.1128/AEM.01786-10

Pool-Zobel, B. L., Neudecker, C., Domizlaff, I., Ji, S., Schillinger, U., Rumney, C., et al. (1996). Lactobacillus- and *bifidobacterium*-mediated antigenotoxicity in the colon of rats. *Nutr. Cancer* 26, 365–380. doi: 10.1080/01635589609514492

Postma, P. W., Lengeler, J. W., and Jacobson, G. R. (1993). Phosphoenolpyruvate:carbohydrate phosphotransferase systems of bacteria. *Microbiol. Rev.* 57, 543–594.

Rafter, J., Bennett, M., Caderni, G., Clune, Y., Hughes, R., Karlsson, P. C., et al. (2007). Dietary synbiotics reduce cancer risk factors in polypectomized and colon cancer patients. *Am. J. Clin. Nutr.* 85, 488–496.

Rastall, R. A., and Gibson, G. R. (2015). Recent developments in prebiotics to selectively impact beneficial microbes and promote intestinal health. *Curr. Opin. Biotechnol.* 32, 42–46. doi: 10.1016/j.copbio.2014.11.002

Ravcheev, D. A., Khoroshkin, M. S., Laikova, O. N., Tsoy, O. V., Sernova, N. V., Petrova, S. A., et al. (2014). Comparative genomics and evolution of regulons of the LacI-family transcription factors. *Front. Microbiol.* 5:294. doi: 10.3389/fmicb.2014.00294

Reichardt, N., Duncan, S. H., Young, P., Belenguer, A., Mcwilliam Leitch, C., Scott, K. P., et al. (2014). Phylogenetic distribution of three pathways for propionate production within the human gut microbiota. *ISME J.* 8, 1323–1335. doi: 10.1038/ismej.2014.14

Rivière, A., Gagnon, M., Weckx, S., Roy, D., and De Vuyst, L. (2015). Mutual cross-feeding interactions between Bifidobacterium longum NCC2705 and eubacterium rectale ATCC 33656 explain the bifidogenic and butyrogenic effects of arabinoxylan-oligosaccharides. *Appl. Environ. Microbiol.* 22, 7767–7781. doi: 10.1128/AEM.02089-15

Riviere, A., Moens, F., Selak, M., Maes, D., Weckx, S., and De Vuyst, L. (2014). The ability of bifidobacteria to degrade arabinoxylan oligosaccharide constituents and derived oligosaccharides is strain dependent. *Appl. Environ. Microbiol.* 80, 204–217. doi: 10.1128/AEM.02853-13

Roberfroid, M. B., Bornet, F., Bouley, C., and Cummings, J. H. (1995). Colonic microflora: nutrition and health. Summary and conclusions of an International Life Sciences Institute (ILSI) [Europe] workshop held in Barcelona, Spain. *Nutr. Rev.* 53, 127–130. doi: 10.1111/j.1753-4887.1995.tb01535.x

Roberfroid, M., Gibson, G. R., Hoyles, L., Mccartney, A. L., Rastall, R., Rowland, I., et al. (2010). Prebiotic effects: metabolic and health benefits. *Br. J. Nutr.* 104 (Suppl. 2), S1–S63. doi: 10.1017/s0007114510003363

Rowland, I. R., Rumney, C. J., Coutts, J. T., and Lievense, L. C. (1998). Effect of *Bifidobacterium longum* and inulin on gut bacterial metabolism and carcinogen-induced aberrant crypt foci in rats. *Carcinogenesis* 19, 281–285. doi: 10.1093/carcin/19.2.281

Ruiz, L., Motherway, M. O., Lanigan, N., and Van Sinderen, D. (2013). Transposon mutagenesis in *Bifidobacterium breve:* construction and characterization of a Tn5 transposon mutant library for *Bifidobacterium breve* UCC2003. *PLoS ONE* 8:e64699. doi: 10.1371/journal.pone.0064699

Ryan, S. M., Fitzgerald, G. F., and Van Sinderen, D. (2005). Transcriptional regulation and characterization of a novel β-fructofuranosidase-encoding gene from *Bifidobacterium* breve UCC2003. *Appl. Environ. Microbiol.* 71, 3475–3482. doi: 10.1128/AEM.71.7.3475-3482.2005

Saier, M. H. Jr. and Ramseier, T. M. (1996). The catabolite repressor/activator (Cra) protein of enteric bacteria. *J. Bacteriol.* 178, 3411–3417.

Sakaguchi, K., He, J., Tani, S., Kano, Y., and Suzuki, T. (2012). A targeted gene knockout method using a newly constructed temperature-sensitive plasmid mediated homologous recombination in *Bifidobacterium* longum. *Appl. Microbiol. Biotechnol.* 95, 499–509. doi: 10.1007/s00253-012-4090-4

Salminen, S., and Wright, A. V. (1998). *Lactic Acid Bacteria : Microbiology And Functional Aspects* New York, NY: Marcel Dekker).

Satokari, R. M., Vaughan, E. E., Smidt, H., Saarela, M., Matto, J., and De Vos, W. M. (2003). Molecular approaches for the detection and identification of *bifidobacteria* and lactobacilli in the human gastrointestinal tract. *Syst. Appl. Microbiol.* 26, 572–584. doi: 10.1078/072320203770865882

Schell, M. A., Karmirantzou, M., Snel, B., Vilanova, D., Berger, B., Pessi, G., et al. (2002). The genome sequence of Bifidobacterium longum reflects its adaptation to the human gastrointestinal tract. *Proc. Natl. Acad. Sci. U.S.A.* 99, 14422–14427. doi: 10.1073/pnas.212527599

Schroder, G., and Lanka, E. (2005). The mating pair formation system of conjugative plasmids-A versatile secretion machinery for transfer of proteins and DNA. *Plasmid* 54, 1–25. doi: 10.1016/j.plasmid.2005.02.001

Sekine, K., Toida, T., Saito, M., Kuboyama, M., Kawashima, T., and Hashimoto, Y. (1985). A new morphologically characterized cell wall preparation (whole peptidoglycan) from *Bifidobacterium infantis* with a higher efficacy on the regression of an established tumor in Mice. *Cancer Res.* 45, 1300–1307.

Sela, D. A. (2011). *Bifidobacterial* utilization of human milk oligosaccharides. *Int. J. Food Microbiol.* 149, 58–64. doi: 10.1016/j.ijfoodmicro.2011.01.025

Sela, D. A., Chapman, J., Adeuya, A., Kim, J. H., Chen, F., Whitehead, T. R., et al. (2008). The genome sequence of *Bifidobacterium* longum subsp. infantis reveals adaptations for milk utilization within the infant microbiome. *Proc. Natl. Acad. Sci. U.S.A.* 105, 18964–18969. doi: 10.1073/pnas.0809584105

Serafini, F., Turroni, F., Guglielmetti, S., Gioiosa, L., Foroni, E., Sanghez, V., et al. (2012). An efficient and reproducible method for transformation of genetically recalcitrant *bifidobacteria*. *FEMS Microbiol. Lett.* 333, 146–152. doi: 10.1111/j.1574-6968.2012.02605.x

Shkoporov, A. N., Efimov, B. A., Khokhlova, E. V., Steele, J. L., Kafarskaia, L. I., and Smeianov, V. V. (2008). Characterization of plasmids from human infant *Bifidobacterium strains*: sequence analysis and construction of *E. coli-Bifidobacterium* shuttle vectors. *Plasmid* 60, 136–148. doi: 10.1016/j.plasmid.2008.06.005

Simon, G. L., and Gorbach, S. L. (1984). Intestinal flora in health and disease. *Gastroenterology* 86, 174–193.

Steed, H., Macfarlane, G. T., Blackett, K. L., Bahrami, B., Reynolds, N., Walsh, S. V., et al. (2010). Clinical trial: the microbiological and immunological effects of synbiotic consumption - a randomized double-blind placebo-controlled study in active Crohn's disease. *Aliment. Pharmacol. Ther.* 32, 872–883. doi: 10.1111/j.1365-2036.2010.04417.x

Sun, Z., Baur, A., Zhurina, D., Yuan, J., and Riedel, C. U. (2012). Accessing the inaccessible: molecular tools for *bifidobacteria*. *Appl. Environ. Microbiol.* 78, 5035–5042. doi: 10.1128/AEM.00551-12

Suzuki, H., Murakami, A., and Yoshida, K. (2013). Motif-guided identification of a glycoside hydrolase family 1 alpha-L-arabinofuranosidase in *Bifidobacterium* adolescentis. *Biosci. Biotechnol. Biochem.* 77, 1709–1714. doi: 10.1271/bbb.130279

Szostková, M., and Horáková, D. (1998). The effect of plasmid DNA sizes and other factors on electrotransformation of *Escherichia coli* JM109. *Bioelectrochem. Bioenerg.* 47, 319–323. doi: 10.1016/S0302-4598(98)00203-7

Takamatsu, D., Osaki, M., and Sekizaki, T. (2001). Thermosensitive suicide vectors for gene replacement in *Streptococcus suis*. *Plasmid* 46, 140–148. doi: 10.1006/plas.2001.1532

Tanaka, R., Takayama, H., Morotomi, M., Kuroshima, T., Ueyama, S., Matsumoto, K., et al. (1983). Effects of administration of TOS and *Bifidobacterium breve* on the human fecal flora. *Bifidobacteria and Microflora*, 2, 17–24. doi: 10.12938/bifidus1982.2.1_17

Tavan, E., Cayuela, C., Antoine, J. M., and Cassand, P. (2002). Antimutagenic activities of various lactic acid bacteria against food mutagens: heterocyclic amines. *J. Dairy Res.* 69, 335–341. doi: 10.1017/S002202990200540X

Teitelbaum, J. E., and Walker, W. A. (2002). Nutritional impact of pre- and probiotics as protective gastrointestinal organisms. *Annu. Rev. Nutr.* 22, 107–138. doi: 10.1146/annurev.nutr.22.110901.145412

Trindade, M. I., Abratt, V. R., and Reid, S. J. (2003). Induction of sucrose utilization genes from *Bifidobacterium* lactis by sucrose and raffinose. *Appl. Environ. Microbiol.* 69, 24–32. doi: 10.1128/AEM.69.1.24-32.2003

Truchado, P., Van Den Abbeele, P., Riviere, A., Possemiers, S., De Vuyst, L., and Van De Wiele, T. (2015). Bifidobacterium longum D2 enhances microbial degradation of long-chain arabinoxylans in an *in vitro* model of the proximal colon. *Benef. Microbes* 6, 1–12. doi: 10.3920/BM2015.0023

Turroni, F., Bottacini, F., Foroni, E., Mulder, I., Kim, J. H., Zomer, A., et al. (2010). Genome analysis of *Bifidobacterium* bifidum PRL2010 reveals metabolic pathways for host-derived glycan foraging. *Proc. Natl. Acad. Sci. U.S.A.* 107, 19514–19519. doi: 10.1073/pnas.1011100107

Turroni, F., Milani, C., Duranti, S., Mancabelli, L., Mangifesta, M., Viappiani, A., et al. (2016). Deciphering bifidobacterial-mediated metabolic interactions and their impact on gut microbiota by a multi-omics approach. *ISME J.* doi: 10.1038/ismej.2015.236. [Epub ahead of print].

Turroni, F., Strati, F., Foroni, E., Serafini, F., Duranti, S., Van Sinderen, D., et al. (2012). Analysis of predicted carbohydrate transport systems encoded by *Bifidobacterium* bifidum PRL2010. *Appl. Environ. Microbiol.* 78, 5002–5012. doi: 10.1128/AEM.00629-12

Van Den Abbeele, P., Gerard, P., Rabot, S., Bruneau, A., El Aidy, S., Derrien, M., et al. (2011). Arabinoxylans and inulin differentially modulate the mucosal and luminal gut microbiota and mucin-degradation in humanized rats. *Environ. Microbiol.* 13, 2667–2680. doi: 10.1111/j.1462-2920.2011.02533.x

Van Den Broek, L. A., Hinz, S. W., Beldman, G., Vincken, J. P., and Voragen, A. G. (2008). Bifidobacterium carbohydrases-their role in breakdown and synthesis of (potential) prebiotics. *Mol. Nutr. Food Res.* 52, 146–163. doi: 10.1002/mnfr.200700121

Van Der Meulen, R., Adriany, T., Verbrugghe, K., and De Vuyst, L. (2006a). Kinetic analysis of bifidobacterial metabolism reveals a minor role for succinic acid in the regeneration of NAD+ through its growth-associated production. *Appl. Environ. Microbiol.* 72, 5204–5210. doi: 10.1128/AEM.00146-06

Van Der Meulen, R., Avonts, L., and De Vuyst, L. (2004). Short fractions of oligofructose are preferentially metabolized by Bifidobacterium animalis DN-173 010. *Appl. Environ. Microbiol.* 70, 1923–1930. doi: 10.1128/AEM.70.4.1923-1930.2004

Van Der Meulen, R., Makras, L., Verbrugghe, K., Adriany, T., and De Vuyst, L. (2006b). *In vitro* kinetic analysis of oligofructose consumption by Bacteroides and Bifidobacterium spp. indicates different degradation mechanisms. *Appl. Environ. Microbiol.* 72, 1006–1012. doi: 10.1128/AEM.72.2.1006-1012.2006

Van Laere, K. M., Beldman, G., and Voragen, A. G. (1997). A new arabinofuranohydrolase from Bifidobacterium adolescentis able to remove arabinosyl residues from double-substituted xylose units in arabinoxylan. *Appl. Microbiol. Biotechnol.* 47, 231–235. doi: 10.1007/s002530050918

Van Laere, K. M., Hartemink, R., Bosveld, M., Schols, H. A., and Voragen, A. G. (2000). Fermentation of plant cell wall derived polysaccharides and their corresponding oligosaccharides by intestinal bacteria. *J. Agric. Food Chem.* 48, 1644–1652. doi: 10.1021/jf990519i

Van Laere, K. M. J., Voragen, C. H. L., Kroef, T., Van Den Broek, L. A. M., Beldman, G., and Voragen, A. G. J. (1999). Purification and mode of action of two different arabinoxylan arabinofuranohydrolases from *Bifidobacterium* adolescentis DSM 20083. *Appl. Microbiol. Biotechnol.* 51, 606–613. doi: 10.1007/s002530051439

Vaughan, E. E., Heilig, H. G., Ben-Amor, K., and De Vos, W. M. (2005). Diversity, vitality and activities of intestinal lactic acid bacteria and *bifidobacteria* assessed by molecular approaches. *FEMS Microbiol. Rev.* 29, 477–490. doi: 10.1016/j.fmrre.2005.04.009

Veereman-Wauters, G., Staelens, S., Van De Broek, H., Plaskie, K., Wesling, F., Roger, L. C., et al. (2011). Physiological and bifidogenic effects of prebiotic supplements in infant formulae. *J. Pediatr. Gastroenterol. Nutr.* 52, 763–771. doi: 10.1097/MPG.0b013e3182139f39

Ventura, M., Canchaya, C., Tauch, A., Chandra, G., Fitzgerald, G. F., Chater, K. F., et al. (2007). Genomics of *Actinobacteria*: tracing the evolutionary history of an ancient phylum. *Microbiol. Mol. Biol. Rev.* 71, 495–548. doi: 10.1128/MMBR.00005-07

Ventura, M., Turroni, F., Lugli, G. A., and Van Sinderen, D. (2014). Bifidobacteria and humans: our special friends, from ecological to genomics perspectives. *J. Sci. Food Agric.* 94, 163–168. doi: 10.1002/jsfa.6356

Ventura, M., Van Sinderen, D., Fitzgerald, G. F., and Zink, R. (2004). Insights into the taxonomy, genetics and physiology of *bifidobacteria*. *Antonie Van Leeuwenhoek* 86, 205–223. doi: 10.1023/B:ANTO.0000047930.11029.ec

Venturi, A., Gionchetti, P., Rizzello, F., Johansson, R., et al. (1999). Impact on the composition of the faecal flora by a new probiotic preparation: preliminary data on maintenance treatment of patients with ulcerative colitis. *Aliment. Pharmacol. Ther.* 13, 1103–1108. doi: 10.1046/j.1365-2036.1999.00560.x

Walton, G. E., Lu, C., Trogh, I., Arnaut, F., and Gibson, G. R. (2012). A randomised, double-blind, placebo controlled cross-over study to determine the gastrointestinal effects of consumption of arabinoxylan-oligosaccharides enriched bread in healthy volunteers. *Nutr. J.* 11, 36. doi: 10.1186/1475-2891-11-36

Wong, J. M., De Souza, R., Kendall, C. W., Emam, A., and Jenkins, D. J. (2006). Colonic health: fermentation and short chain fatty acids. *J. Clin. Gastroenterol.* 40, 235–243. doi: 10.1097/00004836-200603000-00015

**Conflict of Interest Statement:** The authors declare that the research was conducted in the absence of any commercial or financial relationships that could be construed as a potential conflict of interest.

2

# Gut Bifidobacteria Populations in Human Health and Aging

*Silvia Arboleya[1,2], Claire Watkins[1,2,3], Catherine Stanton[1,2] and R. Paul Ross[1,2,4]**

[1] APC Microbiome Institute, University College Cork, Cork, Ireland, [2] Teagasc Food Research Centre, Moorepark, Fermoy, Cork, Ireland, [3] School of Microbiology, University College Cork, Cork, Ireland, [4] School of Science, Engineering and Food Science, University College Cork, Cork, Ireland

The intestinal microbiota has increasingly been shown to have a vital role in various aspects of human health. Indeed, several studies have linked alterations in the gut microbiota with the development of different diseases. Among the vast gut bacterial community, *Bifidobacterium* is a genus which dominates the intestine of healthy breast-fed infants whereas in adulthood the levels are lower but relatively stable. The presence of different species of bifidobacteria changes with age, from childhood to old age. *Bifidobacterium longum*, *B. breve,* and *B. bifidum* are generally dominant in infants, whereas *B. catenulatum*, *B. adolescentis* and, as well as *B. longum* are more prevalent in adults. Increasingly, evidence is accumulating which shows beneficial effects of supplementation with bifidobacteria for the improvement of human health conditions ranging from protection against infection to different extra- and intra-intestinal positive effects. Moreover, bifidobacteria have been associated with the production of a number of potentially health promoting metabolites including short chain fatty acids, conjugated linoleic acid and bacteriocins. The aim of this mini-review is to describe the bifidobacteria compositional changes associated with different stages in life, highlighting their beneficial role, as well as their presence or absence in many disease states.

**Keywords: bifidobacteria, intestinal microbiota, health, aging, probiotics**

***Edited by:***
*Marco Ventura,*
*University of Parma, Italy*

***Reviewed by:***
*Sinead Christine Leahy,*
*AgResearch, New Zealand*
*Simone Guglielmetti,*
*University of Milan, Italy*
*Silvia Turroni,*
*University of Bologna, Italy*

***Correspondence:***
*R. Paul Ross*
*p.ross@ucc.ie*

## INTRODUCTION

The study of the intestinal microbiota, linked with its impact on human health, has become a transforming topic in microbiology research. From the pioneering studies using culture-dependent techniques to the current omics' approach, composition and functionality of the intestinal microbiota has been assessed in a wide variety of studies (Lagier et al., 2015). The complex microbiota and the relationships it plays in the prevention and alleviation of diseases have been a major focus of many works (Turnbaugh et al., 2009; Qin et al., 2010; Karlsson et al., 2013). Members of the genus *Bifidobacterium* have been identified as almost ubiquitous inhabitants of the human host (Biavatti and Mattarelli, 2006), performing an important role in the gut from birth. Due to the traditional interest in bifidobacteria as probiotics this genus has been studied in depth, and changes in the species composition, diversity or relative abundance have been investigated at different stages of life and in several diseases. The aim of this mini-review is to describe the bifidobacteria compositional changes associated with different stages in life, highlighting their beneficial role, as well as the possible role of their presence in the protection against many disease states.

## INTESTINAL MICROBIOTA

The intestinal microbiota plays an important role in human health and has long been associated with such functions as metabolic, protective, and trophic (Guarner and Malagelada, 2003) and more recently functions related to the gut-brain axis or liver-gut axis (Clarke et al., 2014). The current exponential increase in sequencing and the explosion of other "omics" approaches has allowed us to gain a deep knowledge of intestinal microbiota.

The establishment of the intestinal microbiota was initially considered to occur at birth; however, the presence of microorganisms in placenta or amniotic fluid (Collado et al., 2016) suggests a primary fetal colonization. Moreover, the process is determined by different factors which undoubtedly have an effect on microbiota homeostasis (Penders et al., 2006; Faa et al., 2013). Actinobacteria, followed by Proteobacteria and Firmicutes are the main phyla in early childhood, characterized by low diversity and complexity (Koenig et al., 2011; Turroni et al., 2012). The main changes in gut microbiota composition take place in the first stages of life, getting to a relative stability at 1–2 years old (Voreades et al., 2014). The adult-like structure of the gut microbiota is thought to occur after the 3rd year of life (Koenig et al., 2011; Yatsunenko et al., 2012), and reaches a total number of $10^{14}$ microorganisms, comprising of bacteria, eukarya, viruses and archaeal members (Mihajlovski et al., 2010; HMP, 2012). At phylum level, the gut microbiota is made up of 80–90% Firmicutes and Bacteroidetes. At species and strain taxonomic level, the diversity is very high in adult life, and is characterized by high interindividual variability (HMP, 2012). However, the functionality and metabolism of the gut microbiota is highly conserved (Qin et al., 2010). Aging was defined by Imahori (1992) as "the regression of physiological function accompanied by advancement of age." It is a natural process which entails changes in the gastrointestinal tract (GIT), immunosenescence and, in some cases, malnutrition (Woodmansey, 2007; Biagi et al., 2012). In addition, alterations in lifestyle, diet and medication have an unavoidable effect on the elderly microbiota composition and function (Biagi et al., 2012; Claesson et al., 2012). Microbial diversity is reduced in old age; however, Bacteroidetes and Firmicutes still constitute the dominant phyla, with increases in the relative abundance of some other Phyla – most notably Proteobacteria (Biagi et al., 2012; Odamaki et al., 2016).

## BIFIDOBACTERIA AND THEIR RELEVANCE: FROM THE INFANT-TYPE TO THE AGED-TYPE MICROBIOTA

Bifidobacteria are normal inhabitants of the GIT belonging to the Actinobacteria phylum. After the depletion of oxygen by facultative anaerobes, bifidobacterial populations are the most abundant genus present in the healthy infant gut (Favier et al., 2002). During adulthood the levels decrease considerably but remain relatively stable; decreasing again in old age (Odamaki et al., 2016; **Figure** ). Certain strains of *Bifidobacterium* are widely used as probiotics – the safety of which is supported by the long historical consumption in fermented milks and the growing knowledge about their physiology and genomes (Arboleya et al., 2016; O'Callaghan and van Sinderen, 2016). The genus *Bifidobacterium* has been shown to play an important role in the barrier effect, the stimulation of immune system, being associated with a range of beneficial health effects (Picard et al., 2005).

### Early Stages of Life

The initial colonization by bifidobacteria is dependent on a number of extrinsic factors. In terms of vertical transfer, many studies have linked the transmission of bifidobacteria from the mother's vaginal tract, GIT, breast milk, placenta and amniotic fluid to the infant (Makino et al., 2013; Collado et al., 2016). Birth mode, in particular, has a significant impact on this initial colonization, with an increased abundance of bifidobacteria found in infants born vaginally, versus those born by cesarean section (Dominguez-Bello et al., 2010).

Gestational age has been described in terms of its impact on the infant gut, whereby pre-term infants have been characterized in many studies by a dominance of Proteobacteria, with increases in members of *Clostridium* and *Staphylococcus*, and much lower levels of Actinobacteria. In contrast, the full-term infant gut has been correlated with much higher levels of *Bifidobacterium* and *Bacteroides*, which tend to be dominant in the early weeks of life (Arboleya et al., 2012; Barrett et al., 2013; Arboleya et al., 2015).

Numerous studies investigating the effects of breast feeding versus formula feeding have identified specific bifidobacterial species that correlate with the feeding regime (Roger et al., 2010). Using both culture-dependent and molecular methods, studies have found that *Bifidobacterium breve*, *B. bifidum*, *B. longum* ssp. *longum*, and *B. longum* spp. *infantis* are present in both breast- and formula-fed infants (Mevissen-Verhage et al., 1987; Klaassens et al., 2009). However, *B. longum* ssp. *infantis* is more commonly associated with breast fed infants, whereas *B. longum* ssp. *longum* has been found more commonly amongst bottle-fed infants (Guaraldi and Salvatori, 2012). *B. adolescentis*, which is commonly found in adults, was found to be present only in formula-fed babies (Klaassens et al., 2009). Numerous studies have focused on the bifidogenic effect that human breast milk has on the infant gut microbiota (Musilova et al., 2014). Specific glycans found within breast milk are known to be utilized by bifidobacteria; however, it has been shown that breast milk provided by mothers with an inactive allele of the Fucosyltransferase 2 gene (FUT2), an enzyme involved in the transfer of fucose to glycans, delays the establishment of bifidobacteria in the infant (Lewis et al., 2015). Similarly, bifidobacteria provide an important role in the breakdown of human milk oligosaccharides (HMOs; Garrido et al., 2013), creating a clear evolutionary link between the mother, infant and the microbial species present. Indeed, *B. longum* ssp. *infantis* has specifically been studied for its ability to digest different HMO structures (Sela et al., 2008). This species is also associated with anti-inflammatory properties, and the ability to decrease intestinal permeability (Underwood et al., 2015). The growth of infant-derived *Bifidobacterium*, *B. longum* ssp. *infantis*, and *B. bifidum*, in the presence of HMOs has also been shown

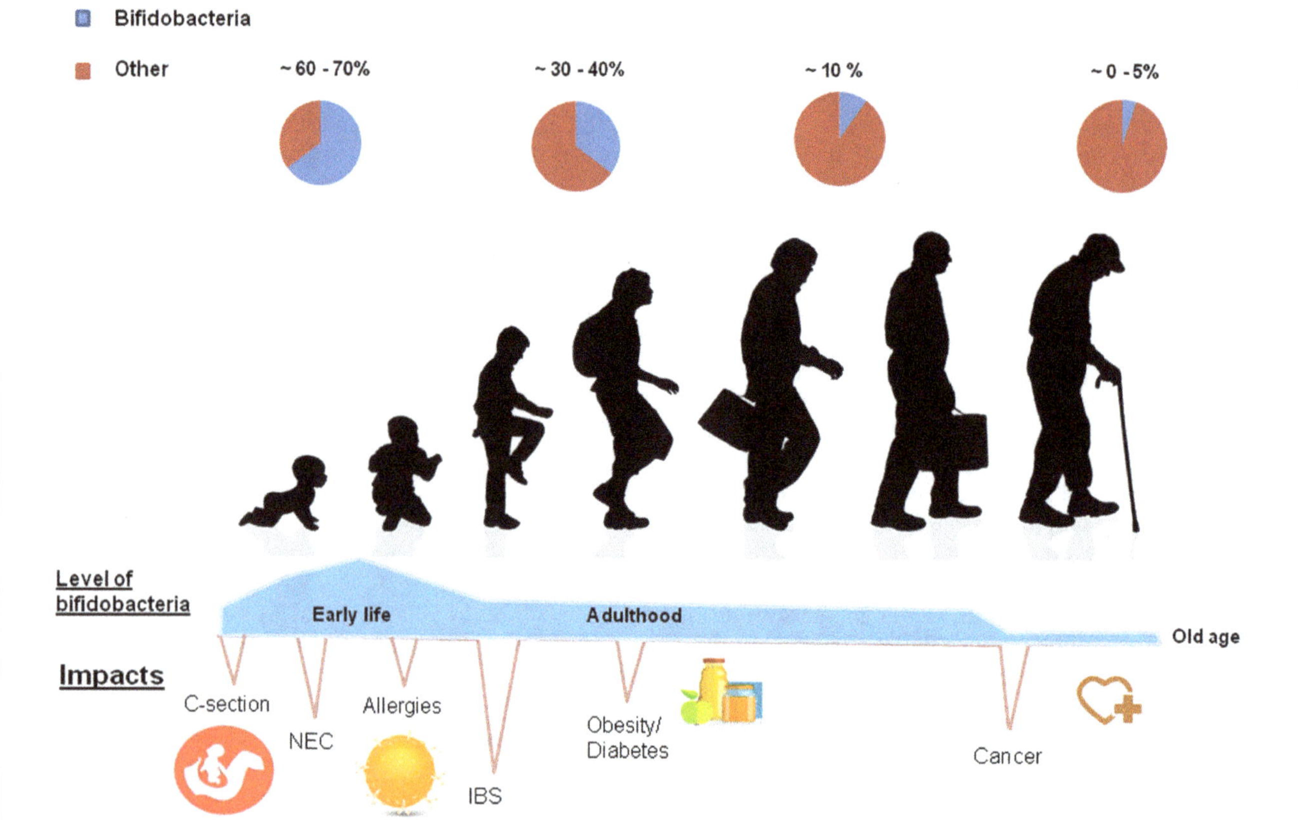


**FIGURE | At birth levels of bifidobacteria are found to be at their highest.** In cases of natural childbirth the numbers are highest at birth. In contrast, they are lower in C sectioned babies. Various diseases such as obesity, diabetes and allergies have been associated with lower numbers of bifidobacteria at various stages of life. When weaned onto solid foods diet is more of an intervening factor and an adult-like (stable) microbiota develops. In this figure the authors hypothesize with regard to the relative abundance of bifidobacteria present at each stage of the life cycle, based on the literature cited in the following review by Voreades et al. (2014).

to promote the adhesive properties of these strains, as well as the expression of anti-inflammatory cytokines and tight junction proteins in eukaryotic cells (Chichlowski et al., 2012). Interestingly, a study published by Del Chierico et al. (2015) focused on the co-correlation between the metabolome and the OTUs occurrence in infant stool samples. Results found that *Bifidobacterium* OTUs were positively correlated with the presence of lactate and alanine during this 1st month of life. Another recent work, negatively correlated the lactase gene with *Bifidobacterium* levels in the gut microbiota of twins, speculating that lactase-persisters babies harbor lower levels of bifidobacteria because of low lactose availability (Goodrich et al., 2016).

In terms of species, *B. longum, B. breve,* and *B. bifidum* are most abundant (**Table** ) in infants (Favier et al., 2002; Gueimonde et al., 2007; Turroni et al., 2012). A recent study described the different *Bifidobacterium* species profile between monozygotic twins (entirely dominated by *B. breve*) and their fraternal sibling (exhibiting higher species diversity) at 1 month of life (Murphy et al., 2015; **Table** ).

## Adulthood and Old Age

In adulthood, the levels of bifidobacteria are lower (2–14% relative abundance) but remain stable (Odamaki et al., 2016). Gueimonde et al. (2007) identified significantly higher levels of *Bifidobacterium* in infants than in adults by q-PCR technique. They postulated that *B. longum* is the most widely abundant species, which is in agreement with other studies (Gavini et al., 2001; Matsuki et al., 2004). However, Matsuki et al. (2004) observed higher levels of *B. adolescentis* and *B. catenulatum* in their adult population. Chaplin et al. (2015) reported a decreased frequency of isolation of *B. bifidum* and *B. breve* with age and an increased trend in *B. adolescentis.*

Currently, there is no agreed definition of old-age-specific gut microbiota profile due to the high inter-individual variability, differences in diet and lifestyle, and the unclear definition of the term "elderly." However, some trends are repetitively observed such as the decrease of bifidobacteria in the elderly population, confirmed by several studies using different technologies (Mitsuoka et al., 1974; Mitsuoka, 1992; Gavini et al., 2001; Hopkins et al., 2001; Gueimonde et al., 2010; Biagi et al., 2012; Salazar et al., 2013). Woodmansey et al. (2004) reported that the decline in bifidobacteria population with aging was accompanied by a decrease in species diversity. This decline was associated with the reduction in adhesion to the intestinal mucosa, but it is not clear if it is due to the changes in the microbiota or in the structure of mucus (He et al., 2001).

**TABLE | (A)** Distribution of the most abundant *Bifidobacterium* species in the intestinal microbiota at different stages of life analyzed using different techniques.

| (A) Human population | *Bifidobacterium* spp. | Techniques | Reference |
|---|---|---|---|
| **Infants** | | | |
| Breast-fed, 22–24 days of age | *B. breve*[a]<br>*B. longum* ssp. *longum*, *B. longum* ssp. *infantis*[b] | PCR | Matsuki et al., 1999 |
| Breast- and Formula fed, 28–90 days of age | *B. longum* ssp. *infantis*<br>*B. breve, B. longum ssp. longum* | PCR | Haarman and Knol, 2005 |
| Breast-fed, 1 month of age | *B. longum*<br>*B. bifidum, B. animalis, B. breve* | PCR | Grönlund et al., 2007 |
| Breast-fed, 3–6 weeks of age | *B. breve*<br>*B. longum ssp. longum, B. longum ssp. infantis* | PCR | Mikami et al., 2009 |
| Full-term, 1 month of age | *B. longum* | q-PCR | Grzeskowiak et al., 2015 |
| Preterm, CS, 1 month of age | *B. longum, B. lactis* | | |
| Preterm, Vaginal, 1 month of age | *B. longum, B. bifidum* | | |
| Twins, 1 month of age | *B. breve* | 16S Metagenomics | Murphy et al., 2015 |
| Fraternal infant, 1 month of age | *B. breve, B. longum*<br>*B. dentium, B. adolescentis* | | |
| **Adults** | | | |
| 23–54 years old, Japanese | *B. catenulatum,*<br>*B. longum, B. adolescentis* | PCR | Matsuki et al., 1999 |
| 25–59 years old, Japanese | *B. longum,*<br>*B. adolescentis, B. catenulatum* | q-PCR | Matsuki et al., 2004 |
| ≤57 years old, Russian | *B. adolescentis* | MALDI-TOF | Chaplin et al., 2015 |
| 20–40 years old, Finnish | *B. longum,*<br>*B. catenulatum* | q-PCR | Gueimonde et al., 2007 |
| 18–39 years old, lean subjects (BMI = $19.83 \pm 0.94$ kg/m$^2$) | *B. longum* | q-PCR | Mayorga Reyes et al., 2016 |
| **Elderly** | | | |
| 69–89 years old, French | *B. adolescentis*<br>*B. longum* | DNA–DNA hybridization | Gavini et al., 2001 |
| 67–75 years old, Scottish | *B. angulatum*<br>*B. longum* | Culture-based analyses | Woodmansey et al., 2004 |
| >70 years old, Finish | *B. catenulatum*<br>*B. longum, B. bifidum* | q-PCR | Gueimonde et al., 2007 |
| 77–95 years old, Spanish | *B. longum*<br>*B. bifidum, B. pseudocatenulatum* | q-PCR | Salazar et al., 2013 |
| **Centenaries** | | | |
| 100–104 years old, Italian | *B. longum*<br>*B. adolescentis, B. bifidum* | Culture-based analyses | Drago et al., 2012 |
| 80–108 years old, Chinese | *B. dentium*<br>*B. longum* | q-PCR | Wang et al., 2015 |

| (B) Disease in human population | *Bifidobacterium spp.* | Techniques | Reference |
|---|---|---|---|
| **Infants** | | | |
| Allergic mothers | ↑ *B. adolescentis* | q-PCR | Grönlund et al., 2007 |
| Coeliac disease (Non-active) | ↓ *B. longum*;↑ *B. dentium* | q-PCR | Collado et al., 2008a |
| Celiac disease | ↓ *B. longum* | q-PCR | Palma et al., 2012 |
| Allergic diseases | ↓ *B. longum*<br>↑ *B. pseudocatenulatum, B. catenulatum* | Culture-based analyses | Akay et al., 2014 |
| **Adults** | | | |
| Allergy | ↑ *B. adolescentis* | PCR-DGGE | Stsepetova et al., 2007 |
| IBS | ↓ *B. catenulatum* | q-PCR | Kerckhoffs et al., 2009 |
| IBS | ↓ *B. catenulatum/pseudocatenulatum* | q-PCR | Lyra et al., 2009 |
| IBS | ↓ *B. pseudocatenulatum, B. gallicum* | HITChip phylogenetic microarray | Rajilic-Stojanovic et al., 2009 |

*(Continued)*

**TABLE | Continued**

| (B) Disease in human population | *Bifidobacterium spp.* | Techniques | Reference |
|---|---|---|---|
| IBS | ↑ *B. adolescentis* | 16S Metagenomics | Jeffery et al., 2012 |
| Hepatitis B virus-related cirrhosis | ↑ *B. dentium*<br>↓ *B. catenulatum, B. longum* | PCR-DGGE and q-PCR | Xu et al., 2012 |
| Obesity | ↓ *B. animalis* | q-PCR | Million et al., 2013 |
| Cystic fibrosis | ↓ *B. catenulatum/pseudocatenulatum, B. longum, B. adolescentis* | PCR-DGGE | Duytschaever et al., 2013 |
| Long-term asthma | ↑ *B. adolescentis* | 16S Metagenomics | Hevia et al., 2016 |

*First row indicates the most abundant species found[a]. Second row indicates other abundant species found[b].* ***(B)*** *Studies focusing on Bifidobacterium species altered in certain diseases.*
*↑ Increased levels; ↓ Decreased levels.*

The bifidobacteria composition in centenary populations was also reported in some studies; however, the results remain somewhat controversial. In a European population, the microbiota composition of centenarians was still similar to that of adults (Biagi et al., 2010), however, higher proportions of bifidobacteria were found in centenarians than in younger elderlies from a region of China (Zhao et al., 2011). Regarding species (**Table** ), *B. longum* was the most abundant in Italian centenarians followed by *B. adolescentis* and *B. bifidum* (Drago et al., 2012), but *B. dentium* was dominant in Chinese centenaries (Wang et al., 2015).

Other extrinsic factors indirectly related to the aging process also affect the bifidobacteria composition. The extended use of antibiotics in the older population undoubtedly has a huge impact on the intestinal microbiota composition, decreasing the bifidobacteria population (Woodmansey et al., 2004; O'Sullivan et al., 2013). While antibiotics remain an essential medical tool, therapies targeted toward the reestablishment of microbiota have been explored, in particular the use of probiotics to correct the imbalance in the bifidobacteria population and the alteration in the intestinal microbiota after antibiotic therapy (Rondanelli et al., 2015). In terms of frailty, van Tongeren et al. (2005) did not find differences in bifidobacteria between elderly people divided into a low and high frailty category. The same trend was observed by Bartosch et al. (2004) between elderly living in the local community and elderly in a hospitalized environment.

## BIFIDOBACTERIA AND DISEASES

Several diseases, both intra- and extra-intestinal, have been associated with alterations in the gut microbiota composition and function (Wu and Lewis, 2013). Although, there is still not a detailed description of "potential alterated-microbiota types," some authors postulate that intestinal microbial alterations could be the prelude to a wide range of disease (Putignani et al., 2014).

### Bifidobacterial Composition in Diseases

Given the widespread use of bifidobacteria as probiotics, they have been studied extensively and as such the aberrancies in bifidobacteria species composition, diversity or changes in their relative abundance have been reported in several diseases.

Obesity is a worldwide disease affecting children and adults, which is commonly associated with alterations in the microbiota. Some studies have shown lower levels of bifidobacteria, linked to higher prevalence of enterobacteria or *Staphylococcus* in obese children (Kalliomäki et al., 2008; Gao et al., 2015). Interestingly, women who gain weight during pregnancy have displayed lower levels of *Bifidobacterium* in contrast to healthy weight pregnant women (8.36 vs. 9.10 log genome equivalents/g feces; Santacruz et al., 2010). These results are correlated with the decreased levels of bifidobacteria in babies whose mother gained significant weight during pregnancy (Collado et al., 2008b), that is over and above the pregnancy itself. In terms of allergic disease, Hevia et al. (2016) observed lower levels of *Bifidobacterium* in patients with long-term asthma. The same study showed a dominance of *B. adolescentis* in both short- and long-term asthmatic individuals, in concordance with other previous studies (Stsepetova et al., 2007). This species was only found in infants of allergic mothers, who displayed lower levels of bifidobacteria in their breast milk (Grönlund et al., 2007). One particular Turkish study described a statistical difference between the increased levels of *B. longum* in healthy children (30.3%) when compared to children with allergic disease (11.1%), suggesting that *B. longum* may play a beneficial role in the disease and thus may be useful as a probiotic for the prevention of allergic pathologies (Akay et al., 2014). Different studies have focused on the relationship between the intestinal microbiota and the pathogenesis of irritable bowel syndrome (IBS), showing an altered microbiota related to IBS patients and lower levels in the *Bifidobacterium* genus (Taverniti and Guglielmetti, 2014).

In the elderly gut microbiota, Hopkins and Macfarlane (2002) focused on the impact of *Clostridium difficile*-associated diarrhea, describing a reduction in numbers of bifidobacteria in elderly people suffering the infection compared to a healthy control group. Decreased numbers of bifidobacteria have also been observed in other illnesses such as cystic fibrosis, hepatitis B and both diabetes Types I and II (Wu et al., 2010; Xu et al., 2012; Duytschaever et al., 2013; Murri et al., 2013).

Overall, there is a repetitive trend between lower bifidobacteria and a variety of common disease states, suggesting a role of bifidobacteria in health. However, whether bifidobacteria numbers have any causal relationship to any of these conditions remains unknown.

## Bifidobacteria as Probiotics in Diseases

Numerous health-promoting effects have been ascribed to strains of the *Bifidobacterium* genus based on their use as probiotics in intervention strategies to address many ill health conditions.

In that regard, the capability of bifidobacteria to stimulate the immune system seems to be species specific as shown in a particular study where the effect of infant derived bifidobacteria on the T-helper $1(T_H1)/T_H2$ balance was examined. Results published by Ménard et al. (2008) demonstrated that *B. bifidum, B. dentium,* and *B. longum* were capable of stimulating systemic and intestinal immunity. It is not surprising that bifidobacteria are so widely used as probiotics in the treatment and prevention of infant disease given their dominance in the infant gut. Their application in pathologies such as allergies, celiac disease, obesity, diarrheas, colic, infections or necrotizing enterocolitis has returned very good results (Di Gioia et al., 2014). They have also been extensively used in adults and elderly, in the treatment of gastrointestinal and respiratory diseases (Biagi et al., 2012; Malaguarnera et al., 2012; Tojo et al., 2014). In an attempt to restore the lipoprotein imbalance found in the blood of children with dyslipidemia, Guardamagna et al. (2014) examined the effects of a probiotic which contained three *Bifidobacterium* strains, selected due to their bile salt hydrolase activity. Their results found a decrease in total cholesterol and low-density lipoprotein cholesterol.

In terms of brain gut disorders, several studies have examined the use of bifidobacteria for their psychobiotic effects in reducing stress, anxiety and depressive like behavior in BALB/c mice. It was concluded that the behavioral effects observed may be partially explained by the differential effects on the immune system, although mechanisms underlying the effects are unknown (Savignac et al., 2014). Furthermore, a particular strain of *B. longum* was shown to have anxiolytic effects in a model of non-infectious colitis through vagal pathways whereby the fermentation products of this species were capable of modifying the excitation of enteric neurons in the gut (Bercik et al., 2011).

Bifidobacteria have also been studied for their ability to specifically localize at tumor sites (Nakamura et al., 2002). Cronin et al. (2010) showed in that *B. breve* UCC 2003 was able to target tumors in athymic MF1 nu/nu mice bearing s.c. B16-F10 murine melanoma tumors, as well as C57 mice bearing s.c. Lewis lung carcinoma tumors. Interestingly, an increasing bacterial load of *B. breve* UCC 2003 was found in the metastatic nodules in the lungs of the B-16 mouse model where oxygen is transferred. Bifidobacteria have also been shown to have an effect on tumor specific T-cell responses in C57BL/6 mice bearing s.c. B16.SIY melanoma growth (Sivan et al., 2015). In this study bifidobacteria were found to have an effect on dendritic cell function and CD8+ T cell responses, reducing tumor cell growth.

*Bifidobacterium* species may have positive effects on human health, while it should be stressed that the increase only in fecal bifidobacteria level cannot in itself a health benefit. However, given the strong association of bifidobacteria to health it may provide a contributory biomarker to disease status lined to some illnesses in the future. Moreover the use of bifidobacteria as delivery vehicles for the administration of therapeutic agents to target tumors due to their anaerobic nature, is a very promising and safe form of therapy for this disease state.

## CONCLUSION

Bifidobacteria are one of the most abundant genera present in the healthy infant gut and represent a significant portion of the microbiota throughout a healthy adult life, playing an important role in gut homeostasis and health. During late adulthood and within several diseases, the levels of *Bifidobacterium* and its species diversity decrease. Nowadays, their association with health and aging is undeniable, however, the jury is still out on whether it is a "cause" or "effect" type relationship. Given the prevalence of bifidobacteria at various stages of a healthy life and the many health promoting attributes associated with their use, it is undoubted that these bacteria play an important role in human health maintenance and protection and also may in the future provide a very important biomarker for certain diseases.

## AUTHOR CONTRIBUTIONS

RR, CS, and SA conceived the manuscript. SA and CW drafted the manuscript. CS and RR reviewed the final version of the manuscript. All the authors approved it for publication.

## ACKNOWLEDGMENTS

This research was supported by the Science Foundation Ireland (SFI)-funded APC Microbiome Institute; by Department of Agriculture Food and Marine (DAFM)-funded INFANTMET (Ref. No. 10FDairy) and ToddlerFood (Ref. No. 14F821) projects; and by European Union's Seventh Framework Program (EUFP7)-funded MyNewGut project (Ref. No. 613979).

## REFERENCES

Akay, H. K., Bahar Tokman, H., Hatipoglu, N., Hatipoglu, H., Siraneci, R., Demirci, M., et al. (2014). The relationship between bifidobacteria and allergic asthma and/or allergic dermatitis: a prospective study of 0–3 years-old children in Turkey. *Anaerobe* 28, 98–103. doi: 10.1016/j.anaerobe.2014.05.006

Arboleya, S., Binetti, A., Salazar, N., Fernandez, N., Solis, G., Hernandez-Barranco, A., et al. (2012). Establishment and development of intestinal microbiota in preterm neonates. *FEMS Microbiol. Ecol.* 79, 763–772. doi: 10.1111/j.1574-6941.2011.01261.x

Arboleya, S., Sánchez, B., Milani, C., Duranti, S., Solís, G., Fernández, N., et al. (2015). Intestinal microbiota development in preterm neonates and effect of perinatal antibiotics. *J. Pediatr.* 166, 538–544. doi: 10.1016/j.jpeds.2014.09.041

Arboleya, S., Stanton, C., Ryan, C. A., Dempsey, E., and Ross, P. R. (2016). Bosom Buddies: the symbiotic relationship between infants and *Bifidobacterium longum* ssp. longum and ssp. infantis. Genetic and probiotic features.

*Annu. Rev. Food Sci. Technol.* 7, 1–21. doi: 10.1146/annurev-food-041715-033151

Barrett, E., Kerr, C., Murphy, K., O'Sullivan, O., Ryan, C. A., Dempsey, E. M., et al. (2013). The individual-specific and diverse nature of the preterm infant microbiota. *Arch. Dis. Child. Fetal Neonatal Ed.* 98, F334–F340. doi: 10.1136/archdischild-2012-303035

Bartosch, S., Fite, A., Macfarlane, G. T., and McMurdo, M. E. T. (2004). Characterization of bacterial communities in feces from healthy elderly volunteers and hospitalized elderly patients by using real-time PCR and effects of antibiotic treatment on the fecal microbiota. *Appl. Environ. Microbiol.* 70, 3575–3581. doi: 10.1128/AEM.70.6.3575-3581.2004

Bercik, P., Park, A. J., Sinclair, D., Khoshdel, A., Lu, J., Huang, X., et al. (2011). The anxiolytic effect of *Bifidobacterium longum* NCC3001 involves vagal pathways for gut-brain communication. *Neurogastroenterol. Motil.* 23, 1132–1139. doi: 10.1111/j.1365-2982.2011.01796.x

Biagi, E., Candela, M., Fairweather-Tait, S., Franceschi, C., and Brigidi, P. (2012). Ageing of the human metaorganism: the microbial counterpart. *Age* 34, 247–267. doi: 10.1007/s11357-011-9217-5

Biagi, E., Nylund, L., Candela, M., Ostan, R., Bucci, L., Pini, E., et al. (2010). Through ageing and beyond: gut microbiota and inflammatory status in seniors and centenarians. *PLoS ONE* 5:e10667. doi: 10.1371/journal.pone.0010667

Biavatti, B., and Mattarelli, P. (2006). "The family Bifidobacteriaceae," in *The Prokaryotes*, 3rd Edn, Vol. 3, eds S. Falkow, E. Rosenberg, K. H. Schleifir, E. Stackebrandt, and M. Dworkin (New York, NY: Springer - Verlag GmbH), 322–382. doi: 10.1007/0-387-30743-5_17

Chaplin, A. V., Brzhozovskii, A. G., Parfenova, T. V., Kafarskaia, L. I., Volodin, N. N., Shkoporov, A. N., et al. (2015). Species diversity of bifidobacteria in the intestinal microbiota studied using MALDI-TOF mass-spectrometry. *Vestn. Ross. Akad. Med. Nauk* 4, 435–440.

Chichlowski, M., De Lartigue, G., German, J. B., Raybould, H. E., and Mills, D. A. (2012). Bifidobacteria isolated from infants and cultured on human milk oligosaccharides affect intestinal epithelial function. *J. Pediatr. Gastroenterol. Nutr.* 55, 321–327. doi: 10.1097/MPG.0b013e31824fb899

Claesson, M. J., Jeffery, I. B., Conde, S., Power, S. E., O'Connor, E. M., Cusack, S., et al. (2012). Gut microbiota composition correlates with diet and health in the elderly. *Nature* 488, 178–184. doi: 10.1038/nature11319

Clarke, G., Stilling, R. M., Kennedy, P. J., Stanton, C., Cryan, J. F., and Dinan, T. G. (2014). Minireview: gut microbiota: the neglected endocrine organ. *Mol. Endocrinol.* 28, 1221–1238. doi: 10.1210/me.2014-1108

Collado, M. C., Donat, E., Ribes-Koninckx, C., Calabuig, M., and Sanz, Y. (2008a). Imbalances in faecal and duodenal *Bifidobacterium* species composition in active and non-active coeliac disease. *BMC Microbiol.* 8:232. doi: 10.1186/1471-2180-8-232

Collado, M. C., Isolauri, E., Laitinen, K., and Salminen, S. (2008b). Distinct composition of gut microbiota during pregnancy in overweight and normal-weight women. *Am. J. Clin. Nutr.* 88, 894–899.

Collado, M. C., Rautava, S., Aakko, J., Isolauri, E., and Salminen, S. (2016). Human gut colonisation may be initiated in utero by distinct microbial communities in the placenta and amniotic fluid. *Sci. Rep.* 6, 23129. doi: 10.1038/srep23129

Cronin, M., Morrissey, D., Rajendran, S., El Mashad, S. M., van Sinderen, D., O'Sullivan, G. C., et al. (2010). Orally administered bifidobacteria as vehicles for delivery of agents to systemic tumors. *Mol. Ther.* 18, 1397–1407. doi: 10.1038/mt.2010.59

Del Chierico, F., Vernocchi, P., Petrucca, A., Paci, P., Fuentes, S., Pratico, G., et al. (2015). Phylogenetic and metabolic tracking of gut microbiota during perinatal development. *PLoS ONE* 10:e0137347. doi: 10.1371/journal.pone.0137347

Di Gioia, D., Aloisio, I., Mazzola, G., and Biavati, B. (2014). Bifidobacteria: their impact on gut microbiota composition and their applications as probiotics in infants. *Appl. Microbiol. Biotechnol.* 98, 563–577. doi: 10.1007/s00253-013-5405-9

Dominguez-Bello, M. G., Costello, E. K., Contreras, M., Magris, M., Hidalgo, G., Fierer, N., et al. (2010). Delivery mode shapes the acquisition and structure of the initial microbiota across multiple body habitats in newborns. *Proc. Natl. Acad. Sci. U.S.A.* 107, 11971–11975. doi: 10.1073/pnas.1002601107

Drago, L., Toscano, M., Rodighiero, V., De Vecchi, E., and Mogna, G. (2012). Cultivable and pyrosequenced fecal microflora in centenarians and young subjects. *J. Clin. Gastroenterol.* 46(Suppl.), S81–S84. doi: 10.1097/MCG.0b013e3182693982

Duytschaever, G., Huys, G., Bekaert, M., Boulanger, L., De Boeck, K., and Vandamme, P. (2013). Dysbiosis of bifidobacteria and *Clostridium* cluster XIVa in the cystic fibrosis fecal microbiota. *J. Cyst. Fibros.* 12, 206–215. doi: 10.1016/j.jcf.2012.10.003

Faa, G., Gerosa, C., Fanni, D., Nemolato, S., van Eyken, P., and Fanos, V. (2013). Factors influencing the development of a personal tailored microbiota in the neonate, with particular emphasis on antibiotic therapy. *J. Matern. Fetal Neonatal Med.* 26(Suppl. 2), 35–43. doi: 10.3109/14767058.2013.829700

Favier, C. F., Vaughan, E. E., De Vos, W. M., and Akkermans, A. D. (2002). Molecular monitoring of succession of bacterial communities in human neonates. *Appl. Environ. Microbiol.* 68, 219–226. doi: 10.1128/AEM.68.1.219-226.2002

Gao, X., Jia, R., Xie, L., Kuang, L., Feng, L., and Wan, C. (2015). Obesity in school-aged children and its correlation with gut *E. coli* and Bifidobacteria: a case-control study. *BMC Pediatr.* 15:64. doi: 10.1186/s12887-015-0384-x

Garrido, D., Dallas, D. C., and Mills, D. A. (2013). Consumption of human milk glycoconjugates by infant-associated bifidobacteria: mechanisms and implications. *Microbiology* 159(Pt 4), 649–664. doi: 10.1099/mic.0.064113-0

Gavini, F., Cayuela, C., Antoine, J.-M., Lecoq, C., Lefebvre, B., Membré, J.-M., et al. (2001). Differences in the distribution of bifidobacterial and enterobacterial species in human faecal microflora of three different (children, adults, elderly) age groups. *Microb. Ecol. Health Dis.* 13, 40–45. doi: 10.1080/089106001750071690

Goodrich, J. K., Davenport, E. R., Beaumont, M., Jackson, M. A., Knight, R., Ober, C., et al. (2016). Genetic determinants of the gut microbiome in UK twins. *Cell Host Microbe* 19, 731–743. doi: 10.1016/j.chom.2016.04.017

Grönlund, M. M., Gueimonde, M., Laitinen, K., Kociubinski, G., Gronroos, T., Salminen, S., et al. (2007). Maternal breast-milk and intestinal bifidobacteria guide the compositional development of the *Bifidobacterium* microbiota in infants at risk of allergic disease. *Clin. Exp. Allergy* 37, 1764–1772. doi: 10.1111/j.1365-2222.2007.02849.x

Grzeskowiak, L., Sales Teixeira, T. F., Bigonha, S. M., Lobo, G., Salminen, S., and Ferreira, C. L. (2015). Gut *Bifidobacterium* microbiota in one-month-old Brazilian newborns. *Anaerobe* 35(Pt B), 54–58. doi: 10.1016/j.anaerobe.2015.07.004

Guaraldi, F., and Salvatori, G. (2012). Effect of breast and formula feeding on gut microbiota shaping in newborns. *Front. Cell. Infect. Microbiol.* 2:94. doi: 10.3389/fcimb.2012.00094

Guardamagna, O., Amaretti, A., Puddu, P. E., Raimondi, S., Abello, F., Cagliero, P., et al. (2014). Bifidobacteria supplementation: effects on plasma lipid profiles in dyslipidemic children. *Nutrition* 30, 831–836. doi: 10.1016/j.nut.2014.01.014

Guarner, F., and Malagelada, J. R. (2003). Gut flora in health and disease. *Lancet* 361, 512–519. doi: 10.1016/S0140-6736(03)12489-0

Gueimonde, M., Debor, L., Tölkkö, S., Jokisalo, E., and Salminen, S. (2007). Quantitative assessment of faecal bifidobacterial populations by real-time PCR using lanthanide probes. *J. Appl. Microbiol.* 102, 1116–1122. doi: 10.1111/j.1365-2672.2006.03145.x

Gueimonde, M., Ouwehand, A., Pitkälä, K., Strandberg, T., Finne-Soveri, H., and Salminen, S. (2010). Fecal *Bifidobacterium* levels in elderly nursing home patients - Are levels as expected? *Biosci. Microflora* 29, 111–113. doi: 10.12938/bifidus.29.111

Haarman, M., and Knol, J. (2005). Quantitative real-time PCR assays to identify and quantify fecal *Bifidobacterium* species in infants receiving a prebiotic infant formula. *Appl. Environ. Microbiol.* 71, 2318–2324. doi: 10.1128/aem.71.5.2318-2324.2005

He, F., Ouwehand, A. C., Isolauri, E., Hosoda, M., Benno, Y., and Salminen, S. (2001). Differences in composition and mucosal adhesion of bifidobacteria isolated from healthy adults and healthy seniors. *Curr. Microbiol.* 43, 351–354. doi: 10.1007/s002840010315

Hevia, A., Milani, C., Lopez, P., Donado, C. D., Cuervo, A., Gonzalez, S., et al. (2016). Allergic patients with long-term asthma display low levels of *Bifidobacterium adolescentis*. *PLoS ONE* 11:e0147809. doi: 10.1371/journal.pone.0147809

HMP (2012). Structure, function and diversity of the healthy human microbiome. *Nature* 486, 207–214. doi: 10.1038/nature11234

Hopkins, M., Sharp, R., and Macfarlane, G. (2001). Age and disease related changes in intestinal bacterial populations assessed by cell culture, 16S rRNA abundance, and community cellular fatty acid profiles. *Gut* 48, 198–205. doi: 10.1136/gut.48.2.198

Hopkins, M. J., and Macfarlane, G. T. (2002). Changes in predominant bacterial populations in human faeces with age and with *Clostridium difficile* infection. *J. Med. Microbiol.* 51, 448–454. doi: 10.1099/0022-1317-51-5-448

Imahori, K. (1992). How I understand aging. *Nutr. Rev.* 50, 351–352. doi: 10.1111/j.1753-4887.1992.tb02477.x

Jeffery, I. B., O'Toole, P. W., Ohman, L., Claesson, M. J., Deane, J., Quigley, E. M., et al. (2012). An irritable bowel syndrome subtype defined by species-specific alterations in faecal microbiota. *Gut* 61, 997–1006. doi: 10.1136/gutjnl-2011-301501

Kalliomäki, M., Collado, M. C., Salminen, S., and Isolauri, E. (2008). Early differences in fecal microbiota composition in children may predict overweight. *Am. J. Clin. Nutr.* 87, 534–538.

Karlsson, F. H., Tremaroli, V., Nookaew, I., Bergstrom, G., Behre, C. J., Fagerberg, B., et al. (2013). Gut metagenome in European women with normal, impaired and diabetic glucose control. *Nature* 498, 99–103. doi: 10.1038/nature12198

Kerckhoffs, A. P., Samsom, M., van der Rest, M. E., de Vogel, J., Knol, J., Ben-Amor, K., et al. (2009). Lower Bifidobacteria counts in both duodenal mucosa-associated and fecal microbiota in irritable bowel syndrome patients. *World J. Gastroenterol.* 15, 2887–2892. doi: 10.3748/wjg.15.2887

Klaassens, E. S., Boesten, R. J., Haarman, M., Knol, J., Schuren, F. H., Vaughan, E. E., et al. (2009). Mixed-species genomic microarray analysis of fecal samples reveals differential transcriptional responses of bifidobacteria in breast- and formula-fed infants. *Appl. Environ. Microbiol.* 75, 2668–2676. doi: 10.1128/AEM.02492-08

Koenig, J. E., Spor, A., Scalfone, N., Fricker, A. D., Stombaugh, J., Knight, R., et al. (2011). Succession of microbial consortia in the developing infant gut microbiome. *Proc. Natl. Acad. Sci. U.S.A.* 108(Suppl. 1), 4578–4585. doi: 10.1073/pnas.1000081107

Lagier, J. C., Hugon, P., Khelaifia, S., Fournier, P. E., La Scola, B., and Raoult, D. (2015). The rebirth of culture in microbiology through the example of culturomics to study human gut microbiota. *Clin. Microbiol. Rev.* 28, 237–264. doi: 10.1128/cmr.00014-14

Lewis, Z. T., Totten, S. M., Smilowitz, J. T., Popovic, M., Parker, E., Lemay, D. G., et al. (2015). Maternal fucosyltransferase 2 status affects the gut bifidobacterial communities of breastfed infants. *Microbiome* 3, 13. doi: 10.1186/s40168-015-0071-z

Lyra, A., Rinttila, T., Nikkila, J., Krogius-Kurikka, L., Kajander, K., Malinen, E., et al. (2009). Diarrhoea-predominant irritable bowel syndrome distinguishable by 16S rRNA gene phylotype quantification. *World J. Gastroenterol.* 15, 5936–5945. doi: 10.3748/wjg.15.5936

Makino, H., Kushiro, A., Ishikawa, E., Kubota, H., Gawad, A., Sakai, T., et al. (2013). Mother-to-infant transmission of intestinal bifidobacterial strains has an impact on the early development of vaginally delivered infant's microbiota. *PLoS ONE* 8:e78331. doi: 10.1371/journal.pone.0078331

Malaguarnera, G., Leggio, F., Vacante, M., Motta, M., Giordano, M., Bondi, A., et al. (2012). Probiotics in the gastrointestinal diseases of the elderly. *J. Nutr. Health Aging* 16, 402–410. doi: 10.1007/s12603-011-0357-1

Matsuki, T., Watanabe, K., Fujimoto, J., Kado, Y., Takada, T., Matsumoto, K., et al. (2004). Quantitative PCR with 16S rRNA-gene-targeted species-specific primers for analysis of human intestinal bifidobacteria. *Appl. Environ. Microbiol.* 70, 167–173. doi: 10.1128/AEM.70.1.167-173.2004

Matsuki, T., Watanabe, K., Tanaka, R., Fukuda, M., and Oyaizu, H. (1999). Distribution of bifidobacterial species in human intestinal microflora examined with 16S rRNA-gene-targeted species-specific primers. *Appl. Environ. Microbiol.* 65, 4506–4512.

Mayorga Reyes, L., Gonzalez Vazquez, R., Cruz Arroyo, S. M., Melendez Avalos, A., Reyes Castillo, P. A., Chavaro Perez, D. A., et al. (2016). Correlation between diet and gut bacteria in a population of young adults. *Int. J. Food Sci. Nutr.* 67, 470–478. doi: 10.3109/09637486.2016.1162770

Ménard, O., Butel, M. J., Gaboriau-Routhiau, V., and Waligora-Dupriet, A. J. (2008). Gnotobiotic mouse immune response induced by *Bifidobacterium* sp. strains isolated from infants. *Appl. Environ. Microbiol.* 74, 660–666. doi: 10.1128/aem.01261-07

Mevissen-Verhage, E. A., Marcelis, J. H., de Vos, M. N., Harmsen-van Amerongen, W. C., and Verhoef, J. (1987). *Bifidobacterium*, *Bacteroides*, and *Clostridium* spp. in fecal samples from breast-fed and bottle-fed infants with and without iron supplement. *J. Clin. Microbiol.* 25, 285–289.

Mihajlovski, A., Doré, J., Levenez, F., Alric, M., and Brugère, J.-F. (2010). Molecular evaluation of the human gut methanogenic archaeal microbiota reveals an age-associated increase of the diversity. *Environ. Microbiol. Rep.* 2, 272–280. doi: 10.1111/j.1758-2229.2009.00116.x

Mikami, K., Takahashi, H., Kimura, M., Isozaki, M., Izuchi, K., Shibata, R., et al. (2009). Influence of maternal bifidobacteria on the establishment of bifidobacteria colonizing the gut in infants. *Pediatr. Res.* 65, 669–674. doi: 10.1203/PDR.0b013e31819ed7a8

Million, M., Angelakis, E., Maraninchi, M., Henry, M., Giorgi, R., Valero, R., et al. (2013). Correlation between body mass index and gut concentrations of *Lactobacillus reuteri*, *Bifidobacterium animalis*, *Methanobrevibacter smithii* and *Escherichia coli*. *Int. J. Obes. (Lond.)* 37, 1460–1466. doi: 10.1038/ijo.2013.20

Mitsuoka, T. (1992). Intestinal flora and aging. *Nutr. Rev.* 50, 438–446. doi: 10.1111/j.1753-4887.1992.tb02499.x

Mitsuoka, T., Hayakawa, K., and Kimura, N. (1974). [The faecal flora of man. II. The composition of Bifidobacterium flora of different age groups (author's transl)]. *Zentralbl. Bakteriol. Orig. A* 226, 469–478.

Murphy, K., O' Shea, C. A., Ryan, C. A., Dempsey, E. M., O' Toole, P. W., Stanton, C., et al. (2015). The gut microbiota composition in dichorionic triplet sets suggests a role for host genetic factors. *PLoS ONE* 10:e0122561. doi: 10.1371/journal.pone.0122561

Murri, M., Leiva, I., Gomez-Zumaquero, J. M., Tinahones, F. J., Cardona, F., Soriguer, F., et al. (2013). Gut microbiota in children with type 1 diabetes differs from that in healthy children: a case-control study. *BMC Med.* 11:46. doi: 10.1186/1741-7015-11-46

Musilova, S., Rada, V., Vlkova, E., and Bunesova, V. (2014). Beneficial effects of human milk oligosaccharides on gut microbiota. *Benef. Microbes* 5, 273–283. doi: 10.3920/bm2013.0080

Nakamura, T., Sasaki, T., Fujimori, M., Yazawa, K., Kano, Y., Amano, J., et al. (2002). Cloned cytosine deaminase gene expression of *Bifidobacterium longum* and application to enzyme/pro-drug therapy of hypoxic solid tumors. *Biosci. Biotechnol. Biochem.* 66, 2362–2366. doi: 10.1271/bbb.66.2362

O'Callaghan, A., and van Sinderen, D. (2016). Bifidobacteria and their role as members of the human gut microbiota. *Front. Microbiol.* 7:925. doi: 10.3389/fmicb.2016.00925

Odamaki, T., Kato, K., Sugahara, H., Hashikura, N., Takahashi, S., Xiao, J. Z., et al. (2016). Age-related changes in gut microbiota composition from newborn to centenarian: a cross-sectional study. *BMC Microbiol.* 16:90. doi: 10.1186/s12866-016-0708-5

O'Sullivan, O., Coakley, M., Lakshminarayanan, B., Conde, S., Claesson, M. J., Cusack, S., et al. (2013). Alterations in intestinal microbiota of elderly Irish subjects post-antibiotic therapy. *J. Antimicrob. Chemother.* 68, 214–221. doi: 10.1093/jac/dks348

Palma, G. D., Capilla, A., Nova, E., Castillejo, G., Varea, V., Pozo, T., et al. (2012). Influence of milk-feeding type and genetic risk of developing coeliac disease on intestinal microbiota of infants: the PROFICEL study. *PLoS ONE* 7:e30791. doi: 10.1371/journal.pone.0030791

Penders, J., Thijs, C., Vink, C., Stelma, F. F., Snijders, B., Kummeling, I., et al. (2006). Factors influencing the composition of the intestinal microbiota in early infancy. *Pediatrics* 118, 511–521. doi: 10.1542/peds.2005-2824

Picard, C., Fioramonti, J., Francois, A., Robinson, T., Neant, F., and Matuchansky, C. (2005). Review article: bifidobacteria as probiotic agents – physiological effects and clinical benefits. *Aliment. Pharmacol. Ther.* 22, 495–512. doi: 10.1111/j.1365-2036.2005.02615.x

Putignani, L., Del Chierico, F., Petrucca, A., Vernocchi, P., and Dallapiccola, B. (2014). The human gut microbiota: a dynamic interplay with the host from birth to senescence settled during childhood. *Pediatr. Res.* 76, 2–10. doi: 10.1038/pr.2014.49

Qin, J., Li, R., Raes, J., Arumugam, M., Burgdorf, K. S., Manichanh, C., et al. (2010). A human gut microbial gene catalogue established by metagenomic sequencing. *Nature* 464, 59–65. doi: 10.1038/nature08821

Rajilic-Stojanovic, M., Heilig, H. G., Molenaar, D., Kajander, K., Surakka, A., Smidt, H., et al. (2009). Development and application of the human intestinal tract chip, a phylogenetic microarray: analysis of universally conserved phylotypes in the abundant microbiota of young and elderly adults. *Environ. Microbiol.* 11, 1736–1751. doi: 10.1111/j.1462-2920.2009.01900.x

Roger, L. C., Costabile, A., Holland, D. T., Hoyles, L., and McCartney, A. L. (2010). Examination of faecal *Bifidobacterium* populations in breast- and formula-fed infants during the first 18 months of life. *Microbiology* 156, 3329–3341. doi: 10.1099/mic.0.043224-0

Rondanelli, M., Giacosa, A., Faliva, M. A., Perna, S., Allieri, F., and Castellazzi, A. M. (2015). Review on microbiota and effectiveness of probiotics use in older. *World J. Clin. Cases* 3, 156–162. doi: 10.12998/wjcc.v3.i2.156

Salazar, N., Lopez, P., Valdes, L., Margolles, A., Suarez, A., Patterson, A. M., et al. (2013). Microbial targets for the development of functional foods accordingly with nutritional and immune parameters altered in the elderly. *J. Am. Coll. Nutr.* 32, 399–406. doi: 10.1080/07315724.2013.827047

Santacruz, A., Collado, M. C., Garcia-Valdes, L., Segura, M. T., Martin-Lagos, J. A., Anjos, T., et al. (2010). Gut microbiota composition is associated with body weight, weight gain and biochemical parameters in pregnant women. *Br. J. Nutr.* 104, 83–92. doi: 10.1017/s0007114510000176

Savignac, H. M., Kiely, B., Dinan, T. G., and Cryan, J. F. (2014). Bifidobacteria exert strain-specific effects on stress-related behavior and physiology in BALB/c mice. *Neurogastroenterol. Motil.* 26, 1615–1627. doi: 10.1111/nmo.12427

Sela, D. A., Chapman, J., Adeuya, A., Kim, J. H., Chen, F., Whiteheadf, T. R., et al. (2008). The genome sequence of *Bifidobacterium longum* subsp. infantis reveals adaptations for milk utilization within the infant microbiome. *Proc. Natl. Acad. Sci. U.S.A.* 105, 18964–18969. doi: 10.1073/pnas.0809584105

Sivan, A., Corrales, L., Hubert, N., Williams, J. B., Aquino-Michaels, K., Earley, Z. M., et al. (2015). Commensal *Bifidobacterium* promotes antitumor immunity and facilitates anti-PD-L1 efficacy. *Science* 350, 1084–1089. doi: 10.1126/science.aac4255

Stsepetova, J., Sepp, E., Julge, K., Vaughan, E., Mikelsaar, M., and de Vos, W. M. (2007). Molecularly assessed shifts of *Bifidobacterium* ssp. and less diverse microbial communities are characteristic of 5-year-old allergic children. *FEMS Immunol. Med. Microbiol.* 51, 260–269. doi: 10.1111/j.1574-695X.2007.00306.x

Taverniti, V., and Guglielmetti, S. (2014). Methodological issues in the study of intestinal microbiota in irritable bowel syndrome. *World J. Gastroenterol.* 20, 8821–8836. doi: 10.3748/wjg.v20.i27.8821

Tojo, R., Suarez, A., Clemente, M. G., de los Reyes-Gavilan, C. G., Margolles, A., Gueimonde, M., et al. (2014). Intestinal microbiota in health and disease: role of bifidobacteria in gut homeostasis. *World J. Gastroenterol.* 20, 15163–15176. doi: 10.3748/wjg.v20.i41.15163

Turnbaugh, P. J., Hamady, M., Yatsunenko, T., Cantarel, B. L., Duncan, A., Ley, R. E., et al. (2009). A core gut microbiome in obese and lean twins. *Nature* 457, 480–484. doi: 10.1038/nature07540

Turroni, F., Peano, C., Pass, D. A., Foroni, E., Severgnini, M., Claesson, M. J., et al. (2012). Diversity of bifidobacteria within the infant gut microbiota. *PLoS ONE* 7:e36957. doi: 10.1371/journal.pone.0036957

Underwood, M. A., German, J. B., Lebrilla, C. B., and Mills, D. A. (2015). *Bifidobacterium longum* subspecies infantis: champion colonizer of the infant gut. *Pediatr. Res.* 77, 229–235. doi: 10.1038/pr.2014.156

van Tongeren, S. P., Slaets, J. P., Harmsen, H. J., and Welling, G. W. (2005). Fecal microbiota composition and frailty. *Appl. Environ. Microbiol.* 71, 6438–6442. doi: 10.1128/aem.71.10.6438-6442.2005

Voreades, N., Kozil, A., and Weir, T. (2014). Diet and the development of the human intestinal microbiome. *Front. Microbiol.* 5:494. doi: 10.3389/fmicb.2014.00494

Wang, F., Huang, G., Cai, D., Li, D., Liang, X., Yu, T., et al. (2015). Qualitative and semiquantitative analysis of fecal *Bifidobacterium* species in centenarians living in Bama, Guangxi, China. *Curr. Microbiol.* 71, 143–149. doi: 10.1007/s00284-015-0804-z

Woodmansey, E. J. (2007). Intestinal bacteria and ageing. *J. Appl. Microbiol.* 102, 1178–1186. doi: 10.1111/j.1365-2672.2007.03400.x

Woodmansey, E. J., McMurdo, M. E., Macfarlane, G. T., and Macfarlane, S. (2004). Comparison of compositions and metabolic activities of fecal microbiotas in young adults and in antibiotic-treated and non-antibiotic-treated elderly subjects. *Appl. Environ. Microbiol.* 70, 6113–6122. doi: 10.1128/AEM.70.10.6113-6122.2004

Wu, G. D., and Lewis, J. D. (2013). Analysis of the human gut microbiome and association with disease. *Clin. Gastroenterol. Hepatol.* 11, 774–777. doi: 10.1016/j.cgh.2013.03.038

Wu, X., Ma, C., Han, L., Nawaz, M., Gao, F., Zhang, X., et al. (2010). Molecular characterisation of the faecal microbiota in patients with type II diabetes. *Curr. Microbiol.* 61, 69–78. doi: 10.1007/s00284-010-9582-9

Xu, M., Wang, B., Fu, Y., Chen, Y., Yang, F., Lu, H., et al. (2012). Changes of fecal *Bifidobacterium* species in adult patients with hepatitis B virus-induced chronic liver disease. *Microb. Ecol.* 63, 304–313. doi: 10.1007/s00248-011-9925-5

Yatsunenko, T., Rey, F. E., Manary, M. J., Trehan, I., Dominguez-Bello, M. G., Contreras, M., et al. (2012). Human gut microbiome viewed across age and geography. *Nature* 486, 222–227. doi: 10.1038/nature 11053

Zhao, L., Qiao, X., Zhu, J., Zhang, X., Jiang, J., Hao, Y., et al. (2011). Correlations of fecal bacterial communities with age and living region for the elderly living in Bama, Guangxi, China. *J. Microbiol.* 49, 186–192. doi: 10.1007/s12275-011-0405-x

**Conflict of Interest Statement:** The authors declare that the research was conducted in the absence of any commercial or financial relationships that could be construed as a potential conflict of interest.

3

# Variations in the Post-Weaning Human Gut Metagenome Profile as a Result of *Bifidobacterium* Acquisition in the Western Microbiome

*Matteo Soverini[1], Simone Rampelli[1]*, Silvia Turroni[1], Stephanie L. Schnorr[2], Sara Quercia[1], Andrea Castagnetti[1], Elena Biagi[1], Patrizia Brigidi[1] and Marco Candela[1]**

[1] *Department of Pharmacy and Biotechnology, University of Bologna, Bologna, Italy,* [2] *Department of Anthropology, University of Oklahoma, Norman, OK, USA*

**Edited by:**
*Francesca Turroni, University College Cork, Ireland*

**Reviewed by:**
*Abelardo Margolles, Consejo Superior de Investigaciones Científicas, Spain*
*Jeffrey David Galley, Baylor College of Medicine, USA*

***Correspondence:**
*Simone Rampelli*
*simone.rampelli@unibo.it*
*Marco Candela*
*marco.candela@unibo.it*

Studies of the gut microbiome variation among human populations revealed the existence of robust compositional and functional layouts matching the three subsistence strategies that describe a trajectory of changes across our recent evolutionary history: hunting and gathering, rural agriculture, and urban post-industrialized agriculture. In particular, beside the overall reduction of ecosystem diversity, the gut microbiome of Western industrial populations is typically characterized by the loss of *Treponema* and the acquisition of *Bifidobacterium* as an abundant inhabitant of the post-weaning gut microbial ecosystem. In order to advance the hypothesis about the possible adaptive nature of this exchange, here we explore specific functional attributes that correspond to the mutually exclusive presence of *Treponema* and *Bifidobacterium* using publically available gut metagenomic data from Hadza hunter-gatherers and urban industrial Italians. According to our findings, *Bifidobacterium* provides the enteric ecosystem with a diverse panel of saccharolytic functions, well suited to the array of gluco- and galacto-based saccharides that abound in the Western diet. On the other hand, the metagenomic functions assigned to *Treponema* are more predictive of a capacity to incorporate complex polysaccharides, such as those found in unrefined plant foods, which are consistently incorporated in the Hadza diet. Finally, unlike *Treponema*, the *Bifidobacterium* metagenome functions include genes that permit the establishment of microbe–host immunological cross-talk, suggesting recent co-evolutionary events between the human immune system and *Bifidobacterium* that are adaptive in the context of agricultural subsistence and sedentary societies.

**Keywords:** ***Bifidobacterium*, *Treponema*, gut microbiota, microbiome, co-evolution, Westernization**

## INTRODUCTION

The gut microbiota (GM) exerts a vital role in human physiology, being strategic for human nutrition, immune protection, and for the preservation of essential metabolic functions (Kau et al., 2011; Nicholson et al., 2012). This raises questions of whether, and to what degree, the human GM can provide the host with extra physiological and metabolic flexibility for the adaptation to different lifestyles and environments. Since 2010, several studies have been conducted with the specific aim to explore GM variation across human populations with different subsistence practices, lifestyles and geographical origin. These human GM surveys revealed the existence of robust GM compositional and functional subgroups, so far generally reflective of the variations in subsistence strategy: hunter-gatherer, rural agricultural, and urban industrial Western lifestyle (Yatsunenko et al., 2012; Schnorr et al., 2014; Martínez et al., 2015; Obregon-Tito et al., 2015; Rampelli et al., 2015).

The findings from this new and emerging field of research, which combines human microbiology ecology and anthropology, resulted in two main conclusions with important implications for both human evolutionary history and human health: first, humans co-evolved with symbiont microbial ecosystems, which have co-adapted along the trajectory of subsistence change across human evolutionary history, from hunter-gatherers to rural agricultural to the most recent development of completely industrialized societies (Quercia et al., 2014; Obregon-Tito et al., 2015); second, despite the considerable variation in rural and traditional life-ways, urban industrial populations stand apart as having a distinctly altered GM profile. Indeed, the GM of urban industrial populations seems to universally share certain compositional qualities, such as: (i) an overall compression of microbial diversity as measured by phylogeny and the number of unique taxa (Segata, 2015), (ii) the loss of the so-called microorganisms "old friends", *Treponema* and *Succinivibrio* (Blaser and Falkow, 2009; Warinner et al., 2015), and (iii) the acquisition of *Bifidobacterium* as typical inhabitant of the adult gut (Schnorr et al., 2014; Martínez et al., 2015; Obregon-Tito et al., 2015).

In agreement with the multiple hit hypothesis (Sonnenburg and Sonnenburg, 2014), dietary changes, sanitization and antibiotic usage are all potential triggers that would explain the reduction of ecosystem diversity and the loss of co-adapted microbial communities commonly observed in the GM profile of urban industrial Western populations. Conversely, the factors that favored the correspondent acquisition of *Bifidobacterium* in the Western microbiome still remain to be defined. Showing a relative abundance that ranges from 3 to 10% of the total ecosystem, bifidobacteria are an abundant bacterial component in the GM of urban industrial, namely Westernized, adults, and also dominates the GM ecosystem of breast-fed infants, where this bacterial family accounts on average for 80% of the total community (Turroni et al., 2012; Bottacini et al., 2014; Milani et al., 2015b). The recent characterization of the bifidobacterial pangenome – 18,181 *Bifidobacterium* specific Cluster of Orthologous Genes (BifCOGs) from 47 sequenced type strains – revealed the saccharolytic functions of this microorganism, and indicated a strong adaptation to the human gut environment (Milani et al., 2015a). Furthermore, recent work has helped advance a possible hypothesis to explain the adoption of *Bifidobacterium* into the Western adult GM. Through comparisons of the GM of Hadza hunter-gatherers and urban industrial Italians, Schnorr et al. (2014) highlighted – for the first time to our knowledge – the substantial lack of *Bifidobacterium* from the GM of some traditional populations. The authors proposed that the lack of bifidobacteria in adult Hadza hunter-gatherers may be "a consequence of the post-weaning GM composition in the absence of agro-pastoral-derived foods", while the continued consumption of dairy into adulthood is one of the possible vectors by which many Westernized populations maintain a relatively large bifidobacterial presence. Obregon-Tito et al. (2015) and Martínez et al. (2015) reached similar conclusions through comparative research on the GM of hunter-gatherers from Peru and rural highlanders from Papua New Guinea, respectively, and also found a low proportion of bifidobacteria relative to individuals from the USA. However, these comparative gut metagenome surveys do not specifically explore the impact of *Bifidobacterium* acquisition on the functional configuration of the GM of Western adults.

Taken together, these findings indicate a scenario of exchange, whereby through Westernization, human populations have lost long-standing commensal microorganisms, in particular *Treponema* and *Succinivibrio*, but have potentially compensated through the adult acquisition of bifidobacteria. In order to examine changes to the GM as a result of these community shifts, we investigate how the loss of *Treponema* and the acquisition of *Bifidobacterium* influenced the human gut metagenome profile. To this aim, we compare gut metagenome functions assigned to *Treponema* and *Bifidobacterium* retrieved from downloadable GM metagenomic data for both Hadza hunter-gatherers and urban Italians. Our findings reveal interesting functional gains in the Western microbiome corresponding to the post-weaning retention of *Bifidobacterium* as a symbiont microorganism, suggesting an opportunistic yet important role of this taxon in our recent history.

## MATERIALS AND METHODS

### Sample Collection and Shotgun Sequencing

The Illumina shotgun sequences used in this study were previously generated (Rampelli et al., 2015), and are publically available at the National Center for Biotechnology Information – Sequence Read Archive (NCBI SRA; SRP056480, Bioproject ID PRJNA278393. Leipzig Ethik-kommission review board, reference number 164-12-21052012).

### *Bifidobacterium* and *Treponema* Species Identification within Italian and Hadza Metagenomes

In order to identify the *Bifidobacterium* and *Treponema* species in Italian and Hadza populations, respectively, the 16S rDNA sequences within the assembled metagenomes were taxonomically selected using the assign_taxonomy.py script of the Qiime pipeline (Caporaso et al., 2010), against the Greengenes database[1]. The assignment at species level was performed by blastn of the *Treponema* and *Bifidobacterium* 16S rDNA sequences against the entire NCBI nucleotide database, in particular the top hit results for each sequence were considered.

[1]http://greengenes.secondgenome.com/downloads

### Characterization of the CAZyme Repertoire Assigned to *Bifidobacterium* and *Treponema* in the Gut Metagenome

Reads from a total of 38 individual GM metagenomes, 27 Hadza and 11 Italians from Rampelli et al. (2015), were downloaded and included in this study. Reads were assembled into contigs using MetaVelvet (Namiki et al., 2012) with 350 bp as insert length. Predicted open reading frames (ORFs) were determined by FragGeneScan (Rho et al., 2010) on assembled contigs, using the "–w 0" option for the fragmented genomic sequences and the parameter "–t complete". From the translated ORFs we detected the CAZymes-coding sequences using hmmscan tool from the HMMER software package (Eddy, 2011) and the dbCAN database (Yin et al., 2012). The outputs were processed by the script hmmscan-parser.sh[2], selecting only the sequences that showed a minimum identity of 30% to the query sequences and an alignment length of at least 100 residues. In order to identify CAZymes derived from *Bifidobacterium* and *Treponema*, we retrieved the nucleotide sequences of the CAZymes detected with hmmscan from the FragGeneScan output, and we blasted them against the NCBI nucleotide database. Only the sequences that showed as best hit an assignment to *Bifidobacterium* for the Italian samples or *Treponema* for the Hadza samples were retained for further analysis. On the basis of the coverage of the contigs, we obtained information concerning the abundance of CAZymes. To compare the data among samples, we obtained a normalized CAZyme abundance by dividing the CAZyme coverages of every correspondent contig for the giga-bases of every correspondent sample. Heatmaps and graphs were generated in R using the packages made4 (Culhane et al., 2005) and stats[3].

### Read-Mapping Approach for the Detection of *Bifidobacterium* and *Treponema* Functions Involved in the Adaptation to the Gut Environment

High quality reads for each sample were aligned to *Bifidobacterium*- or *Treponema*-assigned genes encoding bile acid adaptation, host interaction, and polysaccharide catabolism using bowtie2 and setting the alignment parameters to "–sensitive-local" (See Supplementary Table for the list of genes used for this analysis). As reference for the alignment, two different databases containing orthologous genes from the NCBI genomes of the previously detected *Treponema* or *Bifidobacterium* species were created. Specifically, the databases contain genes for alpha-amylase, beta-galactosidase, mannanase, cellulase, pectinase, and xylanase. Furthermore, the databases were implemented with the sequences of the bile efflux pump, bile salt hydrolase, exopolysaccharide synthase, fimbrial subunit FimQ, sortase, galactosyl transferase, and undecaprenyl-phosphate phosphotransferase, since they were recently reported as genes that facilitate commensal-host cross-talk in *Bifidobacterium* (Gueimonde et al., 2009; Fanning et al., 2012; Turroni et al., 2013; Ferrario et al., 2016). In the event that the *Bifidobacterium* or *Treponema* NCBI genomes did not contain the above-mentioned genes, we supplemented the databases with genes belonging to the taxonomically closest annotated microorganism. The reads that aligned with a reference using bowtie2, were extracted and their taxonomy was further verified by blastn against the entire NCBI nucleotide database. Notably, in the case that the best hits of the blastn search were not assigned to *Bifidobacterium* or *Treponema*, we did not consider those reads for further analysis. The number of hits for each gene was normalized by the number of base pairs in the input file and in the correspondent reference in order to compare the results.

## RESULTS

### Variation in the Gut Microbiome Carbohydrate-Degrading Repertoire as a Result of the Exclusive Presence of *Bifidobacterium* or *Treponema* in the Italian and Hadza Gut Microbial Ecosystems

By studying hunter-gatherers and rural populations, we previously depicted an approximation of the co-evolutional trajectory of the human GM structure in the recent times (Schnorr et al., 2014). In particular, we characterized the taxonomic and functional metabolic potential of the GM in the Hadza hunter-gatherers and urban Italian adults, reporting considerable differences between the two populations that map on to their respective life-ways (Schnorr et al., 2014; Rampelli et al., 2015). Among them, we observed a loss of *Treponema* and an acquisition of *Bifidobacterium* in the Italian GM. Considering the importance of *Bifidobacterium* as a health-promoting commensal of the GM in Western populations, it is important to understand how this beneficial role may have developed through the lens of co-evolution.

We first identified the diversity of the *Bifidobacterium* and *Treponema* species in urban Italian and Hadza GM by reconstructing the full 16S rDNA gene from assembled metagenomes. Italian samples contain sequences belonging to *Bifidobacterium faecale*, *Bifidobacterium pseudocatenulatum*, *Bifidobacterium adolescentis*, *Bifidobacterium coryneforme*, *Bifidobacterium bifidum*, *Bifidobacterium longum*, *Bifidobacterium angulatum*, and *Bifidobacterium dentium*. On the other hand, we detect the 16S rDNA sequences assigned to *Treponema porcinum*, *Treponema bryantii*, *Treponema succinifaciens*, *Treponema parvum*, and *Treponema berlinense* in the GM of the Hadza hunter-gatherers. However, we must acknowledge that these taxonomic assignments are limited by present whole genome for *Treponema* species, most of which have been characterized by work on human pathogens, rather than commensal members of the GM. Ongoing work with species-level variation should be crucial to resolving the disparities in our classification and resolution of these species and their functions in the future (Obregon-Tito et al., 2015). In order

[2]https://github.com/carden24/Bioinformatics_scripts/blob/master/hmmscan-parser.sh

[3]https://stat.ethz.ch/R-manual/R-devel/library/stats/html/00Index.html

to compare the specific saccharolytic functions conferred by *Bifidobacterium* and *Treponema* in the Italian and Hadza microbiomes, we identified a total of 5.4 million of ORFs, of which 14,512 mapped to CAZymes for the Italian samples and 74,651 for the Hadza samples (**Figure** ). Notably, the Hadza metagenomes contain significantly more CAZymes per subject, in terms of ORFs assigned to CAZymes per million of reads, respect to the Italian metagenomes (mean ± SD, Hadza: 233 ± 86, Italians: 137 ± 78), as reported by Rampelli et al. (2015). We then profiled the saccharolytic repertoire of *Bifidobacterium* and *Treponema* in the Italian and Hadza GM as relative abundance at the CAZyme category level based on the coverage of taxonomically assigned contigs (**Figure** ). *Bifidobacterium* showed a higher presence of glycosyl transferase (GT) and carbohydrate esterase (CE), with respect to *Treponema* (GT relative abundance, rel. ab.: 54% for *Bifidobacterium* and 43% for *Treponema*; CE rel. ab.: 6% for *Bifidobacterium* and 5% for *Treponema*). On the other hand, *Treponema* were more enriched in glycoside hydrolase (GH) and carbohydrate binding module (CBM) (GH rel. ab.: 33% for *Treponema* and 31% for *Bifidobacterium*; CBM rel. ab.: 15% for *Treponema* and 8% for *Bifidobacterium*). At the CAZyme family level, we revealed the four families that constitute the core *Bifidobacterium* CAZyme repertoire: GH13 (GH family acting on substrates containing α-glycosidic linkages), GH3 (GH family that groups together exo-acting β-D-glucosidases, α-L-arabinofuranosidase, β-D-xylopyranosidase and *N*-acetyl-β-D-glucosaminidase), GT2 (GT family containing cellulose synthase, mannan synthase, and several monosaccharide-/oligosaccharide-transferases), and GT4 (GT family containing sucrose synthase, glucosyl transferase, and several phosphorylases). The sum of ORFs assigned to these four major families comprises 77% of the total detected CAZyme cohort. The relative abundance of the *Bifidobacterium* and *Treponema* CAZyme families detected in the Italian and Hadza samples reveals several differences in the potential saccharolytic functional contributions of these two microorganisms (**Figure** ). In particular, *Bifidobacterium* have a greater abundance of genes involved in the degradation of lactate, which is produced from pyruvate in the fermentation of simple sugar and commonly found in sour milk as well as in other lacto-fermented foods (family GH2). This finding also corresponds to the exclusive contribution of dairy carbohydrates (~5% of total carbohydrates), in the Italian cohort diet (Supplementary Table ). Emphasis on monosaccharide catabolism is evidenced by enrichment in gene families containing enzymes that metabolize mannose, xylose and arabinose (GH2, GH31, and GH43), which are highly represented in plant and fruit glycans. As a variety of genes were also found that are involved in the degradation of α- and β-glucans (GH3 and GH31), this illustrates an ability of *Bifidobacterium* to retrieve energy also from more complex carbohydrates that are commonly present in the cellulosic biomass of plant foods in the Italian diet: salads, fruits, nuts, cereals and their product derivatives. In addition, *Bifidobacterium* are also enriched in genes involved in the catabolism of sucrose (GH31), which is widely distributed in nature, but robustly manifest in the industrial food products that are consumed daily by most urban populations. Further evidence of these functions comes from detection of a higher abundance of CBM families for lactose, galactose and β-glucans (CBM4, CBM13, and CBM32) in *Bifidobacterium*, with respect to *Treponema*. In contrast, the CAZyme profile of *Treponema* within the Hadza metagenome is mainly devoted to degradation of glucans, galactans, and fructans (GH16, GH32, and GH53), which are sugar polymers that comprise hemicellulose (galactans) and inulin (fructans). The monosaccharide of galactans, galactose, is also expressed in mucilages and glycoproteins that derive from the human host, as well as a number of vegetable-derived carbohydrates. Both sugar polymers are largely implicit in difficult-to-digest plant polysaccharides that escape small intestine absorption and are instead fermented by the colonic microbiota. The Hadza diet is replete with such unrefined plant foods that contain indigestible polysaccharides such as berries, baobab fruit, and particularly tubers (Schnorr et al., 2015). *Treponema* are also enriched in two CAZyme α-amylase families (GH57 and GH77), which are unlike the typical α-amylase GH13 family because they have a conserved trans-glycosylating region (MacGregor et al., 2001). Finally, *Treponema* are better equipped to metabolize peptidoglycans due to a wide range of acetylglucosaminases and peptidoglycan lyases (GH23, GH73, and GH109). These activities were confirmed by the detection of high levels of CBM families for peptidoglycans and α-glucans (CBM50 and CBM48). To specifically investigate the effective differences in polysaccharide-degrading repertoires between *Bifidobacterium* and *Treponema* metagenome functions, we aligned the reads of Italian and Hadza samples to a custom database containing the sequences for alpha-amylase, beta-galactosidase, mannanase, cellulase, pectinase, and xylanase. The taxonomic specificity of the aligned reads was verified by blasting the sequences to the nucleotide database of NCBI (**Figure** ). Unlike *Treponema*, which appears to be functionally equipped to derive energy from a broad spectrum of polysaccharides through increased relative representation of pectinase, xylanase, and cellulase, the polysaccharide-degrading functions assigned to *Bifidobacterium* are enriched in genes coding for β-galactosidase and mannanase (**Figure** ). Interestingly, both genera are configured to hydrolyze 1–4 α-glycosidic bonds (via α-amylase) to a similar degree. The catabolic configurations of *Bifidobacterium* and *Treponema* for polysaccharides are reflective of differences in dietary carbohydrates of Italians and Hadza: pasta, legumes, milk, and dairy products, versus berries, tubers, baobab, and honey, respectively, (Supplementary Table ).

## Differences in the Gut Metagenome Functions Involved in Bacteria/Host Interaction and Gut Adaptation Corresponding to the Presence of *Bifidobacterium* or *Treponema* in the Italian and Hadza Gut Metagenome

We further investigated the presence of the genes involved in host interaction and immune modulation in the ORFs attributed to *Bifidobacterium* and *Treponema*. In this scenario,

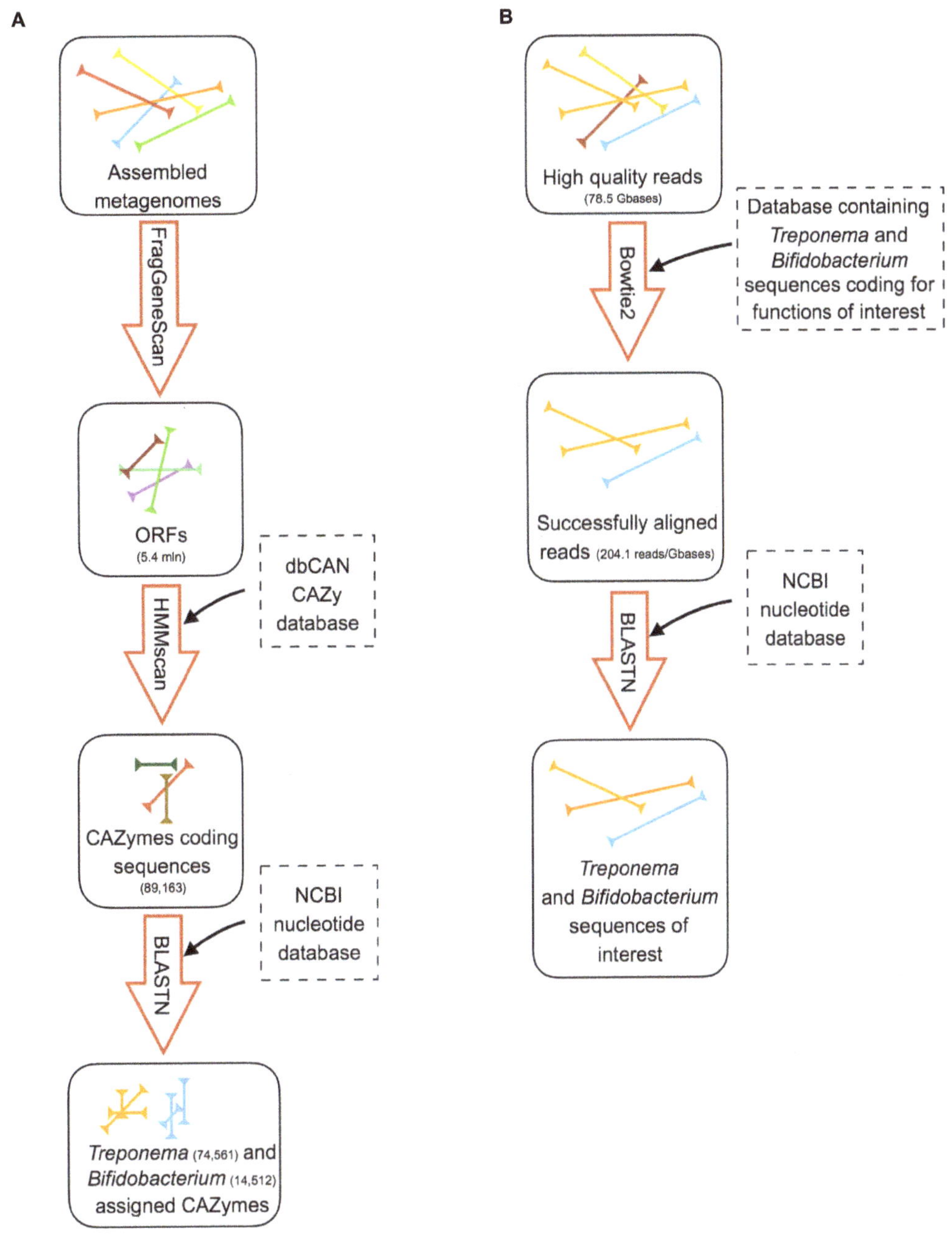


**FIGURE | Schematic representation of the analysis workflow. (A)** Pipeline for the identification and assignment of *Treponema* and *Bifidobacterium* CAZymes on assembled metagenomes: (i) ORFs detection using FragGeneScan; (ii) detection of the CAZyme-coding ORFs by using hmmscan against the dbCAN CAZy database; (iii) taxonomy assignment to CAZyme-coding sequences by blastn against the NCBI nucleotide database. **(B)** Pipeline for the identification of *Treponema* and *Bifidobacterium* sequences coding for functions involved in the adaptation to the gut environment: (i) alignment of high quality reads to databases containing the selected *Treponema* or *Bifidobacterium* functions using bowtie2; (ii) blasting of the successfully aligned reads against the NCBI nucleotide database to confirm the taxonomy.

the sortase-dependent pili are considered key surface molecules in establishing bacterial adherence to the host epithelium and they have been proposed as potential mediator of mucosal immune response (Turroni et al., 2013). Another recognized marker of interaction with the host that affects the immune response is the surface-exopolysaccharide (EPS). Specifically, bacteria producing EPS failed to elicit a strong immune B-cell response compared to EPS-deficient strains (Fanning et al., 2012). Our analyses confirm that the enzymes involved in the production of EPS and pili, namely EPS synthase, undecaprenyl-phosphate phosphotransferase, galactosyl transferase, sortase, and fimbrial subunit FimQ, are typical of the *Bifidobacterium*

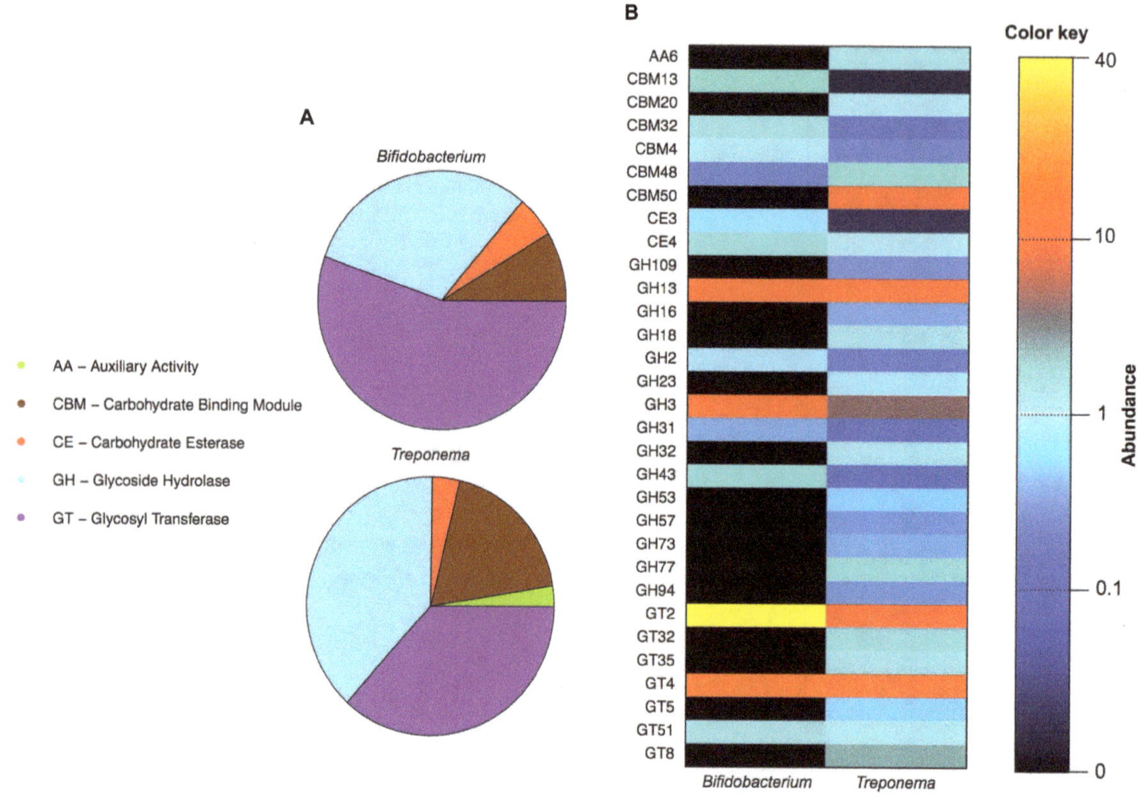


**FIGURE | Overview of *Bifidobacterium* and *Treponema* CAZyme repertoires in the Hadza and Italian samples. (A)** Normalized relative abundance of the CAZyme category levels for *Bifidobacterium* and *Treponema*: auxiliary activity (AA), carbohydrate binding module (CBM), carbohydrate esterase (CE), glycoside hydrolase (GH), and glycosyl transferase (GT) categories. **(B)** Comparison between the *Bifidobacterium* and *Treponema* CAZyme family profiles. The relative abundance of each family is indicated by the color key.

ORFs detected in the Italian metagenome, while virtually absent in the *Treponema* ORFs retrieved from the Hadza GM ecosystem (**Figure** ). Finally, we investigated the presence of genes involved in bile tolerance as mechanisms of bacterial adaptation to the human host. Bile salts are detergent-like compounds with strong antimicrobial activity (Begley et al., 2005), and intestinal bacteria have had to evolve strategies to tolerate physiological concentrations of bile salts to colonize the intestine (Gueimonde et al., 2009). Interestingly, two representative enzymes, which contribute to bile resistance and adaptation to gut environment, the bile-inducible efflux transporters and the bile salt hydrolase, are present in within the *Bifidobacterium* gut metagenome functions, but are not detected in *Treponema* ORFs (**Figure** ).

## DISCUSSION

In our work we explore how the apparent dichotomy between two instance-specific mutually exclusive gut inhabitants, *Bifidobacterium* and *Treponema*, reflects specific functional roles within the human GM ecosystem. We observe that the complete replacement of these microorganisms within the human gut in modern populations follows a curious pattern of lifestyle-associated differences, which stand at opposite ends of an entire subsistence spectrum, from hunting and gathering to post-industrial urban life. This may hint at a more ancient pattern of transition that occurred along more recent human evolutionary history as early settlements permitted the adoption of a fully agro-pastoral regime in place of mobile foraging. As seen in modern rural farmers, these earlier human groups may have harbored both communities of bifidobacteria and treponemes, suggesting an intermediary phase of co-habitation (De Filippo et al., 2010; Obregon-Tito et al., 2015). These patterns are instructive, and we can hypothesize that the stimulus for bifidobacterial acquisition and proliferation in the human GM followed much earlier changes to diet and lifestyle that occurred during the Neolithic transition. Specifically, these changes include permanent settlements, dense population structures, animal co-habitation, plant domestication, and shifts in the type and amount of carbohydrate-based foods (Childe, 1936; Binford, 1968).

While bifidobacteria are absent from the GM ecosystem of modern hunter-gatherers, and thus possibly acquired as a post-weaning symbionts among communities who adopted an agro-pastoral subsistence, *Treponema* has been lost from the human GM along the transition from small-scale rural communities to post-industrial urbanized society. This raises key questions such as 'what are the functional gains and losses in the human GM

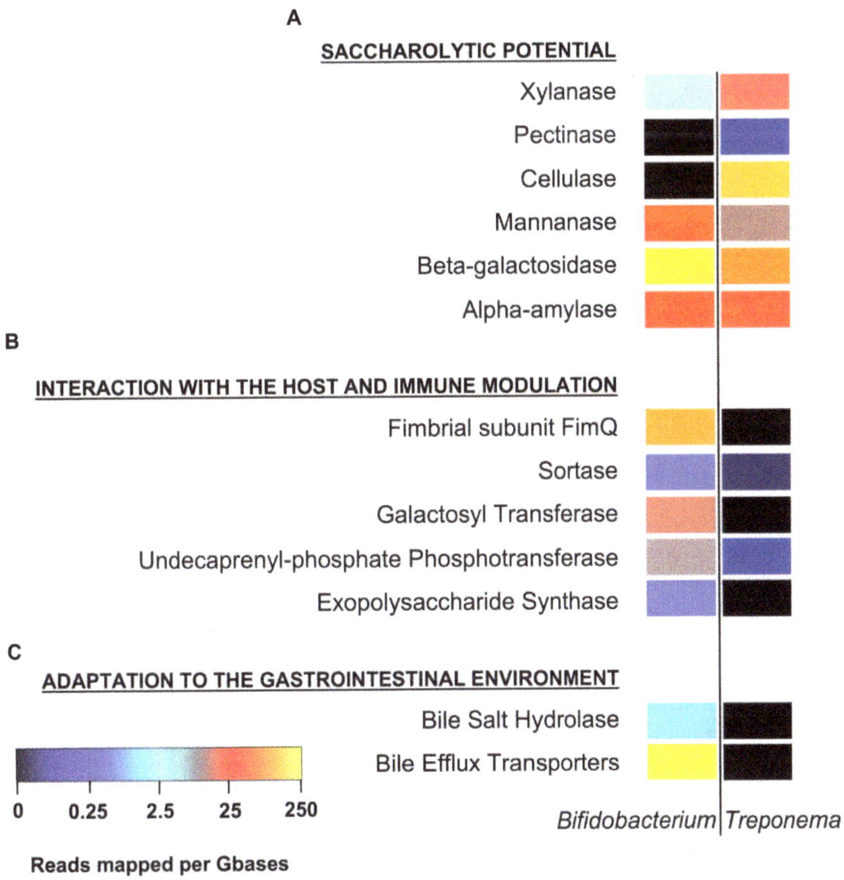


**FIGURE | Profile of *Bifidobacterium* and *Treponema* functions involved in the adaptation to the host environment.** Polysaccharide metabolism **(A)**; interaction with the host and immune modulation **(B)**; adaptation to the human gastrointestinal environment **(C)**. Color key represents reads mapped per giga-bases of the sample of origin. See Supplementary Table for the list of reference genes used.

corresponding to this process?', and 'does an adaptive nature of these changes exist?' To begin answering these questions, we compared metagenome functions assigned to *Bifidobacterium* and *Treponema* in the gut metagenome samples from urban Italians and Hadza hunter-gatherers, respectively. In particular, we focused on functional features key in the context of the GM-human host mutualism, such as the metabolism of complex polysaccharides, which is essential to provide the host with short-chain fatty acids (SCFA), and the microbe–host interaction and immunomodulation processes.

As an initial approach, we described the *Bifidobacterium* and *Treponema* diversity in the urban Italian and Hadza GM. Eight different bifidobacterial species were detected in the Italian GM ecosystem, including the plant polysaccharides degrader *B. adolescentis*, and the milk fermenters *B. bifidum* and *B. longum*. Other bifidobacterial species detected in the Italian GM were *B. faecale*, *B. pseudocatenulatum*, *B. coryneforme*, *B. angulatum*, and *B. dentium*, highlighting considerable bifidobacterial diversity within the Italian GM ecosystem. Given our current limitation in resolving novel *Treponema* taxa, we were still able to identify five species: *T. porcinum*, *T. bryantii*, *T. succinifaciens*, *T. parvum*, and *T. berlinense*. These species were already described in association with the GM of Matses hunter-gatherers from Peru (Obregon-Tito et al., 2015).

According to our findings, *Bifidobacterium* and *Treponema* provide the host metagenome with a different configuration of CAZyme categories. Indeed, the *Bifidobacterium* genes encoding for CAZymes in the Italian gut metagenome are enriched in GT and CE, while the *Treponema* saccharolytic repertoire in the Hadza gut is principally devoted to GH and CBM functions, suggesting the higher propensity of the latter to act as a primary degrader of complex polysaccharides in the gut. Focusing on the CAZyme family level, we observed that *Bifidobacterium* provides the host metagenome with an extremely versatile panel of saccharolytic functions. Bifidobacterial CAZyme families enriched in the Italian samples range from the metabolism of simple sugars, such as lactate, mannose, xylose, and arabinose, to more complex carbohydrates from plant sources, such as α- and β-glucans, lactose, or galactose. These findings are in general agreement with the overall structure of the bifidobacterial glycobiome (Milani et al., 2015a). According to Milani et al. (2015a), when comparing the bifidobacterial GH repertoire with the sequenced members of the human GM, *Bifidobacterium* appears enriched in GH3 and GH43 families for the degradation of plant carbohydrates. These findings are in agreement with our observations that the *Bifidobacterium*-assigned saccharolytic functions emphasize substrate-specific specializations that are not observed among *Treponema*-assigned genes. Instead, the saccharolytic functions supplied by *Treponema* to the Hadza gut metagenome are mainly devoted to the degradation of indigestible polysaccharides, such as galactans, fructans, and glucans, that are present in unrefined and wild plant foods. Other functional peculiarities characteristic of *Treponema* include a completely unique suite of α-amylases and a greater capacity to bind and metabolize peptidoglycan. When we specifically explored gut metagenome differences in polysaccharide-degrading genes, we confirmed the overall greater metabolic polysaccharolytic potential for *Treponema* while *Bifidobacterium* was enriched in beta-galactosidase and mannase.

Interestingly, the specific glycobiomes conferred by *Bifidobacterium* and *Treponema* match the respective dietary habits of the two populations. Hadza consume a heavily plant-based diet, particularly in the rainy season, in which ~70% of kcal are derived from fibrous wild plant foods and the 30% from wild game meat (Marlowe, 2010). Furthermore, Hadza entirely lack dairy in their diet. Correspondingly, the *Treponema*-assigned glycobiome in the Hadza is devoted to the metabolism of the vast array of refractory glucans, galactans, and fructans that are present in the wild Hadza plant foods. The Mediterranean diet, characteristic for the Italian cohort, is abundant in plant foods (salads, fruits, sauces), pasta, bread and olive oil with low to moderate inclusion of dairy, poultry, fish and read meat (Supplementary Table ). Curiously, certain glycobiome functions conferred by bifidobacteria to the Italian GM – e.g., GH2, GH31, and GH43, as well as CBM4, CBM13, and CBM32 – are well suited to deal with dairy carbohydrates and some types of plant glycans, all of which are relatively abundant in the Italian diet.

When we looked for functions associated with host–microbe interactions, we found matches only among *Bifidobacterium*. Conversely, *Treponema* does not provide the Hadza metagenome with known functionalities for direct host interaction. The *Bifidobacterium* functional potential encoded within strains retrieved from the Italian samples included enzymes for the biosynthesis of pili and EPS structures. Both cell components are essential in the context of a mutualistic strategy that permits close interaction and initiation of a tolerated immunological dialog between the bacterium and host (Fanning et al., 2012). Finally, as a functional marker of the bacterial adaptation to the gut environment (Begley et al., 2005), we sought functions involved in bile salts tolerance. Interestingly, only bifidobacteria seem to possess bile-inducible efflux transporters and the bile salt hydrolase required for bile salt mitigation and detoxification. Even if biases in the assignment of *Treponema* functions as a result of the paucity of reference genomes cannot be excluded, our findings could indicate different ecological strategies for *Bifidobacterium* and *Treponema* in the human gut.

## CONCLUSION

Our findings suggest possible co-evolutionary implications for the loss of *Treponema* and the acquisition of *Bifidobacterium* as a stable component of the post-weaning GM ecosystem from post-industrial urban populations. Capable of heterogeneous saccharolytic metabolism, which ranges from complex plant polysaccharides to simpler sugars such as lactose and sucrose, *Bifidobacterium* are well suited to handle the degradative demands imposed by a typical Western diet. Conversely, the progressive loss of more challenging microbiota accessible carbohydrates in the Western diet (Sonnenburg et al., 2016), such as hemicellulose and inulin, would help partially explain the extinction of a more specialized fiber degrader such as *Treponema* from the Western GM ecosystem. Furthermore, unlike *Treponema*, *Bifidobacterium* evolved the capacity to establish an intense microbe–host connection, which may help support a continuous and abundant bifidobacterial presence in adults, allowing this commensal to outcompete other opportunistic, but functionally diverse, microbiota. The acquisition of *Bifidobacterium* as a stable component of the GM ecosystem in small-scale rural agriculturalists, reminiscent of early human farmers, and modern Westernized populations, may therefore engage the functionalities of the host immune system, providing new adaptive solutions in response to changing selective pressures during the restructuring of human diet and society.

## AUTHOR CONTRIBUTIONS

SR and MC conceived the study with the participation of MS in experimental design. MS and SR performed the bioinformatics analysis. MS, SR, SLS, ST, and MC carried out data interpretation and wrote the paper. SQ, AC, EB, and PB participated to the discussion of the results and in revising the final draft of the paper. All authors read and approved the final manuscript.

## SUPPLEMENTARY MATERIAL

The Supplementary Material for this article can be found online at: http://journal.frontiersin.org/article/10.3389/fmicb.2016.01058

## REFERENCES


Begley, M., Gahan, C. G., and Hill, C. (2005). The interaction between bacteria and bile. *FEMS Microbiol. Rev.* 29, 625–651. doi: 10.1016/j.femsre.2004.09.003

Binford, L. R. (1968). "Post pleistocene adaptations," in *New Perspectives in Archaeology*, eds S. Binford and L. R. Binford (Chicago, IL: Adeline), 313–342.

Blaser, M. J., and Falkow, S. (2009). What are the consequences of the disappearing human microbiota? *Nat. Rev. Microbiol.* 7, 887–894. doi: 10.1038/nrmicro2245

Bottacini, F., Ventura, M., Sinderen, D., and Motherway, M. (2014). Diversity, ecology and intestinal function of bifidobacteria. *Microb. Cell Fact.* 13:S4. doi: 10.1186/1475-2859-13-S1-S4

Caporaso, J. G., Kuczynski, J., Stombaugh, J., Bittinger, K., Bushman, F. D., Costello, E. K., et al. (2010). QIIME allows analysis of high-throughput community sequencing data. *Nat. Methods* 7, 335–336. doi: 10.1038/nmeth.f.303

Childe, V. G. (1936). *Man Makes Himself.* London: Watts.

Culhane, A. C., Thioulouse, J., Perrière, G., and Higgins, D. G. (2005). MADE4: an R package for multivariate analysis of gene expression data. *Bioinformatics* 21, 2789–2790. doi: 10.1093/bioinformatics/bti394

De Filippo, C., Cavalieri, D., Di Paola, M., Ramazzotti, M., Poullet, J. B., Massart, S., et al. (2010). Impact of diet in shaping gut microbiota revealed by a comparative study in children from Europe and rural Africa. *Proc. Natl. Acad. Sci. U.S.A.* 107, 14691–14696. doi: 10.1073/pnas.1005963107

Eddy, S. R. (2011). Accelerated profile HMM searches. *PLoS Comput. Biol.* 7:e1002195. doi: 10.1371/journal.pcbi.1002195

Fanning, S., Hall, L. J., Cronin, M., Zomer, A., MacSharry, J., Goulding, D., et al. (2012). Bifidobacterial surface-exopolysaccharide facilitates commensal-host interaction through immune modulation and pathogen protection. *Proc. Natl. Acad. Sci. U.S.A.* 109, 2108–2113. doi: 10.1073/pnas.1115621109

Ferrario, C., Milani, C., Mancabelli, L., Lugli, G. A., Duranti, S., Mangifesta, M., et al. (2016). Modulation of the eps-ome transcription of bifidobacteria through simulation of human intestinal environment. *FEMS Microbiol. Ecol* 92:fiw056. doi: 10.1093/femsec/fiw056

Gueimonde, M., Garrigues, C., van Sinderen, D., de los Reyes-Gavilan, C. G., and Margolles, A. (2009). Bile-inducible efflux transporter from *Bifidobacterium longum* NCC2705, conferring bile resistance. *Appl. Environ. Microbiol.* 75, 3153–3160. doi: 10.1128/AEM.00172-09

Kau, A. L., Ahern, P. P., Griffin, N. W., Goodman, A. L., and Gordon, J. I. (2011). Human nutrition, the gut microbiome and the immune system. *Nature* 474, 327–336. doi: 10.1038/nature10213

MacGregor, E. A., Janecek, S., and Svensson, B. (2001). Relationship of sequence and structure to specificity in the alpha-amylase family of enzymes. *Biochim. Biophys. Acta* 1546, 1–20. doi: 10.1016/S0167-4838(00)00302-2

Marlowe, F. W. (2010). *The Hadza.* Oakland, CA: University of California Press.

Martínez, I., Stegen, J. C., Maldonado-Gómez, M. X., Eren, A. M., Siba, P. M., Greenhill, A. R., et al. (2015). The gut microbiota of rural papua new guineans: composition, diversity patterns, and ecological processes. *Cell Rep.* 11, 527–538. doi: 10.1016/j.celrep.2015.03.049

Milani, C., Lugli, G., Duranti, S., Turroni, F., Mancabelli, L., Ferrario, C., et al. (2015a). Bifidobacteria exhibit social behavior through carbohydrate resource sharing in the gut. *Sci. Rep.* 5, 15782. doi: 10.1038/srep15782

Milani, C., Mancabelli, L., Lugli, G. A., Duranti, S., Turroni, F., Ferrario, C., et al. (2015b). Exploring vertical transmission of bifidobacteria from mother to child. *Appl. Environ. Microbiol.* 81, 7078–7087. doi: 10.1128/AEM.02037-15

Namiki, T., Hachiya, T., Tanaka, H., and Sakakibara, Y. (2012). MetaVelvet: an extension of Velvet assembler to de novo metagenome assembly from short sequence reads. *Nucleic Acids Res.* 40:e155. doi: 10.1093/nar/gks678

Nicholson, J. K., Holmes, E., Kinross, J., Burcelin, R., Gibson, G., Jia, W., et al. (2012). Host-gut microbiota metabolic interactions. *Science* 336, 1262–1267. doi: 10.1126/science.1223813

Obregon-Tito, A. J., Tito, R. Y., Metcalf, J., Sankaranarayanan, K., Clemente, J. C., Ursell, L. K., et al. (2015). Subsistence strategies in traditional societies distinguish gut microbiomes. *Nat. Commun.* 6:6505. doi: 10.1038/ncomms7505

Quercia, S., Candela, M., Giuliani, C., Turroni, S., Luiselli, D., Rampelli, S., et al. (2014). From lifetime to evolution: timescales of human gut microbiota adaptation. *Front. Microbiol.* 5:587. doi: 10.3389/fmicb.2014.00587

Rampelli, S., Schnorr, S. L., Consolandi, C., Turroni, S., Severgnini, M., Peano, C., et al. (2015). Metagenome sequencing of the Hadza hunter-gatherer gut microbiota. *Curr. Biol.* 25, 1682–1693. doi: 10.1016/j.cub.2015.04.055

Rho, M., Tang, H., and Ye, Y. (2010). FragGeneScan: predicting genes in short and error-prone reads. *Nucleic Acids Res.* 38:e191. doi: 10.1093/nar/gkq747

Schnorr, S. L., Candela, M., Rampelli, S., Centanni, M., Consolandi, C., Basaglia, G., et al. (2014). Gut microbiome of the Hadza hunter-gatherers. *Nat. Commun.* 5:3654. doi: 10.1038/ncomms4654

Schnorr, S. L., Crittenden, A. N., Venema, K., Marlowe, F. W., and Henry, A. G. (2015). Assessing digestibility of Hadza tubers using a dynamic in-vitro model. *Am. J. Phys. Anthropol.* 158, 371–385. doi: 10.1002/ajpa.22805

Segata, N. (2015). Gut microbiome: westernization and the disappearance of intestinal diversity. *Curr. Biol.* 25, R611–R613. doi: 10.1016/j.cub.2015.05.040

Sonnenburg, E. D., Smits, S. A., Tikhonov, M., Higginbottom, S. K., Wingreen, N. S., and Sonnenburg, J. L. (2016). Diet-induced extinctions in the gut microbiota compound over generations. *Nature* 529, 212–215. doi: 10.1038/nature16504

Sonnenburg, E. D., and Sonnenburg, J. L. (2014). Starving our microbial self: the deleterious consequences of a diet deficient in microbiota-accessible carbohydrates. *Cell Metab.* 20, 779–786. doi: 10.1016/j.cmet.2014.07.003

Turroni, F., Peano, C., Pass, D. A., Foroni, E., Severgnini, M., Claesson, M. J., et al. (2012). Diversity of bifidobacteria within the infant gut microbiota. *PLoS ONE* 7:e36957. doi: 10.1371/journal.pone.0036957

Turroni, F., Serafini, F., Foroni, E., Duranti, S., O'Connell Motherway, M., Taverniti, V., et al. (2013). Role of sortase-dependent pili of Bifidobacterium bifidum PRL2010 in modulating bacterium-host interactions. *Proc. Natl. Acad. Sci. U.S.A.* 110, 11151–11156. doi: 10.1073/pnas.1303897110

Warinner, C., Speller, C., Collins, M. J., and Lewis, C. M. (2015). Ancient human microbiomes. *J. Hum. Evol.* 79, 125–136. doi: 10.1016/j.jhevol.2014.10.016

Yatsunenko, T., Rey, F. E., Manary, M. J., Trehan, I., Dominguez-Bello, M. G., Contreras, M., et al. (2012). Human gut microbiome viewed across age and geography. *Nature* 486, 222–227. doi: 10.1038/nature11053

Yin, Y., Mao, X., Yang, J., Chen, X., Mao, F., and Xu, Y. (2012). dbCAN: a web resource for automated carbohydrate-active enzyme annotation. *Nucleic Acids Res.* 40, W445–W451. doi: 10.1093/nar/gks479

**Conflict of Interest Statement:** The authors declare that the research was conducted in the absence of any commercial or financial relationships that could be construed as a potential conflict of interest.

# 4

# Why Don't all Infants have Bifidobacteria in their Stool?

*Gerald W. Tannock*[1,2]*, *Pheng Soon Lee*[3,4], *Khai Hong Wong*[3] *and Blair Lawley*[1]

[1] *Department of Microbiology and Immunology, University of Otago, Dunedin, New Zealand,* [2] *Riddet Centre for Research Excellence, Massey University, Palmerston North, New Zealand,* [3] *Mead Johnson Nutrition, Singapore, Singapore,* [4] *Department of Human Nutrition, University of Otago, Dunedin, New Zealand*

**Keywords: bifidobacteria, infants, fecal microbiota, Lachnospiraceae, child development**

**Edited by:**
*Francesca Turroni,*
*University College Cork, Ireland*

**Reviewed by:**
*Sabrina Duranti,*
*University of Parma, Italy*
*Francesca Bottacini,*
*University College Cork, Ireland*

***Correspondence:**
*Gerald W. Tannock*
*gerald.tannock@otago.ac.nz*

Members of the genus *Bifidobacterium* are abundant in the stool of most human infants during the initial exclusively milk-fed period of life, especially at an age of 2–3 months (Harmsen et al., 2000; Favier et al., 2002; Mariat et al., 2009; Coppa et al., 2011; Turroni et al., 2012; Yatsunenko et al., 2012; Tannock et al., 2013; Barrett et al., 2015). Bifidobacteria dominate the stool microbiota regardless of whether the infants are fed human milk or formula based on ruminant milk (cow or goat). However, bifidobacteria have about 20% higher relative abundances in human milk-fed compared to formula-fed babies (Tannock et al., 2013). The greater abundance of bifidobacteria in human-milk-fed infants can, at least in part, be explained by the fact that bifidobacterial species that are enriched in the infant bowel can utilize Human Milk Oligosaccharides (HMO) or their components as growth substrates (Sela et al., 2008; LoCascio et al., 2010; Garrido et al., 2013). It could be anticipated, therefore, that bifidobacteria would be detectable in the stool microbiota of every child nourished at the breast because of the supply of appropriate growth substrates. This expectation is not borne out completely because a proportion of infants have very low abundance or undetectable bifidobacteria as members of the fecal microbiota regardless of breast milk or formula feeding (Young et al., 2004; Gore et al., 2008; Tannock et al., 2013). Antibiotics had not been administered to these infants. How then can the absence of bifidobacteria be explained?

## A GROWTH SUBSTRATE DEFICIT?

The "bifidobacteria-negative" babies have been detected in both human milk and formula-fed infants. Therefore, a bacterial growth substrate effect seems unlikely. While human milk is rich in HMO and ruminant milk lacks these complex molecules (although simpler forms such as sialylated-lactose are present in very small amounts), bifidobacteria are still the most abundant taxon in the feces of infants fed formula un-supplemented with galacto- or fructo-oligosaccharides (Tannock et al., 2013). In this case, lactose and/or glycoproteins and glycolipids are probable growth substrates for bifidobacteria (Turroni et al., 2010; Bottacini et al., 2014; O'Callaghan et al., 2015) in the bowel of exclusively milk fed infants. There is, however, a need to support genomic analysis of bifidobacteria with culture-based investigations of bifidobacterial nutrition based on substrates present in the bowel of exclusively milk-fed babies (other than HMO).

## LACK OF SENSITIVITY OF BIFIDOBACTERIAL DETECTION METHODS?

An obvious reason for bifidobacteria-negative feces is that the detection methods lack sufficient sensitivity. Culture-based methods usually have a lower detection limit of $1 \times 10^3$ per gram, fluorescent in situ hybridization (FISH) $1 \times 10^6 - 10^7$ per gram (manual or digital counts respectively) or $\sim 4 \times 10^4$ by flow cytometry, and denaturing gradient gel electrophoresis of PCR amplicons ~$1 \times 10^5$ – $10^6$ cells (Welling et al., 1997; Jansen et al., 1999; Zoetendal et al., 2001, 2002) or $1 \times 10^4$ using internal transcribed spacer targets (Milani et al., 2014). High throughput DNA sequencing methods, such as Illumina, generate tens of thousands of 16S rRNA gene sequences per DNA sample, but there may be several hundred OTU per sample. Thus, taxa present in very low abundance could be missed. However, reference to rarefaction curves (alpha diversity) during sequence analysis will show whether coverage of the microbiota is near complete or not. Therefore, while lack of sufficient sensitivity of detection methods remains a possibility, it probably does not provide the total explanation.

## BIFIDOBACTERIAL POPULATIONS RISE AND FALL FROM DAY TO DAY?

Most fecal microbiota studies examine a single fecal sample from each participating individual. Comprehensive temporal studies of the fecal microbiota to determine day-to-day variations in composition have not been reported. It is possible that bifidobacteria are present in the feces of all children during early life but that, on some days, the bifidobacterial population falls to undetectable levels. Populations of bifidobacteria in the feces of some adults without diseases are dynamic in terms of strain composition, so there is some support for a concept of temporal instability in the bifidobacterial population of the microbiota (McCartney et al., 1996). **Figure A** shows data from feces collected at intervals from infants during the first 12 weeks of life. In the example, fluctuations in the abundances of bifidobacteria were seen, varying from very low abundance to absence, in feces of individual children. Strikingly, bifidobacteria were not detected in any of the fecal samples of one child. Therefore, bifidobacteria-free infants do seem to be a real phenomenon.

## THE WINDOW OF INFECTIVITY (OPPORTUNITY/COLONIZATION) WAS MISSED?

A window of opportunity is a short time period during which an otherwise unattainable opportunity exists. After the window of opportunity closes, the opportunity ceases to exist. Caufield was the first to describe the "window of infectivity" in the acquisition of commensal bacteria. His example was *Streptococcus mutans* in the oral cavity of children (Caufield et al., 1993; Li and Caufield, 1995). This bacterial species is associated with dental plaque, thus the window of infectivity coincided with the eruption of the first molars. Prior to this, a habitat for *S. mutans* is not available in the oral cavity of children for this species. The Caufield hypothesis reminds us that many factors have to coincide to favor the establishment of a commensal in a body site. Cesarean-delivered babies have lower prevalences of bifidobacteria in their feces in early life (**Figure** ). By analogy to Caufield's studies, this probably relates to a lack of favorable opportunities for bifidobacteria to colonize the bowel relative to the vaginal birth process. Notably, we found that 36% of cesarean-derived babies lacked bifidobacteria, whereas 18% of vaginally delivered infants were bifidobacteria-free at 2 months of age (Tannock et al., 2013).

## OTHER TAXA REPLACE BIFIDOBACTERIA IN SOME BABIES?

If bifidobacteria have not colonized the bowel of certain infants, they are likely to be replaced by other taxa, which may have the requisite metabolic properties to fill the vacant ecological niche. In a study of the fecal microbiotas of Australian babies that were breast milk- or formula-fed, we compared the relative abundances of bacterial taxa in infants that had very low (<10%) or higher (>10%) bifidobacterial content (Tannock et al., 2013). Analysis of the compositions of these microbiotas showed that when *Bifidobacteriaceae* abundance was low, *Lachnospiraceae* abundances tended to be greater in babies in all dietary groups (**Figures C–E**). There was also a tendency for *Erysipelotrichaceae* abundances to be greater in formula-fed babies with low bifidobacterial abundances, being much more evident in the case of goat milk-fed infants. These observations suggest that, yes, other taxa might replace bifidobacteria in the fecal microbiota of some children.

## WHAT ARE THE CONSEQUENCES OF LACKING BIFIDOBACTERIA IN THE BOWEL?

The absence of bifidobacteria in the bowel may be detrimental for infant development. The curious phenomenon whereby mother's milk contains substances not used in the nutrition of the offspring, but which fertilize bifidobacterial growth, is unique to humans. There must be a good reason for this. Enriching bifidobacterial populations in the bowel tends to minimize the abundance of other bacterial species, so a competitive exclusion function could be ascribed to HMO. Additionally, HMO may act as "decoys" in the bowel by binding to pathogens (bacteria and viruses) and their toxins and thus limiting contact with mucosal surfaces (Kunz et al., 2000). The large diversity of HMO structures that is known to occur in human milk suggests a large diversity of decoy functions (Pacheco et al., 2015). Irrespective of where in the World babies live, their gut microbiomes are enriched in genes involved in the *de novo* biosynthesis of folate (Yatsunenko et al., 2012). In contrast, the microbiome of adults favors synthesis of another B vitamin, cobalamin. Folate synthesis is an attribute of bifidobacteria and folate can be absorbed from the large bowel, so enrichment of bifidobacteria in the infant bowel may provide an important contribution to infant nutrition (Aufreiter et al., 2009; D'Aimmo et al., 2012; Lakoff et al., 2014). Folate functions as a coenzyme or co-substrate in single-carbon transfers in the synthesis of nucleic acids and metabolism of amino acids. One of the most important folate-dependent reactions is the conversion of homocysteine to methionine in the synthesis of S-adenosyl-methionine, an important methyl donor. Another folate-dependent reaction, the methylation of deoxyuridylate to thymidylate in the formation of DNA, is required for proper cell division (Crider et al., 2012).

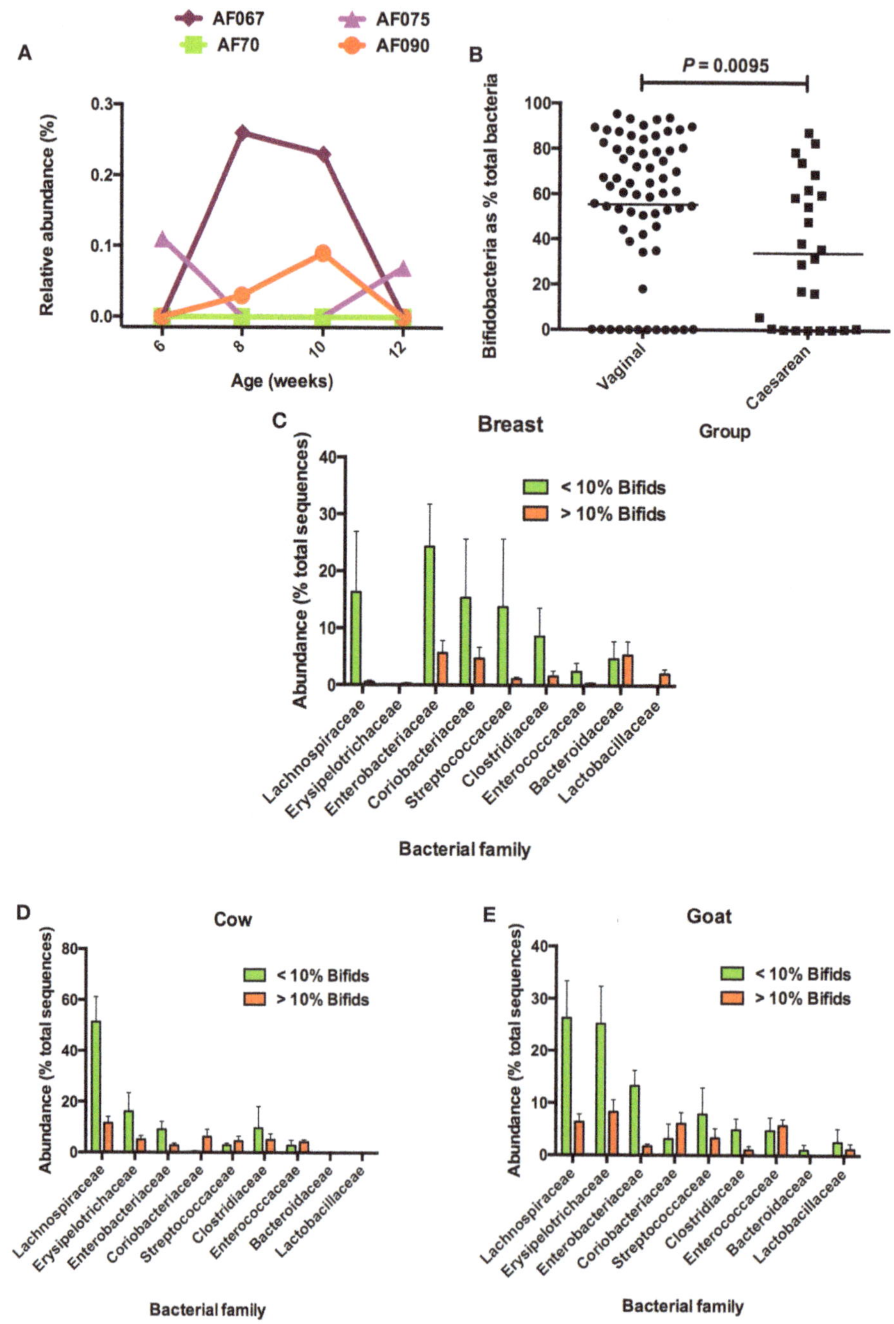


**FIGURE** | **(A)** Examples of babies without bifidobacteria, or very low abundances of bifidobacteria, in feces during the first 12 weeks of life. Bifidobacteria were undetectable by DNA sequencing of 16S rRNA gene amplicons at 6 and 12 weeks in the feces of AF067; 6, 8, 10, and 12 weeks in AF70; 8 and 10 weeks in AF075; and 6 and 12 weeks in AF090. **(B)** Comparison of bifidobacerial abundances in the feces of infants delivered vaginally or by cesarean. Note that in both groups, bifidobacteria were not detected in some infants. **(C)** Comparison of abundances of bacterial families in microbiotas of breast milk-fed infants in relation to abundances of bifidobacteria. **(D)** Comparison of abundances of bacterial families in microbiotas of cow milk formula-fed infants in relation to abundances of bifidobacteria. **(E)** Comparison of abundances of bacterial families in microbiotas of goat milk formula-fed infants in relation to abundances of bifidobacteria. Note that Lachnospiraceae have increased abundances when bifidobacteria have low relative abundance. Figures after Tannock et al. (2013), reproduced with permission.

Neonatal nutrition could, indeed, be the very important reason for the HMO-bifidobacteria-infant paradigm. The foundation of brain structure and function is set early in life through genetic, biological and psychosocial influences. The rate of neonatal brain growth exceeds that of any other organ or body tissue (Wang, 2012). The infant is born with neurons already formed but the synaptic connections between these cells are mostly established and elaborated after birth causing a large nutritional demand for biosynthesis of gangliosides (Svennerholm et al., 1989). Nutrition of the infant in early life affects brain developmental processes including cognition (Uauy and Peirano, 1999; Uauy et al., 2001). While long-chain fatty acids (such as docosahexaenoic acid) have been the focus of much of the research in this field, tantalizing research evidence now indicates that sialic acid (N-acetyl-neuraminic acid), a 9-carbon carbohydrate, is also an essential nutrient for optimal brain development and cognition (Gibson, 1999; Meldrum et al., 2012; Wang, 2012). Strikingly, cortical tissue from human brain contains up to 4 times more sialic acid than that of other mammals tested (Wang et al., 1998). Moreover, the sialic acid concentration in the brain of breast milk-fed babies is higher than in that of formula-fed infants (Wang et al., 2003). These facts correlate with the unique biochemistry of human milk and the unique bacteriology of the infant bowel. Intriguingly, Ruhaak et al. (2014) have reported the detection of sialylated oligosaccharides (3′ sialyl-lactose, 6′ sialyl-lactose, 3′ sialyl-lactosamine, 6′ sialyl-lactosamine) that might result from the hydrolysis of HMO, in the blood of human infants. Thus, bifidobacterial biochemistry in the bowel may have extra-intestinal, nutritional influences important in brain development. However, perhaps the taxa that are abundant in the bowel of infants in the absence of bifidobacteria can carry out these same functions? This interesting possibility remains to be investigated.

## BABIES WITHOUT BIFIDOBACTERIA ARE IMPORTANT SOURCES OF KNOWLEDGE?

Rene Dubos explored in a number of books the interplay between environmental forces and the physical, mental, and spiritual development of humankind. His article published in the journal *Pediatrics* entitled "Biological Freudianism: lasting effects of early environmental influences" encapsulated this theme (Dubos et al., 1966). Drawing on the results of experiments conducted with specific-pathogen-free mice, the authors concluded that "From all points of view, the child is truly the father of the man, and for this reason we need to develop an experimental science that might be called biological Freudianism. Socially and individually the response of human beings to the conditions of the present is always conditioned by the biological remembrance of things past."

Biological Freudianism is clearly of relevance to the concept that the first 1000 days, between conception and the child's second birthday, offer a unique window of opportunity to shape healthier and more prosperous futures. Nutrition during this 1000 day window can have a profound impact on a child's ability to grow, and learn. The influences of the microbiota on the development of the child during early life are potentially very important, and much longitudinal research is required to clarify whether there are continuing, medically important impacts of the microbiota, inlcuding the bifidobacteria, that last throughout the lifetime of humans. Comparisons of the cognitive development and general health status of children that had been bifidobacteria-free, and children that were ex-bifidobacteria-free then intentionally exposed to bifidobacteria, in a longitudinal study extending perhaps 10 or 20 years, would tell us whether these bacteria optimize short and/or long term human development and health.

## AUTHOR CONTRIBUTIONS

GT wrote the article. BL, PL, and KW provided data described in the article.

## REFERENCES


Aufreiter, S., Gregory, J. F., Pfeiffer, C. M., Fazil, Z., Kim, Y.-I., Marcon, N., et al. (2009). Folate is absorbed across the colon of adults: evidence from cecal infusion of $^{13}$C-labelled [6S]-5-formaltetrahydrofolic acid. *Am. J. Clin. Nutr.* 90, 116–123. doi: 10.3945/ajcn.2008.27345

Barrett, E., Deshpandey, A. K., Ryan, C. A., Dempsey, E. M., Murphy, B., O'Sullivan, L., et al. (2015). The neonatal gut harbours distinct bifidobacterial strains. *Arch. Dis. Child. Fetal Neonatal Ed.* 100, F405–F410. doi: 10.1136/archdischild-2014-306110

Bottacini, F., Motherway, M. O., Kucynski, J., O'Connell, K. J., Serafini, F., Duranti, S., et al. (2014). Comparative genomics of the *Bifidobacterium breve* taxon. *BMC Genom.* 15:170. doi: 10.1186/1471-2164-15-170

Caufield, P. W., Cutter, G. R., and Dasanayake, A. P. (1993). Initial acquisition of mutans streptococci by infants: evidence for a discrete window of infectivity. *J. Dent. Res.* 72, 37–45. doi: 10.1177/00220345930720010501

Coppa, G. V., Gabrielli, O., Zampini, L., Galeazzi, T., Ficcadenti, A., Padella, L., et al. (2011). Oligosaccharides in 4 different milk groups, Bifidobacteria and *Ruminococcus obeum*. *J. Pediatr. Gastroenterol. Nutr.* 53, 80–87. doi: 10.1097/MPG.0b013e3182073103

Crider, K. S., Young, T. P., Berry, R. J., and Bailey, L. B. (2012). Folate and DNA methylation: a review of molecular mechanisms and the evidence for folate's role. *Adv. Nutr.* 3, 21–38. doi: 10.3945/an.111.000992

D'Aimmo, M. R., Mattarelli, P., Biavati, B., Carlsson, N. G., and Andlid, T. (2012). The potential of bifidobacteria as a source of natural folate. *J. Appl. Microbiol.* 112, 975–984. doi: 10.1111/j.1365-2672.2012.05261

Dubos, R., Savage, D., Schaedler, R., and Biological Freudianism (1966). Lasting effects of early environmental influences. *Pediatric* 38, 789–800.

Favier, C. F., Vaughan, E. E., De Vos, W. M., and Akkermans, A. D. L. (2002). Molecular monitoring of succession of bacterial communities in human neonates. *Appl. Environ. Microbiol.* 68, 219–226. doi: 10.1128/AEM.68.1.219-226.2002

Garrido, D., Dallas, D. C., and Mills, D. A. (2013). Consumption of human milk glycoconjugates by infant-associated bifidobacteria: mechanisms and implications. *Microbiology* 159, 649–664. doi: 10.1099/mic.0.064113-0

Gibson, R. A. (1999). Long-chain polyunsaturated fatty acids and infant development. *Lancet* 354, 1919–1920. doi: 10.1016/S0140-6736(99)00300-1

Gore, C., Munro, K., Lay, C., Bibiloni, R., Morris, J., Woodcock, A., et al. (2008). *Bifidobacterium pseudocatenulatum* is associated with atopic eczema: a nested case-control study investigating the fecal microbiota of infants. *J. Allergy. Clin. Immunol.* 121, 135–140. doi: 10.1016/j.jaci.2007.07.061

Harmsen, H. J., Wildeboer-Veloo, A. C., Raangs, G. C., Wagendorp, A. A., Klijn, N., Bindels,J. G., et al. (2000). Analysis of intestinal flora development in breast-fed and formula-fed infants by using molecular identification and detection methods. *J. Pediatr. Gastroenterol. Nutr.* 30, 61–67. doi: 10.1097/00005176-200001000-00019

Jansen, G. J., Wildeboer-Veloo, A. C., Tonk, R. H., Franks, A. H., and Welling, G. W. (1999). Development and validation of an automated, microscopy-based method for enumeration of groups of intestinal bacteria. *J. Microbiol. Methods* 37, 215–221. doi: 10.1016/S0167-7012(99)00049-4

Kunz, C., Rudloff, S., Baier, W., Klein, N., and Strobel, S. (2000). Oligosaccharides in human milk: structural, functional, and metabolic aspects. *Annu. Rev. Microbiol.* 20, 699–722. doi: 10.1146/annurev.nutr.20.1.699

Lakoff, A., Fazili, Z., Aufreiter, S., Pfeiffer, C. M., Connolly, B., Gregory, J. F., et al. (2014). Folate is absorbed across the human colon: evidence using enteric-coated caplets containing $^{13}$C-labeled [6S]-5-formyltetrahydrofolate. *Am. J. Clin. Nutr.* 100, 1278–1286. doi: 10.3945/ajcn.114.091785

Li, Y., and Caufield, P. W. (1995). The fidelity of initial acquisition of mutans streptococci by infants from their mothers. *J. Dent. Res.* 74, 681–685. doi: 10.1177/00220345950740020901

LoCascio, R. G., Desai, P., Sela, D. A., Weimer, B., and Mills, D. A. (2010). Broad conservation of milk utilization genes in *Bifidobacterium longum* subsp. *infantis as revealed by comparative genomic hybridization. Appl. Environ. Microbiol.* 76, 7373–7381. doi: 10.1128/AEM.00675-10

Mariat, D., Firmesse, O., Levenez, F., Guimaraes, V. D., Sokol, H., Doré, J., et al. (2009). The *Firmicutes/Bacteroidetes* ratio of the human microbiota changes with age. *BMC Microbiol.* 9:123. doi: 10.1186/1471-2180-9-123

McCartney, A. L., Wenzhi, W., and Tannock, G. W. (1996). Molecular analysis of the composition of the bifidobacterial and lactobacillus microflora of humans. *Appl. Environ. Microbiol.* 62, 4608–4613.

Meldrum, S. J., D'Vaz, N., Simmer, K., Dunstan, J. A., Hird, K., and Prescott, S. L. (2012). Effects of high-dose fish oil supplementation during early infancy on neurodevelopment and language: a randomized controlled trial. *Br. J. Nutr.* 108, 1443–1454. doi: 10.1017/S0007114511006878

Milani, C., Lugli, G. A., Turroni, F., Mancabelli, L., Duranti, S., Viappiani, A., et al. (2014). Evaluation of bifidobacterial community composition in the human gut by means of a targeted amplicon sequencing (ITS)protocol. *FEMS Microbiol. Ecol.* 90, 493–503. doi: 10.1111/1574-6941.12410

O'Callaghan, A., Bottacini, F., Motherway, M. O., and van Sinderen, D. (2015). Pangenome analysis of *Bifidobacterium longum* and site-directed mutagenesis through by-pass of restriction-modification systems. *BMC Genom.* 16:832. doi: 10.1186/s12864-015-1968-4

Pacheco, A. R., Barile, D., Underwood, M. A., and Mills, D. A. (2015). The impact of the milk glycobiome on the neonate gut microbiota. *Ann. Rev. Anim. Sci.* 3, 419–445. doi: 10.1146/annurev-animal-022114-111112

Ruhaak, L. R., Stroble, C., Underwood, M. A., and Lebrilla, C. B. (2014). Detection of milk oligosaccharides in plasma of infants. *Anal. Bioanal. Chem.* 406, 5775–5784. doi: 10.1007/s00216-014-8025-z

Sela, D. A., Chapman, J., Adeuya, A., Kim, J. H., Chen, F., Whitehead, T. R., et al. (2008). The genome sequence of *Bifidobacterium longum* subsp. *infantis reveals adaptations for milk utilization within the infant microbiome. Proc. Natl. Acad. Sci.U.S.A.* 2, 18964–18969. doi: 10.1073/pnas.0809584105

Svennerholm, L., Boström, K., Fredman, P., Mansson, J. E., Rosengren, B., and Rynmark, B. M. (1989). Human brain gangliosides: developmental changes from early fetal stage to advanced age. *Biochim. Biophys. Acta* 1005, 109–117. doi: 10.1016/0005-2760(89)90175-6

Tannock, G. W., Lawley, B., Munro, K., Pathmanathan, S. G., Zhou, S. J., Makrides, M., et al. (2013). Comparison of the compositions of the stool microbiotas of infants fed goat milk formula, cow milk-based formula, or breast milk. *Appl. Environ. Microbiol.* 79, 3040–3048. doi: 10.1128/AEM.03910-12

Turroni, F., Bottacini, F., Foroni, E., Mulder, I., Kim, J.-H., Zomer, A., et al. (2010). Genome analysis of *Bifidobacterium bifidum* PRL2010 reveals metabolic pathways for host-derived glycan foraging. *Proc. Natl. Acad. Sci. U.S.A.* 107, 19514–19519. doi: 10.1073/pnas.1011100107

Turroni, F., Peano, C., Pass, D. A., Foroni, E., Severgnini, M., Claesson, M. J., et al. (2012). Diversity of bifidobacteria within the infant gut microbiota. *PLoS ONE* 7:e36957. doi: 10.1371/journal.pone.0036957

Uauy, R., Mena, P., and Peirano, P. (2001). Mechanisms for nutrient effects on brain development and cognition. *Nestle Nutr. Workshop Ser. Clin. Perform Programme* 5, 41–70. doi: 10.1159/000061845

Uauy, R., and Peirano, P. (1999). Breast is best: human milk is the optimal food for brain development. *Amer. J. Clin. Nutr.* 70, 433–434.

Wang, B. (2012). Molecular mechanism underlying sialic acid as an essential nutrient for brain development and cognition. *Adv. Nutr.* 3, 465S–472S. doi: 10.3945/an.112.001875

Wang, B., McVeagh, P., Petocz, P., and Brand-Miller, J. (2003). Brain ganglioside and glycoprotein sialic acid in breast-fed compared with formula-fed infants. *Am. J. Clin. Nutr.* 78, 1024–1029.

Wang, B., Miller, J. B., McNeil, Y., and McVeagh, P. (1998). Sialic acid concentration of brain gangliosides: variation among eight mammalian species. *Comp. Biochem. Physiol. A Mol. Integr. Physiol.* 19, 435–439. doi: 10.1016/S1095-6433(97)00445-5

Welling, G. W., Elfferich, P., Raangs, G. C., Wildeboer-Veloo, A. C., Jansen, G. J., and Degener, J. E. (1997). 16S ribosomal RNA-taggd oligonucleotide probes for monitoring of intestinal tract bacteria. *Scand. J. Gastroenterol.* 222, 17–19. doi: 10.1080/00365521.1997.11720711

Yatsunenko, T., Rey, F. E., Manary, M. J., Trehan, I., Dominguez-Bello, M. G., Contreras, M., et al. (2012). Human gut microbiome viewed across age and geography. *Nature* 486, 222–227. doi: 10.1038/nature 11053

Young, S. L., Simon, M. A., Baird, M. A., Tannock, G. W., Bibiloni, R., Spencely, K., et al. (2004). Bifidobacterial species differentially affect expression of cell surface markers and cytokines of dendritic cells harvested from cord blood. *Clin. Diagn. Lab. Immunol.* 11, 686–690. doi: 10.1128/cdli.11.4.686-690.2004

Zoetendal, E. G., Ben-Amor, K., Akkermans, A. D., Abee, T., and de Vos, W. M. (2001). DNA isolation protocols affect the detection limit of PCR approaches of bacteria in samples from the human gastrointestinal tract. *Syst. Appl. Microbiol.* 24, 405–410. doi: 10.1078/0723-2020-00060

Zoetendal, E. G., von Wright, A., Vilponen-Samela, T., Ben-Amor, K., Akkermans, A. D., and de Vos, W. M. (2002). Mucosa-associated bacteria in the human gastrointestinal tract are uniformly distributed along the colon and differ from the community recovered from feces. *Appl. Environ. Microbiol.* 68, 3401–3407. doi: 10.1128/AEM.68.7.3401-3407.2002

**Conflict of Interest Statement:** The authors declare that the research was conducted in the absence of any commercial or financial relationships that could be construed as a potential conflict of interest.

5

# Bifidobacteria and Butyrate-Producing Colon Bacteria: Importance and Strategies for their Stimulation in the Human Gut


***Audrey Rivière, Marija Selak, David Lantin, Frédéric Leroy and Luc De Vuyst*** *

*Research Group of Industrial Microbiology and Food Biotechnology, Faculty of Sciences and Bioengineering Sciences, Vrije Universiteit Brussel, Brussels, Belgium*



With the increasing amount of evidence linking certain disorders of the human body to a disturbed gut microbiota, there is a growing interest for compounds that positively influence its composition and activity through diet. Besides the consumption of probiotics to stimulate favorable bacterial communities in the human gastrointestinal tract, prebiotics such as inulin-type fructans (ITF) and arabinoxylan-oligosaccharides (AXOS) can be consumed to increase the number of bifidobacteria in the colon. Several functions have been attributed to bifidobacteria, encompassing degradation of non-digestible carbohydrates, protection against pathogens, production of vitamin B, antioxidants, and conjugated linoleic acids, and stimulation of the immune system. During life, the numbers of bifidobacteria decrease from up to 90% of the total colon microbiota in vaginally delivered breast-fed infants to <5% in the colon of adults and they decrease even more in that of elderly as well as in patients with certain disorders such as antibiotic-associated diarrhea, inflammatory bowel disease, irritable bowel syndrome, obesity, allergies, and regressive autism. It has been suggested that the bifidogenic effects of ITF and AXOS are the result of strain-specific yet complementary carbohydrate degradation mechanisms within cooperating bifidobacterial consortia. Except for a bifidogenic effect, ITF and AXOS also have shown to cause a butyrogenic effect in the human colon, i.e., an enhancement of colon butyrate production. Butyrate is an essential metabolite in the human colon, as it is the preferred energy source for the colon epithelial cells, contributes to the maintenance of the gut barrier functions, and has immunomodulatory and anti-inflammatory properties. It has been shown that the butyrogenic effects of ITF and AXOS are the result of cross-feeding interactions between bifidobacteria and butyrate-producing colon bacteria, such as *Faecalibacterium prausnitzii* (clostridial cluster IV) and *Anaerostipes, Eubacterium,* and *Roseburia* species (clostridial cluster XIVa). These kinds of interactions possibly favor the co-existence of bifidobacterial strains with other bifidobacteria and with butyrate-producing colon bacteria in the human colon.



**Edited by:**
*Marco Ventura,*
*University of Parma, Italy*

**Reviewed by:**
*Sinead Christine Leahy,*
*AgResearch Limited, New Zealand*
*Christian U. Riedel,*
*University of Ulm, Germany*



***Correspondence:**
*Luc De Vuyst*
*ldvuyst@vub.ac.be*



**Keywords: bifidobacteria, butyrate-producing colon bacteria, cross-feeding, prebiotics, probiotics, arabinoxylan-oligosaccharides, inulin-type fructans**

## INTRODUCTION

Whereas, the human gut microbiota has been studied in the past mainly in the context of infectious diseases, it is known today that this enormous number of microorganisms has an indispensable role in the normal development and functioning of the human body (O'Hara and Shanahan, 2006; Sommer and Bäckhed, 2013). Within the adult gastrointestinal tract, the colon contains the most dense ($>10^{11}$ bacteria per mL of luminal content) and metabolically active microbiota (**Figure** ; Whitman et al., 1998; The Human Microbiome Project Consortium, 2012). The immense number of genes ($>100$ times the number of genes of the human genome) encoded by this microbiota, expands the host's biochemical and metabolic capabilities substantially (Bäckhed et al., 2005; The Human Microbiome Project Consortium, 2012). Examples of supporting functions of the human gut microbiota are the degradation of otherwise non-digestible food compounds; the transformation of toxic compounds; and the production of essential vitamins, important metabolic end-products, and defending bacteriocins (Sommer and Bäckhed, 2013). Microbial metabolic end-products, which account for one third of the metabolites present in the human blood, play an important role in gut homeostasis and have an impact on host metabolism and health (Wikoff et al., 2009; Hood, 2012; Louis et al., 2014; Sharon et al., 2014; Richards et al., 2016). The short-chain fatty acids (SCFAs) acetate, butyrate, and propionate (typically occurring in a 3:1:1 ratio) are quantitatively (total concentration of 50–150 mM) and metabolically the most important microbial end-products of the human colon fermentation process (Louis et al., 2014), as they display several physiological effects (**Table** ).

Changes in the gut microbiota composition have been associated with disturbed gut barrier functions, increased gut permeability, and increased plasma lipopolysaccharide concentrations (i.e., metabolic endotoxemia), which causes low-grade inflammation that triggers the development of obesity and metabolic syndrome (Cani et al., 2008). Also other disorders, such as inflammatory bowel disease (IBD, encompassing Crohn's disease and ulcerative colitis), irritable bowel syndrome (IBS), colorectal cancer, and allergies have been linked to changes in the gut microbiota composition (de Vos and de Vos, 2012; Le Chatelier et al., 2013). During the last years, even associations have been made between the gut microbiota composition and behavioral disorders, such as depression, anxiety disorder, regressive autism, and schizophrenia (Collins et al., 2012; Braniste et al., 2014; Dinan et al., 2015). However, whereas increasing numbers of animal studies provide evidence for cause-and-effect relationships between shifts in gut microbiota composition and certain disorders (as in the case of obesity; Ridaura et al., 2013), it has not been proven yet for humans whether changes in the gut microbiota composition can cause disorders or that these changes are a consequence of the disorders themselves (de Vos and de Vos, 2012).

In recent years, a few distinct members of the human gut microbiota have received particular attention because of their dedicated metabolism and central role in gut homeostasis and because their loss adversely affects the remaining microorganisms and/or host's health. *Bifidobacterium* species are one such bacterial species that fulfill important functions within the human colon (Leahy et al., 2005; Rossi and Amaretti, 2011). Decreased numbers of these species in the colon have been associated with several disorders. Moreover, they have shown to interact with other colon bacteria such as butyrate-producing bacteria by cross-feeding interactions. Furthermore, decreased butyrate concentrations and decreased numbers of butyrate producers in the human colon have been associated with disorders. Therefore, this knowledge has encouraged the development of approaches to stimulate the growth and/or activity of bifidobacteria, i.e., the bifidogenic effect, and butyrate-producing colon bacteria, i.e., the butyrogenic effect, in the human colon. The most prevalent approaches to cause bifidogenic and butyrogenic effects involve the consumption of probiotics and prebiotics.

## BIFIDOBACTERIA AND BUTYRATE-PRODUCING COLON BACTERIA

### *Bifidobacterium* Species

#### General Aspects

Bifidobacteria are Gram-positive, anaerobic, saccharolytic bacteria that belong to the phylum Actinobacteria; they mainly occur in the gastrointestinal tract of mammals, birds, and insects, but are present in sewage, human breast milk, fermented milk, cheeses, and water kefir too (Bottacini et al., 2014; Khodayar-Pardo et al., 2014; Laureys and De Vuyst, 2014; Laureys et al., 2016). A typical bifidobacterial genome has an average size ranging from 2.0 to 2.8 Mb and is characterized by a high guanine-plus-cytosine content, with numerous genes involved in the uptake and breakdown of carbohydrates from both diet and host origin (Ventura et al., 2014). Bifidobacteria are among the first bacteria to colonize the human gastrointestinal tract and reach their highest proportion in the colon (up to 90% of the total colon microbiota in vaginally delivered breast-fed infants) during the first 12 months of life (Tannock, 2010; Turroni et al., 2012). This abundance significantly decreases over time to $<5\%$ in adult subjects and decreases even more in the elderly (Arumugam et al., 2011; Duncan and Flint, 2013). At the time of writing, the *Bifidobacterium* genus comprised 51 species (Euzéby, 1997, 2016; Laureys et al., 2016), among which *Bifidobacterium longum*, *Bifidobacterium animalis*, *Bifidobacterium adolescentis*, *Bifidobacterium bifidum*, *Bifidobacterium catenulatum*, *Bifidobacterium pseudocatenulatum*, *Bifidobacterium breve*, *Bifidobacterium pseudolongum*, *Bifidobacterium gallicum*, *Bifidobacterium angulatum*, and *Bifidobacterium faecale* are encountered in the human colon (Turroni et al., 2009; Ventura et al., 2011; Choi et al., 2014). In general, *B. bifidum* and *B. longum* are the dominant species in infants, whereas *B. adolescentis* and *B. longum* dominate the adult gut microbiota (Turroni et al., 2012). Quantitative PCR analyses of fecal samples of 42 Belgian healthy adults have shown that the fecal microbiota of adults contains between zero and four (with an average of two) different bifidobacterial species,

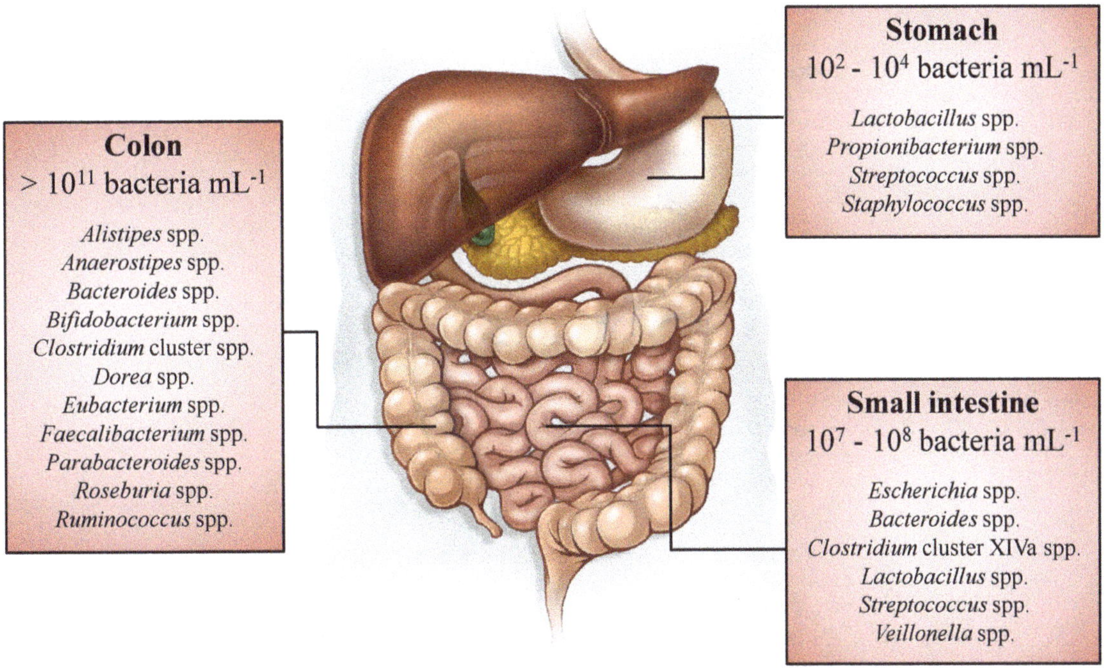


**FIGURE | Spatial distribution and concentrations of bacteria along the gastrointestinal tract of humans (Tuohy and Scott, 2015).** The dominant genera in the stomach, small intestine, and colon are listed, based on 16S rRNA gene sequence studies (Tap et al., 2009; Zoetendal et al., 2012; Delgado et al., 2013; Walker et al., 2014).

among which *B. longum* (present in 90% of the adults), *B. adolescentis* (present in 79% of the adults), and *B. catenulatum* (present in 38% of the adults) are the most frequently detected species (Ishikawa et al., 2013).

## Functional Role in the Colon

From the growing body of scientific evidence associating decreased numbers of bifidobacteria with disorders, it emerges that these species have a disproportionally large impact in the human colon in relation to their relatively low numerical abundance in adults. Hence, a decrease in the relative abundances of *Bifidobacterium* species in the human colon has been associated with antibiotic-associated diarrhea, IBS, IBD, obesity, allergies, and regressive autism (Di Gioia et al., 2014; Grimm et al., 2014). Examples of functions carried out by bifidobacteria include the production and/or liberation of B vitamins, antioxidants, polyphenols, and conjugated linoleic acids; maturation of the immune system during early life and preservation of immune homeostasis during life; preservation of gut barrier functions and protection against pathogens by producing bacteriocins, decreasing luminal pH by the production of acids, and blocking the adhesion of pathogens to the intestinal mucosa (Leahy et al., 2005; Gorissen et al., 2010, 2012; Rossi and Amaretti, 2011; Gagnon et al., 2015). However, these functions are not characteristic for the *Bifidobacterium* genus or certain species, but are rather strain-specific. Another important function of the bifidobacterial genus that contributes to gut homeostasis and host health is the production of acetate and lactate during carbohydrate fermentation, organic acids that in turn can be converted into butyrate by other colon bacteria through cross-feeding interactions (**Table** ; De Vuyst and Leroy, 2011; De Vuyst et al., 2014; Rivière et al., 2015).

## Metabolism

Bifidobacteria display a strictly fermentative metabolism, i.e., they gain energy in the form of ATP by substrate-level phosphorylation during anaerobic carbohydrate breakdown, and play an important role in the human colon with respect to the degradation of carbohydrates that resist digestion and absorption in the upper gastrointestinal tract (Pokusaeva et al., 2011; De Vuyst et al., 2014). Glycoside hydrolases (EC 3.2.1.x) constitute the most important enzyme group that colon bacteria use to degrade poly- and oligosaccharides to fermentable monosaccharides (van den Broek et al., 2008; van den Broek and Voragen, 2008). Compared with the human genome, encoding only 17 glycoside hydrolases for the digestion of food carbohydrates, bifidobacterial genomes possess high numbers of genes encoding these carbohydrases (El Kaoutari et al., 2013). As an example, the genome of *B. longum* NCC2705 contains 56 genes encoding glycoside hydrolases, one gene encoding a carbohydrate esterase (EC 3.1.1.x), but no genes encoding polysaccharide lyases (EC 4.2.2.x; Schell et al., 2002; Lombard et al., 2014). Bifidobacteria are particularly specialized in efficient uptake of short oligosaccharides into the cell, where

**TABLE | Overview of the physiological effects of the short-chain fatty acids (SCFAs) acetate, propionate, and butyrate produced by human colon bacteria (Hamer et al., 2008; Al-Lahham et al., 2010; Havenaar, 2011; Macfarlane and Macfarlane, 2012; Chang et al., 2014; Louis et al., 2014; Tralongo et al., 2014).**

| SCFA | Physiological effect |
|---|---|
| Acetate $CH_3$-$COO^-$ | Reaches the portal vein and is metabolized in various tissues<br>**Intestinal effects**<br>Is a minor energy source for the colon epithelial cells<br>Decreases the pH of the colon (which decreases bile salt solubility, increases mineral absorption, decreases ammonia absorption, and inhibits growth of pathogens)<br>Has anti-inflammatory effects<br>Increases colonic blood flow and oxygen uptake<br>Is used by cross-feeding species as a co-substrate to produce butyrate<br>**Other effects**<br>Is a substrate for cholesterol and fatty acid biosynthesis in the liver<br>Is an energy source for muscle and brain tissue |
| Propionate $CH_3$-$CH_2$-$COO^-$ | Reaches the portal vein and is subsequently taken up by the liver<br>**Intestinal effects**<br>Is a minor energy source for the colon epithelial cells<br>Decreases the pH of the colon (which decreases bile salt solubility, increases mineral absorption, decreases ammonia absorption, and inhibits growth of pathogens)<br>Prevents proliferation and induces apoptosis of colorectal cancer cells<br>Interacts with the immune system<br>Has anti-inflammatory effects<br>**Other effects**<br>Promotes satiety<br>Lowers blood cholesterol levels<br>Decreases liver lipogenesis<br>Improves insulin sensitivity |
| Butyrate $CH_3$-$CH_2$-$CH_2$-$COO^-$ | Is mainly taken up by the colon epithelial cells, only small amounts reach the portal vein and the systemic circulation<br>**Intestinal effects**<br>Is the preferred energy source for the colon epithelial cells<br>Decreases the pH of the colon (which decreases bile salt solubility, increases mineral absorption, decreases ammonia absorption, and inhibits growth of pathogens)<br>Stimulates proliferation of normal colon epithelial cells<br>Prevents proliferation and induces apoptosis of colorectal cancer cells<br>Affects gene expression of colon epithelial cells<br>Plays a protective role against colon cancer and colitis<br>Improves the gut barrier function by stimulation of the formation of mucin, antimicrobial peptides, and tight-junction proteins<br>Interacts with the immune system<br>Has anti-inflammatory effects<br>Stimulates the absorption of water and sodium<br>Reduces oxidative stress in the colon<br>**Other effects**<br>Promotes satiety |

they are further degraded to monosaccharides, i.e., they display a preferential oligosaccharide metabolism, providing them a competitive advantage toward other colon bacteria that degrade carbohydrates extracellularly (Van der Meulen et al., 2004, 2006b; Falony et al., 2009a; De Vuyst and Leroy, 2011; De Vuyst et al., 2014). About 5% of the total bifidobacterial gene content is dedicated to carbohydrate internalization, through either ATP-binding cassette transporters, permeases, or proton symporters (Ventura et al., 2009). For example, *B. longum* NCC2705 contains 15 genes that encode transport systems that could be involved in the transport of oligosaccharides (Schell et al., 2002; Parche et al., 2007). Several laboratory fermentation studies have shown that bifidobacteria can use various non-digestible carbohydrates as energy sources, encompassing plant-derived carbohydrates [such as resistant starch, pectin, inulin, arabinoxylan (AX), cellulose, and their corresponding oligosaccharides] and host-produced carbohydrates (human milk oligosaccharides and mucin), although this ability is strain-dependent too (Klijn et al., 2005; De Vuyst et al., 2014; Rivière et al., 2014; McLaughlin et al., 2015; Selak et al., 2016).

Once internalized into the cytoplasm, hexose monosaccharides (e.g., fructose and glucose) are converted into acetate and lactate by the fructose 6-phosphate phosphoketolase pathway or bifid shunt (De Vuyst et al., 2014). Bifidobacteria will initially cleave, by means of the key enzyme fructose 6-phosphate phosphoketolase, one mole of fructose 6-phosphate into one mole of erythrose 4-phosphate and one mole of acetyl-phosphate (**Figure A**). From erythrose 4-phosphate and an additional mole of fructose 6-phosphate, one mole of ribose 5-phosphate and one mole of xylulose 5-phosphate

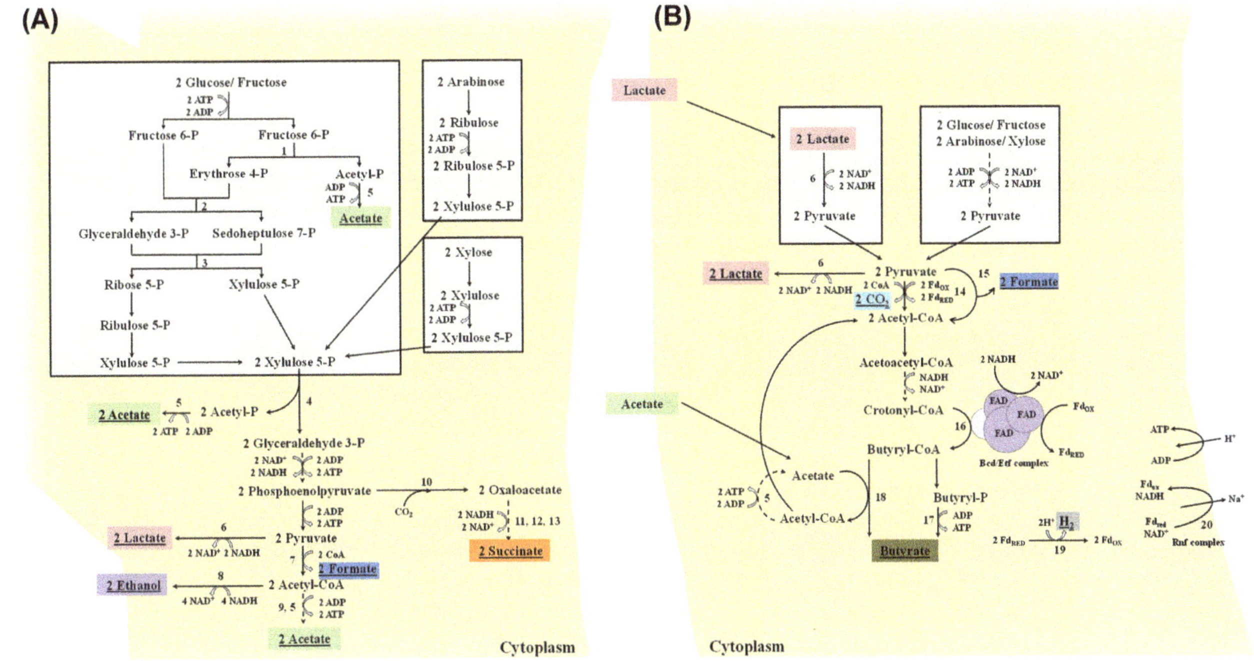


**FIGURE** | **(A)** Schematic representation of the fermentation of hexoses (glucose and fructose) and pentoses (arabinose and xylose) by bifidobacteria through the fructose 6-phosphate phosphoketolase pathway or bifid shunt. **(B)** Schematic representation of the fermentation of hexoses (glucose and fructose) and pentoses (arabinose and xylose) by butyrate-producing colon bacteria through the Embden-Meyerhof-Parnas pathway and pentose-phosphate pathway, respectively, and of lactate. Dashed lines represent different steps. Underlined metabolites are excreted into the extracellular medium. $Fd_{ox}$, oxidized ferredoxin; $Fd_{red}$, reduced ferredoxin; FAD, flavin adenine dinucleotide; enzymes: 1, fructose 6-phosphate phosphoketolase; 2, transaldolase; 3, transketolase; 4, xylulose 5-phosphate phosphoketolase; 5, acetate kinase; 6, lactate dehydrogenase; 7, formate acetyltransferase; 8, bifunctional aldehyde-alcohol dehydrogenase; 9, phosphotransacetylase; 10, phosphoenolpyruvate carboxylase; 11, malate dehydrogenase; 12, fumarase; 13, succinate dehydrogenase; 14, pyruvate:ferredoxin oxidoreductase; 15, pyruvate-formate lyase; 16, butyryl-CoA dehydrogenase/electron-transferring flavoprotein (Bcd/Etf) complex; 17, butyrate kinase; 18, butyryl-CoA:acetate CoA transferase; 19, ferredoxin hydrogenase; and 20, membrane-bound ferredoxin oxidoreductase (Rnf) complex.

are formed by the successive action of a transaldolase and a transketolase. Two moles of xylulose 5-phosphate are subsequently converted into two moles of acetyl-phosphate and two moles of glyceraldehyde 3-phosphate by the action of a xylulose 5-phosphate phosphoketolase. These two moles of acetyl-phosphate plus the additional mole of acetyl-phosphate (produced by the fructose 6-phosphate phosphoketolase) are further converted into three moles of acetate by an acetate kinase, which is accompanied by the production of three moles of ATP. The two moles of glyceraldehyde 3-phosphate are oxidized into two moles of pyruvate by enzymes participating in the Embden-Meyerhof-Parnas pathway, which results in an additional production of two moles of ATP. In a last step, pyruvate can be reduced into lactate by means of a lactate dehydrogenase, which is accompanied by $NAD^+$ recycling. Thus, when fermenting hexose monosaccharides, acetate and lactate are produced in a theoretical molar ratio of 1.5 and three moles of ATP are produced. Pentose monosaccharides (e.g., arabinose and xylose) can also be shuttled into the bifid shunt by their conversion into xylulose 5-phosphate (**Figure A**). However, this is not accompanied by the production of an additional mole of acetate (and thus no additional mole of ATP) as in the case of hexose fermentation, leading to a final theoretical molar ratio of acetate to lactate of 1.0 and two moles of ATP (Pokusaeva et al., 2011; De Vuyst et al., 2014). However, these theoretical ratios are rarely found during bifidobacterial carbohydrate fermentation, due to the production of formate, acetate, and ethanol from pyruvate instead of lactate (**Figure A**), which depends on the available energy source and its consumption rate (Palframan et al., 2003; Van der Meulen et al., 2004, 2006a,b; Falony et al., 2009b; De Vuyst et al., 2014). The production of formate from pyruvate by a formate acetyltransferase, at the expense of lactate, can be explained by the need for additional ATP production by means of the concomitant production of acetate when bifidobacteria are grown on complex carbohydrates to improve their fitness, despite their lower growth rate compared with simple carbohydrates. Bifidobacteria are also able to produce ethanol from acetyl-CoA with a bifunctional aldehyde-alcohol dehydrogenase, at the expense of acetate, to enable the continuation of pyruvate production by regenerating $NAD^+$. This shift in metabolism away from lactate production has been found for the degradation of complex carbohydrates such as inulin-type fructans (ITF; oligofructose and inulin; Van der Meulen et al., 2004; Falony et al., 2009b) and arabinoxylan-oligosaccharides (AXOS; Rivière et al., 2014, 2015). Bifidobacteria can also regenerate $NAD^+$ by the production of succinate from oxaloacetate that is in turn formed from phosphoenolpyruvate (**Figure** ; Van der Meulen et al., 2006a).

## Butyrate-Producing Colon Bacterial Species

### General Aspects

Gene-targeted approaches to investigate the butyrate-producing bacterial communities of the human gut microbiota have led to the consideration that butyrate-producing colon bacteria form a functional group rather than a monophyletic group. Most butyrate producers in the human colon belong to the Firmicutes phylum and in particular clostridial clusters IV and XIVa (Louis and Flint, 2009; Van den Abbeele et al., 2013a; Vital et al., 2014). Clostridial clusters IV and XIVa butyrate producers are Gram-positive, highly oxygen-sensitive, strictly anaerobic, saccharolytic bacteria. The two most dominant bacterial species in the human colon are *Faecalibacterium prausnitzii* (up to 14% of the total fecal gut microbiota, clostridial cluster IV) and *Eubacterium rectale* (up to 13% of the total fecal gut microbiota, clostridial cluster XIVa), and are expected to have a significant contribution to butyrate production (De Vuyst et al., 2014; Walker et al., 2014). Other important butyrate-producing bacterial species in the human colon are *Roseburia* spp. (clostridial cluster XIVa, such as *Roseburia faecis*, *Roseburia inulinivorans*, *Roseburia intestinalis*, and *Roseburia hominis*), *Eubacterium* spp. (clostridial cluster XIVa, such as *Eubacterium hallii*), *Anaerostipes* spp. (clostridial cluster XIVa, such as *Anaerostipes butyraticus*, *Anaerostipes caccae*, and *Anaerostipes hadrus*), and *Butyricicoccus pullicaecorum* (clostridial cluster IV). Some of these species (such as *E. rectale*, *F. prausnitzii*, and *R. intestinalis*) preferentially colonize the mucus layer, and consequently increase the butyrate bioavailability for colon epithelial cells, whereas other species such as *A. caccae* mainly occur in the lumen of the colon (El Aidy et al., 2013; Van den Abbeele et al., 2013a). In contrast to bifidobacteria, clostridial clusters IV and XIVa do not directly colonize the colon in high quantities after birth. In the case of *F. prausnitzii*, it has been shown that fecal numbers in infants younger than 6 months are undetectable, slightly increase between the age of 6 and 24 months, then suddenly increase to reach a peak during late childhood and adolescence, and finally decrease again during adulthood and especially in the elderly (Miquel et al., 2014).

### Functional Role in the Colon

Clostridial clusters IV and XIVa have gained a lot of attention during the last years because of their contribution to gut homeostasis, by preserving gut barrier functions and exerting immunomodulatory and anti-inflammatory properties (Velasquez-Manoff, 2015). In addition to the beneficial properties of the butyrate produced (**Table** ), *F. prausnitzii* produces anti-inflammatory peptides blocking nuclear factor NF-κB activation and cytokine IL-8 production in mice, which provide protection against chemically induced colitis (Qiu et al., 2013; Quévrain et al., 2016). Several studies have shown that the abundance of *B. pullicaecorum*, *E. rectale*, *F. prausnitzii*, and/or *R. intestinalis* is markedly decreased in IBD patients (Morgan et al., 2012; Eeckhaut et al., 2013; Gevers et al., 2014) and that such patients have lower concentrations of butyrate in their feces than healthy individuals (Marchesi et al., 2007; Nemoto et al., 2012). Less butyrate producers were also found in patients with colorectal cancer (Wu et al., 2013). Therefore, methods are being searched to stimulate butyrate-producing human colon bacterial species by diet (prebiotic approach) or by administering these bacteria orally (probiotic approach). In medical applications, pure butyrate by means of tablets or rectal enemas is used as a therapeutic agent for IBD treatment (Geirnaert et al., 2014).

### Metabolism

Like bifidobacteria, members of clostridial clusters IV and XIVa carry out a fermentative metabolism and are often able to degrade a wide range of non-digestible carbohydrates in the human colon anaerobically, encompassing resistant starch, ITF, xylo-oligosaccharides (XOS), and AXOS (Falony et al., 2009c; Louis and Flint, 2009; Scott et al., 2014; Rivière et al., 2015; Moens et al., 2016). As an example, the genome of *E. rectale* ATCC 33656 encodes 52 glycoside hydrolases, encompassing $\beta$-fructofuranosidases, $\alpha$-arabinofuranosidases, $\beta$-xylosidases, exo-oligoxylanases, $\alpha$-amylases, $\alpha$- and $\beta$-glucosidases, $\alpha$- and $\beta$-galactosidases, and cellulases (Lombard et al., 2014). However, inter-genus variations have been found within clostridial clusters IV and XIVa, as not all species and even strains within one species can consume complex carbohydrates to the same extent (Falony et al., 2009c; Scott et al., 2014; Moens et al., 2016). Most of the butyrate-producing colon bacteria use a non-preferential extracellular degradation mechanism for the breakdown of oligo- and polysaccharides, with the release of monosaccharides into the extracellular medium. As illustrated during laboratory batch fermentation experiments, co-cultivation of such butyrate-producing bacteria with bifidobacteria that have a preferential carbohydrate degradation mechanism, can comprise the competitiveness of the butyrate-producing colon bacteria (Falony et al., 2006, 2009c; De Vuyst et al., 2014). For instance, the percentage of oligofructose that was consumed by *F. prausnitzii* DSM 17677$^T$ when co-cultivated with different bifidobacterial strains decreased with an increasing ITF degradation capacity of the latter (Moens et al., 2016).

Once internalized into the cytoplasm, hexoses and pentoses are degraded to pyruvate by the Embden-Meyerhof-Parnas pathway or pentose phosphate pathway, respectively. Like other fermentative bacteria, clostridial clusters IV and XIVa butyrate producers possess several alternative pathways to form different end-metabolites from pyruvate, depending on the bacterial species, carbohydrate source, hydrogen gas pressure, and necessity of redox balancing. Besides butyrate, they can form lactate, formate, hydrogen gas, and carbon dioxide (**Figure** ). Pyruvate can get reduced into lactate by means of a lactate dehydrogenase, which is accompanied by $NAD^+$ recycling (for instance *R. inulinivorans* and *E. rectale*; Falony et al., 2009c; Rivière et al., 2015; Moens et al., 2016). The production of butyrate from pyruvate involves the conversion of pyruvate into acetyl-CoA by a pyruvate:ferredoxin oxidoreductase, with the reduction of ferredoxin and production of carbon dioxide (for instance most clostridial clusters IV and XIVa butyrate-producing colon bacteria; Falony et al., 2009c; Moens et al., 2016) and/or by a pyruvate-formate lyase with the formation of formate (for instance *F. prausnitzii* and *E. rectale*; Rivière et al., 2015; Moens et al., 2016). Two moles of acetyl-CoA are then converted via a four-step pathway into butyryl-CoA, in which the last step is carried out by a butyryl-CoA dehydrogenase/electron-transferring flavoprotein complex that catalyzes the NADH + $H^+$-dependent reduction of crotonyl-CoA coupled to the reduction of ferredoxin. The final step from butyryl-CoA to butyrate is either catalyzed by a butyrate kinase (after phosphorylation of butyryl-CoA) or a butyryl-CoA:acetate CoA transferase (Falony et al., 2009c; Louis and Flint, 2009; Mahowald et al., 2009; De Vuyst and Leroy, 2011; Moens et al., 2016). Only a few butyrate-producing colon bacteria, encompassing *Clostridium butyricum*, *Coprococcus eutactus*, and *Coprococcus comes*, are known to use a butyrate kinase to produce butyrate (Louis and Flint, 2009; Vital et al., 2014). The butyryl-CoA:acetate CoA transferase step involves the consumption of external acetate (coming from for instance bifidobacterial carbohydrate breakdown through cross-feeding) as a co-substrate, thereby producing acetyl-CoA and butyrate. The acetyl-CoA produced can be converted via acetyl-phosphate into acetate, with the production of ATP, by acetate kinase, or recycled into the four-step pathway mentioned above (Falony et al., 2009c). The reduced ferredoxin can be reoxidized via a ferredoxin hydrogenase, with the concomitant production of $H_2$, and/or via a membrane-bound ferredoxin oxidoreductase (Rnf) complex, without production of $H_2$, but with the generation of a proton-motive force that allows additional ATP production (for instance *R. inulinivorans* and *F. prausnitzii*; Falony et al., 2009c; Moens et al., 2016). The production of butyrate thus not only leads to regeneration of $NAD^+$ from NADH + $H^+$ produced in the upper parts of the carbohydrate degradation pathways for ATP production, but also leads to additional ATP production (Falony et al., 2009c; Louis and Flint, 2009; Mahowald et al., 2009; De Vuyst and Leroy, 2011). Some butyrate producers, encompassing *A. caccae*, *A. butyraticus*, *A. hadrus*, and *E. hallii*, can produce butyrate from lactate instead of carbohydrates (**Figure** ; Duncan et al., 2004; Falony et al., 2006, 2009c; Belenguer et al., 2011; De Vuyst and Leroy, 2011; De Vuyst et al., 2014).

## STIMULATION OF BIFIDOBACTERIA AND BUTYRATE-PRODUCING COLON BACTERIA

Since decreased numbers of *Bifidobacterium* species and butyrate-producing bacterial species in the human colon have been reported in patients with diverse disorders and because the SCFAs produced by these species have beneficial effects (**Table** ), these bacteria are potential candidates to be stimulated in the colon to prevent and restore a disturbed gut homeostasis. The most prevalent strategies to stimulate bifidobacteria and butyrate-producing colon bacteria in the human colon involve the consumption of probiotics and prebiotics (Scott et al., 2015).

### Probiotics

According to the international scientific association for probiotics and prebiotics (ISAPP), probiotics are defined as "live microorganisms that, when administered in adequate amounts, confer a health benefit on the host" (Hill et al., 2014). Selected strains of *Bifidobacterium* species are commonly used probiotics and are added to food supplements and foods (especially dairy products). The oral consumption of bifidobacteria has been associated with beneficial effects for different digestive problems and disorders, encompassing acceleration of the gut transit time; improvement of lactose intolerance; prevention of antibiotic-associated diarrhea and necrotizing enterocolitis (in pre-term infants that usually harbor reduced numbers of bifidobacteria); and alleviation of IBS and IBD symptoms (Leahy et al., 2005; Di Gioia et al., 2014; Tojo et al., 2014; Saez-Lara et al., 2015). Also, evidence continues to emerge that bifidobacteria influence immune responses and hence may enhance resistance to infections and allergies (Di Gioia et al., 2014; Frei et al., 2015). Further, bifidobacteria display anti-inflammatory effects and negatively correlate with metabolic endotoxemia (Everard and Cani, 2013). Moreover, interest is growing to use bifidobacterial strains (such as *Bifidobacterium infantis* 35624) as psychobiotics, which are "live organisms that, when ingested in adequate amounts, produce a health benefit in patients suffering from psychiatric illness" (Dinan et al., 2013, 2015). The health effects that bifidobacteria exert are of course strain-related; some bifidobacterial strains are effective, whereas others are not. Moreover, the probiotic health benefits are probably not caused by the bifidobacterial strains consumed solely, but are rather the result of interactions with the resident gut microbiota (Cani and Van Hul, 2015; Scott et al., 2015). Indeed, a recent metagenomic and metatranscriptomic study of feces of 12 healthy individuals has shown that the oral administration of the probiotic strain *Lactobacillus rhamnosus* GG significantly changes the activity of the resident gut microbiota, without influencing the gut microbiota composition itself (Eloe-Fadrosh et al., 2015). Especially genes involved in adhesion, chemotaxis, and/or motility of *Bifidobacterium* spp., *Eubacterium* spp., and *Roseburia* spp. are overexpressed during probiotic consumption, suggesting that the consumption of the probiotic strain promotes interactions between the resident gut microbiota and the host. Nowadays, there is also a growing interest toward the use of other bacterial strains as probiotics, such as *Akkermansia muciniphila* and butyrate-producing colon bacteria, encompassing *B. pullicaecorum*, *E. rectale*, *F. prausnitzii*, and *Roseburia* spp. (Marteau, 2013; Geirnaert et al., 2014; Cani and Van Hul, 2015; Scott et al., 2015). For example, the oral administration of *B. pullicaecorum* 25-3$^T$ and *F. prausnitzii* A2-165 in rodents has shown attenuation of chemically induced colitis (Eeckhaut et al., 2013, 2014; Martín et al., 2015). However, whether these strict anaerobic colon bacteria can survive the harsh industrial production steps and deal with the regulatory hurdles (as these bacteria have no history of safe use) will partly determine their application as probiotics in the future human diet (Figueroa-González et al., 2011; Gosálbez and Ramón, 2015; Kumar et al., 2015; Scott et al., 2015).

Since the implementation of EU legislation on health claims in 2009, no health claims for probiotics in foods have been approved by the European Food Safety Authority (EFSA) neither can the term probiotic further be used as a food label in Europe (Glanville et al., 2015). The only approved health claim is the benefit on lactose digestion when consuming live *Lactobacillus delbrueckii* subsp. *bulgaricus* and *Streptococcus thermophilus* strains present in yogurt or fermented milk (EFSA, 2010).

In severe cases of a disturbed gut homeostasis, whereby probiotic treatments do not suffice, the gut microbiota can be restored by transplanting the complete fecal microbiota from a healthy donor into a diseased person. However, the ISAPP recommends that fecal microbiota transplantations (FMTs) should not be considered as probiotics, as they are uncharacterized mixtures of strains (Hill et al., 2014). FMTs have shown to be very effective for curing *Clostridium difficile* infections, although they have ambiguous outcomes for IBD and IBS (Aroniadis and Brandt, 2014; Pamer, 2014). Furthermore, a step-up FMT strategy has been proposed to treat Crohn's disease and ulcerative colitis, which consists of a FMT, followed by additional FMT steps or standard IBD medications depending on the patient's clinical response to the treatment (Cui et al., 2016). Also, it has been shown that patients with metabolic syndrome display improved insulin sensitivity after being treated with fecal microbiota of healthy individuals (Vrieze et al., 2012). These patients possess increased numbers of butyrate-producing colon bacteria and decreased numbers of Gram-negative bacteria after a FMT. Studies are being performed to see whether FMTs can also cure non-gastrointestinal disorders, such as allergies and behavioral disorders (Xu et al., 2015). However, up to now, few fecal transplants have been performed, as the selection of healthy fecal donors requires a thorough examination to avoid the transfer of pathogens and gut microbiota-associated disorders (Kapel et al., 2014). Therefore, new approaches are being searched to transplant well-defined mixtures of bacteria (de Vos, 2013; Van den Abbeele et al., 2013b). However, an additional challenge in selecting an appropriate healthy donor or bacterial synthetic community is that, despite the large amount of information about the composition and diversity of the human gut microbiota, it is difficult (if not impossible) to define a healthy gut microbiota composition, as each healthy individual harbors a unique gut microbiota (de Vos and de Vos, 2012; Faith et al., 2013; Li et al., 2016).

## Prebiotics

### General

Another strategy to increase bifidobacteria and butyrate-producing bacteria in the human colon is through the consumption of prebiotics, which are defined according to the ISAPP as "a selectively fermented ingredient that results in specific changes in the composition and/or activity of the gastrointestinal microbiota, thus conferring benefit(s) upon host health" (Gibson et al., 2010). To date, all well-known prebiotics are carbohydrates, although other compounds such as, for instance, polyphenols may display prebiotic properties as well (Bindels et al., 2015). Compared with probiotics, prebiotics are more stable and thus can easily be added to foods, such as yogurts, biscuits, breads, cereals, spreads, ice creams, and drinks (Gibson et al., 2010). The criteria for classifying a compound as a prebiotic have been listed as (i) resistance to gastric acidity, hydrolysis by mammalian digestive enzymes, and gastrointestinal absorption; (ii) fermentation by intestinal microbiota; and (iii) selective stimulation of the growth and/or activity of intestinal bacteria associated with health and well-being (Gibson et al., 2004). In the past, the impact of the consumption of prebiotics on the gut microbiota composition was mainly studied regarding species of *Bifidobacterium* and *Lactobacillus* (Verbeke, 2014). However, recent community-wide analyses of the gut microbiota show that prebiotics are not that selective as previously assumed, and that they stimulate other bacteria too (Bindels et al., 2015). It has indeed been shown that butyrate-producing colon bacteria, such as *E. rectale*, *F. prausnitzii*, and *Roseburia* spp., can consume prebiotics such as ITF (Falony et al., 2006, 2009c; Rivière et al., 2015; Moens et al., 2016). Also, the consumption of oligofructose changes the relative abundance of 102 bacterial taxa in mice, of which 16 display a more than 10-fold decrease or increase in abundance (Everard et al., 2011). Therefore, Bindels et al. (2015) proposed to define a prebiotic as "a non-digestible compound that, through its metabolization by microorganisms in the gut, modulates composition and/or activity of the gut microbiota, thus conferring a beneficial physiological effect on the host." Alternatively, the definition of prebiotics has been challenged over time not only according to scientific considerations but also due to its importance for regulators, industry, and consumers (Hutkins et al., 2016). As for probiotics, the term prebiotic cannot be used as a health claim on food products in Europe (Salminen and van Loveren, 2012). Some claims exist for the term fiber (EFSA, 2011a,b, 2015), but not all fibers are prebiotics, whereby the latter are distinguished from the former by the selectivity of their fermentation (Slavin, 2013; Hutkins et al., 2016; Verspreet et al., 2016).

Examples of prebiotic non-digestible carbohydrates that are bifidogenic include poly- and oligosaccharides containing fructose (and a terminal glucose) as in ITF, galactose and glucose (as in galacto-oligosaccharides), glucose (as in isomalto-oligosaccharides), galactose and fructose (as in lactulose), xylose (as in XOS), and arabinose and xylose (as in AX and AXOS) (Roberfroid, 2005; Macfarlane et al., 2008; Broekaert et al., 2011; De Vuyst and Leroy, 2011; De Vuyst et al., 2014). Whereas, in the past the target genus for prebiotic stimulation was *Bifidobacterium* (Gibson et al., 2010), today new prebiotics are searched to stimulate other beneficial bacterial species in the human colon such as butyrate producers. Of special interest are prebiotics that cause both a bifidogenic effect and a butyrogenic effect. ITF, AX, and AXOS are such prebiotics that stimulate both bifidobacteria and the production of butyrate (Falony et al., 2006, 2009b,c; De Vuyst and Leroy, 2011; De Vuyst et al., 2014; Rivière et al., 2015).

### ITF As an Example of Well-Known Prebiotics

Inulin naturally occurs in fruits and plants such as chicory roots, wheat, onion, banana, garlic, and leek, but is generally extracted from chicory roots on an industrial scale (Roberfroid, 2007). Inulin consists of a linear backbone of $\beta$-(2$\rightarrow$1)-linked fructose monomers with a degree of polymerization (DP) between 2 and 65 (average DP of 10), which is often linked to a terminal glucose monomer by an $\alpha$-(1→2)-glycosidic bond (**Figures A,B**). Oligofructose is derived from native inulin by partial enzymatic hydrolysis with an inulinase and has a DP that varies between 2 and 8 (average DP of 4). Given the relative simple structures of ITF, only few bacterial enzymes are required for their degradation in the human colon, encompassing enzymes belonging to the $\beta$-fructofuranosidase (EC 3.2.1.26) superfamily that cleave terminal fructose residues from the non-reducing ends of the fructose polymers (**Figure** Scott et al., 2011). Several $\beta$-fructofuranosidases have been isolated and characterized in colon bacteria, for instance in *Bifidobacterium* species (Warchol et al., 2002; Ehrmann et al., 2003; Omori et al., 2010; Jedrzejczak-Krzepkowska et al., 2011) and *R. inulinivorans* (Scott et al., 2011). Examples of beneficial effects of the consumption of ITF include increased stool frequency, increased colonic absorption of dietary minerals (calcium and magnesium), decreased proteolytic activity, and increased secretion of satiety hormones (Schaafsma and Slavin, 2015).

ITF belong to the most studied prebiotics and their bifidogenic and butyrogenic effects have been well established in various studies (De Vuyst and Leroy, 2011; De Vuyst et al., 2014). For instance, it has been shown that not all bifidobacterial strains benefit in the same way from the presence of ITF in the human colon. A comparative statistical study of 18 bifidobacterial strains, belonging to 10 different species and

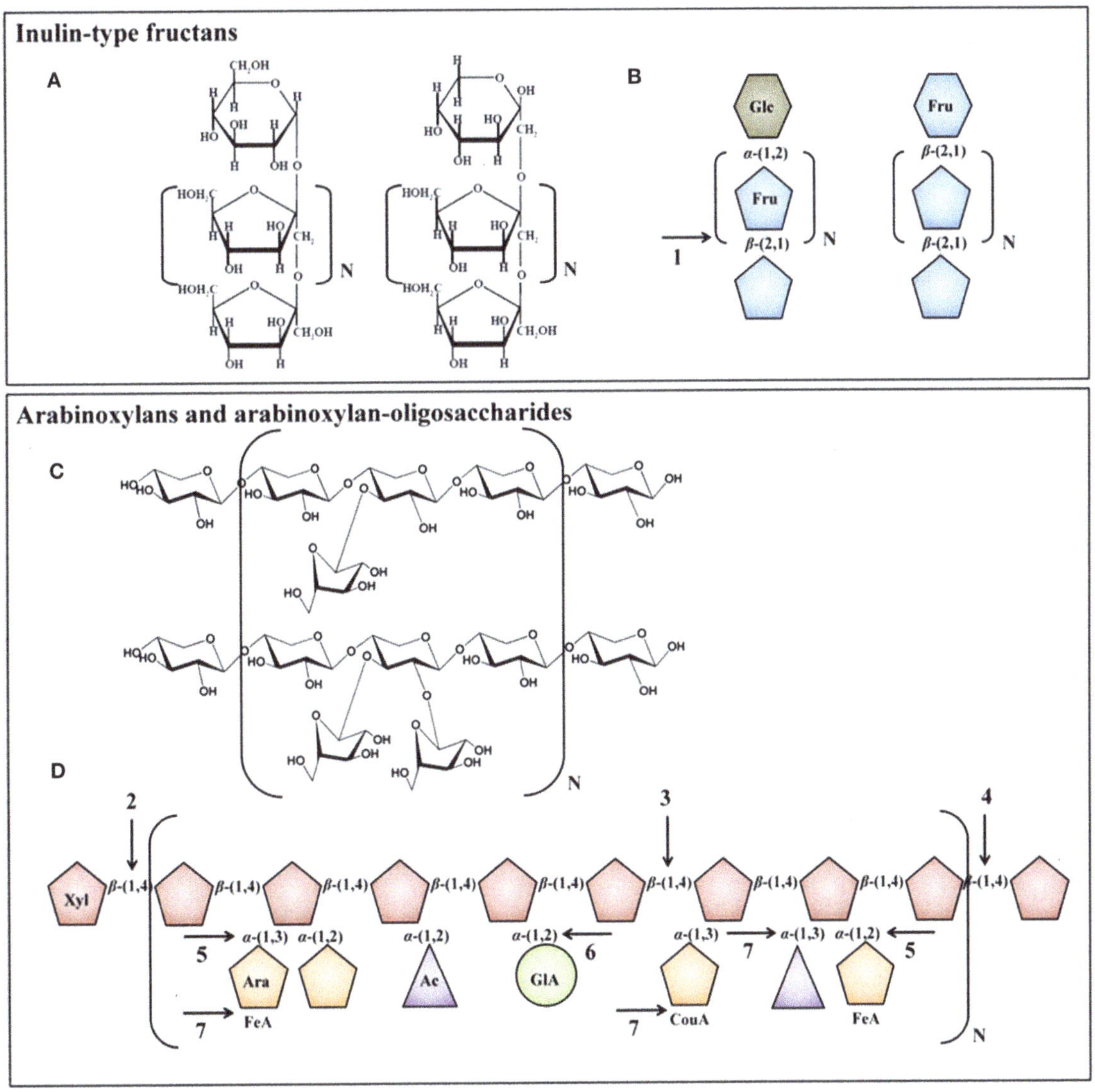


**FIGURE | Chemical structures [(A) and (C)] and schematic representations [(B) and (D)] of ITF, AX, and AXOS molecules.** Glc, glucose; Fru, fructose; Xyl, xylose; Ara, arabinose; FeA, ferulic acid; Ac, acetyl group; GlA, glucuronic acid; CouA, *p*-coumaric acid. Arrows indicate possible hydrolysis of the carbohydrates by bacterial enzymes present in the human colon: 1, $\beta$-fructofuranosidase; 2, $\beta$-xylosidase; 3, $\beta$-endoxylanase; 4, exo-oligoxylanase; 5, $\alpha$-arabinofuranosidase; 6, $\alpha$-glucuronidase; and 7, esterase.

coming from different donors and origins, has shown the existence of four different clusters of strains differing in their mechanisms and capabilities to degrade ITF (Falony et al., 2009b). Some strains only consume fructose (cluster A), whereas others consume both fructose and oligofructose, mainly short oligosaccharides (DP up to seven) after import into the cell, i.e., they display a preferential metabolism (cluster B). Certain strains degrade both oligofructose and inulin (short chain length fractions only) extracellularly, accompanied with the release of fructose into the extracellular medium, i.e., they display a non-preferential metabolism (clusters C and D). A recent study of 190 bifidobacterial strains isolated from different donors and colon regions has shown that these ITF degradation fingerprints are not correlated with the region in the intestine, suggesting that the degradation of ITF is uniform along the human intestine (Selak et al., 2016). Yet, intra-species variability in ITF degradation capacity indicates strain-specific variations. Moreover, within one colon region bifidobacterial strains with different ITF degradation mechanisms occur, which suggests cooperation for the degradation of ITF in the colon, with opportunities for cross-feeding on strain and/or species level. Similar cross-feeding between bifidobacterial strains with complementary degradation mechanisms has also been shown for starch, xylan, and mucin glycoproteins (Egan et al., 2014; Turroni et al., 2015). Also, it has been shown that the consumption of ITF, the bifidogenic effect, and the butyrogenic effect are linked to each other, because of cross-feeding interactions between bifidobacteria and butyrate-producing colon bacteria (**Figure** ; Belenguer et al., 2006; Falony et al., 2006, 2009c; Moens et al., 2016). As an end-metabolite of the bifid shunt and a co-substrate for the production of butyrate (Section AX and AXOS as an Example of Interesting Prebiotics), acetate plays a key role in cross-feeding interactions between bifidobacteria and butyrate-producing colon bacteria in the human colon. In a first type of cross-feeding, both the bifidobacterial and butyrate-producing strains consume ITF (**Figure** ). The consumption of ITF by bifidobacteria provides butyrate-producing colon bacteria with exogenous acetate that is used as a co-substrate to produce butyrate by growing on ITF simultaneously [which is, for instance, the case for *R. intestinalis* DSM 14610 (Falony et al., 2006), *R. inulinivorans* DSM 16841 (Falony et al., 2009c), and *F. prausnitzii* DSM 17677$^{T}$ (Moens et al., 2016)]. However, such cross-feeding interactions can be either a pure commensal or beneficial relationship between these bacteria or can be dominated by competition, depending on the ITF degradation capacities of the bifidobacterial strains involved (Moens et al., 2016). A second type of cross-feeding takes place between bifidobacteria that consume ITF, and concomitantly produce acetate, and acetate-consuming butyrate-producing colon bacteria that are not able to degrade ITF (**Figure** ). Instead of ITF, the latter bacteria consume carbohydrate breakdown products (short-chain oligosaccharides) liberated by the bifidobacterial strain (which is, for instance, the case for *R. hominis* DSM 16839; Belenguer et al., 2006) or lactate (for instance *E. hallii* DSM 17630; Belenguer et al., 2006, and *A. caccae* DSM 14662; Falony et al., 2006).

## AX and AXOS As an Example of Interesting Prebiotics

### *Physiological effects*

Growing interest is devoted to complex non-digestible carbohydrates that ferment slowly and thereby cause bifidogenic and butyrogenic effects along the entire length of the human colon. AX and AXOS, as a broad class of heteropolysaccharides and -oligosaccharides with complex varying structures (**Figures C,D**), belong to these slow-fermenting carbohydrates and hence are able to decrease the production of bacterial toxic metabolites originating from protein and lipid metabolism in the distal colon (Section Bifidobacterium Species; Van Craeyveld et al., 2008; Grootaert et al., 2009; Sanchez et al., 2009; Neyrinck et al., 2011). This is to be explained by a stimulation of saccharolytic activities, an increase in SCFA production, and a lowering of the luminal pH in the distal part of the colon, where carbohydrates are rare and proteolytic bacteria, such as *Bacteroides* spp., are otherwise favored (Duncan et al., 2009). Examples of additional potential benefits of the consumption of AX and AXOS for human health include improved mineral (calcium and magnesium) absorption; increased stool frequency and improved stool consistency; reduced post-prandial glycemic response; reduced blood cholesterol levels; and increased antioxidant capacity (Grootaert et al., 2007; Broekaert et al., 2011; Damen et al., 2011; Mendis and Simsek, 2013). Moreover, the consumption of AXOS, with the increase of bifidobacterial numbers as a result, may help to restore gut barrier functions and cure metabolic endotoxemia in mice (Neyrinck et al., 2012).

### *Occurrence, structural properties, and degradation*

AX naturally occur in the endosperm and bran (pericarp, testa, and aleuron layer) of cereal grains such as wheat, rye, rice, barley, oat, and sorghum, but in varying quantities, depending on the cereal species and the location within the cereal kernel (Izydorczyk and Biliaderis, 2006). For instance, the endosperm of wheat kernels contains *ca.* 2% of AX, whereas the pericarp contains *ca.* 38% of AX (Benamrouche et al., 2002; Maes and Delcour, 2002). AX consist of a linear backbone of 1500 to 15,000 $\beta$-(1→4)-linked xylose monomers, which can randomly be substituted with arabinose monomers on the C-(O)-2 or C-(O)-3 positions (monosubstituted) or on both positions (disubstituted; **Figures C,D**; Izydorczyk and Biliaderis, 1995). Distribution patterns of arabinose substituents on the xylose backbone are not regular for wheat AX; highly branched regions are interlinked by sequences of contiguous non-substituted xylose residues (Gruppen et al., 1993). The number of arabinose substituents bound to the xylose backbone is expressed as the arabinose/xylose ratio (A/X) and depends on the cereal species and the location within the kernel. For instance, in the pericarp, testa, aleuron layer, and endosperm of wheat kernels, different A/Xs are found, namely *ca.* 1.0, 0.1, 0.4, and 0.5, respectively (Izydorczyk and Biliaderis, 1995; Antoine et al., 2003; Barron et al., 2007). The fermentability of AX and AXOS in the human colon is strongly influenced by the complexity of the AXOS molecules and decreases with increasing DP and increasing A/X (Van Craeyveld et al., 2008; Pollet et al., 2012). Additionally, xylose residues can be esterified with glucuronic acid and acetyl groups,

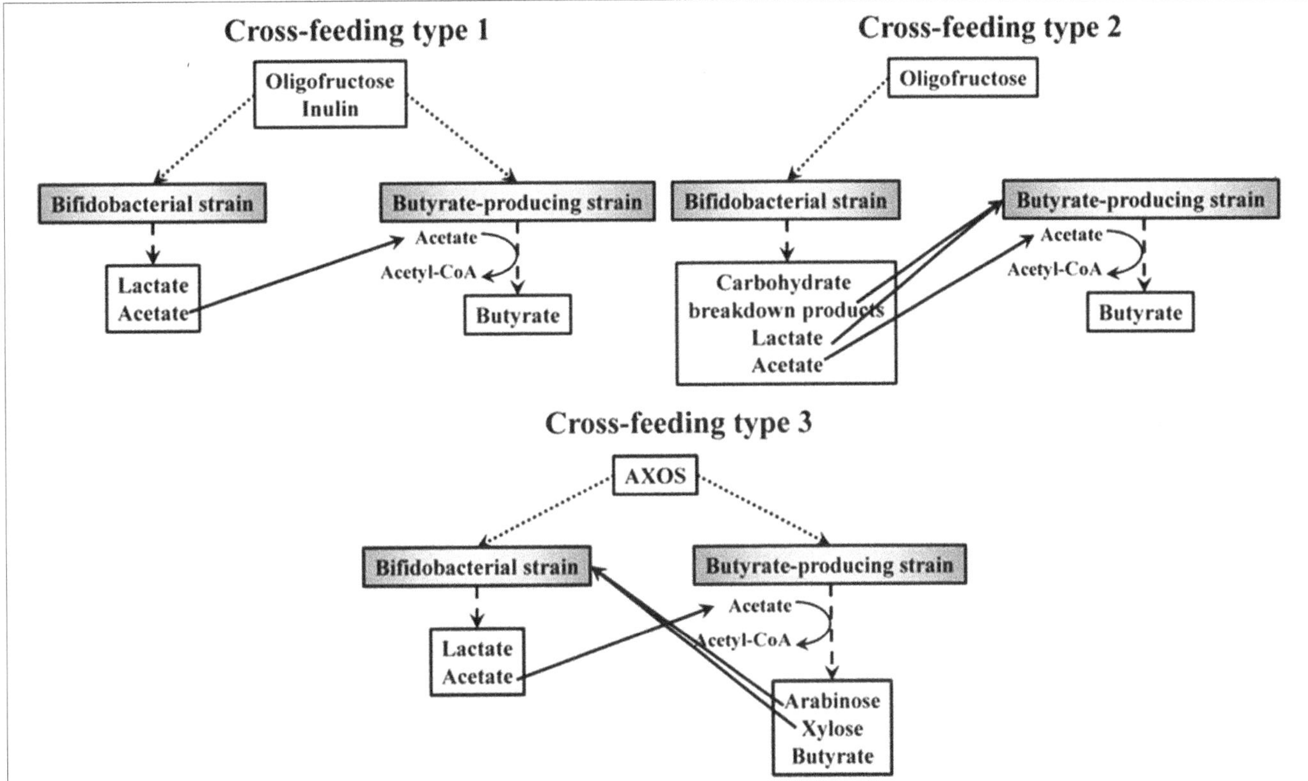


**FIGURE | Different types of cross-feeding that can take place between *Bifidobacterium* spp. and species of butyrate-producing colon bacteria in the human colon.** Arrows indicate consumption of oligofructose, inulin, and AXOS (·····), production of carbohydrate breakdown products and/or metabolic end-products (- - -), and cross-feeding interactions between the bifidobacterial and butyrate-producing strains (—).

whereas arabinose residues can be esterified with ferulic acid and *p*-coumaric acid, although in low numbers (**Figure** ; Izydorczyk and Biliaderis, 1995). These esterifications are of health and physicochemical importance, since ferulic acid and *p*-coumaric acid are antioxidants and potent cross-linking sites for attachment to other AX chains (Bunzel et al., 2001; Ou and Sun, 2014). The presence of feruloylated and diferuloylated arabinose substituents reduces the fermentability of AX and AXOS (Hopkins et al., 2003; Snelders et al., 2014). As cereal whole grains only contain low concentrations of AX (varying between 1.8% of AX in sorghum and 12.1% of AX in rye; Izydorczyk and Biliaderis, 2006), and thus the overall intake of AX is low (especially in modern Western-type diets with high intakes of refined cereal products), AX can be extracted from cereal grains and added to food products in higher concentrations (Broekaert et al., 2009). On an industrial scale, AX are usually extracted from wheat bran that is available in large quantities in Europe (Swennen et al., 2006). AXOS, the hydrolysis products of AX, are formed not only in processed cereal-based food products such as bread and beer (Courtin et al., 2009; Broekaert et al., 2011), but can also be produced on an industrial scale by the enzymatic cleavage of AX with $\beta$-endoxylanases (Broekaert et al., 2009, 2011). This results in various substituted molecules (i.e., AXOS) and non-substituted molecules (i.e., XOS), differing in DP and A/X.

Given their complex structures, the degradation of AX and AXOS in the human colon requires the cooperative action of debranching and depolymerizing bacterial carbohydrate-active enzymes, encompassing $\beta$-endoxylanases (EC 3.2.1.8) that cleave AX into AXOS and XOS; $\beta$-xylosidases (EC 3.2.1.37) that cleave terminal xylose residues from the non-reducing ends of the xylose backbones; exo-oligoxylanases (EC 3.2.1.156) that release terminal xylose residues from the reducing ends of the xylose backbones; and $\alpha$-arabinofuranosidases (EC 3.2.1.55) that remove arabinose substituents from the xylose backbones (**Figure D**). Additional enzymes are needed to cleave glucuronic acid [i.e., $\alpha$-glucuronidase (EC 3.2.1.139)], ferulic acid [i.e., ferulic acid esterase (EC 3.1.1.73)], acetyl groups [i.e., acetyl xylan esterase (EC 3.1.1.72)], and *p*-coumaric acid [i.e., *p*-coumaric acid esterase (EC 3.1.1.-)] from AXOS (**Figure D**; Dodd and Cann, 2009; Lagaert et al., 2014).

To date, AX and AXOS fall under the definition of dietary fiber (European Commission, 2008; Snelders et al., 2014) but are not considered as prebiotics by the EFSA, although they meet the three criteria of prebiotics (see above; Broekaert et al., 2011). AX and AXOS are neither digested nor absorbed in the upper gastrointestinal tract and reach the human colon intact, where they are fermented by the resident colon bacteria and cause bifidogenic and butyrogenic effects (**Table** ). However, as is also the case for other prebiotics, the selective stimulation

**TABLE | Overview of *in vitro* and *in vivo* studies of AX and AXOS.**

| Substrate (avDP-A/X) supplementation+ | Time | *In vitro*/*In vivo* | Significant concentration shift+ | | Method microbial characterization+ | Significant bacterial shift+ | | References |
|---|---|---|---|---|---|---|---|---|
| | | | Butyrate | Propionate | | Increase of | Decrease of | |
| AXOS (Nd-0.87) 13 g day$^{-1}$ | 3 w | *In vivo* Humans | ↑ Most | ↑ | Nd | Nd | Nd | Gråsten et al., 2003 |
| AX (Nd-0.51) 10 g L$^{-1}$ | 48 h | *In vitro* batch fermentation (human fecal inoculum) | ↑ | ↑ Most | 16S rRNA probe hybridization | *Bacteroides-Prevotella-Porphyromonas* spp. | ~ | Hopkins et al., 2003 |
| AX-66 kDa (Nd-0.40) AX-278 kDa (Nd-0.61) AX-354 kDa (Nd-0.61) 1% (m v$^{-1}$) | 12 h | *In vitro* batch fermentation (human fecal inoculum) | ↑ Most Especially AX-66 kDa | ↑ | Fluorescent *in situ* hybridization (FISH) | *Bifidobacterium* spp., *Lactobacillus* spp., and *Bacteroides* spp. *Clostridium coccoides-Eubacterium rectale* spp. (especially AX-66 kDa) | ~ | Hughes et al., 2007 |
| AXOS (61-0.58) (12-0.69) (15-0.27) (5-0.27) (3-0.26) 4% (m m$^{-1}$) | 2 w | *In vivo* Rats | ↑ Only for AXOS (5-0.27) and (3-0.26) in colon | ~ | qPCR | *Bifidobacterium* spp. [only for AXOS (5-0.27, 3-0.26) in cecum] | ~ | Van Craeyveld et al., 2008 |
| AXOS (15-0.27) 3 g L$^{-1}$ | 3 w | *In vitro* SHIME® (human fecal inoculum) | ↓ In proximal colon vessel ↑ In transverse colon vessel | ↓ In proximal colon vessel ↑ Most in transverse colon vessel | qPCR | ~ | *Roseburia* spp. (in proximal colon vessel) | Grootaert et al., 2009 |
| AXOS (29-0.30) 3 g L$^{-1}$ | 3 w | *In vitro* SHIME® (human fecal inoculum) | ↑ Most In proximal, transverse, and distal colon vessels | ↑ In proximal and transverse colon vessels | qPCR | *Bifidobacterium* spp. and *Bacteroides-Prevotella* spp. (in proximal colon vessel) *Lactobacillus* spp. (in proximal and transverse colon vessels) *Cl. coccoides-E. rectale* spp. (in proximal and distal colon vessels) | ~ | Sanchez et al., 2009 |
| AXOS (6-0.26) 10 g day$^{-1}$ | 3 w | *In vivo* Humans | Nd | Nd | qPCR | *Bifidobacterium* spp. and *Bifidobacterium adolescentis* (in some individuals) in feces | *Lactobacillus* spp. in feces | Cloetens et al., 2010 |

*(Continued)*

**TABLE | Continued**

| Substrate (avDP-A/X) supplementation+ | Time | *In vitro*/*In vivo* | Significant concentration shift+ | | Method microbial characterization+ | Significant bacterial shift+ | | References |
|---|---|---|---|---|---|---|---|---|
| | | | Butyrate | Propionate | | Increase of | Decrease of | |
| AXOS (5-0.51) WU-AX (284-0.59) WE-AX (233-0.51) Combinations 5% (m m$^{-1}$) | 2 w | *In vivo* Rats | ↑ Only for WU-AX, WU-AX + AXOS, and WU-AX + AXOS + WE-AX in cecum and colon | ~ | qPCR | *Bifidobacterium* spp. (only for AXOS, WE-AX, WE-AX + AXOS, WU-AX + AXOS, WU-AX + AXOS + WE-AX in cecum and WE-AX, WE-AX + AXOS in colon) *Lactobacillus* spp. (only for WU-AX + AXOS in cecum) *Roseburia-E. rectale* spp. (WU-AX, WU-AX + AXOS, WE-AX + AXOS in cecum) | *Lactobacillus* (for AXOS in cecum) | Damen et al., 2011 |
| AX (60-0.70) 10% (m m$^{-1}$) | 4 w | *In vivo* Mice | Nd | Nd | qPCR | *Bifidobacterium* spp., *Bacteroides-Prevotella* spp., and *Roseburia* spp. in cecum | ~ | Neyrinck et al., 2011 |
| AX (60-0.70) 10% (m m$^{-1}$) | 6 w | *In vivo* Rats | ↑ In cecum and feces | ↑ Most In cecum and feces | High-resolution phylogenetic microarray (HITChip) | Eleven bacterial species (e.g., *Bifidobacterium* spp., *Roseburia intestinalis, E. rectale, Collinsella* spp., *Clostridium colinum, Lachnospira pectinoschiza*) in cecum *Akkermansia muciniphila* (in colon) | Nine bacterial species (e.g., *Ruminococcus bromii, Anaerostipes caccae, Eubacterium limosum,* and *A. muciniphila*) in cecum | Van den Abbeele et al., 2011 |
| AXOS (Nd-Nd) 4.8 g day$^{-1}$ | 3 w | *In vivo* Humans | ↓ | ~ | FISH | *Bifidobacterium* spp. in feces | ~ | Maki et al., 2012 |
| WB (74-0.61) (46-0.63) (42-0.92) (40-0.34) (4-0.22) PSH (300-0.29) (200-0.27) (88-0.16) (72-0.14) 1% (m v$^{-1}$) | 48 h | *In vitro* batch fermentation (SHIME® human fecal inoculum) | ↑ Especially PSH (300-0.29), (200-0.27), (88-0.16) | ↑ Most Especially PSH (200-0.27), (88-0.16), (72-0.14) | Nd | Nd | Nd | Pollet et al., 2012 |

*(Continued)*

**TABLE | Continued**

| Substrate (avDP-A/X) supplementation+ | Time | *In vitro*/*In vivo* | Significant concentration shift+ | | Method microbial characterization+ | Significant bacterial shift+ | | References |
|---|---|---|---|---|---|---|---|---|
| | | | Butyrate | Propionate | | Increase of | Decrease of | |
| β-Endoxylanase-treated bread [containing AXOS (18-Nd)] Normal bread [containing AX (174-Nd)] 2.2 g day$^{-1}$ | 3 w | *In vivo* Humans | ↑ In feces | ~ | FISH | *Bifidobacterium* spp. and *Bacteroides-Prevotella* spp. (for treated and normal bread) in feces *Roseburia-E. rectale* spp. and *E. rectale-Cl. coccoides* spp. (only for normal bread) in feces | *Clostridium histolyticum-Clostridium perfringens* | Walton et al., 2012 |
| AX (Nd-Nd) 10 g L$^{-1}$ | 12 h | *In vitro* batch fermentation (human fecal inoculum) | ~ | ~ | Pyrosequencing | *Bacteroides xylanisolvens* | *Blautia* spp. | Yang et al., 2013 |
| AX (Nd-0.55) 17% (m m$^{-1}$) | 3 w | *In vivo* Pigs | ↑ Most In cecum, proximal colon, transverse colon | ↑ In cecum, proximal colon, transverse colon | qPCR | *Bifidobacterium* spp., *Lactobacillus* spp., *F. prausnitzii*, *R. intestinalis*, and *Blautia coccoides-E. rectale* spp. in feces | ~ | Nielsen et al., 2014 |

+avDP, average degree of polymerization; A/X, arabinose/xylose ratio; Nd, not determined; ↑, increase of concentration; ↓, decrease of concentration; ~, no significant change; qPCR, quantitative PCR; WU-AX, water-unextractable AX; WE-AX, water-extractable AX; WB, AX and AXOS from wheat bran; PSH, AX and AXOS from Psyllium seed husk.

criterion can be questioned. Several *in vivo* and *in vitro* studies have shown that AX and AXOS stimulate, besides bifidobacteria and butyrate-producing colon bacteria, other saccharolytic colon bacteria too, such as *Bacteroides* spp. and *Lactobacillus* spp. (**Table** ). Moreover, a propionogenic effect is supposed to occur. A few studies have shown that AX and AXOS especially stimulate the production of propionate (**Table** ; Hopkins et al., 2003; Van den Abbeele et al., 2011; Pollet et al., 2012). For instance, the mucin-consuming propionate-producing *A. muciniphila* is stimulated in the colon of humanized rats fed with long-chain AX (**Table** ; Van den Abbeele et al., 2011). Whether this is a direct or indirect effect is not known yet.

### *Bifidogenic effects of AX and AXOS*

Several *in vivo* studies (in rodents, pigs, and humans) and *in vitro* studies [during batch and simulator of human intestinal microbial ecosystem (SHIME®) fermentations with fecal slurries] have shown that AX and AXOS are bifidogenic (**Table** ). An *in vivo* study with rats has shown that the bifidogenic effect is only caused by AXOS with low average DPs $\leq 5$ and A/Xs $\leq 0.27$ (Van Craeyveld et al., 2008), whereas other rodent studies have found a stimulation of bifidobacteria by AX and AXOS with high average DPs up to 284 and A/Xs up to 0.70 (**Table** ; Damen et al., 2011; Neyrinck et al., 2011; Van den Abbeele et al., 2011). In the latter study, a 60-fold increase of bifidobacteria in the cecum of rats has been found, caused by the consumption of long-chain AX (average DP of 60, A/X of 0.70; Van den Abbeele et al., 2011). Apart from *in vitro* and animal experiments, human studies have revealed a bifidogenic effect caused by a daily intake of 10 g of AXOS per day (Cloetens et al., 2010), 5.5 g of AXOS per day (Maki et al., 2012), and 2.2 g of AX and AXOS per day (Walton et al., 2012; **Table** ). However, until recently, many fundamental questions remain unanswered. For instance, how can the low numerical abundant bifidobacteria (<5%) compete with other, more abundant, saccharolytic bacteria in the human colon for AX and AXOS? Do bifidobacteria have a preference for certain AX and AXOS molecules? Are all bifidobacterial strains in the human colon stimulated by AX and AXOS? To answer these questions, a detailed knowledge of the carbohydrate-hydrolyzing capacity of bifidobacteria was missing. Indeed, in the past, studies of the degradation of AX and AXOS through mono-culture fermentations with bifidobacterial strains were restricted to monitoring of bacterial growth, pH, and SCFA production (Van Laere et al., 2000; Crittenden et al., 2002), or fermentations of purified short-chain AXOS standards were performed (Pastell et al., 2009) without revealing the complete fermentation capacity of bifidobacteria. Recently, the mechanistic variations in AXOS degradation by 36 bifidobacterial strains from different donors and origins have been investigated (Rivière et al., 2014). The results show that not all bifidobacterial strains are stimulated by AXOS to the same extent. AXOS degradation by bifidobacteria is complex and involves the consumption of arabinose substituents, whether or not followed by the consumption of the xylose backbones of AXOS, either up to xylotetraose or longer and either intracellularly or extracellularly. Several bifidobacterial strains use the arabinose substituents of AXOS solely, whereas others first consume the arabinose substituents and later import the xylose backbones (up to xylotetraose) into the cell. This extracellular arabinose substituent-oriented metabolism of bifidobacteria has been linked to the presence of genes encoding extracellular cell-associated $\alpha$-arabinofuranosidases (Lagaert et al., 2010, 2014; Rivière et al., 2014). The majority of the bifidobacterial strains cannot use xylose backbones longer than xylotetraose, i.e., they display a preferential metabolism, except for one strain among the 36 tested ones, *B. catenulatum* LMG 11043$^{T}$, that also uses longer xylose backbones, i.e., they display a non-preferential metabolism (Rivière et al., 2014). This could explain why the bifidogenic effect is strongly influenced by the complexity of the AXOS molecules and decreases with increasing DP (**Table** ; Van Craeyveld et al., 2008). A multivariate data analysis of the fermentation data of these 36 bifidobacterial strains has revealed five species-independent clusters, representing five different complementary AXOS degradation mechanisms (Rivière et al., 2014). Cluster I strains, albeit not all, consume free arabinose and xylose; cluster II strains have an extracellular arabinose substituent-oriented metabolism; cluster III strains display a preferential metabolism of non-substituted xylose backbones; cluster IV strains combine the degradation mechanisms of clusters II and III; and cluster V strains display a non-preferential AXOS metabolism. The complementary degradation mechanism of bifidobacterial strains and the ability of intracellular and cell-associated degradation of xylose backbones and AXOS, could explain the selective stimulation of bifidobacteria by AXOS in the presence of other saccharolytic colon bacteria in the human colon. Whole-genome sequence annotations have revealed that some bifidobacterial strains contain genes coding for enzymes involved in the debranching of substituents and in the exo-cleavage of the xylose backbones of AX and AXOS (Schell et al., 2002; van den Broek and Voragen, 2008; van den Broek et al., 2008). Indeed, several AXOS-degrading enzymes have been isolated and characterized in bifidobacterial strains, encompassing $\beta$-xylosidases in *B. adolescentis* ATCC 15703 and *B. animalis* subsp. *lactis* BB-12; $\alpha$-arabinofuranosidases in *B. adolescentis* ATCC 15703, *B. adolescentis* DSM 20083, *B. longum* B667, and *B. longum* NCC2705; and exo-oligoxylanases in *B. adolescentis* LMG 10502 (Lagaert et al., 2010, 2011, 2014). However, up to now, no $\beta$-endoxylanases have been found in the genome of bifidobacteria. The only gene (i.e., *BL1543*) that was first annotated as a $\beta$-endoxylanase in *B. longum* NCC2705 (Schell et al., 2002) has shown to be an extracellular membrane-associated $\alpha$-arabinofuranosidase (Lagaert et al., 2010, 2014; Rivière et al., 2014). For the complete utilization of AX, it is likely that most of the *Bifidobacterium* species require cooperation with $\beta$-endoxylanase-producing bacteria, such as *Bacteroides* and *Roseburia* species (Chassard et al., 2007; Dodd et al., 2011). For instance, the genome of *R. intestinalis* L1-82 contains three genes possibly encoding $\beta$-endoxylanases (NCBI Resource Coordinators, 2014).

### *Butyrogenic effects of AX and AXOS*

Besides a bifidogenic effect, AX and AXOS have shown to cause a butyrogenic effect (**Table** ). In seven of the 13 *in vitro* and *in vivo* studies summarized in **Table** , bifidobacteria and butyrate-producing colon bacteria (*F. prausnitzii*, *E. rectale*, and *Roseburia* spp.) are stimulated simultaneously, with a significant increase of butyrate production as a result. As these butyrate-producing colon bacteria are present in high numbers in the colon, a rise in butyrate concentration does not come as a surprise (De Vuyst et al., 2014). In contrast to bifidobacteria, much less is known about the genetic AX- and AXOS-degrading potential of species of butyrate-producing colon bacteria. *In silico* analysis of the genome sequence of, for instance, *E. rectale* ATCC 33656 has shown that there are five genes possibly encoding AXOS-degrading enzymes (exo-oligoxylanase, bifunctional $\beta$-xylosidase/$\alpha$-arabinofuranosidase, $\beta$-xylosidase, and two $\alpha$-arabinofuranosidases; Rivière et al., 2015).

In contrast to ITF, the link between the consumption of AXOS, the bifidogenic effect, and the butyrogenic effect has been assessed only recently (Rivière et al., 2015). It has been shown that a third type of cross-feeding can take place in the presence of AXOS (**Figure** ), for instance in the case of *B. longum* NCC2705 (an arabinose substituent degrader of AXOS) and *E. rectale* ATCC 33656 (a complete AXOS degrader). Both strains consume AXOS (as in cross-feeding type 1), but the bifidobacterial strain is additionally stimulated by consuming the monosaccharides released by the extracellular degradation of AXOS by the *E. rectale* strain, leading to cross-feeding interactions that are mutually beneficial (**Figure** ). It is likely that these kinds of cross-feeding interactions between bifidobacteria and butyrate-producing colon bacteria, caused by prebiotic consumption, will take place *in vivo* in the human colon (Boets et al., 2013). However, the presence of other bacterial strains, with their own mechanisms of carbohydrate degradation (preferential vs. non-preferential) and own cross-feeding interactions within and between species and genera (**Figure** ), complicate the attempts to fully understand the bifidogenic and butyrogenic effects of AX and AXOS in the human colon. Furthermore, the inter-individual variations in bacterial composition make it even more intricate to predict the effects of prebiotic consumption in the colon.

## CONCLUSIONS

Human gut microbiota research has grown tremendously over the last years in terms of technology development and implications for human health. For instance, it has been shown that certain key bacteria within the colon, such as bifidobacteria and butyrate-producing colon bacteria, are negatively correlated with disorders such as IBD and colorectal cancer. Of the same importance is the progress that is being made into the modulation of the gut microbiota through the use of probiotics, prebiotics, and FMTs to improve human health. Whereas, in the past, the focus was on straightforward increase of bifidobacterial cell concentrations, shifts in interests are currently emphasizing that the stimulation of butyrate-producing bacteria in the human colon is of importance too. The consumption of prebiotic ITF and AXOS seems to be a promising approach to counteract decreased numbers of bifidobacteria and butyrate-producing colon bacteria. The challenge for the upcoming years will however be to first find out whether these changes in gut microbiota composition are the cause or the consequence of a disorder.

## AUTHOR CONTRIBUTIONS

AR acted as the main author. MS, DL, FL, and LD all contributed substantially to the writing and critical revision of the manuscript and approved its final version.

## ACKNOWLEDGMENTS

The authors acknowledge their financial support of the Research Council of the Vrije Universiteit Brussel (SRP7, IRP2, and IOF342 projects) and the Hercules Foundation (grant UABR09004). MS is the recipient of a PhD fellowship of the Vrije Universiteit Brussel in the framework of a bilateral agreement with the University of Ljubljana. AR was the recipient of a PhD fellowship of the Research Foundation Flanders (FWO-Vlaanderen). DL is the recipient of a PhD fellowship in the framework of an IWT-SBO project (BRANDING IWT130028).

## REFERENCES

Al-Lahham, S. H., Peppelenbosch, M. P., Roelofsen, H., Vonk, R. J., and Venema, K. (2010). Biological effects of propionic acid in humans; metabolism, potential applications and underlying mechanisms. *Biochim. Biophys. Acta* 1801, 1175–1183. doi: 10.1016/j.bbalip.2010.07.007

Antoine, C., Peyron, S., Mabille, F., Lapierre, C., Bouchet, B., Abecassis, J., et al. (2003). Individual contribution of grain outer layers and their cell wall structure to the mechanical properties of wheat bran. *J. Agric. Food Chem.* 51, 2026–2033. doi: 10.1021/jf0261598

Aroniadis, O. C., and Brandt, L. J. (2014). Intestinal microbiota and the efficacy of fecal microbiota transplantation in gastrointestinal disease. *Gastroenterol. Hepatol.* 10, 230–237.

Arumugam, M., Raes, J., Pelletier, E., Le Paslier, D., Yamada, T., Mende, D. R., et al. (2011). Enterotypes of the human gut microbiome. *Nature* 473, 174–180. doi: 10.1038/nature09944

Bäckhed, F., Ley, R. E., Sonnenburg, J. L., Peterson, D. A., and Gordon, J. I. (2005). Host-bacterial mutualism in the human intestine. *Science* 307, 1915–1920. doi: 10.1126/science.1104816

Barron, C., Surget, A., and Rouau, X. (2007). Relative amounts of tissues in mature wheat (*Triticum aestivum* L.) grain and their carbohydrate and phenolic acid composition. *J. Cereal Sci.* 45, 88–96. doi: 10.1016/j.jcs.2006.07.004

Belenguer, A., Duncan, S. H., Calder, A. G., Holtrop, G., Louis, P., Lobley, G. E., et al. (2006). Two routes of metabolic cross-feeding between *Bifidobacterium adolescentis* and butyrate-producing anaerobes from the human gut. *Appl. Environ. Microbiol.* 72, 3593–3599. doi: 10.1128/AEM.72.5.3593-3599.2006

Belenguer, A., Holtrop, G., Duncan, S. H., Anderson, S. E., Calder, A. G., Flint, H. J., et al. (2011). Rates of production and utilization of lactate by microbial communities from the human colon. *FEMS Microbiol. Ecol.* 77, 107–119. doi: 10.1111/j.1574-6941.2011.01086.x

Benamrouche, S., Crônier, D., Debeire, P., and Chabbert, B. A. (2002). A chemical and histological study on the effect of (1→4)-β-endo-xylanase treatment on wheat bran. *J. Cereal Sci.* 36, 253–260. doi: 10.1006/jcrs.2001.0427

Bindels, L. B., Delzenne, N. M., Cani, P. D., and Walter, J. (2015). Towards a more comprehensive concept for prebiotics. *Nat. Rev. Gastroenterol. Hepatol.* 12, 303–310. doi: 10.1038/nrgastro.2015.47

Boets, E., Houben, E., Windey, K., De Preter, V., Moens, F., Gomand, S., et al. (2013). "*In vivo* evaluation of bacterial cross-feeding in the colon using stable isotope techniques: a pilot study," in *Digestive Disease Week*, Orlando, FL. Gastroenterology 144.

Bottacini, F., Ventura, M., van Sinderen, D., and O'Connell Motherway, M. (2014). Diversity, ecology and intestinal function of bifidobacteria. *Microb. Cell Fact.* 13, S4. doi: 10.1186/1475-2859-13-S1-S4

Braniste, V., Al-Asmakh, M., Kowal, C., Anuar, F., Abbaspour, A., Tóth, M., et al. (2014). The gut microbiota influences blood-brain barrier permeability in mice. *Sci. Transl. Med.* 6, 263ra158. doi: 10.1126/scitranslmed.3009759

Broekaert, W. F., Courtin, C., and Delcour, J. (2009). *(Arabino)xylan Oligosaccharide Preparation.* WO 2009117790 A2. PCT International Publication.

Broekaert, W. F., Courtin, C. M., Verbeke, K., Van de Wiele, T., Verstraete, W., and Delcour, J. A. (2011). Prebiotic and other health-related effects of cereal-derived arabinoxylans, arabinoxylan-oligosaccharides, and xylooligosaccharides. *Crit. Rev. Food Sci. Nutr.* 51, 178–194. doi: 10.1080/10408390903044768

Bunzel, M., Ralph, J., Marita, J. M., Hatfield, R. D., and Steinhart, H. (2001). Diferulates as structural components in soluble and insoluble cereal dietary fibre. *J. Sci. Food Agric.* 81, 653–660. doi: 10.1002/jsfa.861

Cani, P. D., Bibiloni, R., Knauf, C., Waget, A., Neyrinck, A. M., Delzenne, N. M., et al. (2008). Changes in gut microbiota control metabolic endotoxemia-induced inflammation in high-fat diet-induced obesity and diabetes in mice. *Diabetes* 57, 1470–1481. doi: 10.2337/db07-1403

Cani, P. D., and Van Hul, M. (2015). Novel opportunities for next-generation probiotics targeting metabolic syndrome. *Curr. Opin. Biotechnol.* 32, 21–27. doi: 10.1016/j.copbio.2014.10.006

Chang, P. V., Hao, L., Offermanns, S., and Medzhitov, R. (2014). The microbial metabolite butyrate regulates intestinal macrophage function via histone deacetylase inhibition. *Proc. Natl. Acad. Sci. U.S.A.* 111, 2247–2252. doi: 10.1073/pnas.1322269111

Chassard, C., Goumy, V., Leclerc, M., Del'homme, C., and Bernalier-Donadille, A. (2007). Characterization of the xylan-degrading microbial community from human faeces. *FEMS Microbiol. Ecol.* 61, 121–131. doi: 10.1111/j.1574-6941.2007.00314.x

Choi, J. H., Lee, K. M., Lee, M. K., Cha, C. J., and Kim, G. B. (2014). *Bifidobacterium faecale* sp. nov., isolated from human *faeces*. *Int. J. Syst. Evol. Microbiol.* 64, 3134–3139. doi: 10.1099/ijs.0.063479-0

Cloetens, L., Broekaert, W. F., Delaedt, Y., Ollevier, F., Courtin, C. M., Delcour, J. A., et al. (2010). Tolerance of arabinoxylan-oligosaccharides and their prebiotic activity in healthy subjects: a randomised, placebo-controlled cross-over study. *Br. J. Nutr.* 103, 703–713. doi: 10.1017/S0007114509992248

Collins, S. M., Surette, M., and Bercik, P. (2012). The interplay between the intestinal microbiota and the brain. *Nat. Rev. Microbiol.* 10, 735–742. doi: 10.1038/nrmicro2876

Courtin, C. M., Broekaert, W. F., Swennen, K., Aerts, G., Van Craeyveld, V., and Delcour, J. A. (2009). Occurrence of arabinoxylo-oligosaccharides and arabinogalactan peptides in beer. *J. Am. Soc. Brew. Chem.* 67, 112–117. doi: 10.1094/asbcj-2009-0323-01

Crittenden, R., Karppinen, S., Ojanen, S., Tenkanen, M., Fagerstrom, R., Matto, J., et al. (2002). *In vitro* fermentation of cereal dietary fibre carbohydrates by probiotic and intestinal bacteria. *J. Sci. Food Agric.* 82, 781–789. doi: 10.1002/jsfa.1095

Cui, B., Li, P., Xu, L., Peng, Z., Xiang, J., He, Z., et al. (2016). Step-up fecal microbiota transplantation (FMT) strategy. *Gut Microbes.* doi: 10.1080/19490976.2016.1151608. [Epub ahead of print].

Damen, B., Verspreet, J., Pollet, A., Broekaert, W. F., Delcour, J. A., and Courtin, C. M. (2011). Prebiotic effects and intestinal fermentation of cereal arabinoxylans and arabinoxylan oligosaccharides in rats depend strongly on their structural properties and joint presence. *Mol. Nutr. Food. Res.* 55, 1862–1874. doi: 10.1002/mnfr.201100377

Delgado, S., Cabrera-Rubio, R., Mira, A., Suárez, A., and Mayo, B. (2013). Microbiological survey of the human gastric ecosystem using culturing and pyrosequencing methods. *Microb. Ecol.* 63, 763–772. doi: 10.1007/s00248-013-0192-5 de Vos, W. M. (2013). Fame and future of faecal transplantations - developing next-generation therapies with synthetic microbiomes. *Microb. Biotechnol.* 6, 316–325. doi: 10.1111/1751-7915.12047

de Vos, W. M., and de Vos, E. A. (2012). Role of the intestinal microbiome in health and disease: from correlation to causation. *Nutr. Rev.* 1, S45–S56. doi: 10.1111/j.1753-4887.2012.00505.x

De Vuyst, L., and Leroy, F. (2011). Cross-feeding between bifidobacteria and butyrate-producing colon bacteria explains bifidobacterial competitiveness, butyrate production, and gas production. *Int. J. Food Microbiol.* 149, 73–80. doi: 10.1016/j.ijfoodmicro.2011.03.003

De Vuyst, L., Moens, F., Selak, M., Rivière, A., and Leroy, F. (2014). Summer meeting 2013: growth and physiology of bifidobacteria. *J. Appl. Microbiol.* 116, 477–491. doi: 10.1111/jam.12415

Di Gioia, D., Aloisio, I., Mazzola, G., and Biavati, B. (2014). Bifidobacteria: their impact on gut microbiota composition and their applications as probiotics in infants. *Appl. Microbiol. Biotechnol.* 98, 563–577. doi: 10.1007/s00253-013-5405-9

Dinan, T. G., Stanton, C., and Cryan, J. F. (2013). Psychobiotics: a novel class of psychotropic. *Biol. Psychiat.* 74, 720–726. doi: 10.1016/j.biopsych.2013.05.001

Dinan, T. G., Stilling, R. M., Stanton, C., and Cryan, J. F. (2015). Collective unconscious: how gut microbes shape human behavior. *J. Psychiat. Res.* 63, 1–9. doi: 10.1016/j.jpsychires.2015.02.021

Dodd, D., and Cann, I. K. (2009). Enzymatic deconstruction of xylan for biofuel production. *Glob. Change Biol. Bioenergy* 18, 2–17. doi: 10.1111/j.1757-1707.2009.01004.x

Dodd, D., Mackie, R., and Cann, I. K. (2011). Xylan degradation, a metabolic property shared by rumen and human colonic Bacteroidetes. *Mol. Microbiol.* 79, 292–304. doi: 10.1111/j.1365-2958.2010.07473.x

Duncan, S. H., and Flint, H. J. (2013). Probiotics and prebiotics and health in ageing populations. *Maturitas* 75, 44–50. doi: 10.1016/j.maturitas.2013.02.004

Duncan, S. H., Louis, P., and Flint, H. J. (2004). Lactate-utilizing bacteria, isolated from human feces, that produce butyrate as a major fermentation product. *Appl. Environ. Microbiol.* 70, 5810–5817. doi: 10.1128/AEM.70.10.5810-5817.2004

Duncan, S. H., Louis, P., Thomson, J. M., and Flint, H. J. (2009). The role of pH in determining the species composition of the human colonic microbiota. *Environ. Microbiol.* 11, 2112–2122. doi: 10.1111/j.1462-2920.2009.01931.x

Eeckhaut, V., Ducatelle, R., Sas, B., Vermeire, S., and Van Immerseel, F. (2014). Progress towards butyrate-producing pharmabiotics: *Butyricicoccus pullicaecorum* capsule and efficacy in TNBS models in comparison with therapeutics. *Gut* 63, 367. doi: 10.1136/gutjnl-2013-305293

Eeckhaut, V., Machiels, K., Perrier, C., Romero, C., Maes, S., Flahou, B., et al. (2013). *Butyricicoccus pullicaecorum* in inflammatory bowel disease. *Gut* 62, 1745–1752. doi: 10.1136/gutjnl-2012-303611

EFSA (2010). Scientific opinion on the substantiation of health claims related to live yoghurt cultures and improved lactose digestion (ID 1143, 2976), pursuant to Article 13 (1) of regulation (EC) No 1924/20061. *EFSA J.* 8, 1763. doi: 10.2903/j.efsa.2010.1763

EFSA (2011a). Scientific opinion on the substantiation of health claims related to resistant starch and reduction of post-prandial glycaemic responses (ID 681), "digestive health benefits" (ID 682) and "favours a normal colon metabolism" (ID 783) pursuant to Article 13 (1) of Regulation (EC) No 1924/2006. *EFSA J.* 9, 2024. doi: 10.2903/j.efsa.2011.2024

EFSA (2011b). Scientific opinion on the substantiation of health claims related to arabinoxylan produced from wheat endosperm and reduction of post-prandial glycaemic responses (ID 830) pursuant to Article 13 (1) of Regulation (EC) No 1924/2006. *EFSA J.* 9, 2205. doi: 10.2903/j.efsa.2011.2205

EFSA (2015). Scientific opinion on the substantiation of a health claim related to "native chicory inulin" and maintenance of normal defecation by increasing stool frequency pursuant to Article 13.5 of Regulation (EC) No 1924/2006. *EFSA J.* 13, 3951. doi: 10.2903/j.efsa.2015.3951

Egan, M., O'Connell Motherway, M., Kilcoyne, M., Kane, M., Joshi, L., Ventura, M., et al. (2014). Cross-feeding by *Bifidobacterium breve* UCC2003 during co-cultivation with *Bifidobacterium bifidum* PRL2010 in a mucin-based medium. *BMC Microbiol.* 14:282. doi: 10.1186/s12866-014-0282-7

Ehrmann, M. A., Korakli, M., and Vogel, R. F. (2003). Identification of the gene for beta-fructofuranosidase of *Bifidobacterium lactis* DSM10140$^T$ and characterization of the enzyme expressed in *Escherichia coli*. *Curr. Microbiol.* 46, 391–397. doi: 10.1007/s00284-002-3908-1

El Aidy, S., Van den Abbeele, P., Van de Wiele, T., Louis, P., and Kleerebezem, M. (2013). Intestinal colonization: how key microbial players become established in this dynamic process. *Bioessays* 35, 913–923. doi: 10.1002/bies.201300073

El Kaoutari, A., Armougom, F., Gordon, J. I., Raoult, D., and Henrissat, B. (2013). The abundance and variety of carbohydrate-active enzymes in the human gut microbiota. *Nat. Rev. Microbiol.* 11, 497–504. doi: 10.1038/nrmicro3050

Eloe-Fadrosh, E. A., Brady, A., Crabtree, J., Drabek, E. F., Ma, B., Mahurkar, A., et al. (2015). Functional dynamics of the gut microbiome in elderly people during probiotic consumption. *MBio* 6:e00231. doi: 10.1128/mBio.00231-15

European Commission (2008). Commission directive 2008/100/EC. *Official Journal European Union, p. L 285/9.*

Euzéby, J. P. (1997). List of bacterial names with standing in nomenclature: a folder available on the internet. *Int. J. Syst. Bacteriol.* 47, 590–592. doi: 10.1099/00207713-47-2-590

Euzéby, J. P. (2016). *"Bifidobacterium". List of Prokaryotic Names with Standing in Nomenclature.* Available online at: http://www.bacterio.net/bifidobacterium.html (Accessed May 19, 2016).

Everard, A., and Cani, P. D. (2013). Diabetes, obesity and gut microbiota. *Best Pract. Res. Clin. Gastroenterol.* 27, 73–78. doi: 10.1016/j.bpg.2013.03.007

Everard, A., Lazarevic, V., Derrien, M., Girard, M., Muccioli, G. G., Neyrinck, A. M., et al. (2011). Responses of gut microbiota and glucose and lipid metabolism to prebiotics in genetic obese and diet-induced leptin-resistant mice. *Diabetes* 60, 2775–2786. doi: 10.2337/db11-0227

Faith, J. J., Guruge, J. L., Charbonneau, M., Subramanian, S., Seedorf, H., Goodman, A. L., et al. (2013). The long-term stability of the human gut microbiota. *Science* 341:1237439. doi: 10.1126/science.1237439

Falony, G., Calmeyn, T., Leroy, F., and De Vuyst, L. (2009a). Coculture fermentations of *Bifidobacterium* species and *Bacteroides thetaiotaomicron* reveal a mechanistic insight into the prebiotic effect of inulin-type fructans. *Appl. Environ. Microbiol.* 75, 2312–2319. doi: 10.1128/AEM.02649-08

Falony, G., Lazidou, K., Verschaeren, A., Weckx, S., Maes, D., and De Vuyst, L. (2009b). *In vitro* kinetic analysis of fermentation of prebiotic inulin-type fructans by *Bifidobacterium* species reveals four different phenotypes. *Appl. Environ. Microbiol.* 75, 454–461. doi: 10.1128/AEM.01488-08

Falony, G., Verschaeren, A., De Bruycker, F., De Preter, V., Verbeke, K., Leroy, F., et al. (2009c). *In vitro* kinetics of prebiotic inulin-type fructan fermentation by butyrate-producing colon bacteria: implementation of online gas chromatography for quantitative analysis of carbon dioxide and hydrogen gas production. *Appl. Environ. Microbiol.* 75, 5884–5892. doi: 10.1128/AEM.00876-09

Falony, G., Vlachou, A., Verbrugghe, K., and De Vuyst, L. (2006). Cross-feeding between *Bifidobacterium longum* BB536 and acetate-converting, butyrate-producing colon bacteria during growth on oligofructose. *Appl. Environ. Microbiol.* 72, 7835–7841. doi: 10.1128/AEM.01296-06

Figueroa-González, I., Quijano, G., Ramírez, G., and Cruz-Guerrero, A. (2011). Probiotics and prebiotics - perspectives and challenges. *J. Sci. Food Agric.* 91, 1341–1348. doi: 10.1002/jsfa.4367

Frei, R., Akdis, M., and O'Mahony, L. (2015). Prebiotics, probiotics, synbiotics, and the immune system: experimental data and clinical evidence. *Curr. Opin. Gastroenterol.* 31, 153–158. doi: 10.1097/MOG.0000000000000151

Gagnon, M., Savard, P., Rivière, A., LaPointe, G., and Roy, D. (2015). Bioaccessible antioxidants in milk fermented by *Bifidobacterium longum* subsp. *longum* strains. *Biomed. Res. Int.* 2015:169381. doi: 10.1155/2015/169381

Geirnaert, A., Steyaert, A., Eeckhaut, V., Debruyne, B., Arends, J. B., Van Immerseel, F., et al. (2014). *Butyricicoccus pullicaecorum*, a butyrate producer with probiotic potential, is intrinsically tolerant to stomach and small intestine conditions. *Anaerobe* 30, 70–74. doi: 10.1016/j.anaerobe.2014.08.010

Gevers, D., Kugathasan, S., Denson, L. A., Vázquez-Baeza, Y., Van Treuren, W., Ren, B., et al. (2014). The treatment-naive microbiome in new-onset Crohn's disease. *Cell Host Microbe* 15, 382–392. doi: 10.1016/j.chom.2014.02.005

Gibson, G. R., Probert, H. M., Loo, J. V., Rastall, R. A., and Roberfroid, M. B. (2004). Dietary modulation of the human colonic microbiota: updating the concept of prebiotics. *Nutr. Res. Rev.* 17, 259–275. doi: 10.1079/NRR200479

Gibson, G. R., Scott, K. P., Rastall, R. A., Tuohy, K. M., Hotchkiss, A., Dubert-Ferrandon, A., et al. (2010). Dietary prebiotics: current status and new definition. *Food Sci. Technol. Bull.* 7, 1–19. doi: 10.1616/1476-2137.15880

Glanville, J., King, S., Guarner, F., Hill, C., and Sanders, M. E. (2015). A review of the systematic review process and its applicability for use in evaluating evidence for health claims on probiotic foods in the European Union. *Nutr. J.* 14, 16. doi: 10.1186/s12937-015-0004-5

Gorissen, L., De Vuyst, L., Raes, K., De Smet, S., and Leroy, F. (2012). Conjugated linoleic and linolenic acid production kinetics by bifidobacteria differ among strains. *Int. J. Food Microbiol.* 155, 234–240. doi: 10.1016/j.ijfoodmicro.2012.02.012

Gorissen, L., Raes, K., Weckx, S., Dannenberger, D., Leroy, F., De Vuyst, L., et al. (2010). Production of conjugated linoleic acid and conjugated linolenic acid isomers by *Bifidobacterium* species. *Appl. Microbiol. Biotechnol.* 87, 2257–2266. doi: 10.1007/s00253-010-2713-1

Gosálbez, L., and Ramón, D. (2015). Probiotics in transition: novel strategies. *Trends Biotechnol.* 33, 195–196. doi: 10.1016/j.tibtech.2015.01.006

Gråsten, S., Liukkonen, K. H., Chrevatidis, A., El-Nezami, H., Poutanen, K., and Mykkänen, H. (2003). Effects of wheat pentosan and inulin on the metabolic activity of fecal microbiota and on bowel function in healthy humans. *Nutr. Res.* 23, 1503–1514. doi: 10.1016/S0271-5317(03)00164-7

Grimm, V., Westermann, C., and Riedel, C. U. (2014). Bifidobacteria-host interactions - an update on colonisation factors. *Biomed. Res. Int.* 2014:960826. doi: 10.1155/2014/960826

Grootaert, C., Delcour, J. A., Courtin, C. M., Broekaert, W. F., Verstraete, W., and Van de Wiele, T. (2007). Microbial metabolism and prebiotic potency of arabinoxylan oligosaccharides in the human intestine. *Trends Food Sci. Technol.* 18, 64–71. doi: 10.1016/j.tifs.2006.08.004

Grootaert, C., Van den Abbeele, P., Marzorati, M., Broekaert, W. F., Courtin, C. M., Delcour, J. A., et al. (2009). Comparison of prebiotic effects of arabinoxylan oligosaccharides and inulin in a simulator of the human intestinal microbial ecosystem. *FEMS Microbiol. Ecol.* 69, 231–242. doi: 10.1111/j.1574-6941.2009.00712.x

Gruppen, H., Kormelink, F. J. M., and Voragen, A. G. J. (1993). Enzymic degradation of water-unextractable cell wall material and arabinoxylans from wheat flour. *J. Cereal Sci.* 18, 129–143. doi: 10.1006/jcrs.1993.1041

Hamer, H. M., Jonkers, D., Venema, K., Vanhoutvin, S., Troost, F. J., and Brummer, R. J. (2008). Review article: the role of butyrate on colonic function. *Aliment. Pharmacol. Ther.* 27, 104–119. doi: 10.1111/j.1365-2036.2007.03562.x

Havenaar, R. (2011). Intestinal health functions of colonic microbial metabolites: a review. *Benef. Microbes* 2, 103–114. doi: 10.3920/BM2011.0003

Hill, C., Guarner, F., Reid, G., Gibson, G. R., Merenstein, D. J., Pot, B., et al. (2014). Expert consensus document. The International Scientific Association for Probiotics and Prebiotics consensus statement on the scope and appropriate use of the term probiotic. *Nat. Rev. Gastroenterol. Hepatol.* 11, 506–514. doi: 10.1038/nrgastro.2014.66

Hood, L. (2012). Tackling the microbiome. *Science* 336, 1209. doi: 10.1126/science.1225475

Hopkins, M. J., Englyst, H. N., Macfarlane, S., Furrie, E., Macfarlane, G. T., and McBain, A. J. (2003). Degradation of cross-linked and non-cross-linked arabinoxylans by the intestinal microbiota in children. *Appl. Environ. Microbiol.* 69, 6354–6360. doi: 10.1128/AEM.69.11.6354-6360.2003

Hughes, S. A., Shewry, P. R., Li, L., Gibson, G. R., Sanz, M. L., and Rastall, R. A. (2007). *In vitro* fermentation by human fecal microflora of wheat arabinoxylans. *J. Agric. Food Chem.* 55, 4589–4595. doi: 10.1021/jf070293g

Hutkins, R. W., Krumbeck, J. A., Bindels, L. B., Cani, P. D., Fahey, G., Goh, Y. J., et al. (2016). Prebiotics: why definitions matter. *Curr. Opin. Biotechnol.* 37, 1–7. doi: 10.1016/j.copbio.2015.09.001

Ishikawa, E., Matsuki, T., Kubota, H., Makino, H., Sakai, T., Oishi, K., et al. (2013). Ethnic diversity of gut microbiota: species characterization of *Bacteroides fragilis* group and genus *Bifidobacterium* in healthy Belgian adults, and comparison with data from Japanese subjects. *J. Biosci. Bioeng.* 116, 265–270. doi: 10.1016/j.jbiosc.2013.02.010

Izydorczyk, M. S., and Biliaderis, C. G. (1995). Cereal arabinoxylans: advances in structure and physicochemical properties. *Carbohydr. Polym.* 28, 33–48. doi: 10.1016/0144-8617(95)00077-1

Izydorczyk, M. S., and Biliaderis, C. G. (2006). "Arabinoxylans: technology and nutritionally functional plant polysaccharides," in *Functional Food Carbohydrates*, eds C. G. Biliaderis and M. S. Izydorczyk (Boca Raton, FL: CRC Press), 249–290.

Jedrzejczak-Krzepkowska, M., Tkaczuk, K. L., and Bielecki, S. (2011). Biosynthesis, purification and characterization of $\beta$-fructofuranosidase from *Bifidobacterium longum* KN29.1. *Proc. Biochem.* 46, 1963–1972. doi: 10.1016/j.procbio.2011.07.005

Kapel, N., Thomas, M., Corcos, O., Mayeur, C., Barbot-Trystram, L., Bouhnik, Y., et al. (2014). Practical implementation of faecal transplantation. *Clin. Microbiol. Infec.* 20, 1098–1105. doi: 10.1111/1469-0691.12796

Khodayar-Pardo, P., Mira-Pascual, L., Collado, M. C., and Martínez-Costa, C. (2014). Impact of lactation stage, gestational age and mode of delivery on breast milk microbiota. *J. Perinatol.* 34, 599–605. doi: 10.1038/jp.2014.47

Klijn, A., Mercenier, A., and Arigoni, F. (2005). Lessons from the genomes of bifidobacteria. *FEMS Microbiol. Rev.* 29, 491–509. doi: 10.1016/j.fmrre.2005.04.010

Kumar, H., Salminen, S., Verhagen, H., Rowland, I., Heimbach, J., Bañares, S., et al. (2015). Novel probiotics and prebiotics: road to the market. *Curr. Opin. Biotechnol.* 32, 99–103. doi: 10.1016/j.copbio.2014.11.021

Lagaert, S., Pollet, A., Courtin, C. M., and Volckaert, G. (2014). β-Xylosidases and α-L-arabinofuranosidases: accessory enzymes for arabinoxylan degradation. *Biotechnol. Adv.* 32, 316–332. doi: 10.1016/j.biotechadv.2013.11.005

Lagaert, S., Pollet, A., Delcour, J. A., Lavigne, R., Courtin, C. M., and Volckaert, G. (2010). Substrate specificity of three recombinant α-L-arabinofuranosidases from *Bifidobacterium adolescentis* and their divergent action on arabinoxylan and arabinoxylan oligosaccharides. *Biochem. Biophys. Res. Commun.* 26, 644–650. doi: 10.1016/j.bbrc.2010.10.075

Lagaert, S., Pollet, A., Delcour, J. A., Lavigne, R., Courtin, C. M., and Volckaert, G. (2011). Characterization of two β-xylosidases from *Bifidobacterium adolescentis* and their contribution to the hydrolysis of prebiotic xylooligosaccharides. *Appl. Microbiol. Biotechnol.* 92, 1179–1185. doi: 10.1007/s00253-011-3396-y

Laureys, D., Cnockaert, M., De Vuyst, L., and Vandamme, P. (2016). *Bifidobacterium aquikefiri* sp. nov., isolated from water kefir. *Int. J. Syst. Evolut. Microbiol.* 66, 1281–1286. doi: 10.1099/ijsem.0.000877

Laureys, D., and De Vuyst, L. (2014). Microbial species diversity, community dynamics, and metabolite kinetics of water kefir fermentation. *Appl. Environ. Microbiol.* 80, 2564–2572. doi: 10.1128/AEM.03978-13

Leahy, S. C., Higgins, D. G., Fitzgerald, G. F., and van Sinderen, D. (2005). Getting better with bifidobacteria. *J. Appl. Microbiol.* 98, 1303–1315. doi: 10.1111/j.1365-2672.2005.02600.x

Le Chatelier, E., Nielsen, T., Qin, J., Prifti, E., Hildebrand, F., Falony, G., et al. (2013). Richness of human gut microbiome correlates with metabolic markers. *Nature* 500, 541–546. doi: 10.1038/nature12506

Li, S., Zhu, A., Benes, V., Costea, P. I., Hercog, R., Hildebrand, F., et al. (2016). Durable coexistence of donor and recipient strains after fecal microbiota transplantation. *Science* 352, 586–589. doi: 10.1126/science.aad8852

Lombard, V., Ramulu, H. G., Drula, E., Coutinho, P. M., and Henrissat, B. (2014). The carbohydrate-active enzymes database (CAZy) in 2013. *Nucleic Acids Res.* 42, D490–D495. doi: 10.1093/nar/gkt1178

Louis, P., and Flint, H. J. (2009). Diversity, metabolism and microbial ecology of butyrate-producing bacteria from the human large intestine. *FEMS Microbiol. Lett.* 294, 1–8. doi: 10.1111/j.1574-6968.2009.01514.x

Louis, P., Hold, G. L., and Flint, H. J. (2014). The gut microbiota, bacterial metabolites and colorectal cancer. *Nat. Rev. Microbiol.* 12, 661–672. doi: 10.1038/nrmicro3344

Macfarlane, G. T., and Macfarlane, S. (2012). Bacteria, colonic fermentation, and gastrointestinal health. *J. AOAC Int.* 95, 50–60. doi: 10.5740/jaoacint.SGE_Macfarlane

Macfarlane, G. T., Steed, H., and Macfarlane, S. (2008). Bacterial metabolism and health-related effects of galacto-oligosaccharides and other prebiotics. *J. Appl. Microbiol.* 104, 305–344. doi: 10.1111/j.1365-2672.2007.03520.x

Maes, C., and Delcour, J. A. (2002). Structural characterisation of water-extractable and water-unextractable arabinoxylans in wheat bran. *J. Cereal Sci.* 35, 315–326. doi: 10.1006/jcrs.2001.0439

Mahowald, M. A., Rey, F. E., Seedorf, H., Turnbaugh, P. J., Fulton, R. S., Wollam, A., et al. (2009). Characterizing a model human gut microbiota composed of members of its two dominant bacterial phyla. *Proc. Natl. Acad. Sci. U.S.A.* 106, 5859–5864. doi: 10.1073/pnas.0901529106

Maki, K. C., Gibson, G. R., Dickmann, R. S., Kendall, C. W., Chen, C. Y., Costabile, A., et al. (2012). Digestive and physiologic effects of a wheat bran extract, arabino-xylan-oligosaccharide, in breakfast cereal. *Nutrition* 28, 1115–1121. doi: 10.1016/j.nut.2012.02.010

Marchesi, J. R., Holmes, E., Khan, F., Kochhar, S., Scanlan, P., Shanahan, F., et al. (2007). Rapid and noninvasive metabonomic characterization of inflammatory bowel disease. *J. Proteome Res.* 6, 546–551. doi: 10.1021/pr060470d

Marteau, P. (2013). Butyrate-producing bacteria as pharmabiotics for inflammatory bowel disease. *Gut* 62, 1673. doi: 10.1136/gutjnl-2012-304240

Martín, R., Miquel, S., Chain, F., Natividad, J. M., Jury, J., Lu, J., et al. (2015). *Faecalibacterium prausnitzii* prevents physiological damages in a chronic low-grade inflammation murine model. *BMC Microbiol.* 15:67. doi: 10.1186/s12866-015-0400-1

McLaughlin, H. P., Motherway, M. O., Lakshminarayanan, B., Stanton, C., Paul Ross, R., Brulc, J., et al. (2015). Carbohydrate catabolic diversity of bifidobacteria and lactobacilli of human origin. *Int. J. Food Microbiol.* 203, 109–121. doi: 10.1016/j.ijfoodmicro.2015.03.008

Mendis, M., and Simsek, S. (2013). Arabinoxylans and human health. *Food Hydrocoll.* 42, 239–243. doi: 10.1016/j.foodhyd.2013.07.022

Miquel, S., Martín, R., Bridonneau, C., Robert, V., Sokol, H., Bermúdez-Humarán, L. G., et al. (2014). Ecology and metabolism of the beneficial intestinal commensal bacterium *Faecalibacterium prausnitzii*. *Gut Microbes* 5, 146–151. doi: 10.4161/gmic.27651

Moens, F., Weckx, S., and De Vuyst, L. (2016). Bifidobacterial inulin-type fructan degradation capacity determines cross-feeding interactions between bifidobacteria and *Faecalibacterium prausnitzii*. *Int. J. Food Microbiol.* 231, 76–85 doi: 10.1016/j.ijfoodmicro.2016.05.015

Morgan, X. C., Tickle, T. L., Sokol, H., Gevers, D., Devaney, K. L., Ward, D. V., et al. (2012). Dysfunction of the intestinal microbiome in inflammatory bowel disease and treatment. *Genome Biol.* 13:R79. doi: 10.1186/gb-2012-13-9-r79

NCBI Resource Coordinators (2014). Database resources of the National Center for Biotechnology Information. *Nucleic Acids Res.* 42, D7–D17. doi: 10.1093/nar/gkt1146

Nemoto, H., Kataoka, K., Ishikawa, H., Ikata, K., Arimochi, H., Iwasaki, T., et al. (2012). Reduced diversity and imbalance of fecal microbiota in patients with ulcerative colitis. *Dig. Dis. Sci.* 57, 2955–2964. doi: 10.1007/s10620-012-2236-y

Neyrinck, A. M., Possemiers, S., Druart, C., van de Wiele, T., De Backer, F., Cani, P. D., et al. (2011). Prebiotic effects of wheat arabinoxylan related to the increase in bifidobacteria, *Roseburia* and *Bacteroides/Prevotella* in diet-induced obese mice. *PLoS ONE* 6:e20944. doi: 10.1371/journal.pone.0020944

Neyrinck, A. M., Van Hée, V. F., Piront, N., De Backer, F., Toussaint, O., Cani, P. D., et al. (2012). Wheat-derived arabinoxylan oligosaccharides with prebiotic effect increase satietogenic gut peptides and reduce metabolic endotoxemia in diet-induced obese mice. *Nutr. Diabetes* 2, e28. doi: 10.1038/nutd.2011.24

Nielsen, T. S., Lærke, H. N., Theil, P. K., Sørensen, J. F., Saarinen, M., Forssten, S., et al. (2014). Diets high in resistant starch and arabinoxylan modulate digestion processes and SCFA pool size in the large intestine and faecal microbial composition in pigs. *Br. J. Nutr.* 112, 1837–1849. doi: 10.1017/S000711451400302X

O'Hara, A. M., and Shanahan, F. (2006). The gut flora as a forgotten organ. *EMBO Rep.* 7, 688–693. doi: 10.1038/sj.embor.7400731

Omori, T., Ueno, K., Muramatsu, K., Kikuchi, M., Onodera, S., and Shiomi, N. (2010). Characterization of recombinant $\beta$-fructofuranosidase from *Bifidobacterium adolescentis* G1. *Chem. Centr. J.* 4:9. doi: 10.1186/1752-153X-4-9

Ou, J., and Sun, Z. (2014). Feruloylated oligosaccharides: structure, metabolism and function. *J. Funct. Foods* 7, 90–100. doi: 10.1016/j.jff.2013.09.028

Palframan, R. J., Gibson, G. R., and Rastall, R. A. (2003). Carbohydrate preferences of *Bifidobacterium* species isolated from the human gut. *Curr. Issues Intest. Microbiol.* 4, 71–75.

Pamer, E. G. (2014). Fecal microbiota transplantation: effectiveness, complexities, and lingering concerns. *Mucosal Immunol.* 7, 210–214. doi: 10.1038/mi.2013.117

Parche, S., Amon, J., Jankovic, I., Rezzonico, E., Beleut, M., Barutcu, H., et al. (2007). Sugar transport systems of *Bifidobacterium longum* NCC2705. *J. Mol. Microbiol. Biotechnol.* 12, 9–19. doi: 10.1159/000096455

Pastell, H., Westermann, P., Meyer, A. S., Tuomainen, P., and Tenkanen, M. (2009). *In vitro* fermentation of arabinoxylan-derived carbohydrates by bifidobacteria and mixed faecal microbiota. *J. Agric. Food Chem.* 57, 8598–8606. doi: 10.1021/jf901397b

Pokusaeva, K., Fitzgerald, G. F., and van Sinderen, D. (2011). Carbohydrate metabolism in Bifidobacteria. *Genes Nutr.* 6, 285–306. doi: 10.1007/s12263-010-0206-6

Pollet, A., Van Craeyveld, V., Van de Wiele, T., Verstraete, W., Delcour, J. A., and Courtin, C. M. (2012). *In vitro* fermentation of arabinoxylan oligosaccharides and low molecular mass arabinoxylans with different structural properties from wheat (*Triticum aestivum* L.) bran and psyllium (*Plantago ovata* Forsk) seed husk. *J. Agric. Food Chem.* 60, 946–954. doi: 10.1021/jf203820j

The Human Microbiome Project Consortium (2012). Structure, function and diversity of the healthy human microbiome. *Nature* 486, 207–214. doi: 10.1038/nature11234

Qiu, X., Zhang, M., Yang, X., Hong, N., and Yu, C. (2013). *Faecalibacterium prausnitzii* upregulates regulatory T cells and anti-inflammatory cytokines in treating TNBS-induced colitis. *J. Crohns Colitis* 7, e558–e568. doi: 10.1016/j.crohns.2013.04.002

Quévrain, E., Maubert, M. A., Michon, C., Chain, F., Marquant, R., Tailhades, J., et al. (2016). Identification of an anti-inflammatory protein from *Faecalibacterium prausnitzii*, a commensal bacterium deficient in Crohn's disease. *Gut* 65, 415–425. doi: 10.1136/gutjnl-2014-307649

Richards, L. B., Li, M., van Esch, B. C. A. M., Garssen, J., and Folkerts, G. (2016). The effects of short-chain fatty acids on the cardiovascular system. *Pharma Nutr.* 4, 68–111. doi: 10.1016/j.phanu.2016.02.001

Ridaura, V. K., Faith, J. J., Rey, F. E., Cheng, J., Duncan, A. E., Kau, A. L., et al. (2013). Gut microbiota from twins discordant for obesity modulate metabolism in mice. *Science* 341:1241214. doi: 10.1126/science.1241214

Rivière, A., Gagnon, M., Weckx, S., Roy, D., and De Vuyst, L. (2015). Mutual cross-feeding interactions between *Bifidobacterium longum* NCC2705 and *Eubacterium rectale* ATCC 33656 explain the bifidogenic and butyrogenic effects of arabinoxylan-oligosaccharides. *Appl. Environ. Microbiol.* 81, 7767–7781. doi: 10.1128/AEM.02089-15

Rivière, A., Moens, F., Selak, M., Maes, D., Weckx, S., and De Vuyst, L. (2014). The ability of bifidobacteria to degrade arabinoxylan oligosaccharide constituents and derived oligosaccharides is strain dependent. *Appl. Environ. Microbiol.* 80, 204–217. doi: 10.1128/AEM.02853-13

Roberfroid, M. B. (2005). Introducing inulin-type fructans. *Br. J. Nutr.* 93, S13–S25. doi: 10.1079/bjn20041350

Roberfroid, M. B. (2007). Inulin-type fructans: functional food ingredients. *J. Nutr.* 137, 2493–2502. doi: 10.1201/9780203504932

Rossi, M., and Amaretti, A. (2011). "Probiotic properties of bifidobacteria" in *Bifidobacteria, Genomics and Molecular Aspects*, eds B. Mayo and D. van Sinderen (Norwich: Caister Academic Press), 97–123.

Saez-Lara, M. J., Gomez-Llorente, C., Plaza-Diaz, J., and Gil, A. (2015). The role of probiotic lactic acid bacteria and bifidobacteria in the prevention and treatment of inflammatory bowel disease and other related diseases: a systematic review of randomized human clinical trials. *Biomed. Res. Int.* 2015:505878. doi: 10.1155/2015/505878

Salminen, S., and van Loveren, H. (2012). Probiotics and prebiotics: health claim substantiation. *Microb. Ecol. Health Dis.* 23:18568. doi: 10.3402/mehd.v23i0.18568

Sanchez, J. I., Marzorati, M., Grootaert, C., Baran, M., Van Craeyveld, V., Courtin, C. M., et al. (2009). Arabinoxylan-oligosaccharides (AXOS) affect the protein/carbohydrate fermentation balance and microbial population dynamics of the simulator of human intestinal microbial ecosystem. *Microb. Biotechnol.* 2, 101–113. doi: 10.1111/j.1751-7915.2008.00064.x

Schaafsma, G., and Slavin, J. L. (2015). Significance of inulin fructans in the human diet. *Compr. Rev. Food Sci. Food Saf.* 14, 37–47. doi: 10.1111/1541-4337.12119

Schell, M. A., Karmirantzou, M., Snel, B., Vilanova, D., Berger, B., Pessi, G., et al. (2002). The genome sequence of *Bifidobacterium longum* reflects its adaptation to the human gastrointestinal tract. *Proc. Natl. Acad. Sci. U.S.A.* 99, 14422–14427. doi: 10.1073/pnas.212527599

Scott, K. P., Antoine, J. M., Midtvedt, T., and van Hemert, S. (2015). Manipulating the gut microbiota to maintain health and treat disease. *Microb. Ecol. Health Dis.* 26, 25877. doi: 10.3402/mehd.v26.25877

Scott, K. P., Martin, J. C., Chassard, C., Clerget, M., Potrykus, J., Campbell, G., et al. (2011). Substrate-driven gene expression in *Roseburia inulinivorans*: importance of inducible enzymes in the utilization of inulin and starch. *Proc. Natl. Acad. Sci. U.S.A.* 1, 4672–4679. doi: 10.1073/pnas.1000091107

Scott, K. P., Martin, J. C., Duncan, S. H., and Flint, H. J. (2014). Prebiotic stimulation of human colonic butyrate-producing bacteria and bifidobacteria, *in vitro*. *FEMS Microbiol. Ecol.* 87, 30–40. doi: 10.1111/1574-6941.12186

Selak, M., Rivière, A., Moens, F., Van den Abbeele, P., Geirnaert, A., Rogelj, I., et al. (2016). Inulin-type fructan fermentation by bifidobacteria depends on the strain rather than the species and region in the human intestine. *Appl. Microbiol. Biotechnol.* 100, 4097–4107. doi: 10.1007/s00253-016-7351-9

Sharon, G., Garg, N., Debelius, J., Knight, R., Dorrestein, P. C., and Mazmanian, S. K. (2014). Specialized metabolites from the microbiome in health and disease. *Cell Metab.* 20, 719–730. doi: 10.1016/j.cmet.2014.10.016

Slavin, J. (2013). Fiber and prebiotics: mechanisms and health benefits. *Nutrients* 5, 1417–1435. doi: 10.3390/nu5041417

Snelders, J., Olaerts, H., Dornez, E., Van de Wiele, T., Aura, A. M., Vanhaecke, L., et al. (2014). Structural features and feruloylation modulate the fermentability and evolution of antioxidant properties of arabinoxylanoligosaccharides during *in vitro* fermentation by human gut derived microbiota. *J. Funct. Foods* 10, 1–12. doi: 10.1016/j.jff.2014.05.011

Sommer, F., and Bäckhed, F. (2013). The gut microbiota - masters of host development and physiology. *Nat. Rev. Microbiol.* 11, 227–238. doi: 10.1038/nrmicro2974

Swennen, K., Courtin, C. M., Lindemans, G. C. J. E., and Delcour, J. A. (2006). Large-scale production and characterisation of wheat bran arabinoxylooligosaccharides. *J. Sci. Food Agric.* 86, 1722–1731. doi: 10.1002/jsfa.2470

Tannock, G. W. (2010). "Analysis of bifidobacterial populations in bowel ecology studies" in *Bifidobacteria, Genomics and Molecular Aspects*, eds B. Mayo and D. van Sinderen (Norwich: Caister Academic Press), 1–15.

Tap, J., Mondot, S., Levenez, F., Pelletier, E., Caron, C., Furet, J. P., et al. (2009). Towards the human intestinal microbiota phylogenetic core. *Environ. Microbiol.* 11, 2574–2584. doi: 10.1111/j.1462-2920.2009.01982.x

Tojo, R., Suárez, A., Clemente, M. G., de los Reyes-Gavilán, C. G., Margolles, A., Gueimonde, M., et al. (2014). Intestinal microbiota in health and disease: role of bifidobacteria in gut homeostasis. *World J. Gastroenterol.* 20, 15163–15176. doi: 10.3748/wjg.v20.i41.15163

Tralongo, P., Tomasello, G., Sinagra, E., Damiani, P., Leone, A., Palumbo, V. D., et al. (2014). The role of butyric acid as a protective agent against inflammatory bowel diseases. *Euromediterranean Biomed. J.* 9, 24–35. doi: 10.3269/1970-5492.2014.9.4

Tuohy, K. M., and Scott, K. P. (2015). "The microbiota of the human gastrointestinal tract: a molecular view" in *Diet-Microbe Interactions in the Gut*, eds K. M. Tuohy and D. Del Rio (London: Elsevier), 1–15.

Turroni, F., Özcan, E., Milani, C., Mancabelli, L., Viappiani, A., van Sinderen, D., et al. (2015). Glycan cross-feeding activities between bifidobacteria under *in vitro* conditions. *Front. Microbiol.* 6:1030. doi: 10.3389/fmicb.2015.01030

Turroni, F., Foroni, E., Pizzetti, P., Giubellini, V., Ribbera, A., Merusi, P., et al. (2009). Exploring the diversity of the bifidobacterial population in the human intestinal tract. *Appl. Environ. Microbiol.* 75, 1534–1545. doi: 10.1128/AEM.02216-08

Turroni, F., Peano, C., Pass, D. A., Foroni, E., Severgnini, M., Claesson, M. J., et al. (2012). Diversity of bifidobacteria within the infant gut microbiota. *PLoS ONE* 7:e36957. doi: 10.1371/journal.pone.0036957

Van Craeyveld, V., Swennen, K., Dornez, E., Van de Wiele, T., Marzorati, M., Verstraete, W., et al. (2008). Structurally different wheat-derived arabinoxylooligosaccharides have different prebiotic and fermentation properties in rats. *J. Nutr.* 138, 2348–2355. doi: 10.3945/jn.108.094367

Van den Abbeele, P., Belzer, C., Goossens, M., Kleerebezem, M., De Vos, W. M., Thas, O., et al. (2013a). Butyrate-producing *Clostridium* cluster XIVa species specifically colonize mucins in an *in vitro* gut model. *ISME J.* 7, 949–961. doi: 10.1038/ismej.2012.158

Van den Abbeele, P., Gerard, P., Rabot, S., Bruneau, A., El Aidy, S., Derrien, M., et al. (2011). Arabinoxylans and inulin differentially modulate the mucosal and luminal gut microbiota and mucin-degradation in humanized rats. *Environ. Microbiol.* 13, 2667–2680. doi: 10.1111/j.1462-2920.2011.02533.x

Van den Abbeele, P., Verstraete, W., El Aidy, S., Geirnaert, A., and Van de Wiele, T. (2013b). Prebiotics, faecal transplants and microbial network units to stimulate biodiversity of the human gut microbiome. *Microb. Biotechnol.* 6, 335–340. doi: 10.1111/1751-7915.12049

van den Broek, L. A. M., Hinz, S. W. A., Beldman, G., Vincken, J. P., and Voragen, A. G. J. (2008). *Bifidobacterium* carbohydrases - their role in breakdown and synthesis of (potential) prebiotics. *Nutr. Food. Res.* 52, 146–163. doi: 10.1002/mnfr.200700121

van den Broek, L. A. M., and Voragen, A. G. J. (2008). *Bifidobacterium* glycoside hydrolases and (potential) prebiotics. *Innov. Food Sci. Emerg. Technol.* 9, 401–407. doi: 10.1016/j.ifset.2007.12.006

Van der Meulen, R., Adriany, T., Verbrugghe, K., and De Vuyst, L. (2006a). Kinetic analysis of bifidobacterial metabolism reveals a minor role for succinic acid in the regeneration of $NAD^+$ through its growth-associated production. *Appl. Environ. Microbiol.* 72, 5204–5210. doi: 10.1128/AEM.00146-06

Van der Meulen, R., Avonts, L., and De Vuyst, L. (2004). Short fractions of oligofructose are preferentially metabolized by *Bifidobacterium animalis* DN-173 010. *Appl. Environ. Microbiol.* 70, 1923–1930. doi: 10.1128/AEM.70.4.1923-1930.2004

Van der Meulen, R., Makras, L., Verbrugghe, K., Adriany, T., and De Vuyst, L. (2006b). *In vitro* kinetic analysis of oligofructose consumption by *Bacteroides* and *Bifidobacterium* spp. indicates different degradation mechanisms. *Appl. Environ. Microbiol.* 72, 1006–1012. doi: 10.1128/AEM.72.2.1006-1012.2006

Van Laere, K. M. J., Hartemink, R., Bosveld, M., Schols, H. A., and Voragen, A. G. J. (2000). Fermentation of plant cell wall derived polysaccharides and their corresponding oligosaccharides by intestinal bacteria. *J. Agric. Food Chem.* 48, 1644–1652. doi: 10.1021/jf990519i

Velasquez-Manoff, M. (2015). Gut microbiome: the peacekeepers. *Nature* 518, S3–S11. doi: 10.1038/518S3a

Ventura, M., O'Flaherty, S., Claesson, M. J., Turroni, F., Klaenhammer, T. R., van Sinderen, D., et al. (2009). Genome-scale analyses of health-promoting bacteria: probiogenomics. *Nat. Rev. Microbiol.* 7, 61–71. doi: 10.1038/nrmicro2047

Ventura, M., Turroni, F., Bottacini, F., Giubellini, V., and van Sinderen, D. (2011). "Bifidobacterial ecology and comparative genomics: perspectives and prospects," in *Bifidobacteria, Genomics and Molecular Aspects*, eds B. Mayo and D. van Sinderen (Norwich: Caister Academic Press), 31–44.

Ventura, M., Turroni, F., Lugli, G. A., and van Sinderen, D. (2014). Bifidobacteria and humans: our special friends, from ecological to genomics perspectives. *J. Sci. Food Agric.* 94, 163–168. doi: 10.1002/jsfa.6356

Verbeke, K. (2014). "Prebiotics and synbiotics: how do they affect health?" in *Clinical Insights: Probiotics, Prebiotics and Gut Health*, eds M. H. Floch and A. Kim (London: Future Medicine Ltd.), 47–61.

Verspreet, J., Damen, D., Broekaert, W. F., Verbeke, K., Delcour, J. A., and Courtin, C. M. (2016). A critical look at prebiotics within the dietary fiber concept. *Annu. Rev. Food Sci. Technol.* 7, 167–190. doi: 10.1146/annurev-food-081315-032749

Vital, M., Howe, A. C., and Tiedje, J. M. (2014). Revealing the bacterial butyrate synthesis pathways by analyzing (meta)genomic data. *MBio* 5, e00889–e00814. doi: 10.1128/mBio.00889-14

Vrieze, A., Van Nood, E., Holleman, F., Salojärvi, J., Kootte, R. S., Bartelsman, J. F., et al. (2012). Transfer of intestinal microbiota from lean donors increases insulin sensitivity in individuals with metabolic syndrome. *Gastroenterology* 143, 913–916. doi: 10.1053/j.gastro.2012.06.031

Walker, A. W., Duncan, S. H., Louis, P., and Flint, H. J. (2014). Phylogeny, culturing, and metagenomics of the human gut microbiota. *Trends Microbiol.* 22, 267–274. doi: 10.1016/j.tim.2014.03.001

Walton, G. E., Lu, C., Trogh, I., Arnaut, F., and Gibson, G. R. (2012). A randomised, double-blind, placebo controlled cross-over study to determine the gastrointestinal effects of consumption of arabinoxylan-oligosaccharides enriched bread in healthy volunteers. *Nutr. J.* 11:36. doi: 10.1186/1475-2891-11-36

Warchol, M., Perrin, S., Grill, J. P., and Schneider, F. (2002). Characterization of a purified beta-fructofuranosidase from *Bifidobacterium infantis* ATCC 15697. *Lett. Appl. Microbiol.* 35, 462–467. doi: 10.1046/j.1472-765X.2002.01224.x

Whitman, W. B., Coleman, D. C., and Wiebe, W. J. (1998). Prokaryotes: the unseen majority. *Proc. Natl. Acad. Sci. U.S.A.* 95, 6578–6583. doi: 10.1073/pnas.95.12.6578

Wikoff, W. R., Anfora, A. T., Liu, J., Schultz, P. G., Lesley, S. A., Peters, E. C., et al. (2009). Metabolomics analysis reveals large effects of gut microflora on mammalian blood metabolites. *Proc. Natl. Acad. Sci. U.S.A.* 106, 3698–3703. doi: 10.1073/pnas.0812874106

Wu, N., Yang, X., Zhang, R., Li, J., Xiao, X., Hu, Y., et al. (2013). Dysbiosis signature of fecal microbiota in colorectal cancer patients. *Microb. Ecol.* 66, 462–470. doi: 10.1007/s00248-013-0245-9

Xu, M. Q., Cao, H. L., Wang, W. Q., Wang, S., Cao, X. C., Yan, F., et al. (2015). Fecal microbiota transplantation broadening its application beyond intestinal disorders. *World J. Gastroenterol.* 21, 102–111. doi: 10.3748/wjg.v21.i1.102

Yang, J., Martínez, I., Walter, J., Keshavarzian, A., and Rose, D. J. (2013). *In vitro* characterization of the impact of selected dietary fibers on fecal microbiota composition and short chain fatty acid production. *Anaerobe* 23, 74–81. doi: 10.1016/j.anaerobe.2013.06.012

Zoetendal, E. G., Raes, J., van den Bogert, B., Arumugam, M., Booijink, C. C., Troost, F. J., et al. (2012). The human small intestinal microbiota is driven by rapid uptake and conversion of simple carbohydrates. *ISME J.* 6, 1415–1426. doi: 10.1038/ismej.2011.212

**Conflict of Interest Statement:** The authors declare that the research was conducted in the absence of any commercial or financial relationships that could be construed as a potential conflict of interest.

# 6

# Cell-Free Spent Media Obtained from *Bifidobacterium bifidum* and *Bifidobacterium crudilactis* Grown in Media Supplemented with 3′-Sialyllactose Modulate Virulence Gene Expression in *Escherichia coli* O157:H7 and *Salmonella* Typhimurium

**Edited by:**
Francesca Turroni,
University College Cork, Ireland

**Reviewed by:**
Alberto Amaretti,
University of Modena and Reggio Emilia, Italy
Frédéric Borges,
École Nationale Supérieure d'Agronomie et des Industries Alimentaires - Université de Lorraine, France

***Correspondence:**
Véronique Delcenserie
veronique.delcenserie@ulg.ac.be

*Pauline Bondue[1], Sébastien Crèvecoeur[1], François Brose[1], Georges Daube[1], Marie-Christine Seghaye[2], Mansel W. Griffiths[3], Gisèle LaPointe[3] and Véronique Delcenserie[1]**

*[1] Department of Food Science, Fundamental and Applied Research for Animal and Health, Faculty of Veterinary Medicine, University of Liège, Liège, Belgium, [2] Department of Pediatrics, University Hospital Liège-Notre-Dame des Bruyères, Belgium, [3] Canadian Research Institute for Food Safety, University of Guelph, Guelph, Canada*

Complex oligosaccharides from human milk (HMO) possess an antimicrobial activity and can promote the growth of bifidobacteria such as *Bifidobacterium bifidum* and *Bifidobacterium longum* subsp. *infantis*. In addition, fermentation of carbohydrates by bifidobacteria can result in the production of metabolites presenting an antivirulence effect on several pathogenic bacteria. Whey is rich in complex bovine milk oligosaccharides (BMO) structurally similar to HMO and *B. crudilactis*, a species of bovine origin, is able to metabolize some of those complex carbohydrates. This study focused on the ability of *B. bifidum* and *B. crudilactis* to grow in a culture medium supplemented in 3′-sialyllactose (3′SL) as the main source of carbon, a major BMO encountered in cow milk. Next, the effects of cell-free spent media (CFSM) were tested against virulence expression of *Escherichia coli* O157:H7 and *Salmonella enterica* serovar Typhimurium. Both strains were able to grow in presence of 3′SL, but *B. crudilactis* showed the best growth (7.92 ± 0.3 log cfu/ml) compared to *B. bifidum* (6.84 ± 0.9 log cfu/ml). Then, CFSM were tested for their effects on virulence gene expression by *ler* and *hilA* promoter activity of luminescent mutants of *E. coli* and *S.* Typhimurium, respectively, and on wild type strains of *E. coli* O157:H7 and *S.* Typhimurium using RT-qPCR. All CFSM resulted in significant under expression of the *ler* and *hilA* genes for the luminescent mutants and *ler* (ratios of −15.4 and −8.1 respectively) and *qseA* (ratios of −2.1 and −3.1) for the wild type strain of *E. coli* O157:H7. The 3′SL, a major BMO, combined with some bifidobacteria strains of bovine or human origin could therefore be an interesting synbiotic to maintain or restore the intestinal health of young children. These effects observed *in vitro* will be further investigated regarding the overall phenotype of pathogenic agents and the exact nature of the active molecules.

**Keywords: *Bifidobacterium bifidum*, *Bifidobacterium crudilactis*, bovine milk oligosaccharide, *Escherichia coli* enterohemorragic O157:H7, *Salmonella enterica* serovar Typhimurium, virulence expression, 3′-sialyllactose, whey**

## INTRODUCTION

Due to the influence on the microbiota of carbohydrate source present in food, breast-fed children are generally in better health than children fed with formula (Arrieta et al., 2014; Smilowitz et al., 2014; Scott et al., 2015). Human milk oligosaccharides (HMO) are complex oligosaccharides found in human milk. Their concentration can reach 15 g/l and more than 500 structures have been identified (Pacheco et al., 2015). These HMO, produced in mammary glands, cannot be metabolized either by the host or most bacteria, while bifidobacteria species have enzymatic activity able to degrade specific α- and β-bonds (Garrido et al., 2013). These bacteria such as *Bifidobacterium bifidum* subsp. *infantis* or *Bifidobacterium bifidum* are mainly found in the feces of breast-fed children. Indeed, the fecal microbiota of breast-fed child contain more than 75% bifidobacteria (Di Gioia et al., 2014). In addition to facilitating the growth of beneficial bacteria such as bifidobacteria, HMO can bind with pathogenic virus or bacteria, limiting adhesion to intestinal epithelium and therefore decreasing pathogens colonization by promoting natural excretion (Smilowitz et al., 2014). Also, other metabolites produced by lactic acid bacteria and bifidobacteria are able to inhibit virulence gene expression of *Escherichia coli* O157:H7 (Medellin-Pena and Griffiths, 2009; Zeinhom et al., 2012), *Salmonella enterica* serovar Typhimurium SA 941256 (Medellin-Pena et al., 2007; Bayoumi and Griffiths, 2012; *S.* Typhimurium) and *Campylobacter jejuni* (Mundi et al., 2013).

Five monosaccharides can be found in different HMO structures: glucose (Glc), galactose (Gal), N-acetylglucosamine (GlcNAc), fucose (Fuc), and sialic acid, also called N-acetylneuraminic (NeuAc). High quantities of lacto-N-biose type I (LNB: Galβ1-3GlcNAc) and fucosylated HMO are an important characteristic of human milk (Chichlowski et al., 2011; Garrido et al., 2013; Dotz et al., 2014; Smilowitz et al., 2014). *B. infantis* is able to fully degrade HMO intracellularly, contrary to *B. longum*, *B. breve,* and *B. bifidum* (Sela and Mills, 2010; Underwood et al., 2015). It possesses genes encoding specific transporters and four enzymes necessary for HMO degradation (α-fucosidase, α-sialidase, β-galactosidase, and β-N-hexosaminidase) (Sela et al., 2008; Sela, 2011; Smilowitz et al., 2014). HMO degradation by *B. bifidum* occurs outside the cells. Indeed, *B. bifidum* possesses a lacto-N-biosidase, which cleaves LNB from HMO. LNB is internalized using a specific transport system and is then degraded using LNB-phosphorylase (Sela, 2011; Smilowitz et al., 2014). Therefore, *B. infantis* and *B. bifidum*, despite their different gene clusters, both grow very well in the presence of HMO as sole source of carbon (Asakuma et al., 2011; Barile and Rastall, 2013).

Common oligosaccharides used in infant formula are galacto-oligosaccharides (GOS) and fructo-oligosaccharides (FOS), including inulin. GOS are composed of Glc and Gal, while FOS are composed of fructose and Glc. Their structures are very simple and linear. They are also bifidogenic, but because of their simple structure, they can also be consumed by other members of the intestinal microbiota such as *Bacteroides* spp. or *Clostridium* spp. (Chichlowski et al., 2011; Scholtens, 2014). This is probably why the fecal microbiota of formula-fed children contains only 30% bifidobacteria (Di Gioia et al., 2014) and the species that are present are different from those observed in the feces of breast-fed children. The predominant bifidobacteria in formula-fed children are those encountered in adult feces, such as *Bifidobacterium longum* subsp. *longum* and *Bifidobacterium adolescentis,* which present a less diverse enzymatic arsenal (De Vuyst et al., 2013).

Bovine milk oligosaccharides (BMO) can have similar composition and branching as HMO, so they could share some common properties. In addition, *B. infantis* or *B. bifidum* can grow in the presence of these carbohydrates (Sela, 2011; Milani et al., 2014). More than 60 BMO have been identified so far (Pacheco et al., 2015) and whey, a by-product of the dairy industry, is an important low cost source of BMO (Barile et al., 2009; Zivkovic and Barile, 2011). Furthermore, *Lactobacillus acidophilus* La-5 grown in medium supplemented with dairy ingredients such as whey, presented a protective effect in EHEC-infected mice (Zeinhom et al., 2012). However, BMO concentration in bovine milk is lower than HMO concentration in human milk (Barile et al., 2009; Tao et al., 2009; Kelly et al., 2013). Also, fucosylation occurs at very low frequency compared to sialylation, which is contrary to human milk (Tao et al., 2008). Even if the degree of polymerization in BMO is lower than in HMO, they are also protected by α- and β-bonds which are less accessible to other bacteria (Chichlowski et al., 2011). One of the most important BMO found in cow milk is 3′-sialyllactose (3′SL: NeuAcα2-3Galβ1-4Glc) and its concentration in colostrum can reach 0.85 mg/ml (Nakamura et al., 2003; Urashima et al., 2013). The majority of bifidobacteria grows only in anaerobic conditions, an environment very difficult to reproduce on an industrial scale. In addition, they have to survive

**Abbreviations:** 3′SL, 3′-sialyllactose; AE, attaching and effacing; BHI, brain heart infusion; BMO, bovine milk oligosaccharide; CFSM, cell-free spent medium; EHEC, enterohaemorragic *Escherichia coli;* FOS, fructo-oligosaccharide; Fuc, fucose; Gal, galactose; Glc, glucose; GlcNAc, N-acetylglucosamine; GOS, galacto-oligosaccharide; HMO, human milk oligosaccharide; LacNac, N-acetyllactosamine; LB, Luria Bertani; LNB, lacto-N-biose; LNnT, lacto-N-neotetraose; LNT, lacto-N-tetraose; MRS, De Man, Rogosa and Sharpe medium; NeuAc, N-acetylneuraminic acid or sialic acid; NeuGc, N-glycolylneuraminic acid; OD, optical density; T3SS, type III secretion system.

the acidity of the stomach, bile salts and pancreatic enzymatic activity of the digestive tract in order to reach the colon where they will consume oligosaccharides. *Bifidobacterium crudilactis* FR/62/B/3, a species isolated from raw cow's milk cheese, is oxygen tolerant (Daube et al., 2006; Delcenserie et al., 2013). The genome encodes enzymes degrading BMO (Delcenserie et al., 2007; Milani et al., 2014). These strains from bovine milk could be an interesting source of probiotics for formula supplementation (Delcenserie et al., 2013).

The type III secretion system (T3SS) has a major role in virulence expression of *S.* Typhimurium and *E. coli* O157:H7 by injecting effector proteins in intestinal cells and forming attaching and effacing (AE) lesions on host enterocytes (Bayoumi and Griffiths, 2012). The proteins produced by *S.* Typhimurium are controlled and activated by *hilA* and *sopD* genes and the *ssrB2* gene is a major regulator of the T3SS (Guri et al., 2016). Regarding *E. coli* O157:H7, AE lesions are controlled by a pathogenicity island named locus of enterocyte effacement (LEE) through *ler* gene, which is regulated by *qseA* gene also implicated in quorum sensing (Medellin-Pena et al., 2007). The gene *luxS*, major regulator of quorum sensing and influencing the *qseA* and *ler* genes, is also involved in expression of genes encoding flagella and biofilm formation (*fliC*) or synthesis of shiga-toxin ( *B2*) (Kaper et al., 2004; Wood et al., 2006).

The aims of this work were therefore to study the growth potential of *B. crudilactis* FR/62/B/3 compared to *B. bifidum* BBA1 in culture media supplemented with whey or 3′SL and to evaluate the effects of CFSM on virulence expression of *Escherichia coli* O157:H7 and *Salmonella* Typhimurium. Because of its bovine origin, the hypothesis is that *B. crudilactis* FR/62/B/3 can metabolize components present in whey, especially BMO, as explained previously. *B. bifidum* BBA1, a strain isolated from breastfed children feces, was chosen to provide a comparison with a strain of human origin as this strain should be able to use BMO or 3′SL as a source of carbon, due to their similarity with HMO (Zivkovic and Barile, 2011). Tanimomo et al. (2016) developed a culture medium answering to the specificities of *B. crudilactis* FR/62/B/3. This formula, more suited to bifidobacteria from bovine origin, was used in this study as an optimized medium in which different sources of carbohydrates have been tested. The effects of CFSM obtained from theses cultures on intestinal pathogens virulence were firstly investigated using rapid-testing on a luminescent reporter mutants. Next, virulence gene expression was more deeply investigated on several virulence genes using RT-qPCR on wild pathogenic strains. A special attention was given to the controls (unfermented media). Indeed, some nutrients have previously been shown to have a repressive effect on virulence gene expression of *E. coli* O157:H7 (Delcenserie et al., 2012) and therefore, it was important to exclude that potential effect from the observed results with the fermented media.

## MATERIALS AND METHODS

### Bacterial Strains and Growth Conditions

*Bifidobacterium bifidum* BBA1 was isolated from feces from a breast-fed child (CHU - Hôpital des Bruyères, Liège, Belgium) and *B. crudilactis* FR/62/B/3 from Saint-Marcellin, a raw cow milk cheese from Vercors (France). Both strains were stored at −80°C and grown on De Man, Rogosa, and Sharpe (MRS) medium (Oxoid, Hampshire, UK) supplemented with cysteine-HCl (0.5 g/l) and mupirocin (0.08 g/l) at 37°C for 48 h in an anaerobic workstation (Led Techno, Heusden-Zolder, Belgium) containing 10% $H_2$, 10% $CO_2$, and 80% $N_2$. Several successive cultures, in the same conditions as described previously, have been realized in MRS broth, prior to use. Pathogenic enterohaemorrhagic *E. coli* (EHEC) strain O157:H7 ATCC 35150 ($stx_2^+$) and *S. enterica* serovar Typhimurium strain ATCC 14028 were stored at −80°C and grown in Luria Bertani (LB) media (Sigma-Aldrich, Diegem, Belgium). Two reporter mutants, *E. coli* O157:H7 ATCC 43888 ($stx^-$, LEE:*lux*) containing plasmid LEE1-luxCDABE and resistant to ampicillin ($Amp^r$) and kanamycin ($Kan^r$) and *S.* Typhimurium SA 941 256 containing plasmid pSB377 (*hil*A::*lux*CDABE; $Amp^r$) were designed by Medellin-Pena et al. (2007) and Bayoumi and Griffiths (2010), respectively. Both strains were from the Canadian Research Institute for Food Safety Collection and were grown under aerobic conditions at 37°C in brain heart infusion (BHI) broth (Bio-Rad, Marnes-la-coquette, France) supplemented with ampicillin (50 mg/l). A medium optimized for *B. crudilactis* FR/62/B/3, called MRS2 (Tanimomo et al., 2016) was considered as the reference medium for this study (**Table** ) and was modified by removing or replacing glucose: MRS2 without any glucose (MRS2 G) (control), MRS2 with a mix of glucose and whey (MRS2-Wh) and MRS2 with 3′SL (MRS2-3′SL) as the only source of carbohydrate (**Table** ). Whey was collected at the beginning of a curdling process of a Belgian cheese factory (Liège area, Belgium). The quantity of lactose in MRS2-Wh medium was estimated to 25 g/l, based on lactose concentration of sweet whey (50 g/l of lactose; Food and Agriculture Organization/Organisation Mondiale de la Santé, 1998). However, mature bovine milk contains only traces of BMO (Kelly et al., 2013). The 3′SL, added to MRS2-3′SL, was provided by Carbosynth laboratory (Berkshire, UK). The concentration of 0.85 g/l was chosen to be close to natural concentrations found in colostrum (Nakamura

**TABLE** | **Composition of modified MRS2 media.**

| | MRS2 G | MRS2 | MRS2-Wh | MRS2-3′SL |
|---|---|---|---|---|
| Yeast extract (g/l) | 15.5 | 15.5 | 15.5 | 15.5 |
| Peptone of casein (g/l) | 15.5 | 15.5 | 15.5 | 15.5 |
| $K_2HPO_4$ (g/l) | 0.9 | 0.9 | 0.9 | 0.9 |
| $KH_2PO_4$ (g/l) | 0.9 | 0.9 | 0.9 | 0.9 |
| NaCl (g/l) | 0.009 | 0.009 | 0.009 | 0.009 |
| $MnSO_4.H_2O$ (g/l) | 0.17 | 0.17 | 0.17 | 0.17 |
| $MgSO_4.7H_2O$ (g/l) | 0.007 | 0.007 | 0.007 | 0.007 |
| $FeSO_4.7H_2O$ (g/) | 0.009 | 0.009 | 0.009 | 0.009 |
| Tween 80 (ml/l) | 0.9 | 0.9 | 0.9 | 0.9 |
| Cysteine (g/l) | 0.4 | 0.4 | 0.4 | 0.4 |
| Glucose (g/l) | – | 20 | 10 | – |
| Whey (ml/l) | – | – | 500 | – |
| 3′-sialyllactose (g/l) | – | – | – | 0.85 |

et al., 2003). *B. bifidum* BBA1 and *B. crudilactis* FR/62/B/3 were grown in three independent experiments under the same anaerobic conditions as previously at 37°C for 48 h. Five log/ml of bifidobacteria from a fresh 48 h culture of bifidobacteria were inoculated into the fresh media (1% v/v). The concentration of 5 log/ml was confirmed by plating several dilutions of bifidobacteria at day 0 post inoculation. Bacterial growth was determined by viable counts after 48 h incubation. Cell free spent media (CFSM) were obtained after two centrifugation steps at 5000 × (Eppendorf Centrifuge 5804, Hamburg, Germany) for 10 min. Supernatants were then sterilized by filtration (Minisart® 0.45 μm and 0.2 μm, Sartorius, Vilvoorde, Belgium). Next, CFSM were freeze-dried (Virtis Benchtop 3.3 EL, SP Scientific, Suffolk, United-Kingdom) and rehydrated with sterile distilled water to obtain a 10x concentration. The same treatment was applied to non-fermented culture media (controls). The pH of rehydrated CFSM was adjusted to 7 using 1 M NaOH.

## Measurement of LEE and *hilA* Promoter Activity

Both *E. coli* reporter strains were grown overnight in BHI broth supplemented with ampicillin (BHI-Amp). Each overnight culture was diluted 1:100 with fresh BHI-Amp broth supplemented (test samples) or not (control samples) with 10% concentrated CFSM obtained from fermented MRS2-3′SL. Two hundred microliters of each sample were distributed into triplicate wells of a sterile, opaque 96-well microliter plate (Corning 3610, Fisher Scientific, Ottawa, Ontario, Canada) and incubated at 30°C. Luminescence was measured every hour for 24 h, with a Victor multilabel counter (Wallac, PerkinElmer Life Sciences Canada, Woodbridge, Ontario, Canada). Luminescence was expressed in counts per second. Optical density (OD) was determined using a Genesys 20 spectrophotometer (Thermo scientific, Erembodegen, Belgium) adjusted to 600 nm.

## Contact between Concentrated CFSM and Wild Type Pathogenic Strains

*E. coli* O157:H7 ATCC 43890 was grown in LB agar and a single colony was taken from the plate and incubated overnight in LB broth at 37°C with aeration. The same procedure was applied for *S.* Typhimurium ATCC 14028 using BHI medium. The cultures were homogenized and 50 μl were diluted in 4.5 ml of LB broth for *E. coli* and BHI broth for *S.* Typhimurium. Then, 450 μl of each concentrated CFSM was added to the bacterial suspensions. Triplicate cultures were incubated at 37°C for 4 h on a shaker at 150 rpm. Bacterial growth was determined by OD measurement at 600 nm. *E. coli* O157:H7 and *S.* Typhimurium were grown in LB and BHI broth alone, respectively, as controls.

## Gene Expression Analysis by RT-qPCR

The method was adapted with some modifications from Tellez et al. (2012), Mith et al. (2014) and Guri et al. (2016). After 4 h of incubation (Delcenserie et al., 2012), cells were collected by centrifugation at 5000 × g for 10 min at room temperature (Eppendorf Centrifuge 5804, Hamburg, Germany). The pellet was suspended in 100 μl TE buffer (10 mM Tris and 1 mM EDTA) containing 1% lysozyme (Roche, Mannheim, Germany). Samples were stored at −20°C overnight. On the next day, RNA was extracted using the RNeasy® Mini Kit (Qiagen, Antwerp, Belgium). DNA contamination was eliminated from each sample using the DNase I Recombinant RNase-free Kit (Roche Diagnostics GmbH, Mannheim, Germany). To inactivate the DNase, samples were heated at 75°C for 10 min. The quantity of RNA was determined by measuring the absorbance at 260 nm using a Nanodrop 2000 Spectrophotometer (Thermo Scientific, USA). The purity and quality of RNA were verified by measuring the ratio of absorbance (260 nm/280 nm) and by using agarose gel electrophoresis (Eurogentec, Seraing, Belgium). The concentration of RNA used for reverse transcription was normalized to 100 ng/μl for *E. coli* and to 50 ng/μl for *S.* Typhimurium. Next, the RNA was subjected to reverse transcription polymerase chain reaction (RT-PCR) using a high-capacity cDNA Reverse Transcription Kit (Applied Biosystems, Ghent, Belgium). Briefly, 1 μg of RNA was reverse transcribed with 0.8 μl of desoxyribonucleoside triphosphate (dNTP; 100 mM), 1 μl of Multiscribe® Reverse Transcriptase (50 U/μl), 2 μl of 10X RT Random Primers and 2 μl of 10X RT Buffer in an adjusted total volume of 20 μl. For each sample, a no-RT control was included to confirm the absence of DNA contamination. Synthesis of cDNA was performed in a Mastercycler Gradient Thermocycler (Flexigene, Cambridge, UK) under the following conditions: 25°C for 10 min, 37°C for 120 min, 85°C for 5 min and a cooling step at 4°C. Then, cDNA was stored at −20°C.

To study the effects of bioactive molecules present in culture supernatant on gene expression of *E. coli* O157:H7 ATCC 43890, the expression of genes *ler* (involved in attaching effacing lesions), *fliC* (involved in mobility), *stxB2* (encoding subunit B of Shiga-toxin 2), *luxS* (major regulator of quorum sensing and producing AI-2), and *qseA* (involved in quorum sensing and regulator of LEE expression) was determined using qPCR. The same method was used to study effects on *S.* Typhimurium virulence gene expression of *hilA* (invasion protein regulator), *ssrB2* (Type III secretion system regulator), and *sopD* (secreted effector protein). Quantitative PCR amplification was conducted using the GoTaq® qPCR Master Mix (Promega, Leiden, Netherlands) and using the ABI 7300 Real Time PCR System (Applied Biosystems, Singapore) for *E. coli* or the Light Cycler 480 (Roche Diagnostics, Mannheim, Germany) for *S.* Typhimurium. The primers were synthesized by Eurogentec (Liège, Belgium) and have been used in previous studies (**Table** ). The RT-qPCR was performed in a total volume of 20 μl, containing 10 μl of GoTaq® Master Mix, 5.75 μl of molecular grade water, 1 μl of forward primer (10 μM), 1 μl of reverse primer (10 μM), 0.25 μl of carboxy-X-rhodamine (30 μM), and 2 μl of diluted cDNA. The qPCR conditions for *E. coli* were: initial denaturation at 95°C for 3 min; denaturation, annealing and elongation repeated 45 times: 95°C for 15 s, 58°C for 30 s and 72°C for 45 s; melting curve program: 60–95°C with a heating rate of 0.1°C/s. The qPCR conditions for *S.* Typhimurium were: denaturation at 95°C for 10 min; 40 cycles of amplification and quantification: 95°C for 30 s, 56°C for 30 s and 72°C for 30 s; melting curve program: 60–95°C with a heating rate of 0.1°C s. The annealing temperature, optimized at 56°C, was determined experimentally. Each specific amplicon was validated for the presence of a single melting temperature peak and a single band of expected size on a 2% agarose gel after electrophoresis. Cycle threshold ($C_t$) values were determined using the ABI 7300 System SDS Software for *E. coli* and the Light Cycler Software 480 version 1.5 for *S.* Typhimurium. Four housekeeping genes were tested for *E. coli*: *gnd* (6-phosphogluconate deshydrogenase), *gst* (glutathione S-transferase), 16S gene (ribosomal RNA) and *recA* (recombinase A). Three housekeeping genes were tested for *S.* Typhimurium: *gmk* (guanylate kinase), *rpoD* (sigma factor) and 16S gene (ribosomal RNA gene). Because *recA* and *gmk* were the most stable under different treatments, they were selected for normalizing transcript expression levels. The experiments were replicated three independent times. To determine relative changes in gene expression, the formula described by Pfaffl (Pfaffl, 2001) was used: ratio = $(E_{target})^{\Delta Ct\ target(control\text{-}sample)}/(E_{reference})^{\Delta Ct\ reference(control\text{-}sample)}$, where E is the efficiency of the qPCR, calculated according to the equation: $E = 10^{(-1/slope)}$.

**TABLE | Primers of virulence genes used for qPCR (F: forward; R: reverse).**

| Primer | Sequence of PCR primers (5′-3′)[a] | References |
|---|---|---|
| ***E. coli* Housekeeping and Virulence Genes** | | |
| *gnd* | F: 5′-GGTAATACCTTCTTCCAGGACACC-3′<br>R: 5′-TAGTGCGCCCTCCTCACC-3′ | Rashid et al., 2006 |
| *gst* | F: 5′-CTTTGCCGTTAACCCTAAGGG-3′<br>R: 5′-GCTGCAATGTGCTCTAACCC-3′ | Pfaffl, 2001 |
| *recA* | F: 5′-CAATATTCCCCACTGCTGCC-3′<br>R: 5′-CACCTAGGCGACGATCCCT-3′ | Takle et al., 2007 |
| *16S* | F: 5′-GGTGAGCTGGTTGATCTGGG-3′<br>R: 5′-GCATTCGCTTTACCCTGACC-3′ | Takle et al., 2007 |
| *ler* | F: 5′-TTTCTTCTTCAGTGTCCTTCA-3′<br>R: 5′-TGCGGAGATTATTTATTATGA-3′ | Medellin-Pena et al., 2007 |
| *fliC* | F: 5′-TACCATCGCAAAAGCAACTCC-3′<br>R: 5′-GTCGGCAACGTTAGTGATACC-3′ | Medellin-Pena et al., 2007 |
| *luxS* | F: 5′-GATCATACCCGGATGGAAG-3′<br>R: 5′-AGAATGCTACGCGCAATATC-3′ | Medellin-Pena et al., 2007 |
| *stxB2* | F: 5′-AGATGTTTATGGCGGTTTTA-3′<br>R: 5′-TTAAACTGCACTTCAGCAAA-3′ | Medellin-Pena et al., 2007 |
| *qseA* | F: 5′-CGCGGATCCCGTTGGCACAGGTTTGTACA-3′<br>R: 5′-CGCGGATCCCGTTGGCACAGGTTTGTACA-3′ | Medellin-Pena et al., 2007 |
| ***S.* Typhimurium Housekeeping and Virulence Genes** | | |
| *gmk* | F: 5′-TTGGCAGGGAGGCGTTT-3′<br>R: 5′-GCGCGAAGTGCCGTAGTAAT-3′ | Rashid et al., 2006 |
| *rpoD* | F: 5′-ACATGGGTATTCAGGTAATGGAAGA-3′<br>R: 5′-CGGTGGGTATTCAGGTAATGGAAGA-3′ | Botteldoorn et al., 2006 |
| *16S* | F: 5′-AGGCCTTCGGGTTGTAAAGT-3′<br>R: 5′-GTTAGCCGGTGCTTCTTCTG-3′ | Xu et al., 2010 |
| *hilA* | F: 5′-TGTCGGAAGATAAAGAGCAT-3′<br>R: 5′-AAGGAAGTATCGCCAATGTA-3′ | Guri et al., 2016 |
| *sopD* | F: 5′-ATTAATGCCGGTAACTTTGA-3′<br>R: 5′-CTCTGAAAACGGTGAATAGC-3′ | Guri et al., 2016 |
| *ssrB2* | F: 5′-TGGTTTACACAGCATACCAA-3′<br>R: 5′-GGTCAATGTAACGCTTGTTT-3′ | Guri et al., 2016 |

## Statistical Analysis

The data are means ± standard error of three replicates. A Student's *t*-test was used to assess the statistical significance of the differences between test and control groups, where $p \leq 0.05$ was considered as significant.

# RESULTS

## Growth of *Bifidobacterium bifidum* and *Bifidobacterium crudilactis*

*B. crudilactis* FR/62/B/3 showed increase in viable counts in MRS2, MRS2-Wh and MRS2-3′SL compared to MRS2 G, but the highest counts were observed on MRS2-Wh (8.9 ± 0.6 log cfu/ml, **Table** ). The same trend was observed for *B. bifidum* BBA1 with slightly lower counts (**Table** ) compared to *B. crudilactis* FR/62/B/3. The highest counts were also observed for MRS2-Wh (8.1 ± 0.3 log cfu/ml).

## Activity of CFSM from MRS2-3′SL Fermented by Bifidobacteria on Bioluminescent Reporter Gene Expression

Luminescence expression of the plasmids LEE::*lux*CDABE and *hilA*:*lux*CDABE reached its maximum at 4 h for the *E. coli* mutant and at 13 h for the *S.* Typhimurium mutant, respectively. In the presence of supernatants from fermented MRS2-3′SL medium, bioluminescence induction decreased for both mutants (**Figures** ) showing a decrease in promoter expression of *ler* and *hilA*. These results, statistically significant for both strains, were more pronounced for *hilA* gene expression of *S.* Typhimurium (**Figure** ).

## Effect of CFSM on *E. coli* O157:H7 Virulence Gene Expression

After incubation of 4 h, the different CFSM had no negative impact on growth. The average OD of *E. coli* O157:H7 at 600 nm after 4 h exposure was around 1.286 ± 0.119. From every tested housekeeping gene, *recA* was the most stable and was chosen to normalize the results according to the efficiency of each pair of primers (virulence genes *ler*, *fliC*, *stxB2*, *luxS*, and *qseA*) monitored using qPCR (Pfaffl, 2001; Tellez et al., 2012) (**Table** ). The fermented MRS2 G, MRS2, and MRS2-Wh media did not show significant impact on gene expression

**TABLE | Counts of *B. bifidum* and *B. crudilactis* after 48 h of incubation in MRS2 G, MRS2, MRS2-Wh, and MRS2-3′SL media.**

| | Final concentrations after 48 h incubation (log cfu/ml) | |
|---|---|---|
| | ***B. bifidum*** | ***B. crudilactis*** |
| MRS2 G | 6.9 ± 0.3 | 5.5 ± 0.5 |
| MRS2 | 7.3 ± 0.8 | 7.8 ± 1.4 |
| MRS2-Wh | 8.1 ± 0.3 | 8.9 ± 0.6 |
| MRS2-3′SL | 6.8 ± 0.9 | 7.9 ± 0.3 |

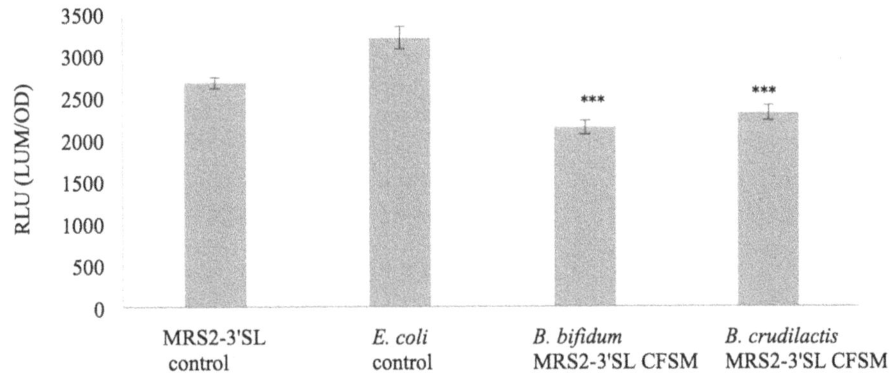


**FIGURE | Effect of CFSM from MRS2-3′SL medium fermented with *B. bifidum* and *B. crudilactis* on *E. coli* O157:H7 (stx⁻, LEE1:*lux*) ATCC 43888 expression.** The *E. coli* control is *E. coli* grown on BHI only. The MRS2-3′SL control is *E. coli* grown on BHI and CFSM from MRS2-3′SL medium unfermented. Data are the means ± the standard deviations derived from triplicate and expressed as relative light units (RLU) defined as counts per seconds, adjusted to OD600 (RLU/OD600) and where OD is fixed at 0.806 and was taken after 4 h of incubation. LUM: luminescence; OD: optical density. ***$P \leq 0.005$.

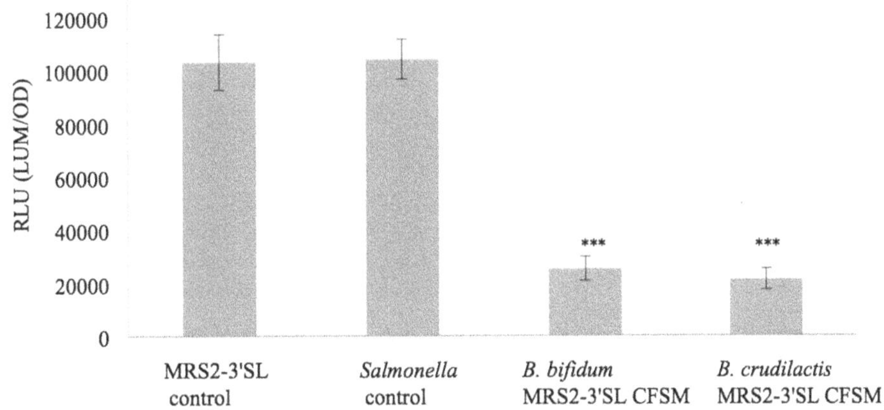


**FIGURE | Effect of CFSM from MRS2-3′SL medium fermented with *B. bifidum* and *B. crudilactis* on *S.* Typhimurium (*hilA::lux*) SA 941256 expression.** The *Salmonella* control is *Salmonella* grown on BHI only. The MRS2-3′SL control is *S.* Typhimurium grown on BHI and CFSM from MRS2-3′SL medium unfermented. Data are the means ± the standard deviations derived from triplicate and expressed as relative light units (RLU) defined as counts per seconds, adjusted to OD600 (RLU/OD600) and where OD is fixed at 0.909 and was taken after 13 h of incubation. LUM: luminescence; OD: optical density. ***$P \leq 0.005$.

(**Figures A–C**) compared to non-fermented control media, meaning that CFSM had no effects. However, significant down-regulation of virulence genes of *E. coli* O157:H7 was observed in the presence of fermented MRS2-3′SL medium (**Figure** ). The medium fermented by *B. bifidum* BBA1 and *B. crudilactis* FR/62/B/3 induced a down-regulation of the *E. coli ler* gene (ratios of −15.4; $P \leq 0.01$ and −8.1; $P \leq 0.05$, respectively). A down-regulation of the *qseA* gene was also observed (ratios of −2.1; $P \leq 0.01$ and −3.1; $P \leq 0.05$, respectively). A non-significant trend for up-regulation of the *fliC* gene (ratios of 25.8 and +20.8, respectively) was noted while a non-significant trend for down-regulation of the *stxB2* gene (ratios of −4.6 and −4.2, respectively) was observed. In the case of fermentation by *B. crudilactis* FR/62/B/3, a slight non-significant trend for down-regulation of the *luxS* gene was observed (ratio of −2.1).

## Effect of CFSM on *S.* Typhimurium Virulence Gene Expression

After incubation of 4 h, the OD measurements showed no negative impact on growth. The average OD of *S.* Typhimurium at 600 nm and after 4 h of exposure was 0.862 ± 0.078. From every tested housekeeping gene, *gmk* was the most stable and was chosen to normalize and adjust the results according to the efficiency of each pair of primers (virulence genes *hilA*, *ssrB2*, and *sopD*) monitored using qPCR (Guri et al., 2016, **Table** ). The CFSM of MRS2-3′SL medium fermented by *B. bifidum* induced a slight down-regulation of the *ssrB2* gene (−2.1; $P \leq 0.05$). The same trend was observed for the genes *hilA* and *sopD* (−2.5 and −1.9, respectively). MRS2-3′SL CFSM fermented by *B. crudilactis* FR/62/B/3 did not show any significant effect on virulence gene expression (**Figure** ). A significant increase of *sopD* expression is observed but too light to be biologically meaningful (1.1; $P \leq$

**TABLE | Effect of CFSM on expression (cycle threshold values ± standard error) of virulence genes of enterohaemorragic *E. coli* O157:H7 after 4 h incubation.**

| Gene | PCR efficiency[a] (%) | LB control[b] | Test supernatants from unfermented media[c] | | | | CFSM from fermented media by *B. bifidum*[d] | | | | CFSM from fermented media by *B. crudilactis*[e] | | | |
|---|---|---|---|---|---|---|---|---|---|---|---|---|---|---|
| | | | MRS2 G | MRS2 | MRS2-Wh | MRS2-3′SL | MRS2 G | MRS2 | MRS2-Wh | MRS2-3′SL | MRS2 G | MRS2 | MRS2-Wh | MRS2-3′SL |
| *recA* | 100 | 20 | 21.3 ± 0.6 | 18.2 ± 0.5 | 21.0 ± 0.7 | 20.3 ± 0.3 | 21.2 ± 0.4 | 21.4 ± 1.2 | 23.6 ± 1.9 | 19.9 ± 0.9 | 20.9 ± 0.5 | 23.7 ± 0.2 | 21.7 ± 0.3 | 19.9 ± 1.0 |
| *ler* | 93 | 18.9 | 25.2 ± 0.4 | 20.3 ± 1.3 | 26.4 ± 1.6 | 19.9 ± 0.8 | 25.6 ± 1.2 | 25.4 ± 0.3 | 28.4 ± 1.6 | 23.8 ± 0.4 | 24.2 ± 1.4 | 25.6 ± 1.0 | 27.8 ± 2.0 | 22.8 ± 2.0 |
| *fliC* | 97 | 20.8 | 18.2 ± 4.0 | 22.8 ± 1.4 | 24.1 ± 1.1 | 23.4 ± 2.0 | 19.1 ± 0.3 | 24.9 ± 4.2 | 26.0 ± 4.9 | 18.7 ± 1.5 | 19.8 ± 1.6 | 21.7 ± 3.7 | 25.3 ± 3.1 | 19.3 ± 0.6 |
| *luxS* | 91 | 23.6 | 26.8 ± 0.9 | 20.7 ± 1.6 | 30.1 ± 1.9 | 25.0 ± 1.3 | 27.7 ± 1.0 | 28.3 ± 0.9 | 32.6 ± 4.7 | 25.4 ± 1.5 | 27.4 ± 0.3 | 26.8 ± 3.1 | 29.8 ± 3.3 | 25.8 ± 2.2 |
| *stxB2* | 95 | 26.1 | 28.8 ± 1.0 | 25.7 ± 1.0 | 33.4 ± 2.4 | 26.5 ± 2.6 | 29.3 ± 3.0 | 30.4 ± 1.5 | 35.3 ± 1.0 | 27.6 ± 0.5 | 29.7 ± 2.5 | 30.1 ± 2.3 | 33.8 ± 3.8 | 27.8 ± 2.0 |
| *qseA* | 91 | 22.5 | 24.1 ± 0.4 | 26.3 ± 0.4 | 24.4 ± 0.8 | 21.9 ± 0.5 | 24.7 ± 1.4 | 25.3 ± 2.0 | 27.0 ± 2.1 | 22.7 ± 1.2 | 24.1 ± 0.9 | 23.5 ± 3.0 | 25.2 ± 1.0 | 23.3 ± 1.0 |

[a] *PCR efficiency:* $E = [(10^{(-1/slope)})/2] \times 100\%$.
[b] *E. coli O157:H7 grown in LB broth for 4 h.*
[c] *E. coli O157:H7 grown in LB broth supplemented with unfermented culture media CFSM for 4 h.*
[d] *E. coli O157:H7 grown in LB broth supplemented with fermented concentrated culture media CFSM from B. bifidum for 4 h.*
[e] *E. coli O157:H7 grown in LB broth supplemented with fermented concentrated culture media CFSM from B. crudilactis for 4 h.*

0.05), and the same but non-significant trend is observed with *hilA* gene (1.6). Interestingly, a down-regulation of virulence genes was observed with CFSM obtained from MRS2 without glucose. In the case of fermentation by *B. crudilactis* FR/62/B/3, the genes *hilA*, *ssrB2* and *sopD* were down regulated (−8.3, −10.9, and −6.2, respectively; $P \leq 0.05$). The same but non-significant trend was observed for *B. bifidum* BBA1 (−8.5, −8.0, and −2.6, respectively, **Figure** ).

## DISCUSSION

*B. crudilactis* FR/62/B/3 presented the best growth potential compared to *B. bifidum* BBA1, particularly with whey or 3′SL instead of glucose. This could be explained by the fact that *B. crudilactis* FR/62/B/3 was originally isolated from raw cow milk and raw milk cheese. This species possesses the genetic machinery suggesting some ability to synthetize specific enzymes for BMO degradation, as highlighted by the presence of genes coding for β-galactosidase and α- or β-glucosidase, genes also present in other bacteria such as *B. bifidum*, *B. longum* subsp. *infantis*, *B. mongoliense*, *B. biavatii*, *B. kashiwanohense*, and *B. stellenboschense* (Delcenserie et al., 2007; Milani et al., 2014, 2015; Bondue and Delcenserie, 2015). *B. bifidum* BBA1 grew as well on medium containing 3′SL as the main carbohydrate source. This species possesses genes encoding some of the enzymes cleaving BMO bonds, so their expression would lead to growth. However, the growth results of bifidobacteria on this media were similar to those observed with MRS2 G (control). This could mean that those genes may not be expressed efficiently under the conditions tested, or another function is necessary. A next step would be to measure the expression of genes involved in carbohydrate metabolism such as genes coding for β-galactosidase and α- or β-glucosidase to validate the expression of those genes in the presence of BMO. Another hypothesis could be attributed to the presence of residual glucose from MRS culture allowing bifidobacteria to grow in medium exempt of carbohydrate. Indeed, the glucose present in initial MRS medium was in excess (Tanimomo et al., 2016), but a negligible part of it (maximum 1 mg) has been transferred with the inoculum. Another source of glucose could be the presence of residual carbohydrate in the yeast extracts or peptone extracts used in MRS media. The positive effects of media supplemented with milk products on growth of probiotics has been demonstrated previously (Champagne et al., 2014). This is confirmed in the present study as the best levels of growth were reached on MRS2-Wh media for both strains studied. In addition to BMO, whey is rich in lactose (Food and Agriculture Organization/Organisation Mondiale de la Santé, 1998), a carbohydrate source that is easily consumed by bifidobacteria (Delcenserie et al., 2007).

As demonstrated previously, the CFSM obtained from some lactic acid bacteria or bifidobacteria can induce a decrease in virulence gene expression of pathogenic bacteria such as *C. jejuni* (Mundi et al., 2013). Likewise, *B. bifidum* ATCC 29521, and *L. acidophilus* La-5 CFSM were able to produce metabolites inhibiting virulence gene expression of *E. coli* O157:H7 (Medellin-Pena et al., 2007) and *S.* Typhimurium

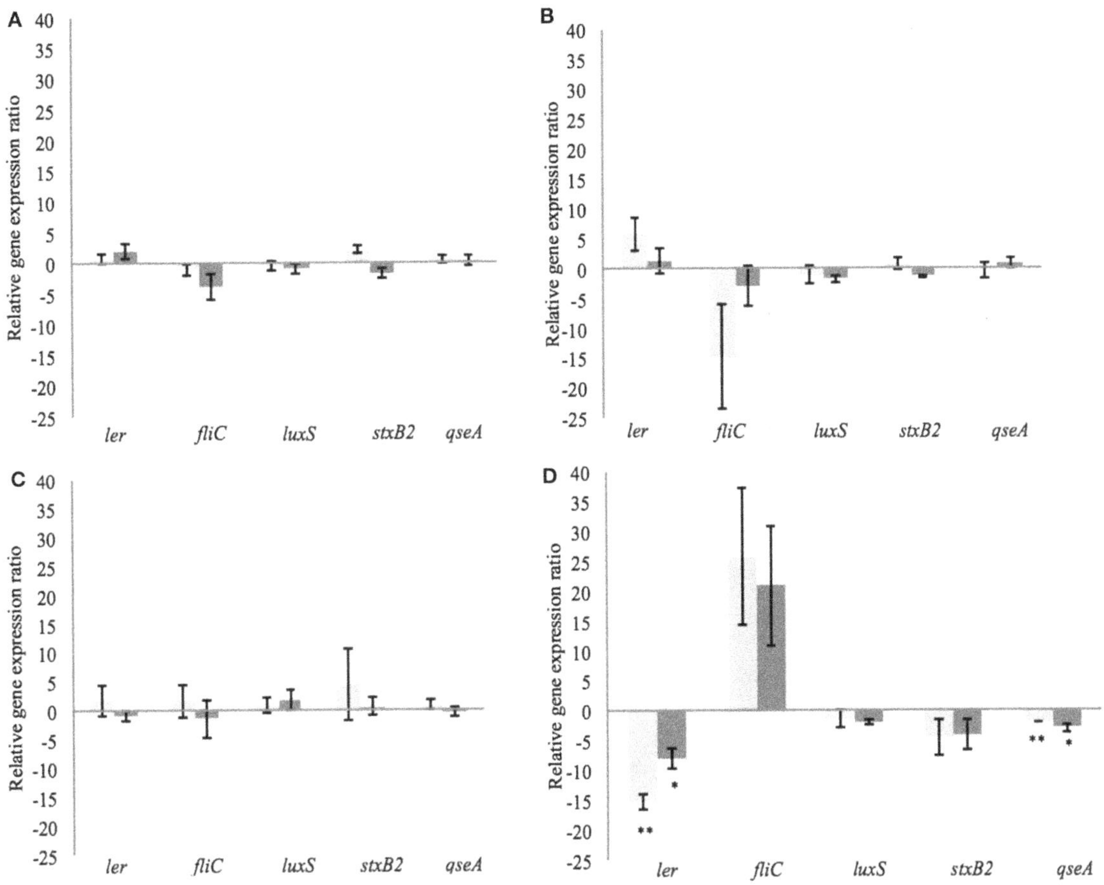


**FIGURE | Effect of CFSM from MRS2 G medium (A), MRS2 medium (B), MRS2-Wh medium (C), and MRS2-3′ SL medium (D) fermented by *B. bifidum* (light gray) and *B. crudilactis* (dark gray) on virulence gene expression of enterohaemorrhagic *E. coli* O157:H7 (EHEC) after 4 h of incubation.** Gene expression ratios of *E. coli* O157:H7 were normalized to the expression of the housekeeping gene *recA* and compared with those of the unfermented media. Negative values represent down-regulation of genes and positive values represent up-regulation of genes. $^{*}P \leq 0.05$; $^{**}P \leq 0.01$.

(Bayoumi and Griffiths, 2012). In our study, CFSM collected from MRS2-3′SL medium fermented by *B. bifidum* BBA1 and *B. crudilactis* FR/62/B/3 down-regulated most of the virulence genes tested in *E. coli* O157:H7, except the *fliC* gene, which tended to be up-regulated. This is not surprising according to the fact that *fliC* gene is not coded by the LEE operon and therefore not necessarily regulated as other virulence genes involved in T3SS and situated within the LEE operon (Falcao et al., 2004). No significant effect has been observed with CFSM from MRS2 and MRS2-Wh medium. In addition, CFSM obtained from bifidobacteria grown in media enriched in 3′SL were able to affect virulence gene expression of *E. coli* O157:H7 without having any impact on its growth, at least during the first 4 h of incubation.

In those media, higher in carbohydrates, more fermentation products such as lactate or acetate are synthetized and could have an inhibiting effect on pathogenic bacteria, as well as acidifying the media. However, all CFSM were neutralized before testing them against *E. coli* or *Salmonella*, meaning that the pH did not exert any effect on *E. coli* O157:H7 growth. Furthermore, under neutral pH, the organic acids were under dissociated form and should not present any bactericidal or bacteriostatic action, contrary to un-dissociated forms (Momose and Hirayama, 2008).

The genes involved in virulence expression such as *ler* but also *fliC* genes are regulated by *luxS*, involved in quorum sensing. However, nutrients can interfere with quorum sensing mechanisms (Henke and Bassler, 2004; Kaper et al., 2004; Nakanishi et al., 2006; Mellies et al., 2007) and induce a decrease in virulence gene expression through a decrease in *luxS* expression. Delcenserie et al. (2012) previously demonstrated the effects of glucose in down-regulating virulence gene expression of *E. coli* O157:H7. The present study brought out similar observations with lactose instead of glucose (data not shown). The *ler* gene was the most affected by the presence of those carbohydrates and the effect was dose-dependent. Media used as controls and containing glucose or lactose (MRS2 and MRS2-Wh) down-regulated this gene but no effect was observed with medium containing mainly 3′SL as a source of carbohydrate (MRS2-3′SL).

To be able to metabolize 3′SL, *B. bifidum* and *B. crudilactis* have to secrete sialidases through which NeuAc (sialic acid) can

**TABLE | Effect of CFSM on expression (cycle threshold values ± standard error) of virulence gene expression of *S.* Typhimurium after 4 h incubation.**

| Gene | PCR efficiency[a] (%) | BHI control[b] | Test supernatants from unfermented media[c] | | CFSM from fermented media by *B. bifidum*[d] | | CFSM from fermented media by *B. crudilactis*[e] | |
|---|---|---|---|---|---|---|---|---|
| | | | MRS2 G | MRS2-3′SL | MRS2 G | MRS2-3′SL | MRS2 G | MRS2-3′SL |
| *gmk* | 100 | 23 | 23.3 ± 0.3 | 24.4 ± 1.2 | 23.4 ± 0.4 | 24.4 ± 1.6 | 21.8 ± 0.2 | 25.0 ± 1.6 |
| *hilA* | 91 | 31.5 | 28.6 ± 0.2 | 31.2 ± 0.7 | 31.4 ± 2.8 | 32.5 ± 1.2 | 30.2 ± 0.4 | 31.3 ± 2.0 |
| *ssrB2* | 115 | 31.2 | 28.7 ± 1.1 | 30.8 ± 0.1 | 30.9 ± 1.6 | 31.6 ± 1.6 | 30.2 ± 0.6 | 32 ± 2.0 |
| *sopD* | 91 | 30.1 | 27.7 ± 0.5 | 30.2 ± 1.0 | 29.3 ± 1.6 | 31.2 ± 1.5 | 28.9 ± 0.6 | 30.9 ± 2.1 |

[a]*PCR efficiency:* $E = [(10^{(-1/slope)})/2] \times 100\%$.
[b]*S. Typhimurium grown in BHI broth for 4 h.*
[c]*S. Typhimurium grown in BHI broth supplemented with concentrated supernatants from culture media unfermented for 4 h.*
[d]*S. Typhimurium grown in BHI broth supplemented with concentrated supernatants from culture media fermented B. bifidum for 4 h.*
[e]*S. Typhimurium grown in BHI broth supplemented with concentrated supernatants from culture media fermented B. crudilactis for 4 h.*

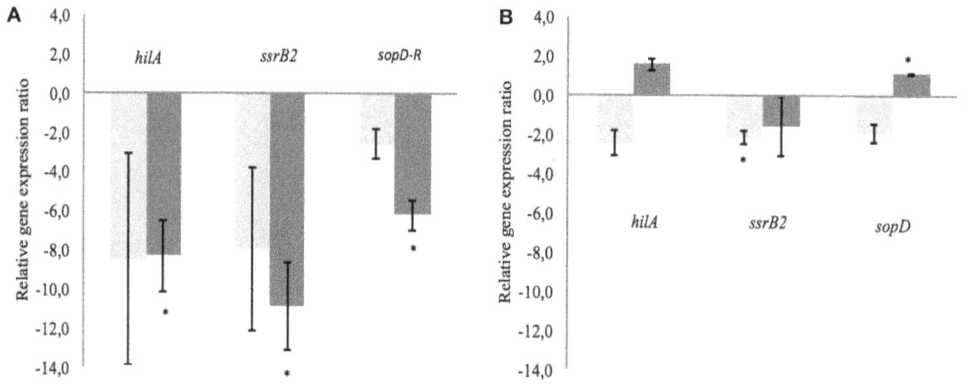


**FIGURE | Effect of CFSM from MRS2 G medium (A) and MRS2-3′ SL medium (B) fermented by *B. bifidum* (light gray) and *B. crudilactis* (dark gray) on virulence gene expression of *S.* Typhimurium after 4 h of incubation.** Gene expression ratios of *S.* Typhimurium were normalized to the expression of the housekeeping gene *gmk* and compared with those of the unfermented media. Negative values represent down-regulation of genes and positive values represent up-regulation of genes. **P* ≤ 0.05.

be produced. *B. bifidum* does not use this sialic acid, which is available for other bacteria such as *B. breve* (Milani et al., 2015). Therefore, if 3′SL is metabolized by *B. bifidum*, free sialic acid was probably present in fermented CFSM from 3′SL medium. Usually, pathogenic bacteria are able to bind this free sialic acid to their cell surface and use it to improve their resistance to the host's innate immune response, or can consume it as a nutrient (Vimr et al., 2004; Severi et al., 2007). NeuAc also exerts a major role in *Salmonella enterica* subsp. *enterica* serovar Typhi adhesion to intestinal epithelium (Sakarya et al., 2010). This means that in theory, sialic acid could have an impact on *S.* Typhimurium and *E. coli* growth, but no effect on growth as measured by OD has been observed. In addition, NeuAc had probably no impact on virulence gene expression in our study, when supplied in the medium.

The non-significant trend for up-regulation of *fliC* observed in our study should be clarified. Indeed, this trend seems higher when *E. coli* O157:H7 was exposed to CFSM from fermented 3′SL. A hypothesis could be that the presence of residual complex carbohydrates affects gene expression of *fliC*. The presence or absence of some nutrients, or stress, could play a role in virulence expression, including *fliC* (Mei et al., 2015). Several studies investigated the effects of some stress (oxidative stress, heat shock, long storage) on down-regulation of *fliC* gene while other virulence genes were upregulated (Carey et al., 2009; Mei et al., 2015; Singh and Jiang, 2015). The experimental protocol of this study submitted *E. coli* O157:H7 to some stress, which may influence virulence gene expression. Genes involved in general stress (*uspA* and *rpoS*), in starvation (*phoA* and *dpS*), in cold shock (*cspA*, *cspC*, and *cspE*) and in acid resistance (*gadW*) have been investigated and the results have shown that the different treatments did not affect expression of stress-related genes (data not shown). This leads us to suppose that virulence factor expression has not been influenced by the experimental conditions.

The results observed using RT-qPCR with CFSM from fermented MRS2-3′SL on *S.* Typhimurium virulence gene expression did not confirm the results observed with luminescent reporter strains. However, CFSM obtained after fermentation of MRS2 without any glucose fermented by *B. crudilactis* FR/62/B/3 down-regulated several virulence genes. Regarding *B. bifidum* BBA1 CFSM, a trend to down-regulation was observed as well. These down-regulations could be caused by other non-carbohydrate metabolites produced by bifidobacteria and these bioactive molecules could originate from the degradation of proteins. Most known bioactive molecules, such as subpeptin JM4-A and subpeptin JM4-B are antimicrobial peptides synthetized by *Bacillus subtilis* and active against *Salmonella*, *Staphylococcus aureus*, and *Bacillus cereus* (Sumi et al., 2015). Nisin, a bacteriocin well known, is produced by *Lactococcus lactis* and has a negative impact on *Listeria* or *Clostridium* (Ebbensgaard et al., 2015). The results of our study suggest that the CFSM activity is not due to an antimicrobial effect but due to an antivirulent effect.

In conclusion, this study provides the information that CFSM obtained from MRS2-3′SL medium fermented by *B. bifidum* BBA1 and *B. crudilactis* FR/62/B/3 down-regulated LEE1 expression of the luminescent *E. coli* reporter strain and *hilA* expression of luminescent *S.* Typhimurium reporter strain. These results agree with the decreasing virulence gene expression of *ler* and *qseA* for *E. coli*, but not for *S.* Typhimurium. The contact between *S.* Typhimurium and CFSM from fermented MRS without glucose showed down-regulation of genes *hilA*, *ssrB2*, and *sopD*. According to this *in vitro* study, the antivirulent metabolites issuing from fermentation by bifidobacteria could have a negative impact on T3SS of both pathogens, decreasing expression of genes mainly implicated in this virulence mechanism (*ler* and *qseA* genes for *E. coli* O157:H7; *hilA*, *ssrB2*, and *sopD* genes for *S.* Typhimurium). The potential upregulation of *fliC* in *E. coli* O157:H7 could increase the motility as well as biofilm formation. A phenotypic analysis of the pathogens under the experimental conditions could bring more insights about its virulence pattern. Information is lacking about the nature of the active molecules, but the activity of those CFSM might be due to small peptides or proteins with low molecular weight and resistant to pH modification and heat, or products obtained from carbohydrate metabolism. Size exclusion chromatography could contribute to separating and isolating these bioactive molecules in order to identify them. In the future, the effects of these metabolites will be investigated in a human gastrointestinal model to study the impact on microbiota to mimic *in vivo* conditions.

## AUTHOR CONTRIBUTIONS

PB did the experiments, interpreted the results and wrote the manuscript. SC and FB participated to the experiments. MS, GD, GL, and MG were involved in the design of the study and provided help for interpretation of the results. VD participated to the design of the study, interpretation of the results and writing of the manuscript.

## ACKNOWLEDGMENTS

This study was funded by the "Fond Opportunité" from Ulg. The author wishes to acknowledge Nicolas Korda, Emilie Arnould, staff from the Bruyères hospital (Liège, Belgium) and Herve society (Liège, Belgium) for their valuable help on this project.

## REFERENCES

Arrieta, M. C., Stiemsma, L. T., Amenyogbe, N., Brown, E. M., and Finlay, B. (2014). The intestinal microbiome in early life: health and disease. *Front. Immunol.* 5:427. doi: 10.3389/fimmu.2014.00427

Asakuma, S., Hatakeyama, E., Urashima, T., Yoshida, E., Katayama, T., Yamamoto, K., et al. (2011). Phisiology of the consumption of human milk oligosaccharides by infant-gut associated bifidobacteria. *J. Biol. Chem.* 40, 34583–34592. doi: 10.1074/jbc.M111.248138

Barile, D., and Rastall, R. A. (2013). Human milk and related oligosaccharides as prebiotics. *Curr. Opin. Biotechnol.* 24, 214–219. doi: 10.1016/j.copbio.2013.01.008

Barile, D., Tao, N., Lebrilla, C. B., Coisson, J. D., Arlorio, M., and German, J. B. (2009). Permeate from cheese whey ultrafiltration is a source of milk oligosaccharides. *Int. Dairy J.* 19, 524–530. doi: 10.1016/j.idairyj.2009.03.008

Bayoumi, M. A., and Griffiths, M. W. (2010). Probiotics down-regulate genes in *Salmonelle enterica* serovar Typhimurium pathogenicity islands 1 and 2. *J. Food Prot.* 73, 452–460.

Bayoumi, M. A., and Griffiths, M. W. (2012). *In vitro* inhibition of expression of virulence genes responsible for colonization and systemic spread of enteric pathogens using *Bifidobacterium bifidum* secreted molecules. *Int. J. Food Microbiol.* 156, 255–263. doi: 10.1016/j.ijfoodmicro.2012.03.034

Bondue, P., and Delcenserie, V. (2015). Genome of bifidobacteria and carbohydrate metabolism. *Korean J. Food Sci. Anim. Resour.* 35, 1–9. doi: 10.5851/kosfa.2015.35.1.1

Botteldoorn, N., Van Coillie, E., Grijspeerrdt, K., Werbrouck, H., Haesebrouck, F., Donné, E., et al. (2006). Real-time reverse transcription PCR for the quantification of the mntH expression of *Salmonella enterica* as a functiun of growth phase and phagosome-like conditions. *J. Microbiol. Methods* 66, 125–135. doi: 10.1016/j.mimet.2005.11.003

Carey, C. M., Kostrzynska, M., and Thompson, S. (2009). *Escherichia coli* O157:H7 stress and virulence gene expression on romaine lettuce using comparative real-time PCR. *J. Microbiol. Methods* 2, 235–242. doi: 10.1016/j.mimet.2009.02.010

Champagne, C. P., Raymond, Y., Pouliot, Y., Gauthier, S. F., and Lessard, M. (2014). Effect of bovine colostrum, cheese whey, and spray-dried porcine plasma on the *in vitro* growth of probiotic bacteria and *Escherichia coli*. *Can. J. Microbiol.* 60, 287–295. doi: 10.1139/cjm-2014-0130

Chichlowski, M., German, J. B., Lebrilla, C. B., and Mills, D. A. (2011). The influence of milk oligosaccharides on microbiota of infants: opportunities for formulas. *Annu. Rev. Food Sci. Technol.* 2, 331–351. doi: 10.1146/annurev-food-022510-133743

Daube, G., Delcenserie, V., and Gavini, F. (2006). *Probiotic Bifidobacterial Species*. PCT/EP2006/061247, US 20080274085, WO 2006/122850, 31-03-2006. European Patent Office, European Union.

Delcenserie, V., Gavini, F., Beerens, H., Tresse, O., Franssen, C., and Daube, G. (2007). Description of a new species, *Bifidobacterium crudilactis* sp. nov., isolated from raw milk and raw milk cheeses. *Syst. Appl. Microbiol.* 30, 381–389. doi: 10.1016/j.syapm.2007.01.004

Delcenserie, V., Lapointe, G., Charaslertrangsi, T., Rabalski, A., and Griffiths, M. W. (2012). Glucose decreases virulence gene expression of *Escherichia coli* O17:H7. *J. Food Prot.* 75. 748–752. doi: 10.4315/0362-028X.JFP-11-384

Delcenserie, V., Taminiau, B., Gavini, F., de Schaetzen, M. A., Cleenwerck, I., Theves, M., et al. (2013). Detection and characterization of *Bifidobacterium crudilactis and B. mongoliense* able to grow during the manufacturing process of French raw milk cheeses. *BMC Microbiol.* 13:239. doi: 10.1186/1471-2180-13-239

De Vuyst, L., Moens, F., Selak, M., Riviere, A., and Leroy, F. (2013). Summer Meeting 2013: growth and physiology of bifidobacteria. *J. Appl. Microbiol.* 116, 477–491. doi: 10.1111/jam.12415

Di Gioia, D., Aloisio, I., Mazzola, G., and Biavati, B. (2014). Bifidobacteria: their impact on gut microbiota composition and their applications as probiotics in infants. *Appl. Microbiol. Biotechnol.* 98, 563–577. doi: 10.1007/s00253-013-5405-9

Dotz, V., Rudloff, S., Meyer, C., Lochnit, G., and Kunz, C. (2014). Metabolic fate of neutral human milk oligosaccharides in exclusively breastfed infants. *Mol. Nutr. Food Res.* 59, 355–364. doi: 10.1002/mnfr.201400160

Ebbensgaard, A., Mordhorst, H., Overgaard, M. T., Nielsen, C. G., Aarestrup, F. M., and Hansen, E. B. (2015). Comparative evaluation of antimicrobial activity of different antimicrobial peptides against a range of pathogenic bacteria. *PLoS ONE* 10:e0144611. doi: 10.1371/journal.pone.0144611

Falcao, J. P., Falcao, D. P., and Gomes, T. A. T. (2004). Ice as a vehicle for diarrheagenic *Escherichia coli*. *Int. J. Food Microbiol.* 91, 99–103. doi: 10.1016/S0168-1605(03)00327-1

Food and Agriculture Organization/Organisation Mondiale de la Santé (FAO/OMS) (1998). *Le lait et les Produits Laitiers dans la Nutrition Humaine.* Available online at: http://www.fao.org/docrep/t4280f/t4280f0h.htm (Accessed June 16, 2016).

Garrido, D., Dallas, D. C., and Mills, D. A. (2013). Consumption of human milk glycoconjugates by infant-associated bifidobacteria: mechanisms and implications. *Microbiology* 159, 649–664. doi: 10.1099/mic.0.064113-0

Guri, A., Paligot, M., Crèvecoeur, S., Piedboeuf, B., Claes, J., Daube, G., et al. (2016). *In vitro* screening of mare's milk antimicrobial effect and antiproliferative activity. *FEMS Microbiol. Lett.* 363, 1–7. doi: 10.1093/femsle/fnv234

Henke, J. B., and Bassler, B. (2004). Quorum sensing regulates type III secretion in *Vibrio harveyi* and *Vibrio parahaemolyticus*. *J. Bacteriol.* 186, 3794–3805. doi: 10.1128/JB.186.12.3794-3805.2004

Kaper, J. B., Nataro, J. P., and Mobley, H. L. (2004). Pathogenic *Escherichia coli*. *Nat. Rev. Microbiol.* 2. 123–140. doi: 10.1038/nrmicro818

Kelly, V., Davis, S., Berry, S., Melis, J., Spelman, R., Snell, R., et al. (2013). Rapid, quantitative analysis of 3′- and 6′-sialyllactose in milk by flow-injection analysis-mass spectrometry: screening of milks for naturally elevated sialyllactose concentration. *J. Dairy Sci.* 96, 7684–7691. doi: 10.3168/jds.2013-6972

Medellin-Pena, M. J., and Griffiths, M. W. (2009). Effect of molecules secreted by *Lactobacillus acidophilus* strain La-5 on *Escherichia coli* O157:H7 colonization. *Appl. Environ. Microbiol.* 75, 1165–1172. doi: 10.1128/AEM.01651-08

Medellin-Pena, M. J., Wang, H., Johnson, R., Anand, S., and Griffiths, M. W. (2007). Probiotics affect virulence-related gene expression in *Escherichia coli* O157:H7. *Appl. Environ. Microbiol.* 73, 4259–4267. doi: 10.1128/AEM.00159-07

Mei, G. Y., Tang, J., Carey, C., Bach, S., and Kostrzynska, M. (2015). The effect of oxidative stress on gene expression of Shiga toxin-producing Escherichia coli (STEC) O157:H7 and non-O157 serotypes. *Int. J. Food Microbiol.* 215, 7–15. doi: 10.1016/j.ijfoodmicro.2015.07.029

Mellies, J. L., Barron, A. M. S., and Carmona, A. M. (2007). Enteropathogenic and enterrohemorrhagic *Escherichia coli* virulence gene regulation. *Infect. Inmmun.* 75. 4199–4210. doi: 10.1128/IAI.01927-06

Milani, C., Lugli, G. A., Duranti, S., Turroni, F., Bottacini, F., Mangifesta, M., et al. (2014). Genomic encyclopedia of type strains of the genus bifidobacterium. *Appl. Environ. Microbiol.* 80, 6290–6302. doi: 10.1128/AEM.02308-14

Milani, C., Lugli, G. A., Duranti, S., Turroni, F., Mancabelli, L., Ferrario, C., et al. (2015). Bifidobacteria exhibit social behavior through carbohydrate resource sharing in the gut. *Sci. Rep.* 5:15782. doi: 10.1038/srep15782

Mith, H., Clinquart, A., Zhiri, A., Daube, G., and Delcenserie, V. (2014). The impact of oregano (*Origanum heracleoticum*) essential oil and carvacrol on virulence gene transcription by *Escherichia coli* O157:H7. *FEMS Microbiol. Lett.* 362. 1–7. doi: 10.1093/femsle/fnu021

Momose, Y., and Hirayama, K. (2008). Effect of organic acids on inhibition of *Escherichia coli* O157:H7 colonization in gnobiotic mice associated with infant intestinal microbiota. *Anton. Leeuw. Int. J. G.* 93. 141–149. doi: 10.1007/s10482-007-9188-9

Mundi, A., Delcenserie, V., Amiri-Jami, M., Moorhead, S., and Griffiths, M. W. (2013). Cell-free preparations of *Lactobacillus acidophilus* strain La-5 and *Bifidobacterium longum* strain NCC2705 affect virulence gene expression in *Campylobacter jejuni*. *J. Food Prot.* 76, 1740–1746. doi: 10.4315/0362-028X.JFP-13-084

Nakamura, T., Kawase, H., Kimura, K., Watanabe, Y., Ohtani, M., Arai, I., et al. (2003). Concentrations of sialyloligosaccharides in bovine colostrum and milk during the prepartum and early lactation. *J. Dairy Sci.* 86, 1315–1320. doi: 10.3168/jds.S0022-0302(03)73715-1

Nakanishi, N., Abe, H., Ogura, Y., Hayashi, T., Tashiro, K., Kuhara, S., et al. (2006). ppGpp with DksA controls gene expression in the locus of enterocyte effacement (LEE) pathogenicity island of enterohaemorrhagic *Escherichia coli* through activation of two virulence regulatory genes. *Mol. Microbiol.* 61. 194–205. doi: 10.1111/j.1365-2958.2006.05217.x

Pacheco, A. R., Barile, D., Underwood, M. A., and Mills, D. A. (2015). The impact of the milk glycobiome on the neonate gut microbiota. *Annu. Rev. Anim. Biosci.* 3, 419–445. doi: 10.1146/annurev-animal-022114-111112

Pfaffl, M. W. (2001). A new mathematical model for relative quantification in real-time RT-PCR. *Nucleic Acids Res.* 29, 2002–2007. doi: 10.1093/nar/29.9.e45

Rashid, R. A., Tabata, T. A., Oatley, M. J., Besser, T. E., Tarr, P. I., and Moseley, S. L. (2006). Expression of putative virulence factors of *Escherichia coli* O157:H7 differs in bovine and human infections. *Infect. Immun.* 74, 4142–4148. doi: 10.1128/IAI.00299-06

Sakarya, S., Göktürk, C., Öztürk, T., and Ertugrul, M. B. (2010). Sialic acid is required for nonspecific adherence of *Salmonella enterica* ssp. *enterica* serovar Typhi on Caco-2cells. *FEMS Immunol. Med. Microbiol.* 58. 330–335. doi: 10.1111/j.1574-695X.2010.00650.x

Scholtens, P. A. (2014). Stool characteristics of infants receiving short-chain galacto-oligosaccharides and long-chain fructo-oligosaccharides: a review. *World, J. Gastroenterol.* 20:13446. doi: 10.3748/wjg.v20.i37.13446

Scott, K. P., Antoine, J., Midtvedt, T., and Van Hemert, S. (2015). Manipulating the gut microbiota to maintain health and treat disease. *Microb. Ecol. Health Dis.* 1, 1–10. doi: 10.3402/mehd.v26.25877

Sela, D. A. (2011). Bifidobacterial utilization of human milk oligosaccharides. *Int. J. Food Microbiol.* 149, 58–64. doi: 10.1016/j.ijfoodmicro.2011.01.025

Sela, D. A., Chapman, J., Adeuya, A., Kim, J. H., Chen, F., Whitehead, T. R., et al. (2008). The genome sequence of *Bifidobacterium longum* subsp. *infantis* reveals adaptations for milk utilization within the infant microbiome. *Proc. Natl. Acad. Sci. U.S.A.* 105, 18964–18969. doi: 10.1073/pnas.0809584105

Sela, D. A., and Mills, D. A. (2010). Nursing our microbiota: molecular linkages between bifidobacteria and milk oligosaccharides. *Trends Microbiol.* 18, 298–307. doi: 10.1016/j.tim.2010.03.008

Severi, E., Hood, D. W., and Thomas, G. H. (2007). Sialic acid utilization by bacterial pathogens. *Microbiology* 153, 2817–2822. doi: 10.1099/mic.0.2007/009480-0

Singh, R., and Jiang, X. (2015). Expression of stress and virulence genes in *Escherichia coli* O157:H7 heat shocked in fresh dairy compost. *J. Food Prot.* 78, 31–41. doi: 10.4315/0362-028X.JFP-13-529

Smilowitz, J. T., Lebrilla, C. B., Mills, D. A., German, J. B., and Freeman, S. L. (2014). Breast milk oligosaccharides: structure-function relationships in the neonate. *Annu. Rev. Nutr.* 34, 1–27. doi: 10.1146/annurev-nutr-071813-105721

Sumi, C. D., Yang, B. W., Yeo, I. C., and Hahm, Y. T. (2015). Antimicrobial peptides of the genus *Bacillus*: a new era for antibiotics. *Can. J. Microbiol.* 61, 93–103. doi: 10.1139/cjm-2014-0613

Takle, G. W., Toth, I. K., and Brurberg, M. B. (2007). Evaluation of reference genes for real-time RT-PCR expression studies in the plant pathogen *Pectobacterium atrosepticum*. *BMC Plant Biol.* 7:50. doi: 10.1186/1471-2229-7-50

Tanimomo, J., Delcenserie, V., Taminiau, B., Daube, G., Saint-Hubert, C., and Durieux, A. (2016). Growth and freeze-drying optimization of *Bifidobacterium crudilactis*. *Food Nutr. Sci.* 7, 616–626. doi: 10.4236/fns.2016.77063

Tao, N., DePeters, E. J., Freeman, S., German, J. B., Grimm, R., and Lebrilla, C. B. (2008). Bovine milk glycome. *J. Dairy Sci.* 91, 3768–3778. doi: 10.3168/jds.2008-1305

Tao, N., DePeters, E. J., German, J. B., Grimm, R., and Lebrilla, C. B. (2009). Variations in bovine milk oligosaccharides during early and middle lactation stages analyzed by high-performance liquid chromatography-chip/mass spectrometry. *J. Dairy Sci.* 92, 2991–3001. doi: 10.3168/jds.2008-1642

Tellez, A., Corredig, M., Guri, A., Zanabria, R., Griffiths, M. W., and Delcenserie, V. (2012). Bovine milk fat globule membrane affects virulence expression in *Escherichia coli* O157:H7. *J. Dairy Sci.* 95, 6313–6319. doi: 10.3168/jds.2012-5560

Underwood, M. A., German, J. B., Lebrilla, C. B., and Mills, D. A. (2015). *Bifidobacterium longum* subespecies *infantis*: champion colonizer in the infant gut. *Pediatr. Res.* 77, 229–235. doi: 10.1038/pr.2014.156

Urashima, T., Taufik, E., Fukuda, K., and Asakuma, S. (2013). Recent advances in studies on milk oligosaccharides of cows and other domestic farm animals. *Biosci. Biotechnol. Biochem.* 77, 455–466. doi: 10.1271/bbb.120810

Vimr, E. R., Kalivoda, K. A., Deszo, E. L., and Steenbergen, S. M. (2004). Diversity of microbial sialic acid metabolism. *Microbiol. Mol. Biol. Rev.* 68, 132–153. doi: 10.1128/MMBR.68.1.132-153.2004

Wood, K. T., Gonzalez Barrios, A. F., Herzberg, M., and Lee, J. (2006). Motility influences biofilm architecture in *Escherichia coli*. *Appl. Microbiol. Biotechnol.* 72, 361–367. doi: 10.1007/s00253-005-0263-8

Xu, H., Lee, H. Y., and Ahn, J. (2010). Growth and virulence properties of biofilm-forming *Salmonella enterica* serovar Typhimurium under different acidic conditions. *Appl. Environ. Microbiol.* 76, 7910–7917. doi: 10.1128/AEM.01508-10

Zeinhom, M., Tellez, A. M., Delcenserie, V., El-Kholy, A. M., El-Shinawy, and Griffiths, M. W. (2012). Yoghurt containing bioactive molecules produced by *Lactobacillus acidophilus* La-5 exerts a protective effect against enterohaemorrhagic *Escherichia coli* (EHEC) in mice. *J. Food Prot.* 10, 1796–1805. doi: 10.4315/0362-028X.JFP-11-508

Zivkovic, A. M., and Barile, D. (2011). Bovine milk as a source of functional oligosaccharides for improving human health. *Adv. Nutr.* 2, 284–289. doi: 10.3945/an.111.000455

**Conflict of Interest Statement:** The authors declare that the research was conducted in the absence of any commercial or financial relationships that could be construed as a potential conflict of interest.

# 7

# Exploring Amino Acid Auxotrophy in *Bifidobacterium bifidum* PRL2010

*Chiara Ferrario[1], Sabrina Duranti[1], Christian Milani[1], Leonardo Mancabelli[1], Gabriele A. Lugli[1], Francesca Turroni[1], Marta Mangifesta[2], Alice Viappiani[2], Maria C. Ossiprandi[3], Douwe van Sinderen[4] and Marco Ventura[1]**

*[1] Laboratory of Probiogenomics, Department of Life Sciences, University of Parma, Parma, Italy, [2] GenProbio Ltd., Parma, Italy, [3] Department of Medical-Veterinary Science, University of Parma, Parma, Italy, [4] APC Microbiome Institute and School of Microbiology, University College Cork, National University of Ireland, Cork, Ireland*

The acquisition and assimilation strategies followed by members of the infant gut microbiota to retrieve nitrogen from the gut lumen are still largely unknown. In particular, no information on these metabolic processes is available regarding bifidobacteria, which are among the first microbial colonizers of the human intestine. Here, evaluation of amino acid auxotrophy and prototrophy of *Bifidobacterium bifidum*, with particular emphasis on *B. bifidum* strain PRL2010 (LMG S-28692), revealed a putative auxotrophy for cysteine. In addition, we hypothesized that cysteine plays a role in the oxidative stress response in *B. bifidum*. The use of glutathione as an alternative reduced sulfur compound did not alleviate cysteine auxotrophy of this strain, though it was shown to stimulate expression of the genes involved in cysteine biosynthesis, reminiscent of oxidative stress response. When PRL2010 was grown on a medium containing complex substrates, such as whey proteins or casein hydrolysate, we noticed a distinct growth-promoting effect of these compounds. Transcriptional analysis involving *B. bifidum* PRL2010 cultivated on whey proteins or casein hydrolysate revealed that the biosynthetic pathways for cysteine and methionine are modulated by the presence of casein hydrolysate. Such findings support the notion that certain complex substrates may act as potential prebiotics for bifidobacteria in their ecological niche.

**Keywords: bifidobacteria, chemically defined medium, genomics, microbiota**

***Edited by:***
*David Berry,*
*University of Vienna, Austria*

***Reviewed by:***
*Thomas Clavel,*
*Technische Universität München, Germany*
*Clarissa Schwab,*
*ETH Zurich, Switzerland*

****Correspondence:***
*Marco Ventura*
*marco.ventura@unipr.it*

***Specialty section:***
*This article was submitted to Microbial Symbioses, a section of the journal Frontiers in Microbiology*

## INTRODUCTION

The gut lumen contains a very complex mixture of compounds from alimentary and endogenous origins together with living microorganisms. The intestinal microbiota is metabolically active and plays a significant role in host physiology and metabolism (Hamer et al., 2012). The ability to metabolize peptides and amino acids is shared by a large number of bacteria ranging from saccharolytic bacteria to obligate amino acid fermenters present in gut microbiota (Davila et al., 2013). Peptides are the preferred substrates over free amino acids for many colonic bacteria, probably due to kinetic advantages of peptide uptake systems. Moreover, nitrogen source, such as amino acids, are fermented to short-chain fatty acids and organic acids, representing energy fuel for the colonic mucosa (Davila et al., 2013).

Milk proteins and peptides such as lactoferrin, lactoperoxidase, and lysozyme are reported to provide a non-immune defense against microbial infections (Schanbacher et al., 1997). In addition, they are known to stimulate growth of several members of the human infant microbiota such

*Lactobacillus* and *Bifidobacterium* (Liepke et al., 2002; McCann et al., 2006). In this latter ecological context, the bacterial population is dominated by bifidobacteria, which remain a prominent component of the gut microbiota until weaning (Turroni et al., 2012a; Duranti et al., 2015; Underwood et al., 2015). Member of the genus *Bifidobacterium* are anaerobic microorganisms, typically resident in the gastro intestinal tract of mammals and insects (Lugli et al., 2014), where they are known to interact with their hosts using various genetic strategies (O'Connell Motherway et al., 2011; Fanning et al., 2012; Ventura et al., 2012; Turroni et al., 2014).

Among host-derived nutrients, milk proteins significantly influence the composition of the gut microbiota, supplying these microorganisms with nitrogen and amino acids (Liepke et al., 2002). Enhancement of (bifido)bacterial growth is frequently associated with milk proteins and the peptides that arise from the hydrolysis of these proteins (Nagpal et al., 2011; Lonnerdal, 2013).

Compared to carbon metabolism, for which a large body of scientific data is available (Pokusaeva et al., 2011; Marcobal et al., 2013), only very limited knowledge is available on the acquisition and assimilation processes that are used by members of the infant gut microbiota to retrieve nitrogen from the gut lumen (Liepke et al., 2002). For Gram positive bacteria, nitrogen metabolism has been investigated in *Lactobacillus delbrueckii* subsp. *bulgaricus* (Liu et al., 2012), *Lactobacillus rhamnosus* (Lebeer et al., 2007) and *Bacillus* sp. (Fisher, 1999; Even et al., 2006).

Recently, specific interest has been directed toward sulfur-containing amino acids and global control of cysteine and methionine metabolism in both Gram positive and negative bacteria, such as *Lactococcus lactis*, *Salmonella* sp., *Vibrio fischeri* and *Clostridium perfringens* (Fernandez et al., 2002; Andre et al., 2010; Alvarez et al., 2015; Singh et al., 2015). Cysteine biosynthesis is the key mechanism by which inorganic sulfur is reduced and incorporated into organic compounds (Kredich, 1992), where it plays an essential role in the formation of the catalytic sites of several enzymes, or protein folding and assembly via the formation of disulfide bonds (Mihara and Esaki, 2002). Sulfur-containing compounds that are used for the synthesis of cysteine and methionine are transported into the bacterial cell through different mechanisms: the first involves sulfate permease related to inorganic phosphate transporters (CysC) and then the reduction of sulfate to sulfide (Mansilla and de Mendoza, 2000) (**Figure** ). The second involves aliphatic sulfonate ATP-binding cassette (ABC) transporters (SsuBD) (van der Ploeg et al., 1998) and the subsequent conversion into sulfide by an FMNH monooxygenase (**Figure** ). The following reaction of sulfide with *O*-acetyl-L-serine (OAS) results in cysteine synthesis by the action of an *O*-acetylserine thiol-lyase (Bogicevic et al., 2012). Alternatively, cysteine can be directly transported inside the cell by symporter proteins (TcyBCP) (Burguiere et al., 2004). Methionine biosynthesis is closely linked to cysteine production by the action of serine acetyltransferase, which uses cysteine and an *O*-acetylhomoserine to generate cystathionine, where the latter compound is then converted to homocysteine and methionine (Fernandez et al., 2002) (**Figure** ).

In this study, in order to understand the role of the bifidobacterial population in the utilization of nitrogen available in the human gut, we evaluated the amino acid metabolism of the infant stool isolate *B. bifidum* PRL2010 (LMG S-28692), a bifidobacterial prototype for analysis of interactions between microbes and the intestinal mucosa (Turroni et al., 2013), by coupling physiological data on a chemically defined medium (CDM) with transcriptional analysis. Specific emphasis was placed on sulfur amino acids/metabolism of PRL2010 since these amino acids are particularly important for the bacterial cells of gut commensals in coping against gut related stresses (e.g., oxidative stress) (Even et al., 2006).

Furthermore, PRL2010 metabolism of complex substrates from milk such as casein hydrolysate and whey proteins was investigated.

## MATERIALS AND METHODS

### Bacterial Strains and DNA Extraction

Bifidobacterial strains used in this study are reported in **Table** . Strains were grown anaerobically in de Man, Rogosa, Sharpe (MRS) medium (Scharlau, Spain), which was supplemented with 0.05% L-cysteine-HCl and incubated at 37°C for 16 h. Anaerobic conditions were achieved by the use of an anaerobic cabinet (Ruskin), in which the atmosphere consisted of 10% $CO_2$, 80% $N_2$, and 10% $H_2$.

### *Bifidobacterium* CDM Development

For amino acid auxotrophy and prototrophy tests, a CDM was employed based on a previously described formulation (Petry et al., 2000; Cronin et al., 2012) for *Lactobacillus* and *Bifidobacterium*, with some modifications. Briefly, to the already reported CDM, 50 mg/l of guanine and 4.0 mg/l of thiamine was added. Several simple sugars were screened including glucose, fructose, galactose, lactose, ribose, xylose, fucose, mannose, and rhamnose. All carbohydrates were added at 2% (w/w). The medium was sterilized by filtration (0.22 μm). When the CDM was prepared without amino acids it is termed basal CDM (bCDM). All components of CDM were purchased from Sigma (USA).

### Amino Acid and Nitrogen Growth Assay

Cell growth on CDM was monitored by measuring the optical density of cultures at 600 nm (OD 600) using a plate reader (Biotek, Winooski, VT, USA). The plate reader was run in discontinuous mode, with absorbance readings performed after 24 h of incubation and preceded by 30 s of shaking at medium speed. Bacteria were cultivated in the wells of a 96-well microtiter plate, with each well containing a different amino acid, and incubated in an anaerobic cabinet.

For all growth tests, cells were recovered from an overnight MRS broth culture, centrifuged at 3000 rpm for 5 min in anaerobiosis, and washed with bCDM to remove protein and sugar residues. In each of the 96-wells of the microplate, 135 μl of medium was inoculated with 15 μl of washed cells diluted to OD 1.0 with bCDM, obtaining a final

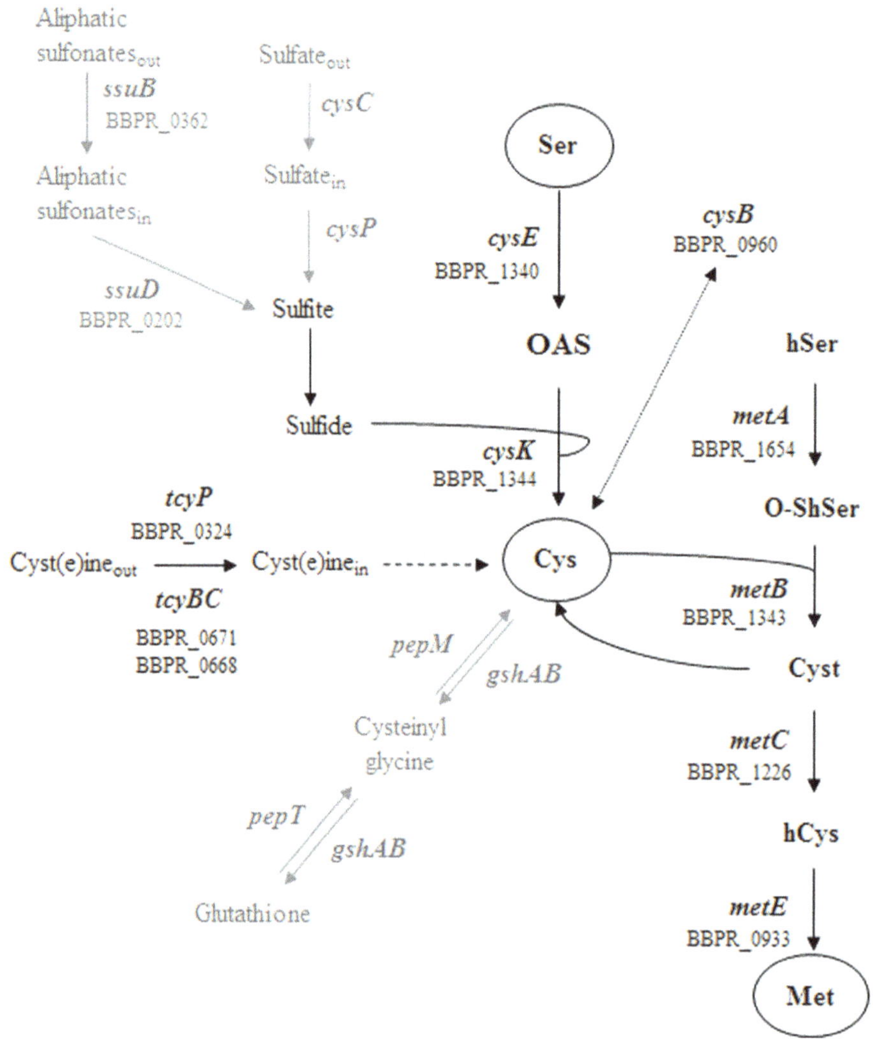


**FIGURE | Schematic representation of the metabolic pathways for sulfur amino acid in Gram positive bacteria.** The different ORFs of *Bifidobacterium bifidum* PRL2010 encoding the predicted enzymes are indicated. The metabolic steps present in Gram positive bacteria but absent in PRL2010 (sulfate assimilation and glutathione synthesis) are indicated by gray arrows. Molecule involved are reported as follow: serine Ser, *O*-acetyl-L-serine OAS, cysteine Cys, homoserine hSer, *O*-succinylhomoserine *O*-ShSer, cystathionine Cyst, homocysteine hCys and methionine Met.

inoculum OD of 0.1. Wells were covered with 30 μl of sterile mineral oil in order to maintain anaerobic conditions. Cultures were grown in biologically independent triplicates and the resulting growth data were expressed as the means from these replicates.

Once it had been established that CDM supports bifidobacterial growth, amino acid assays were performed using CDM in which individual amino acids were omitted or with bCDM supplemented with 0.2 g/l of a particular amino acid. To understand amino acid metabolism and in particular the metabolism of sulfur-containing amino acids derived from complex substrates, bCDM was supplemented with 2–0.5% (w/w) of whey protein or casein hydrolysate (Sigma). To test the influence of reduced sulfur substrate instead cysteine to PRL2010 growth, 5 mM of reduced glutathione (Sigma) was added to bCDM formulation without any other amino acid (bCDM + Glut). To evaluate the utilization of source of sulfur and nitrogen available in the gut environment, 0.5 g/l of taurine was added to bCDM.

## Identification of Genes Involved in Sulfur-containing Amino Acid Metabolism

The identification of genes involved in cysteine and methionine metabolism in PRL2010 and other *B. bifidum* strains was performed by using the BLASTP program (Gish and States, 1993). For the BLAST search, previously identified genes involved in sulfur metabolism of lactic acid bacteria were used (Liu et al., 2012). Twenty bp oligonucleotides for RT-qPCR experiments were manually designed on identified putative genes to obtain

**TABLE | Bifidobacterial strains used in this study.**

| Bacteria | Strains[a] | Genome accession numbers |
|---|---|---|
| *B. actinocoloniiforme* | DSM 22766 | JGYK00000000 |
| *B. adolescentis* | ATCC 15703 | AP009256.1 |
| *B. angulatum* | LMG 11039 | JGYL00000000 |
| *B. animalis* subsp. *animalis* | LMG 10508 | JGYM00000000 |
| *B. animalis* subsp. *lactis* | DSM 10140 | CP001606.1 |
| *B. asteroides* | LMG 10735 (PRL2011) | CP003325.1 |
| *B. biavatii* | DSM 23969 | JGYN00000000 |
| *B. bifidum* | LMG 11041 | JGYO00000000 |
| *B. bifidum* | PRL2010 | CP001840 |
| *B. bifidum* | 85B | JSDU00000000 |
| *B. bifidum* | 324B | JSDT00000000 |
| *B. bifidum* | 156B | JSDS00000000 |
| *B. bifidum* | LMG 11583 | JSDZ00000000 |
| *B. bifidum* | LMG 11582 | JSDY00000000 |
| *B. bifidum* | LMG 13200 | JSEB00000000 |
| *B. bifidum* | LMG 13195 | JSEA00000000 |
| *B. bohemicum* | DSM 22767 | JGYP00000000 |
| *B. bombi* | DSM 19703 | ATLK00000000 |
| *B. boum* | LMG 10736 | JGYQ00000000 |
| *B. breve* | LMG 13208 | JGYR00000000 |
| *B. callitrichos* | DSM 23973 | JGYS00000000 |
| *B. catenulatum* | LMG 11043 | JGYT00000000 |
| *B. choerinum* | LMG 10510 | JGYU00000000 |
| *B. coryneforme* | LMG 18911 | CP007287 |
| *B. crudilactis* | LMG 23609 | JHAL00000000 |
| *B. cuniculi* | LMG 10738 | JGYV00000000 |
| *B. dentium* | LMG 11405 (Bd1) | CP001750.1 |
| *B. gallicum* | LMG 11596 | JGYW00000000 |
| *B. gallinarum* | LMG 11586 | JGYX00000000 |
| *B. indicum* | LMG 11587 | CP006018 |
| *B. kashiwanohense* | DSM 21854 | JGYY00000000 |
| *B. longum* subsp. *infantis* | ATCC 15697 | AP010889.1 |
| *B. longum* subsp. *longum* | LMG 13197 | JGYZ00000000 |
| *B. longum* subsp. *suis* | LMG 21814 | JGZA00000000 |
| *B. magnum* | LMG 11591 | JGZB00000000 |
| *B. merycicum* | LMG 11341 | JGZC00000000 |
| *B. minimum* | LMG 11592 | JGZD00000000 |
| *B. mongoliense* | DSM 21395 | JGZE00000000 |
| *B. moukalabense* | DSM 27321 | AZMV00000000.1 |
| *B. pseudocatenulatum* | LMG 10505 | JGZF00000000 |
| *B. pseudolongum* subsp. *globosum* | LMG 11596 | JGZG00000000 |
| *B. pseudolongum* subsp. *pseudolongum* | LMG 11571 | JGZH00000000 |
| *B. psychraerophilum* | LMG 21775 | JGZI00000000 |
| *B. pullorum* | LMG 21816 | JGZJ00000000 |
| *B. reuteri* | DSM 23975 | JGZK00000000 |
| *B. ruminantium* | LMG 21811 | JGZL00000000 |
| *B. saeculare* | LMG 14934 | JGZM00000000 |
| *B. saguini* | DSM 23967 | JGZN00000000 |
| *B. scardovii* | LMG 21589 | JGZO00000000 |
| *B. stellenboschense* | DSM 23968 | JGZP00000000 |
| *B. stercoris* | DSM 24849 | JGZQ00000000 |
| *B. subtile* | LMG 11597 | JGZR00000000 |
| *B. thermacidophilum* subsp. *porcinum* | LMG 21689 | JGZS00000000 |
| *B. thermacidophilum* subsp. *thermacidophilum* | LMG 21395 | JGZT00000000 |
| *B. thermophilum* | JCM 1207 | JGZV00000000 |
| *B. tsurumiense* | JCM 13495 | JGZU00000000 |

[a]*ATCC, American Type Culture Collection, USA. LMG, Belgian Co-ordinated Collection of Microorganisms-Bacterial Collection, Belgium. DSM, German Collection of Microorganism and Cell Cultures, Germany. JCM, Japan Collection of Microorganisms, Japan.*

amplicons with a size ranging from 150 to 200 bp. Primers were checked with Primer Blast (Ye et al., 2012) and listed in **Table**

## RNA Isolation, Reverse Transcription and RT-qPCR

Total RNA was isolated from PRL2010 cultures grown in CDM, bCDM supplemented with cysteine, or bCDM supplemented with cysteine and whey protein or casein hydrolysate (2% w/w). PRL2010 cells grown in MRS was used as a control condition. Cultures were grown in biologically independent triplicates. Cells were harvested by centrifugation step at 4000 × *g* for 5′ at 4°C when cells had reached late exponential phase (OD values of 0.8–1.0, except for bCDM supplemented with glutathione where cells were harvested at OD 0.35). Cell pellets were resuspended in 500 μl of RNAprotect reagent (Qiagen, UK) and mechanically lysed by inclusion of 0.1 mm zirconium-silica beads (Biospec Products, Bartlesville, OK, USA) and by subjecting the sample to three 2 min pulses at maximum speed in a bead beater (Biospec Products, Bartlesville, OK, USA) with intervals of 3 min on ice. RNA was extracted with the RNeasy mini kit (Qiagen) as reported in the manufacturer's instructions. Quality and integrity of the RNA was checked by Tape station 2200 (Agilent Technologies, USA) analysis and only samples displaying a RIN value above seven were used. RNA concentration and purity was then determined with a Picodrop microlitre Spectrophotometer (Picodrop). Reverse transcription to cDNA was performed with the iScript Select cDNA synthesis kit (Biorad) using the following thermal cycle: 5 min at 25°C, 30 min at 42°C, 10 min at 45°C, 10 min at 50°C and 5 min at 85°C.

The mRNA expression levels of these genes were analyzed with SYBR green technology in quantitative real-time PCR (qRT-PCR) using SoFast EvaGreen Supermix (Biorad) on a Bio-Rad CFX96 system according to the manufacturer's instructions. Quantitative PCR was carried out according to the following cycle: initial hold at 96°C for 30 s and then 40 cycles at 96°C for 2 s and 60°C for 5 s. Gene expression was normalized relative to a housekeeping genes as previously described (Turroni et al., 2011) and reported in **Table** . The amount of template cDNA used for each sample was 12.5 ng.

**TABLE | Primers used for RT-qPCR experiments.**

| Target | ORF | | Primer Fw 5′-3′ | | Primer Rv 5′-3′ | Size (bp) |
|---|---|---|---|---|---|---|
| *cysE* | BBPR_1340 | cysE-fw | CGCGACCATGCGCGACTACC | cysE-rv | GAGGATGCGCTCGTGTCCGC | 187 |
| *cysK* | BBPR_1344 | cysK-fw | CGAACCAGTACGACAACCCC | cysK-rv | GATGGAGCCTTCCGGATCGG | 203 |
| *cysB* | BBPR_0960 | cysB-fw | GACGACCTCAAGCCGTTCCC | cysB-rv | GTCGCCGTTGTCGATGCCGG | 189 |
| *metB* | BBPR_1343 | metB-fw | GGAGCCCGACCCGACCACCG | metB-rv | CAGCAGCACGTCAATCGCGG | 214 |
| *metC* | BBPR_1226 | metC-fw | CATGGGTGTGGGAAGCGAGG | metC-rv | TCGATGTCCCAGTTGTGCCG | 189 |
| *metE* | BBPR_0933 | metE-fw | GATGCTGGACACCGCGATCC | metE-rv | GGCGGATCTCGGTGCTCTCC | 206 |
| *metA* | BBPR_1654 | metA-fw | GTTCGCTCTCGGCCATTGGG | metA-rv | CGGCGTGGTCTGATACACCC | 205 |
| *rpoB*[a] | | BBP-rpo-for | GTGCAGACCGACAGCTTCGAC | BBP-rpo-rev | GAGATCTCGTTGAAGAACTCGTC | |
| *ldh*[a] | | BBP-ldh-for | CACCATGAACAGGAACAAAGTTG | BBP-ldh-rev | GAATGATCGATGAGTACGAGCTC | |
| *atpD*[a] | | BBP-atp-uni | CAGAGCCGATCAATGGACGTG | BBP-atp-rev | GTGCTGCTCGACCTCAAGCGTGAT | |

[a]Turroni et al. (2011).

## Statistical Analyses

Statistical significance between means was analyzed using the two way ANOVA. Statistically different means were determined using the Bonferroni *post hoc* test at 5% ($P$-value $< 0.05$). Values are expressed as the means ± standard errors from three experiments. Statistical calculations were performed using the software program GraphPad Prism 5 (La Jolla, CA, USA).

# RESULTS

## Development of a CDM for *B. bifidum* PRL2010

We modified the previously described CDM formulations (Petry et al., 2000; Cronin et al., 2012) based on the nutrient requirements of *B. bifidum* PRL2010. Several growth attempts on CDM minimal modifications (Petry et al., 2000; Cronin et al., 2012), i.e., where various compounds were omitted one after the other, allowed the identification of a number of components that were either essential or non-essential for growth of PRL2010 cells. Notably, folic acid and pyridoxal were eliminated from $CDM_{PRL2010}$ composition, while guanine and thiamine were supplemented. When testing different sugars it was observed that PRL2010 exhibits the best growth performance with lactose, consistent with previous studies (Turroni et al., 2010, 2012b), and this sugar was therefore used for $CDM_{PRL2010}$ formulation.

## Evaluation of Amino Acids Auxotrophy and Prototrophy of PRL2010

When PRL2010 cells were cultivated on $CDM_{PRL2010}$, they exhibited reduced growth (OD600 value of 1.21 ± 0.3) compared to that observed when grown on a nutrient-rich medium such as MRS (OD600 value of 2.9 ± 0.2). In order to assess PRL2010 amino acid prototrophy/auxotrophy, growth experiments were performed using $CDM_{PRL2010}$ where an individual amino acid had been omitted at time, and $bCDM_{PRL2010}$ medium with the inclusion of one amino acid at time. The achieved growth yield was compared to that obtained for complete $CDM_{PRL2010}$ or $bCDM_{PRL2010}$ respectively (**Figure** ).

In both experiments, only when cysteine was removed or supplied to the medium a significant decrease or increase (ranging from four- to sixfold, $P < 0.05$) of the obtained growth yield was observed, respectively, suggesting that PRL2010 is auxotrophic for cysteine. Furthermore, PRL2010 seems unable to grow on sulfate as its sole sulfur source, such as when this strain is grown in $bCDM_{PRL2010}$ (a medium that contains $MnSO_4$, $MgSO_4$, and $FeSO_4$).

Another reducing compound, glutathione, was added to $bCDM_{PRL2010}$ and only very limited growth was detected when PRL2010 cells were cultivated for 24 h (OD600 value of 0.31 ± 0.038). Furthermore, we decided to investigate if taurine, which is a common nitrogen and sulfur sources present in the gut environment (Carbonero et al., 2012) influence the growth yields of PRL2010. However, we did achieved any significant grow (OD600 = 0.10 ± 0.01) of PRL2010 when taurine was used as the unique nitrogen and sulfur sources.

## Assessing Cysteine Auxotrophy of Members of the Genus *Bifidobacterium*

We further investigated the behavior of other strains belonging to the *B. bifidum* species (see **Table** ), when cultivated under similar growth conditions (**Figure** ). These experiments showed that strains LMG11041, 156B, 85B, 324B, and LMG13195 were unable to grow on $CDM_{PRL2010}$, (OD600 values ≤ 0.3) after 24 h of incubation (**Figure** ). The other *B. bifidum* strains investigated (i.e., LMG13200, LMG11582, and LMG11583) reached OD600 values of 0.7–0.9 and exhibited an identical auxotrophic behavior as PRL2010 for cysteine (**Figure** ).

Within the genus *Bifidobacterium*, the same auxotrophic behavior for cysteine appears to be widely distributed. In fact, of the currently recognized 48 (sub)species harboring the genus *Bifidobacterium*, only *B. boum* LMG10736, *B. minimum* LMG11592, *B. pullorum* LMG21816, *B. ruminantium* LMG21811, *B. saguini* DSM23967 and *B. scardovii* LMG21589 were shown to be able to grow in $CDM_{PRL2010}$ without cysteine, though such strains reached OD600 values of just 0.5 ± 0.1 (**Figure** ). No bifidobacterial strain was able to grow in $bCDM_{PRL2010}$.

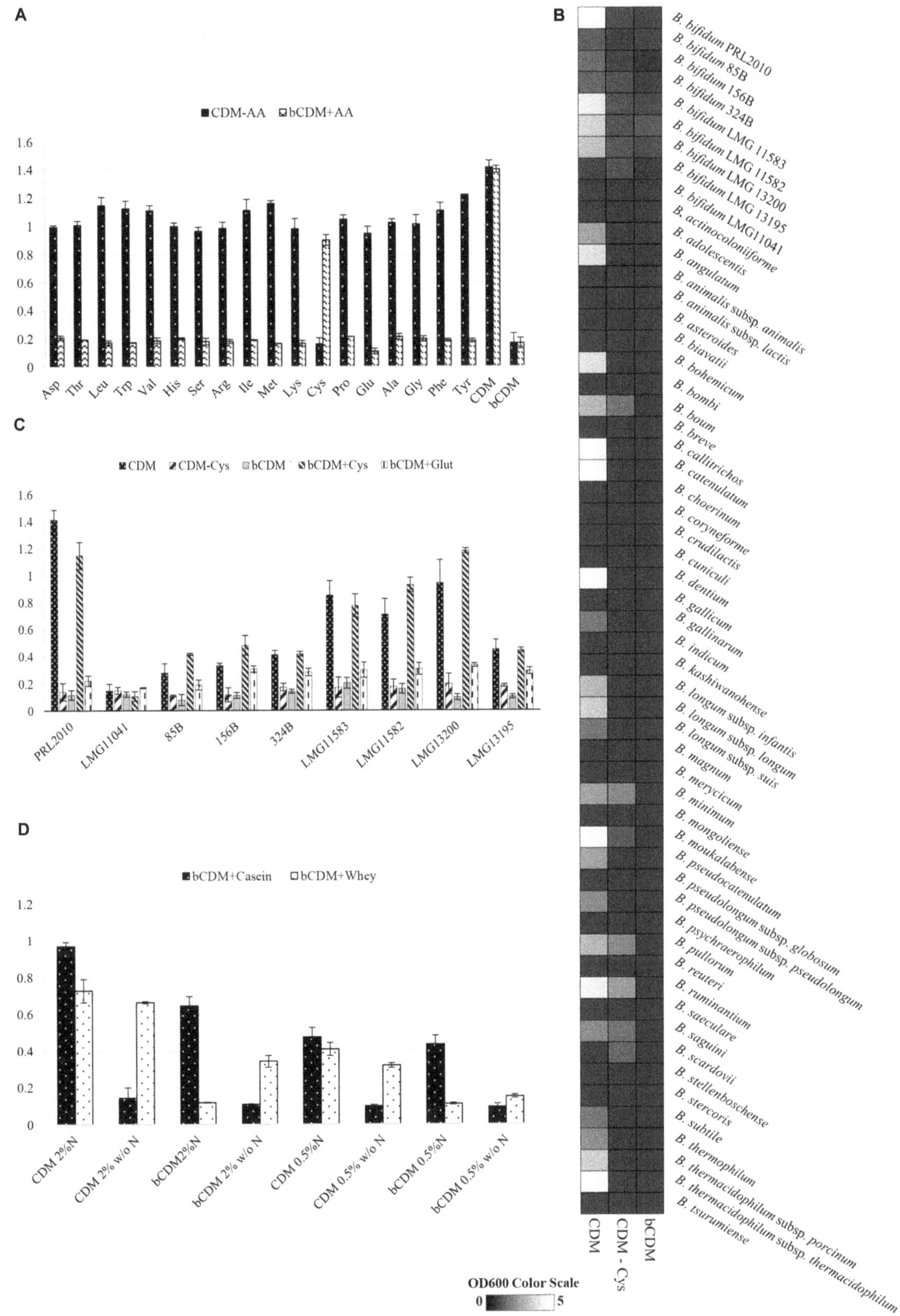


**FIGURE** | Continued

**FIGURE | Continued**
**Growth of *B. bifidum* strains.** Growth was measured as the optical density of the medium at 600 nm (OD600). Cultures were grown in triplicates. **(A)** Reports the growth of *B. bifidum* PRL2010 in $CDM_{PRL2010}$. In these tests, one amino acid at time was removed (CDM-AA) or supplied (bCDM + AA) to the medium. Amino acids are reported in the horizontal axis as follows: aspartic acid Asp, threonine Thr, leucine Leu, tryptophan Trp, valine Val, histidine His, serine Ser, arginine Arg, isoleucine Iso, methionine Met, lysine Lys, cysteine Cys, proline Pro, glutamine Gln, alanine Ala, glycine Gly, phenylalanine Phe and tyrosine Tyr. **(B)** Shows an heat map representing the growth performance of all of the type strains of the currently recognized 48 (sub)species belonging to the genus *Bifidobacterium* on $CDM_{PRL2010}$, CDM-$Cys_{PRL2010}$ and $bCDM_{PRL2010}$. The different shading represents the optical density reached by the various cultures. **(C)** Displays the growth of *B. bifidum* strains LMG11041, 85B, 156B, 324B, LMG11583, LMG11582 LMG13200 and LMG13195 in comparison with PRL2010 in $CDM_{PRL2010}$, $CDM_{PRL2010}$ without cysteine (CDM – Cys), basal $CDM_{PRL2010}$ (bCDM), basal $CDM_{PRL2010}$ with cysteine (bCDM + Cys) and basal $CDM_{PRL2010}$ with glutathione (bCDM + Glut).
**(D)** Illustrates the growth of *B. bifidum* PRL2010 in $CDM_{PRL2010}$ supplemented with complex substrates like whey proteins or casein hydrolysate. For both substrates two concentration were tested, 0.5 and 2% (wt/wt). For every concentration was evaluated the presence of amino acid (CDM or bCDM) or other nitrogen sources (N or w/o N).

## Sulfur Amino Acid Metabolism of *B. bifidum* PRL2010

A general prediction based on genomic data about nitrogen metabolism within the genus *Bifidobacterium* was previously reported by (Milani et al., 2014). The presence of genes involved in amino acid biosynthesis appears to be conserved among the seven phylogenetic groups of the genus *Bifidobacterium* (Lugli et al., 2014). However, the genes that are predicted to be involved in sulfur-containing amino acid metabolism were shown to be variably present within bifidobacterial genomes. In this context, an *in silico* analysis of the *B. bifidum* PRL2010 genome (Turroni et al., 2010) for putative genes involved in sulfur-containing amino acid transport did not reveal any positive match.

Aliphatic sulfonates can be used as alternative sulfur sources for the synthesis of cysteine (van der Ploeg et al., 1998). Bioinformatics analyses revealed the occurrence of two genes (BBPR_0202 and BBPR_0362) encoding two putative ABC-type permeases, in the chromosome of PRL2010. A low level of homology with genes involved in sulfonate transport (Even et al., 2006) was detected (Supplementary Table ), possibly explaining why *B. bifidum* PRL2010 cells are unable to grow with sulfate as the only sulfur source (bCDM condition, see **Figure** ). Another mechanism to achieve sulfur from the environment is based on the intake of cysteine by symporter proteins. This type of symporter may participate in the uptake of cysteine (Vitreschak et al., 2008). In this context, a putative sodium dicarboxylate symporter gene (BBPR_0324) was identified in PRL2010 (see Supplementary Table S1). Moreover, two putative genes (BBPR_0668 and BBPR_0671) predicted to encode two carriers involved in glutamate transport system (GluA and GluD), exhibited 53 and 26% homology, respectively, with the genes that encode the L-cysteine uptake system of *B. subtilis* (Supplementary Table S1).

As mentioned above, *B. bifidum* PRL2010 cells were shown to be unable to grow in presence of reduced glutathione (and in the absence of cysteine). Such physiological findings are in agreement with *in silico* analyses of PRL2010 chromosome sequences. In fact, the *pepT* and *pepM* genes, which are constituting the pathway for degradation of this compound (Andre et al., 2010) to generate cysteine, are absent in PRL2010 genome. Furthermore, a homolog of the *gshAB* gene, which specifies the glutamate–cysteine ligase/glutathione synthase, is also absent in chromosome of PRL2010 (**Figure** ).

Genes predicted to be involved in the cysteine biosynthesis I/homocysteine degradation pathway and methionine biosynthesis I pathway were identified in PRL2010 (**Figure** ). *In silico* analyses of PRL2010 genome revealed the occurrence of the *cysE* (BBPR_1340) and *cysK* (BBPR_ 1344) genes, which encode the predicted serine acetyltransferase that transfers an acetyl group to serine, and the cysteine synthase, respectively (Liu et al., 2012) (**Figure** ). In the same genomic region, we also identified the *metB* gene (BBPR_1343) predicted to encode a cystathionine-γ-synthase, which is catalyzing the conversion of cysteine to cystathionine, as well as the *luxS* gene (BBPR_1341), encoding an S-ribosylhomocysteinase involved in the production of homocysteine, and the *recQ* gene (BBPR_1342) encoding an ATP-dependent DNA helicase. When the presence of these genes was investigated in the genomes of other *B. bifidum* strains (Duranti et al., 2015) included in this study, a high level of homology (higher than 98% at nucleotide level) was found. Furthermore, in the genome sequences of four *B. bifidum* strains, i.e., LMG13200, LMG13195, LMG11582, and LMG11583 (Duranti et al., 2015), an additional acetyltransferase-encoding gene was identified (**Figure** ).

Other genes such as the *cysB* gene (BBPR_0960), *metC* (BBPR_1226) and *metA* (BBPR_1654) that are predicted to be involved in cysteine and methionine metabolism (Fernandez et al., 2002; Liu et al., 2012) are scattered across the PRL2010 genome.

## Growth Evaluation in Complex Substrates

The effects of complex substrates, such as whey proteins or casein hydrolysate, on PRL2010 growth were tested and are reported in **Figure** . Whey proteins and casein hydrolysate were dissolved in $CDM_{PRL2010}$ and $bCDM_{PRL2010}$ with or without other nitrogen sources at 0.5 or 2% concentration (wt/wt), respectively. Casein hydrolysate better supports PRL2010 growth in presence of nitrogen, compared to what was observed when this strain was cultivated on whey proteins. In CDM or bCDM without other nitrogen sources, PRL2010 cells seemed to metabolize whey proteins more efficiently as displayed by the higher OD 600 values that were reached (**Figure** ).

## Targeted Gene Expression Analyses of PRL2010 with Different Sulfur Substrate

Transcription of genes involved in sulfur metabolism, such as those of cysteine (*cysE, cysK* and *cysB*) and methionine metabolism (*met*A, *metE, metB,* and *metC*), were investigated

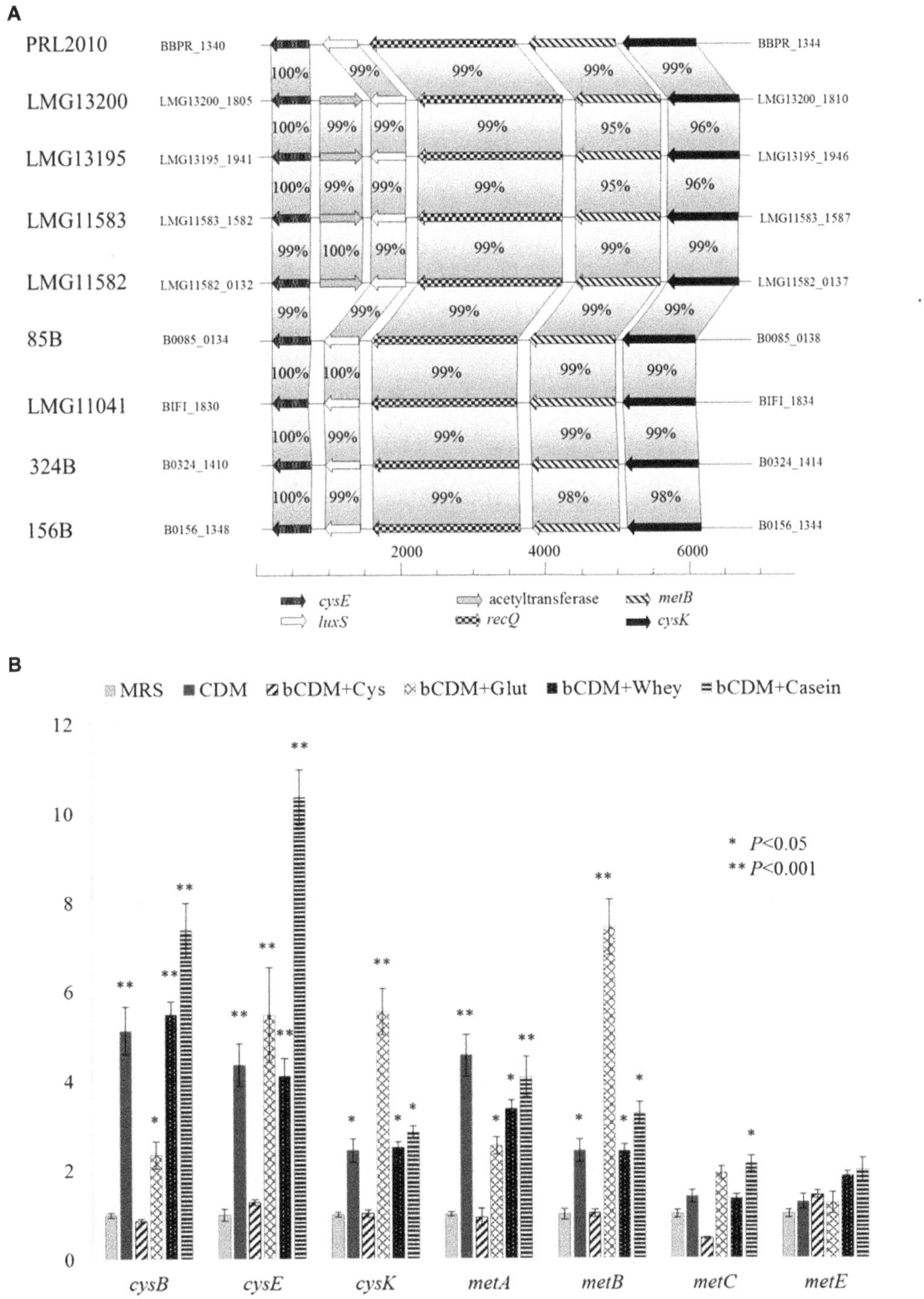


**FIGURE | Schematic representation of genes involved in cysteine and methionine metabolism in *B. bifidum* species and transcriptional analysis in *B. bifidum* PRL2010. (A)** Shows the genetic map of the predicted cysteine/methionine metabolism gene region identified in the genome of *B. bifidum* PRL2010 and compared with other *B. bifidum* strains. Each individual gene is represented by an arrow and is colored or marked according to the predicted function as indicated in the figure. **(B)** Reports the relative transcription levels of cysteine and methionine metabolism genes from *B. bifidum* PRL2010 upon cultivation in complete $CDM_{PRL2010}$, $bCDM_{PRL2010}$ supplemented with cysteine (bCDM + Cys), $bCDM_{PRL2010}$ supplemented with glutathione (bCDM + Glut), $bCDM_{PRL2010}$ with cysteine and 2% (wt/wt) whey protein (bCDM + Whey) and $bCDM_{PRL2010}$ with cysteine and 2% (wt/wt) casein hydrolysate as analyzed by quantitative real-time PCR assays. The histograms indicate the relative amounts of the *cysE, cysK, cysB, metA, metE, metB,* and *metC* genes mRNAs for the specific samples. The *y* axis indicates the logarithmic fold induction of the investigated gene compared to the reference condition (MRS). The *x* axis represents the different gene tested. Asterisks indicate statistically significant differences compared to the control. The error bar for each column represent the standard deviation calculated from three replicates.

using a qRT-PCR approach, the results of which are reported in **Figure B**.

When PRL2010 cells were cultivated in the complete $CDM_{PRL2010}$, *cysB, cysE, cysK, metA* and *metB* were overexpressed ($P < 0.05$) (**Figure B**). The occurrence of cysteine in the basal $CDM_{PRL2010}$ (bCDM + Cys) does not seem to modulate expression of genes involved in sulfur amino acid metabolism. Glutathione (bCDM + Glut) does not allow a significant growth of PRL2010 (OD600 values of 0.31 ± 0.038). However, it enhanced the transcription of the *cysB, cysE, cysK, metA* and *metB* genes ($P < 0.05$).

Regarding complex substrates, *cys* genes appear to be less induced when PRL2010 cells are cultivated in whey protein compared to basal $CDM_{PRL2010}$ in the presence of casein hydrolysate (bCDM + Whey and bCDM + Casein respectively in **Figure B**), although at significant level ($P < 0.05$). Moreover, casein hydrolysate significantly increases the transcription level of *metC*, the cystathionine β-liase ($P < 0.05$).

In all conditions tested, no significant transcriptional changes were detected for the *metC* and *metE* genes predicted to encode for cystathionine β-liase and homocysteine methyltransferase, respectively, except for *metC* when PRL2010 was grown in $bCDM_{PRL2010}$ with casein hydrolysate.

## DISCUSSION

Following birth, the human intestine is rapidly colonized by a vast array of microorganisms. *Bifidobacterium*, and in particular, *B. bifidum* strains are abundant in breast-fed infants, due to their capacity to grow on mucin and on human milk oligosaccharides (Turroni et al., 2010). The bifidogenic effect of breast milk due to bioactive peptides presence is well known (Liepke et al., 2002). A similar bifidogenic effect was shown for milk-derived k-caseins with loss of activity when the disulfide bonds were oxidized (Poch and Bezkorovainy, 1991).

Here, we investigated for the first time sulfur-containing amino acid metabolism in *B. bifidum* PRL2010 through the development of a CDM called $CDM_{PRL2010}$ and by the molecular characterization of the putative auxotrophic behavior of this strain for cysteine. Data indicates that $bCDM_{PRL2010}$ does not support *B. bifidum* PRL2010 growth, unless cysteine addition. The same behavior was extended to three other *B. bifidum* strains, i.e., LMG13200, LMG11582 and LMG11583. Furthermore, cysteine auxotrophy is not a common feature of all the (sub)species harboring the genus *Bifidobacterium*, since representatives of some species such as *B. boum, B. minimum, B. pullorum, B. ruminantium, B. saguini* and *B. scardovii* are able to grow without cysteine, although rather poorly.

*In silico* analyses of PRL2010 genome did not reveal the presence of the genetic arsenal needed to sulfate transport and reduction to sulfide. Growth experiments showed that cysteine is the only amino acid necessary to sustain PRL2010 growth but when the strain is cultivated in basal $CDM_{PRL2010}$ with cysteine (bCDM + Cys) the transcription of genes involved in cysteine and methionine metabolism was not stimulated by the availability of these amino acid residues. Similar results were reported previously for other bacterial species, such as *Escherichia coli* (Kredich, 1992), *Bacillus subtilis* (Mansilla and de Mendoza, 2000), and *Lactococcus lactis* (Fernandez et al., 2002). Another reduced sulfur compound was used to understand if the role of cysteine in PRL2010 is linked to the reducing effect that it exploits on the redox potential (Even et al., 2006). However, reduced glutathione does not sustain any appreciable strain growth, yet enhanced the transcription of genes predicted to be involved in sulfur amino acid metabolism (*cysB, cysE, cysK, metA* and *metB*). Similar behavior was previously reported for *E. coli* (Kredich, 1992) and *B. subtilis* (Mansilla and de Mendoza, 2000).

Complex substrates from dairy industry such as whey proteins and casein hydrolysate act as a reservoir of amino acid, peptides and free protein. Transcriptional analysis showed that whey proteins and casein hydrolysate increased the transcriptions of genes involved in serine degradation and/or conversion to cysteine and methionine (*cysB, cysE, cysK metA* and *metB*).

## CONCLUSION

This study provides new insights into the amino acid utilization ability of the *B. bifidum* species. This work also suggested the existence of a relationship between the sulfur amino acid metabolism and the redox state of the cell. The use of complex nitrogen sources available in the infant gut revealed an enhancement of growth yield and expression of genes involved in sulfur amino acid metabolism in PRL2010. These results could open a new avenue of research for the development of novel functional foods based on milk caseins and whey proteins with high content of cysteine or cysteine precursor's compounds that could act as prebiotics for *Bifidobacterium* enrichment.

## AUTHOR CONTRIBUTIONS

CF performed the work and wrote the manuscript, SD performed the work, CM, performed bioinformatics analyses, LM performed bioinformatics analyses, GL performed bioinformatics analyses, MM performed the work, AV performed the work, MO contributed data, DvS wrote the manuscript, MV wrote the manuscript.

## ACKNOWLEDGMENTS

We thank Science Foundation Ireland for financial support to DvS. This study was funded by the Irish Government's National Development Plan (Grant number SFI/12/RC/2273) to DvS.

## SUPPLEMENTARY MATERIAL

The Supplementary Material for this article can be found online at: http://journal.frontiersin.org/article/10.3389/fmicb.2015.01331

## REFERENCES


Alvarez, R., Neumann, G., Fravega, J., Diaz, F., Tejias, C., Collao, B., et al. (2015). CysB-dependent upregulation of the *Salmonella* Typhimurium cysJIH operon in response to antimicrobial compounds that induce oxidative stress. *Biochem. Biophys. Res. Commun.* 458, 46–51. doi: 10.1016/j.bbrc.2015.01.058

Andre, G., Haudecoeur, E., Monot, M., Ohtani, K., Shimizu, T., Dupuy, B., et al. (2010). Global regulation of gene expression in response to cysteine availability in *Clostridium perfringens*. *BMC Microbiol.* 10:234. doi: 10.1186/1471-2180-10-234

Bogicevic, B., Berthoud, H., Portmann, R., Meile, L., and Irmler, S. (2012). CysK from Lactobacillus casei encodes a protein with O-acetylserine sulfhydrylase and cysteine desulfurization activity. *Appl. Microbiol. Biotechnol.* 94, 1209–1220. doi: 10.1007/s00253-011-3677-5

Burguiere, P., Auger, S., Hullo, M. F., Danchin, A., and Martin-Verstraete, I. (2004). Three different systems participate in L-cystine uptake in *Bacillus* subtilis. *J. Bacteriol.* 186, 4875–4884. doi: 10.1128/JB.186.15.4875-4884.2004

Carbonero, F., Benefiel, A. C., Alizadeh-Ghamsari, A. H., and Gaskins, H. R. (2012). Microbial pathways in colonic sulfur metabolism and links with health and disease. *Front. Physiol.* 3:448. doi: 10.3389/fphys.2012.00448

Cronin, M., Zomer, A., Fitzgerald, G. F., and van Sinderen, D. (2012). Identification of iron-regulated genes of *Bifidobacterium breve* UCC2003 as a basis for controlled gene expression. *Bioeng. Bugs* 3, 157–167. doi: 10.4161/bbug.18985

Davila, A. M., Blachier, F., Gotteland, M., Andriamihaja, M., Benetti, P. H., Sanz, Y., et al. (2013). Intestinal luminal nitrogen metabolism: role of the gut microbiota and consequences for the host. *Pharmacol. Res.* 68, 95–107. doi: 10.1016/j.phrs.2012.11.005

Duranti, S., Milani, C., Lugli, G. A., Turroni, F., Mancabelli, L., Sanchez, B., et al. (2015). Insights from genomes of representatives of the human gut commensal *Bifidobacterium bifidum*. *Environ. Microbiol.* 17, 2515–2531. doi: 10.1111/1462-2920.12743

Even, S., Burguiere, P., Auger, S., Soutourina, O., Danchin, A., and Martin-Verstraete, I. (2006). Global control of cysteine metabolism by CymR in *Bacillus* subtilis. *J. Bacteriol.* 188, 2184–2197. doi: 10.1128/Jb.188.6.2184-2197.2006

Fanning, S., Hall, L. J., Cronin, M., Zomer, A., MacSharry, J., Goulding, D., et al. (2012). Bifidobacterial surface-exopolysaccharide facilitates commensal-host interaction through immune modulation and pathogen protection. *Proc. Natl. Acad. Sci. U.S.A.* 109, 2108–2113. doi: 10.1073/pnas.1115621109

Fernandez, M., Kleerebezem, M., Kuipers, O. P., Siezen, R. J., and van Kranenburg, R. (2002). Regulation of the metC-cysK operon, involved in sulfur metabolism in *Lactococcus lactis*. *J. Bacteriol.* 184, 82–90. doi: 10.1128/JB.184.1.82-90.2002

Fisher, S. H. (1999). Regulation of nitrogen metabolism in *Bacillus* subtilis: vive la difference! *Mol. Microbiol.* 32, 223–232. doi: 10.1046/j.1365-2958.1999.01333.x

Gish, W., and States, D. J. (1993). Identification of protein coding regions by database similarity search. *Nat. Genet.* 3, 266–272. doi: 10.1038/ng0393-266

Hamer, H. M., De Preter, V., Windey, K., and Verbeke, K. (2012). Functional analysis of colonic bacterial metabolism: relevant to health? *Am. J. Physiol. Gastrointest. Liver Physiol.* 302, G1–G9. doi: 10.1152/ajpgi.00048.2011

Kredich, N. M. (1992). The molecular-basis for positive regulation of Cys promoters in *Salmonella*-Typhimurium and *Escherichia-Coli*. *Mol. Microbiol.* 6, 2747–2753. doi: 10.1111/j.1365-2958.1992.tb01453.x

Lebeer, S., De Keersmaecker, S. C. J., Verhoeven, T. L. A., Fadda, A. A., Marchal, K., and Vanderleyden, J. (2007). Functional analysis of luxS in the probiotic strain *Lactobacillus rhamnosus* GG reveals a central metabolic role important for growth and Biofilm formation. *J. Bacteriol.* 189, 860–871. doi: 10.1128/Jb.01394-06

Liepke, C., Adermann, K., Raida, M., Magert, H. J., Forssmann, W. G., and Zucht, H. D. (2002). Human milk provides peptides highly stimulating the growth of bifidobacteria. *Eur. J. Biochem.* 269, 712–718. doi: 10.1046/j.0014-2956.2001.02712.x

Liu, M. J., Prakash, C., Nauta, A., Siezen, R. J., and Francke, C. (2012). Computational analysis of cysteine and methionine metabolism and its regulation in dairy starter and related bacteria. *J. Bacteriol.* 194, 3522–3533. doi: 10.1128/JB.06816-11

Lonnerdal, B. (2013). Bioactive proteins in breast milk. *J. Paediatr. Child Health* 49, 1–7. doi: 10.1111/jpc.12104

Lugli, G. A., Milani, C., Turroni, F., Duranti, S., Ferrario, C., Viappiani, A., et al. (2014). Investigation of the evolutionary development of the genus *Bifidobacterium* by comparative genomics. *Appl. Environ. Microbiol.* 80, 6383–6394. doi: 10.1128/AEM.02004-14

Mansilla, M. C., and de Mendoza, D. (2000). The *Bacillus* subtilis cysP gene encodes a novel sulphate permease related to the inorganic phosphate transporter (Pit) family. *Microbiology* 146(Pt 4), 815–821. doi: 10.1099/00221287-146-4-815

Marcobal, A., Southwick, A. M., Earle, K. A., and Sonnenburg, J. L. (2013). A refined palate: bacterial consumption of host glycans in the gut. *Glycobiology* 23, 1038–1046. doi: 10.1093/glycob/cwt040

McCann, K. B., Shiell, B. J., Michalski, W. P., Lee, A., Wan, J., Roginski, H., et al. (2006). Isolation and characterisation of a novel antibacterial peptide from bovine alpha(s1)-casein. *Int. Dairy J.* 16, 316–323. doi: 10.1016/j.idairyl.2005.05.005

Mihara, H., and Esaki, N. (2002). Bacterial cysteine desulfurases: their function and mechanisms. *Appl. Microbiol. Biotechnol.* 60, 12–23. doi: 10.1007/s00253-002-1107-4

Milani, C., Lugli, G. A., Duranti, S., Turroni, F., Bottacini, F., Mangifesta, M., et al. (2014). Genomic encyclopedia of type strains of the genus *Bifidobacterium*. *Appl. Environ. Microbiol.* 80, 6290–6302. doi: 10.1128/AEM.02308-14

Nagpal, R., Behare, P., Rana, R., Kumar, A., Kumar, M., Arora, S., et al. (2011). Bioactive peptides derived from milk proteins and their health beneficial potentials: an update. *Food Funct.* 2, 18–27. doi: 10.1039/c0fo00016g

O'Connell Motherway, M., Zomer, A., Leahy, S. C., Reunanen, J., Bottacini, F., Claesson, M. J., et al. (2011). Functional genome analysis of *Bifidobacterium breve* UCC2003 reveals type IVb tight adherence (Tad) pili as an essential and conserved host-colonization factor. *Proc. Natl. Acad. Sci. U.S.A.* 108, 11217–11222. doi: 10.1073/pnas.1105380108

Petry, S., Furlan, S., Crepeau, M. J., Cerning, J., and Desmazeaud, M. (2000). Factors affecting exocellular polysaccharide production by *Lactobacillus delbrueckii* subsp bulgaricus grown in a chemically defined medium. *Appl. Environ. Microbiol.* 66, 3427–3431. doi: 10.1128/AEM.66.8.3427-3431.2000

Poch, M., and Bezkorovainy, A. (1991). Bovine-milk kappa-casein trypsin digest is a growth enhancer for the genus *Bifidobacterium*. *J. Agric. Food Chem.* 39, 73–77. doi: 10.1021/jf00001a013

Pokusaeva, K., Fitzgerald, G. F., and van Sinderen, D. (2011). Carbohydrate metabolism in Bifidobacteria. *Genes Nutr.* 6, 285–306. doi: 10.1007/s12263-010-0206-6

Schanbacher, F. L., Talhouk, R. S., and Murray, F. A. (1997). Biology and origin of bioactive peptides in milk. *Livestock Prod. Sci.* 50, 105–123. doi: 10.1016/j.ijfoodmicro.2013.06.019

Singh, P., Brooks, J. F. II, Ray, V. A., Mandel, M. J., and Visick, K. L. (2015). CysK plays a role in biofilm formation and colonization by *Vibrio fischeri*. *Appl. Environ. Microbiol.* 81, 5223–5234. doi: 10.1128/AEM.00157-15

Turroni, F., Bottacini, F., Foroni, E., Mulder, I., Kim, J. H., Zomer, A., et al. (2010). Genome analysis of *Bifidobacterium bifidum* PRL2010 reveals metabolic pathways for host-derived glycan foraging. *Proc. Natl. Acad. Sci. U.S.A.* 107, 19514–19519. doi: 10.1073/pnas.1011100107

Turroni, F., Foroni, E., Montanini, B., Viappiani, A., Strati, F., Duranti, S., et al. (2011). Global genome transcription profiling of *Bifidobacterium bifidum* PRL2010 under in vitro conditions and identification of reference genes for quantitative real-time PCR. *Appl. Environ. Microbiol.* 77, 8578–8587. doi: 10.1128/AEM.06352-11

Turroni, F., Peano, C., Pass, D. A., Foroni, E., Severgnini, M., Claesson, M. J., et al. (2012a). Diversity of bifidobacteria within the infant gut microbiota. *PLoS ONE* 7:e36957. doi: 10.1371/journal.pone.0036957

Turroni, F., Strati, F., Foroni, E., Serafini, F., Duranti, S., van Sinderen, D., et al. (2012b). Analysis of predicted carbohydrate transport systems encoded by *Bifidobacterium bifidum* PRL2010. *Appl. Environ. Microbiol.* 78, 5002–5012. doi: 10.1128/AEM.00629-12

Turroni, F., Serafini, F., Foroni, E., Duranti, S., O'Connell Motherway, M., Taverniti, V., et al. (2013). Role of sortase-dependent pili of *Bifidobacterium bifidum* PRL2010 in modulating bacterium-host interactions. *Proc. Natl. Acad. Sci. U.S.A.* 110, 11151–11156. doi: 10.1073/pnas.1303897110

Turroni, F., Taverniti, V., Ruas-Madiedo, P., Duranti, S., Guglielmetti, S., Lugli, G. A., et al. (2014). *Bifidobacterium bifidum* PRL2010 modulates the host innate immune response. *Appl. Environ. Microbiol.* 80, 730–740. doi: 10.1128/AEM.03313-13

Underwood, M. A., German, J. B., Lebrilla, C. B., and Mills, D. A. (2015). *Bifidobacterium* longum subspecies infantis: champion colonizer of the infant gut. *Pediatr. Res.* 77, 229–235. doi: 10.1038/pr.2014.156

van der Ploeg, J. R., Cummings, N. J., Leisinger, T., and Connerton, I. F. (1998). *Bacillus* subtilis genes for the utilization of sulfur from aliphatic sulfonates. *Microbiology* 144(Pt 9), 2555–2561. doi: 10.1099/00221287-144-9-2555

Ventura, M., Turroni, F., Motherway, M. O., MacSharry, J., and van Sinderen, D. (2012). Host-microbe interactions that facilitate gut colonization by commensal bifidobacteria. *Trends Microbiol.* 20, 467–476. doi: 10.1016/j.tim.2012.07.002

Vitreschak, A. G., Mironov, A. A., Lyubetsky, V. A., and Gelfand, M. S. (2008). Comparative genomic analysis of T-box regulatory systems in bacteria. *RNA* 14, 717–735. doi: 10.1261/rna.819308

Ye, J., Coulouris, G., Zaretskaya, I., Cutcutache, I., Rozen, S., and Madden, T. L. (2012). Primer-BLAST: a tool to design target-specific primers for polymerase chain reaction. *BMC Bioinform.* 13:134. doi: 10.1186/1471-2105-13-134

**Conflict of Interest Statement:** The Guest Associate Editor David Berry declares that despite having hosted a Frontiers Research Topic with the authors Marco Ventura and Francesca Turroni, the review process was handled objectively. The authors declare that the research was conducted in the absence of any commercial or financial relationships that could be construed as a potential conflict of interest.

# 8

# Hypocholesterolemic and Prebiotic Effects of a Whole-Grain Oat-Based Granola Breakfast Cereal in a Cardio-Metabolic "At Risk" Population

*Michael L. Connolly*[1,2], *Xenofon Tzounis*[1,2], *Kieran M. Tuohy*[1,3]* *and Julie A. Lovegrove*[1,2]*

[1] *Hugh Sinclair Unit of Human Nutrition, University of Reading, Reading, UK,* [2] *Institute for Cardiovascular and Metabolic Research, University of Reading, Reading, UK,* [3] *Department of Food Quality and Nutrition, Research and Innovation Centre - Fondazione Edmund Mach, Trento, Italy*

**Edited by:**
Francesca Turroni,
University College Cork, Ireland

**Reviewed by:**
Marco Ventura,
University of Parma, Italy
Francesca Scazzina,
University of Parma, Italy

***Correspondence:**
Kieran M. Tuohy
k.m.tuohy@fmach.it
Julie A. Lovegrove
j.a.lovegrove@reading.ac.uk

Meta-analyses of randomized controlled trials (RTC) have confirmed the hypocholesterolaemic effect of oats and oat based fibers. However, the mechanisms by which oats or oat fractions lower cholesterol is not totally clear. Recognizing the important role of the gut microbiome in metabolism and metabolic disease risk, we examined the impact of whole grain oat Granola (WGO) on the human gut microbiota and cardio-metabolic risk factors using a randomized crossover dietary intervention in at risk individuals (ClinicalTrials.gov Identifier: NCT01925365). We randomized 32 individuals at risk of developing cardio-metabolic disease by virtue of mild hypercholesterolaemia or glucose intolerance, into two groups consuming either 45 g of WGO or non-whole grain (NWG) breakfast cereals daily for two 6-week intervention periods separated by a 4-week wash out period in a randomized, controlled, crossover, double-blinded design. Confirming the cholesterol lowering effect of WGO, we observed a significant time by treatment interaction, for total cholesterol (TC) ($P = 0.0001$) and LDL-cholesterol (LDL-C) ($P = 0.02$) compared to NWG. A significant time by treatment interaction was also observed for the relative abundance of fecal bifidobacteria ($P = 0.0001$), lactobacilli ($P = 0.001$) and total bacterial count ($P = 0.008$), which were all elevated after consumption of WGO. Daily consumption of WGO resulted in a prebiotic effect on the human gut microbiota composition and significant reductions in TC and LDL-C concentrations. Prebiotic modulation of the human gut microbiota may thus constitute a previously unrecognized mechanism contributing to the hypocholesterolaemic effects of whole grain oat Granola.

**Keywords: whole-grain oat granola, prebiotic, cholesterol, cardiovascular risk, *Bifidobacterium***

## INTRODUCTION

Several large epidemiological studies and a number of meta-analyses of nutritional interventions have reported a positive association between increased whole grain intake and reduced risk of developing a range of chronic diseases (Chatenoud et al., 1998; Jacobs et al., 1999; Montonen et al., 2003; Mellen et al., 2008; He et al., 2010; Ye et al., 2012). Consumption of oats or oat based products by individuals with various metabolic disease risk factors (e.g., hypercholesterolemia, obesity, and diabetes) and in different ethnic groups, has been shown to mediate an appreciable normalization of plasma cholesterol levels (Ripsin et al., 1992; Queenan et al., 2007; Wolever et al., 2011; Charlton et al., 2012; Zhang et al., 2012). The cholesterol lowering activity of oats is usually attributed to its ability to reduce intestinal absorption of cholesterol and/or inhibit the enterohepatic circulation of bile acids by increasing carriage of cholesterol and/or bile acids into the colon and facilitating their excretion in feces (Ryan et al., 2007; Gunness and Gidley, 2010; Borneo and León, 2012). Whole grain oats contain a number of potentially bioactive components capable of modulating cholesterol metabolism in mammals, including unsaturated fatty acids, fibers, such as beta-glucan, arabinoxylans, arabinogalactans, and resistant starch. Some of these polysaccharides can form viscous gels in aqueous solutions, and/or directly bind cholesterol or bile acids, while all are fermentable by the gut microbiota into short chain fatty acids (SCFA). Oats, and whole grain oat Granola also contains polyphenolic compounds and phytoeostrogens which may also modulate the gut microbiota and impact on host metabolic parameters (Ryan et al., 2007; Borneo and León, 2012). However, it is the gel forming nature of beta-glucans which is most commonly attributed to the cholesterol lowering effect of oats. Tiwari and Cummins (2011) performed a meta-analysis on 126 studies and concluded that there was a significant dose-dependent inverse relationship between beta-glucan from oats and barley and blood total cholesterol, LDL-cholesterol, and triacylglycerol concentrations with 3 g/day being sufficient to lower blood total cholesterol by −0.30 mmol/L (Tiwari and Cummins, 2011). While the ability of oats and beta-glucans to increase excretion of bile acids and cholesterol in feces appears to be well established, the underlying mechanism still remains to be fully elucidated. Although the gel forming nature of beta-glucans reducing bile acid/cholesterol absorption is the most commonly proposed mechanism, recent studies with probiotic microorganisms raise the possibility of bile salt hydrolase (BSH) activity as another possible mechanism by which oats and fermentable fibers can lower plasma cholesterol (Gunness and Gidley, 2010; Ooi et al., 2010; Ejtahed et al., 2011; Jones et al., 2012a,b). Studies on non-gel forming prebiotic fibers, which modulate gut bacteria and increase BSH active lactobacilli and bifidobacteria, support the hypothesis that an increased BSH activity due to gut bacterial modulation could reduce plasma cholesterol (Tanaka et al., 1999; Kim et al., 2004). Yet few studies have examined the impact of whole grain oat based food products on human gut microbiota and no study to date has measured the ability of dietary oats to modulate the composition and relative abundance of commensal bacteria *in vivo*.

The present study aimed to address this knowledge gap by determining the effectiveness of a whole grain oat (WGO) Granola breakfast cereal, compared to a refined breakfast cereal to beneficially modulate gut microbiota and its metabolic output, plasma lipids, gut satiety hormones and inflammation markers in a randomized, controlled, double-blind, crossover dietary intervention study. It tested the hypothesis that whole grain oat based cereal can mediate a prebiotic modulation of the human gut microbiota, particularly increased bifidobacteria abundance and concomitant reduction of plasma LDL cholesterol.

## MATERIALS AND METHODS

### Subjects

Men and women (age range 23–64 year), recruited from the general population, attended the Hugh Sinclair Unit of Human Nutrition, at the University of Reading in a fasted state for measurement of height, weight, plasma glucose, total cholesterol concentrations, and hematology. To qualify for entry into the study subjects required a BMI of 18–30 kg/m$^2$ and either glucose intolerance (fasting glucose > 5.5 but < 6.9 mmol/L) or mild to moderate hypercholesterolaemic (total cholesterol > 5.2 but < 7.8 mmol/L). Exclusion criteria for the study were as follows: medical history of heart disease, diabetes mellitus, cancer, pancreatitis or renal disease, use of lipid lowering drugs, systemic corticosteroids or drugs for regulating hemostasis, exposure to any investigational agent < 42 d before the study; presence of gastrointestinal disorder or use of a drug likely to alter gastrointestinal motility or nutrient absorption, a history of substance misuse or alcoholism, a current pregnancy, planned pregnancy, or given birth in the past 12 months, antibiotic treatment 6 weeks previous to study start date, an allergy or intolerance to intervention breakfast cereals components or smoking. The study was conducted according to the guidelines laid down in the Declaration of Helsinki and was given a favorable ethical opinion for conduct by the University of Reading's Research Ethics Committee. All participants gave written informed consent before participation. The study was registered as a clinical trial on ClinicalTrial.gov (NCT01925365).

### Study Design

The dietary intervention study was a randomized, double blind, controlled, crossover design. Thirty-two volunteers (20 women, 12 men) were recruited onto the study. For a 2-week period prior to dietary intervention, volunteers followed their habitual diet but were required not to consume confirmed prebiotics (such as inulin), probiotics (such as live yogurts or fermented milk drinks), and whole grain products. The subjects were randomly allocated into one of two groups using a random number generator. One group consumed the WGO Granola breakfast cereal (45 g/day) for 6 weeks, and then after a 4-week washout period, consumed the NWG breakfast cereal (45 g/day) for 6 weeks. The other group received the breakfast cereals in the opposite order. During each washout period no breakfast cereal was consumed. All test materials were packaged, labeled and randomized by Jordans Cereals (Biggleswade., UK) prior to the study, neither the investigators nor study subjects were aware of which cereal was allocated.

Study subjects were asked to keep diaries while ingesting the breakfast cereals, to record stool frequency and consistency (constipation, hard, formed, soft, or diarrhea), abdominal pain, intestinal bloating and flatulence (none, mild, moderate, or severe) on a daily basis. Any concomitant medication and adverse events were recorded. Faecal samples, saliva samples, and 20 ml fasting venous blood samples were collected from each volunteer at six time points before and after each treatment and 14 days after each of the treatment periods (i.e., 0, 42, 56, 70, 112, and 126 days).

## Composition of Breakfast Cereals

The nutritional composition of WGO and NWG breakfast cereals were provided by Jordans Cereals (Biggleswade., UK) except for the β-glucan content which was analyzed by Leatherhead Food Research (using the Megazyme kit). The energy content of the WGO and NWG cereals was 417 kcal/100 g and 380 kcal/100 g, respectively. WGO, in the form of Granola breakfast cereal, comprised mainly of whole grain oats with small amounts of almond and dried fruit, and contained (per 100 g) 67.8 g carbohydrates, of which 26.8 g sugars; 8.3 g protein, 12.4 g fat, 6.3 g fiber and 2.9 g β-glucan. NWG, in the form of flaked corn cereal, contained (per 100 g) 84.4 g carbohydrate, of which 8.8 g sugars; 7.4 g protein, 1.1 g fat, 3.0 g fiber and no detectable β-glucan.

## Culture Independent Enumeration of Fecal Bacteria

Fecal bacteria enumeration was the primary outcome measure. Fecal samples were stored in an anaerobic cabinet (10% $H_2$; 10% $CO_2$; 80% $N_2$) for no longer than 2 h prior to processing. Changes in fecal bacteria populations upon consumption of the test and control cereals were monitored using 16S rRNA probes labeled with Cy3 for specific bacterial groups or the nucleic acid stain DAPI for total bacterial counts and Fluorescence *in situ* Hybridisation (FISH). The bacterial groups selected for enumeration were *Bifidobacterium* spp., *Bacteroides/Prevotella* spp., *Lactobacillus/ Enterococcus* spp., *Clostridium coccoides-Eubacterium rectale* group, *Clostridium histolyticum* group, and *Atopobium* cluster including most *Coriobacteriaceae* species, using the specific probes Bif164, Bac303, Lab158, Erec482, His150, and Ato291, respectively (Langendijk et al., 1995; Manz et al., 1996; Franks et al., 1998; Harmsen et al., 2000). FISH was performed essentially as described by Rycroft et al. (2001) and Daims et al. (2005). Briefly, fecal samples (375 μl) fixed in 4% paraformaldehyde (pH 7.2) overnight at 4°C were then centrifuged at 1500 × g for 5 min, washed twice with phosphate buffer saline (PBS 0.1 M, pH 7.0), re-suspended in a mixture of PBS/99% ethanol (1:1 v/v) and stored at −20°C for up to 3 months.

For the hybridisations, 20 μl of each sample was pipetted onto Teflon- and poly-l-lysine-coated, six-well (10 mm diameter each) slides (Tekdon Inc., Myakka City, FL, USA). Slides were dried at 46°C for 15 min and then submerged in a series of ethanol solutions (50, 80, and 96%, 3 min each). This process was used for all samples, except those where the Lab158 probe was used. Sample slides probed with Lab158 were subjected to an additional step with 50 μl of lysozyme (1 mg/mL in 100 mM Tris-HCl, pH 8.0) at 37°C for 15 min prior to being submerged in the ethanol solutions. A probe/hybridization buffer mixture (5 μl of a 50 ng $\mu l^{-1}$ stock of probe plus 45 μl of hybridization buffer) was applied to the surface of each well. Hybridisations were performed for 4 h in an ISO20 oven (Grant Boekel). Slides were stored in the dark at 4°C (for a maximum of 3 days) until cells were counted. Slides were enumerated using a Nikon E400 Eclipse microscope fitted with an EPI-fluorescence attachment, 15 randomized views were counted for each sample.

## SCFA Analysis

High-performance liquid chromatography (HPLC) was performed to determine fecal SCFA concentration. Aliquots (1 ml) of 10% (w/v) fecal suspension were centrifuged at 13000 g for 10 min and the supernatant was stored at −20°C for up to 3 months. Supernatants were then filtered using 0.2 mm polycarbonate syringe filters (Whatman, Maidstone, Kent, UK) and injected (20 μl) into an HPLC system (Merck, Whitehouse Station, NJ, USA) equipped with refractive index detection. For the preparation of the external standard containing the SCFA; acetic, propionic and butyric acid were added to give a final concentration of 25 mM to HCl (6 M) and HPLC gradient water. Dilutions of the external standards were prepared and added to the internal standard (ratio 4:1) to give a final concentration for the internal standard of 20 mM ethyl butyric acid, and a final concentration of external standards as 80, 40, 20, 10, 5.0, 1.0, and 0.5 mM. The column used was an ion-exclusion REZEX-ROA organic acid column (Phenomenex, Inc., Torrance, CA, USA) maintained at 85°C. $H_2SO_4$ in HPLC-grade $H_2O$ (0.0025 mmol/l) was used as the eluent, and the flow rate was maintained at 0.5 ml/min. Quantification of the samples was obtained through calibration curves of acetic, propionic and butyric acids.

## Blood Samples Collection and Analysis

Blood samples were collected into a 10 ml EDTA tube (BD vacutainer EDTA tube, BD, Cowley, Oxon., UK) for the analysis of fasting total cholesterol (TC), HDL-cholesterol (HDL-C), LDL-cholesterol (LDL-C), triacylglycerol (TAG), CRP, IL-6, TNF-α, PYY, GLP-1 and insulin concentrations; into a 10 ml fluoride/oxalate tube (BD vacutainer fluoride/oxalate tube) for the analysis of fasting glucose concentration. These blood analytes were the secondary outcome measures. Following collection all blood samples were kept on ice until centrifugation. Plasma samples were recovered by centrifugation at 1700 g for 10 min, dispensed into 1.5 ml microcentrifuge tubes and frozen at −20°C within 1 h from collection. Plasma samples were defrosted and centrifuged for 5 min at 1500 g prior to analysis.

Plasma TAG, glucose, CRP, TC, LDL-C, and HDL-C concentrations were determined on a Monarch Automatic Analyzer ILab 600 using enzymatic kits (Instrumentation Laboratories Ltd, Cheshire, UK). Two quality control samples, (Wako Control Serum I and II. Alpha Laboratories Ltd, Eastleigh, Hants., UK), containing known normal and abnormal concentrations were included at the beginning and end of each batch analysis. Results were accepted if the quality control values were within the range specified by the manufacturers.

Determination of plasma insulin (DAKO Diagnostic Ltd, Cambridgeshire, UK), IL-6 and TNF-α (R&D Systems, Abingdon, UK), PYY and GLP-1 (Yanaihara Institute Inc. Shizuoka, Japan), feacal and saliva sIgA (Immundianostik AG, Bensheim, Germany) calprotectin (NovaTec Immundiagnostica GmbH, Dietzenbach, Germany) concentrations were performed using specific commercial ELISA kits according to manufacturer's instructions. Concentrations of the samples and quality controls were determined automatically by reading from the standard curve using MasterPlex (version EX 2010) computer software.

## Dietary Analysis

Volunteers were asked on week 5 of each of the intervention arms to complete a 3-day estimated diet diary (2 week and one weekend day). The volunteers received both verbal and written instructions on how to complete the diet diaries. These were analyzed using Dietplan 6.60 (Forestfield Software Ltd), to determine the macro and micronutrient content of the participant's diets during each intervention period.

## Statistical Analysis

Normality testing was carried out to determine if data was normally distributed and transformed if required. Linear mixed model analysis of variance was performed for testing effects of visit, treatment, and visit by treatment interaction. Tukey's post-test with 95% confidence was used for multiple comparisons after a visit by treatment interaction was established. Minitab 16 for Windows was used for all analysis. $P < 0.05$ was taken as significant. The sample size of 28 was required to detect 0.5 $Log_{10}$ change in bifidobacterial counts with power set at 0.9, and a significance level of 0.05 based on previous prebiotic studies in healthy volunteers conducted with the same microbiological techniques investigating whole grain maize (Carvalho-Wells et al., 2010). To allow for a 10% dropout rate a total of 32 individuals were recruited.

**TABLE | Mean (SEM) of daily macronutrient intake during the WGO and NWG intervention periods[a].**

| Daily intake (*n* = 24) | WGO treatment Mean ± SEM | NWG treatment mean ± SEM | *P*-value |
|---|---|---|---|
| Energy, *kJ* | 7904 ± 404 | 7512 ± 349 | 0.312 |
| Energy, *kcal* | 1916 ± 78 | 1787 ± 83 | 0.301 |
| Fat, *%E* | 34.0 ± 1.1 | 30.6 ± 1.0 | 0.009 |
| SFA, *%E* | 11.9 ± 0.5 | 10.9 ± 0.7 | 0.162 |
| MUFA, *%E* | 12.5 ± 0.6 | 10.4 ± 0.5 | 0.002 |
| PUFA, *%E* | 5.9 ± 0.4 | 5.5 ± 0.4 | 0.238 |
| Total TFA, *%E* | 0.7 ± 0.1 | 0.8 ± 0.1 | 0.158 |
| Protein, *%E* | 16.4 ± 0.8 | 17.0 ± 0.8 | 0.407 |
| Carbohydrate %E | 49.8 ± 1.3 | 52.3 ± 1.2 | 0.073 |
| Starch, *%E* | 27.1 ± 1.2 | 31.4 ± 1.4 | 0.007 |
| Total Sugar, *%E* | 21.2 ± 1.1 | 18.7 ± 1.4 | 0.080 |
| NSP, *g* | 14.4 ± 0.7 | 13.9 ± 0.9 | 0.615 |
| Total fiber (AOAC method), *g* | 18.8 ± 0.9 | 18.2 ± 1.1 | 0.626 |

[a]*Determined from estimated 3-day dietary diaries. SFA, saturated fatty acids; MUFA, monounsaturated fatty acids; PUFA, polyunsaturated fatty acids; TFA, trans fatty acids; NMES, non-milk extrinsic sugars; NSP, non-starch polysaccharide; AOAC, Association of Official Analytical Chemists.*

## RESULTS

## Subjects Characteristics

Of the 32 participants recruited, one individual withdrew before the start of the study due to antibiotic treatment, and another withdrew after visit 1 due to personal reasons. Samples from these individuals were excluded from analysis. Thus, a total of 30 participants were included (women $n = 19$, men $n = 11$). The mean age was 42 years with a range between 19 to 60 years. The participant's baseline mean weight was 75.2 kg (SD ± 20.4), BMI 26.4 (± 5.7), were mildly hypercholesterolamic TC 5.4 mmol/L (± 1.0) or glucose intolerant (glucose 5.6± 0.6 mmol/L).

## Dietary Analysis

Nutrient analysis of the diets during the two intervention periods is shown in **Table** . The nutrient intakes in both intervention periods were similar except for total fat as % of total food energy (%E) ($P = 0.009$) and MUFA (%E) ($P = 0.002$) which were higher, and starch (%E) ($P = 0.007$), which was lower while participants consumed the WGO compared with the NWG cereal. These differences broadly reflected compliance to the intervention foods.

## Faecal Bacteria and SCFA Analysis

The data for the fecal SCFA analysis is shown in **Table** . No significant changes were detected either between visits or groups. Fluorescence *in situ* hybridization was used to enumerate the main gut bacterial groups (**Table** ). Baseline levels of all bacterial groups were comparable between the two treatments. A time by treatment interaction for bifidobacteria ($P = 0.000$; **Figure A**), lactobacilli ($P = 0.001$; **Figure B**) and total bacterial population ($P = 0.008$; **Figure C**) was observed. After consumption of WGO for a 6-week period the numbers of bifidobacteria ($Log_{10}$ bacteria cells/g feces) ($P = 0.000$), lactobacilli ($P = 0.000$), and total bacteria population ($P = 0.017$) significantly increased compared to the respective baseline. A significant decrease in bifidobacteria ($P = 0.03$) and total bacteria population ($P = 0.034$) was observed on NWG consumption after the 6-week feeding time. A significant difference at 6 weeks between bifidobacteria ($P = 0.000$) and total population ($P = 0.003$) size after WGO and NWG was observed, however they returned to near baseline levels after the 4-week washout. No significant changes in the population size of *Bacteroides* spp., *Atopobium* spp., and *Clostridium* group were observed.

## Blood Lipid Parameters

Baseline concentrations of all lipid parameters were comparable between the two treatments (**Table** ). A significant time by treatment interaction for TC ($P = 0.000$), and LDL-C ($P = 0.002$) was observed. After consumption of WGO for a 6-week period, TC concentrations significantly decreased ($P = 0.0016$), in contrast after NWG a significant increase in TC levels was observed ($P = 0.0016$; **Figure A**). Plasma LDL-C was seen to significantly increase ($P = 0.0055$) after consumption of the NWG cereal (**Figure B**). A significant difference at 6 weeks between TC ($P = 0.000$) and LDL-C ($P = 0.009$) after WGO and NWG was observed, however they returned to near baseline levels after the 4-week washout. No significant changes in HDL-cholesterol, glucose or TAG concentrations were observed.

**TABLE | Mean (±SEM) concentrations of fecal short chain fatty acids (SCFA) before during and after WGO and NWG treatment arms.**

| SCFA | Pre-WGO Mean ± SEM | Post-WGO Mean ± SEM | Wash-WGO Mean ± SEM | Pre-NWG Mean ± SEM | Post-NWG Mean ± SEM | Wash-NWG Mean ± SEM | *P*-value Time by treat |
|---|---|---|---|---|---|---|---|
| Acetate, mM/g | 53.7 ± 12.5 | 67.4 ± 12.9 | 70.3 ± 15.2 | 59.4 ± 15.9 | 66.8 ± 19.1 | 57.2 ± 18.4 | 0.213 |
| Propionate, mM/g | 6.5 ± 2.5 | 21.5 ± 4.1 | 19.5 ± 3.2 | 21.3 ± 3.5 | 24.5 ± 4.9 | 24.3 ± 5.5 | 0.572 |
| Butyrate, mM/g | 14.6 ± 2.0 | 18.7 ± 5.2 | 22.1 ± 4.5 | 13.4 ± 3.1 | 18.2 ± 5.3 | 15.7 ± 4.6 | 1.33 |

*(n = 30), visit 1 represents baseline.*

**TABLE | Mean (± SEM) fecal bacterial numbers (*n* = 30) over the trial period.**

| Bacteria Group, $log_{10}$ cells/g feces | Pre-WGO Mean ± SEM | Post-WGO Mean ± SEM | Wash-WGO Mean ± SEM | Pre-NWG Mean ± SEM | Post-NWG Mean ± SEM | Wash-NWG Mean ± SEM | *P*-value Time by treat |
|---|---|---|---|---|---|---|---|
| *Bifidobacterium* | $7.97 \pm 0.05^{a}$ | $8.35 \pm 0.05^{ab}$ | $8.09 \pm 0.05^{a}$ | $8.12 \pm 0.05^{bc}$ | $8.00 \pm 0.04^{b}$ | $7.93 \pm 0.05^{bc}$ | 0.000 |
| *Bacteroides* and *Prevotella* | 8.97 ± 0.05 | 8.96 ± 0.05 | 8.95 ± 0.05 | 8.96 ± 0.05 | 8.97 ± 0.05 | 8.94 ± 0.05 | 0.990 |
| *Lactobacillus* | $8.12 \pm 0.06^{ab}$ | $8.28 \pm 0.06^{abe}$ | $8.20 \pm 0.05^{abf}$ | $8.20 \pm 0.05^{c}$ | $8.14 \pm 0.05^{d}$ | $8.02 \pm 0.05^{cdef}$ | 0.001 |
| *Ruminococcus* | 8.77 ± 0.06 | 8.89 ± 0.06 | 8.83 ± 0.07 | 8.85 ± 0.06 | 8.70 ± 0.07 | 8.67 ± 0.07 | 0.059 |
| *Clostridium histolyticum/perfringens* | 7.98 ± 0.07 | 7.92 ± 0.06 | 7.92 ± 0.06 | 7.95 ± 0.05 | 7.98 ± 0.06 | 7.95 ± 0.07 | 0.632 |
| *Atopobium* | 8.01 ± 0.06 | 7.97 ± 0.06 | 8.00 ± 0.05 | 8.01 ± 0.06 | 8.05 ± 0.06 | 8.09 ± 0.05 | 0.618 |
| *Total Population* | $9.87 \pm 0.03^{a}$ | $9.98 \pm 0.04^{abe}$ | $9.85 \pm 0.03^{b}$ | $9.98 \pm 0.03^{cd}$ | $9.87 \pm 0.05^{c}$ | $9.78 \pm 0.04^{de}$ | 0.008 |

*Bacterial counts in stool samples as determined by fluorescence in situ hybridisation are shown expressed as mean $log_{10}$ cells/g feces. WGO, Whole Oat Grain; NWG, Non-Whole Grain; [abcdef] All values in one row with a common letter are significantly different from each other (P < 0.05, Tukey's post-test).*

## Insulin Sensitivity/Resistance and Marker of Inflammation

The homeostasis model for insulin resistance (HOMA IR), as well as QUICKI (model for insulin sensitivity) was calculated for both treatments (**Table** ). There was no significant effect for either treatment. However, there was a near significant time by treatment interaction for fasting glucose ($p = 0.066$). After the WGO cereal fasting plasma glucose reduced and was lower at the end of treatment (week 6) than after the NWG. Biomarkers of inflammation including; C-Reactive Protein (CRP), IL-6, TNF-alpha from blood plasma, IgA in stool and saliva and calprotectin levels in stool samples are shown in **Table** . No significant time by treatment interaction was recorded for any of the inflammation biomarkers after consumption of either cereal treatments.

## Anthropometric Measures and Satiety Gut Hormones

The data for changes in volunteer's GLP-1 and PYY concentrations and BMI, weight and fat mass are shown in **Table** . No significant changes were detected between visits or groups for either of the cereal treatments.

## Biomarkers of Gut Health

Stool frequency and consistency, qualitatively graded by volunteers as hard, formed, or soft, varied considerably between individuals. No significance differences were observed between treatments. The severity and frequency of reported changes in digestive tolerance varied greatly between volunteers, with neither treatment resulting in adverse symptomology (data not shown).

# DISCUSSION

This novel study was designed to test the hypothesis that a whole grain oat based Granola breakfast cereal can mediate a prebiotic modulation of the human gut microbiota typified by increased relative abundance of bifidobacteria in particular, and concomitant reduction of plasma LDL cholesterol. These data also support a significant body of evidence that ingestion of oats, and oat derived fractions, most notably, β-glucan, at 3 g/day, is associated with a significant reduction in blood TC and LDL-C in hypercholesterolemic groups (EFSA Panel on Dietetic Products, Nutrition and Allergies (NDA), 2010; Tiwari and Cummins, 2011). Although, the bile acid and possibly cholesterol binding abilities of β-glucans have been suggested to be responsible for the hypocholesterolemic effects *in vivo*, other mechanisms may also be involved, including those linked to the human gut microbiota which have not been addressed to date. In this study the WGO group consumed 1.3 g β-glucan daily, which is under half of the daily recommendation for significant TC and LDL-C reduction. However, despite the low β-glucan dose, WGO consumption resulted in 0.94 mmol/l and 0.4 mmol/l lower concentrations of TC and LDL-c , respectively, compared with the NWG at the end of the 6-week intervention period. These data suggest that other mechanisms of action, other than β-glucan, could have contributed to the observed cholesterol reduction. Similar to β-glucan, other fibers or prebiotics, which do not form viscous gels within the intestinal tract, have been shown to lower blood cholesterol, though these prebiotics including

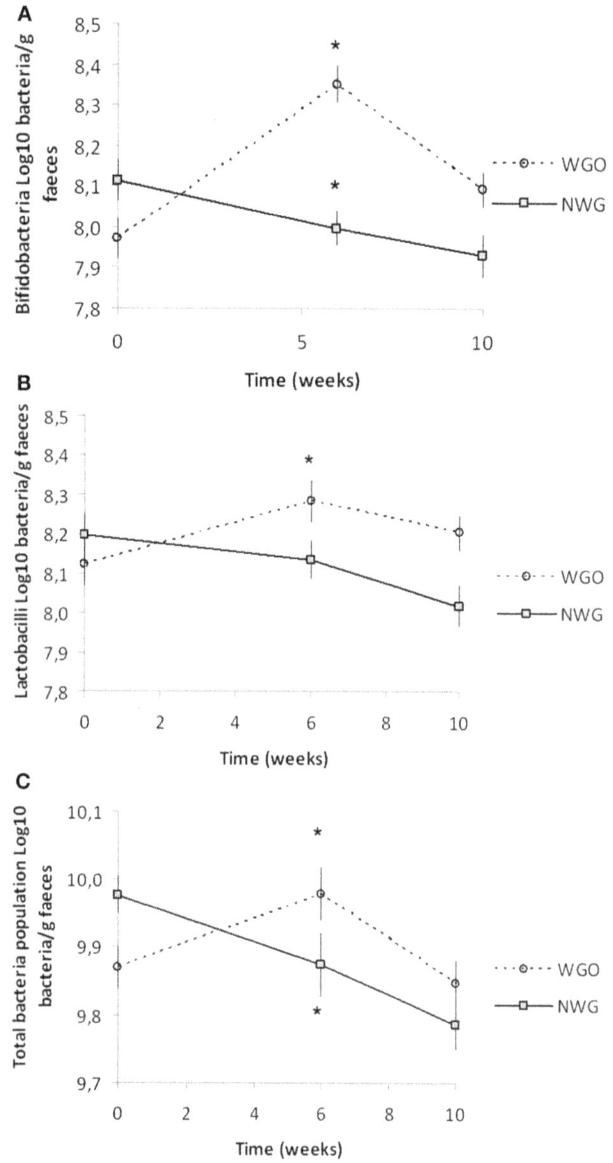


**FIGURE | Mean (± SEM) fecal bacteria changes over the trial period for both cereal treatments ($n$ = 30). (A)** Bifidobacteria, **(B)** Lactobacilli, and **(C)** Total Population in stool samples as determined by fluorescence *in situ* hybridization. * Significantly different from baseline ($P$ < 0.05, Tukey's post-test).

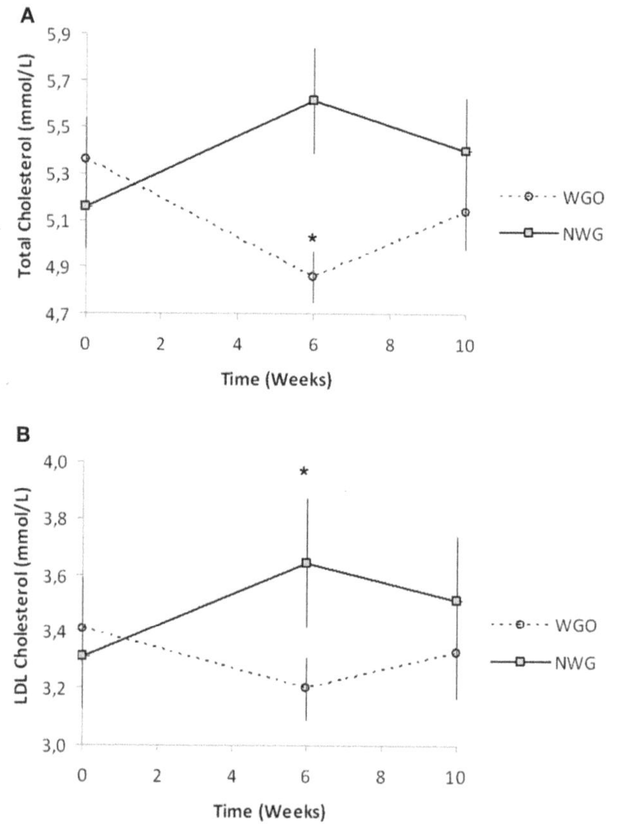


**FIGURE | TC and LDL-C concentrations in fasted blood plasma over the trial period for both cereal treatments ($n$ = 30). (A)** TC and **(B)** LDL-C. * Significantly different from baseline ($P$ < 0.05, Tukey's post-test).

the fructans, xylooligosaccharides, galactooligosaccharides and resistant starch have not received the same attention from the scientific community as β-glucans (Park et al., 2004; Beylot, 2005; Sheu et al., 2008; Vulevic et al., 2013). In parallel, a number of human feeding studies with well-powered cohorts of hypercholesteroleamic individuals have demonstrated that probiotics selected for bile salt hydrolase (BSH) activity lower plasma cholesterol levels to a similar extent to oats and β-glucan (Jones et al., 2012a,b). Such observations raise the intriguing possibility, that along with the viscous gel forming activities of β-glucan, oats may also possess other biological activities which could contribute to their cholesterol lowering ability; not least the ability to stimulate bacteria within the gut with BSH activity thereby increasing bile acid deconjugation, facilitating bile acid excretion in feces and triggering *de novo* hepatic bile acid synthesis from circulating cholesterol. Therefore, the current study was designed to investigate whether WGO mediate a prebiotic modulation of the human gut microbiota with specific increases in bifidobacteria, a group of bacteria with well recognized BSH activity and whether this was associated with reducing plasma cholesterol (Tanaka et al., 1999).

Currently few studies, and none containing oats, have measured the ability of whole grain cereals to modulate bacterial relative abundance in human feces (Costabile et al., 2008; Carvalho-Wells et al., 2010). Andersson et al. (2013) observed gut microbiota modulation in response to a dietary hypocholesterolic intervention with oats in two strains of C57BL/6 mice. We have previously shown that different whole grain cereals, including whole grain oat preparations, have prebiotic potential *in vitro*, determined by significant stimulation of both bifidobacteria and lactobacilli bacteria (Connolly et al., 2010, 2012). Using the same WGO cereal that induced the highest modulation in bifidobacteria and lactobacilli bacteria *in vitro* and also had the lowest GI (37), we have shown for the first time in humans that WGO consumption for 6 weeks

**TABLE | Mean (±SEM) fasting concentrations of plasma lipids and lipoprotein concentrations, glucose, insulin, markers of insulin resistance, and inflammatory markers at baseline, during and following ingestion of WGO and NWG.**

| Biomarker | Pre-WGO Mean ± SEM | Post-WGO Mean ± SEM | Wash-WGO Mean ± SEM | Pre-NWG Mean ± SEM | Post-NWG Mean ± SEM | Wash-NWG Mean ± SEM | *P*-value Time by treat |
|---|---|---|---|---|---|---|---|
| TC, *mmol/L* | 5.38 ± 0.18 | 4.86 ± 0.11∞ | 4.99 ± 0.17 | 5.31 ± 0.19 | 5.80 ± 0.23∞ | 5.4 ± 0.24 | 0.000 |
| HDL, *mmol/L* | 1.46 ± 0.08 | 1.48 ± 0.08 | 1.52 ± 0.08 | 1.51 ± 0.08 | 1.50 ± 0.08 | 1.43 ± 0.07 | 0.158 |
| LDL, *mmol/L* | 3.41 ± 0.92 | 3.22 ± 1.14 | 3.18 ± 1.08 | 3.31 ± 0.16 | 3.62 ± 0.14∞ | 3.51 ± 0.15 | 0.002 |
| TAG, *mmol/L* | 1.21 ± 0.11 | 1.13 ± 0.11 | 1.06 ± 0.11 | 1.38 ± 0.20 | 1.46 ± 0.20 | 1.38 ± 0.20 | 0.518 |
| Glucose, *mmol/L* | 5.44 ± 0.10 | 5.33 ± 0.19 | 5.33 ± 0.12 | 5.71 ± 0.13 | 5.82 ± 0.11 | 5.88 ± 0.12 | 0.066 |
| Insulin, *uIU/mL* | 8.42 ± 0.92 | 8.66 ± 1.13 | 7.72 ± 1.00 | 7.91 ± 1.01 | 10.23 ± 1.47 | 8.95 ± 1.08 | 0.296 |
| HOMA-IR ∇ | 2.17 ± 0.27 | 2.21 ± 0.32 | 2.00 ± 0.31 | 1.99 ± 0.27 | 2.63 ± 0.39 | 2.31 ± 0.30 | 0.259 |
| QUICKI ∇ | 0.35 ± 0.01 | 0.36 ± 0.01 | 0.36 ± 0.01 | 0.36 ± 0.01 | 0.35 ± 0.01 | 0.35 ± 0.01 | 0.137 |
| CRP, *mg/L* | 1.69 ± 0.35 | 2.45 ± 0.92 | 1.82 ± 0.47 | 1.80 ± 0.47 | 2.36 ± 0.49 | 2.02 ± 0.48 | 0.934 |
| IgA Saliva, *ng/mL* | 265 ± 27 | 245 ± 44 | 288 ± 45 | 229 ± 24 | 340 ± 47 | 299 ± 41 | 0.208 |
| IgA Stool, *ng/mL* | 157 ± 31 | 154 ± 39 | 156 ± 36 | 162 ± 31 | 173 ± 42 | 194 ± 48 | 0.197 |
| TNF-alpha, *pg/mL* | 20.2 ± 4.0 | 36.5 ± 15.7 | 32.9 ± 10.9 | 46.3 ± 26.0 | 42.2 ± 14.8 | 27.4 ± 8.4 | 0.519 |
| Calprotectin, *mg/kg* | 25.2 ± 8.2 | 25.9 ± 9.6 | 20.0 ± 4.9 | 24.5 ± 8.9 | 27.6 ± 10.3 | 24.9 ± 8.4 | 0.066 |
| IL-6, *pg/mL* | 4.13 ± 1.47 | 5.88 ± 178 | 3.85 ± 2.00 | 4.09 ± 1.71 | 7.16 ± 3.46 | 4.85 ± 1.74 | 0.925 |
| PYY, *ng/mL* | 0.88 ± 0.07 | 0.79 ± 0.08 | 0.93 ± 0.07 | 0.77 ± 0.06 | 0.78 ± 0.03 | 0.91 ± 0.06 | 0.551 |
| GLP-1, *ng/mL* | 1.04 ± 0.04 | 1.01 ± 0.06 | 1.12 ± 0.03 | 1.01 ± 0.04 | 1.02 ± 0.03 | 1.08 ± 0.03 | 0.641 |
| BMI, *kg/m²* | 26.3 ± 1.1 | 26.1 ± 1.1 | 26.2 ± 1.1 | 26.2 ± 1.0 | 26.3 ± 1.0 | 26.3 ± 1.0 | 0.061 |
| Fat Mass, *kg* | 23.0 ± 2.0 | 22.6 ± 2.0 | 22.8 ± 2.0 | 22.8 ± 1.9 | 23.2 ± 2.0 | 22.9 ± 1.2 | 0.059 |

*BMI, fat mass and weight and concentrations of satiety gut hormones GLP-1 and PYY of participants before, after and following washout of both test cereals (n= 30).*
*∞Significantly different from baseline (P < 0.05, Tukey's post-test).*
*∇ Insulin sensitivity calculated using Quantitative Insulin Sensitivity Check Index (QUICKI), and insulin resistance calculated using homeostatic model assessment (HOMA) measured before and after 42 wk of either WGO or NWG treatment. All values are average of volunteer's measurements ± SEM, (n=30). All measurements were taken in the morning after the subjects had fasted for 12 h.*

can stimulate a prebiotic modulation of the gut microbiota selectively increasing fecal bifidobacteria, lactobacilli, and total bacteria compared to a control treatment (Connolly et al., 2010, 2012). These bacterial numbers returned to baseline levels after 4 weeks washout in the group of metabolically at risk participants. No significant changes in *Bacteroides* spp., *C. histolyticum/perfringens* group, or *Atopobium* spp. was detected in the fecal samples collected, as also found in previous studies with other prebiotics. Furthermore, we observed a significant reduction in TC and LDL-C concentrations after the WGO treatment compared with NWG supporting other findings (EFSA Panel on Dietetic Products, Nutrition and Allergies (NDA), 2010; Tiwari and Cummins, 2011). Although this was associated with a significantly higher monounsaturated fatty acid (MUFA) and lower starch intake in the WGO group, this small absolute change in macronutrient intake is unlikely to have impacted on the circulating lipids, based on the predicted changes reported by Mensink et al. (1% dietary energy exchange of carbohydrate with MUFA was associated with a non-significant reduction in total cholesterol of 0.006 mmol/l and a significant, but relatively small reduction in LDL-C of 0.009 mmol/l) (Mensink et al., 2003). A number of well controlled animal and human feeding studies with strains of bacteria which possess bile salt hydrolase (BSH) activity, like *Bifidobacterium* and *Lactobacillus* spp., have reported significant reductions in circulating TC and LDL-C levels to a similar degree as oats, β-glucans and other gel forming fibers like psylleum (Begley et al., 2006; Ooi et al., 2010; Ejtahed et al., 2011; Jones et al., 2012a,b). In addition *in vitro* fermentation studies have shown that various oat fractions, including β-glucans can both increase SCFA production and modulate relative abundance of key gut microbiota bacteria (Hughes et al., 2008; Kim and White, 2009; Connolly et al., 2012; Nordlund et al., 2012). The ability of oats and β-glucans to increase SCFA production by the gut microbiota *in vitro* and in animal studies has been shown, while few studies have examined the ability of oats or β-glucans to increase fermentation in humans (Nilsson et al., 2008; Turunen et al., 2011). It has been postulated that possible mechanisms of action of prebiotics on cholesterol reduction may include physiological actions of SCFA, particularly propionate, to reduce *de novo* hepatic cholesterol synthesis by down regulation of HMG CoA reductase; increased cholesterol excretion from cholesterol assimilation, deconjugation of bile acids and cholesterol binding to bacterial or plant cell walls (Shapiro and Rodwell, 1971; Levrat et al., 1994).

The mechanisms involved with the observed cholesterol reduction in the present study are not clear. A possible limitation of the study was that plasma bile salts and SCFA were not measured. Although it was observed that fecal SCFA concentrations were similar between treatments, these data may not reflect plasma SCFA concentrations and this mechanism cannot be ruled out. Recent studies have shown that SCFA induce their own active uptake transporters on the gut wall, hindering the use of fecal SCFA concentrations as an indicator of colonic SCFA production from fermentation (Borthakur et al., 2012). Further mechanistic studies are required to examine how prebiotic modulation of the gut microbiota may be linked to cholesterol metabolism, and specifically whether a prebiotic type microbiota modulation increases bile acid deconjugation impacting on bile acid excretion or signaling within the intestine and at other body sites specifically in the liver and brain. In summary we have confirmed that dietary WGO ingestion had an appreciable impact on the composition of the human gut microbiota, and significantly reduces plasma TC and LDL-C, a step toward understanding their gut mediated communication with host energy metabolism.

## AUTHOR CONTRIBUTIONS

JL, KT, and MC designed the study. MC conducted the study. MC analyzed the data and drafted the paper and XT performed the dietary analysis. JL, KT, and XT edited the paper and all authors have read and approved the final manuscript. JL and KT had primary responsibility for final content.

## FUNDING

The study was funded by the Reading Endowment Trust Fund (RETF) and Jordans Cereals (Biggleswade, UK) who kindly supplied the breakfast cereals, however the sponsor had no input into the study hypothesis and design, data analysis and interpretation.

## ACKNOWLEDGMENTS

The authors would like to thank Jan Luff for her assistance with subject recruitment and the study volunteers for their participation in the study.

## REFERENCES

Andersson, K. E., Axling, U., Xu, J., Swärd, K., Ahrné, S., Molin, G., et al. (2013). Diverse effects of oats on cholesterol metabolism in C57BL/6 mice correlate with expression of hepatic bile acid-producing enzymes. *Eur. J. Nutr.* 52, 1755–1769. doi: 10.1007/s00394-012-0479-1

Begley, M., Hill, C., and Gahan, C. G. (2006). Bile salt hydrolase activity in probiotics. *Appl. Environ. Microbiol.* 72, 1729–1738. doi: 10.1128/AEM.72.3.1729-1738.2006

Beylot, M. (2005). Effects of inulin-type fructans on lipid metabolism in man and in animal models. *Br. J. Nutr.* 93(Suppl. 1), S163–S168. doi: 10.1079/BJN20041339

Borneo, R., and León, A. E. (2012). Whole grain cereals: functional components and health benefits. *Food Funct.* 3, 110–119. doi: 10.1039/C1FO10165J

Borthakur, A., Priyamvada, S., Kumar, A., Natarajan, A. A., Gill, R. K., Alrefai, W. A., et al. (2012). A novel nutrient sensing mechanism underlies substrate-induced regulation of monocarboxylate transporter-1. *Am. J. Physiol. Gastrointest. Liver Physiol.* 303, 1126–1133. doi: 10.1152/ajpgi.00308.2012

Carvalho-Wells, A. L., Helmolz, K., Nodet, C., Molzer, C., Leonard, C., McKevith, B., et al. (2010). Determination of the *in vivo* prebiotic potential of a maize-based whole grain breakfast cereal: a human feeding study. *Br. J. Nutr.* 104, 1353–1356. doi: 10.1017/S0007114510002084

Charlton, K. E., Tapsell, L. C., Batterham, M. J., O'Shea, J., Thorne, R., Beck, E., et al. (2012). Effect of 6 weeks' consumption of β-glucan-rich oat products on cholesterol levels in mildly hypercholesterolaemic overweight adults. *Br. J. Nutr.* 107, 1037–1047. doi: 10.1017/S0007114511003850

Chatenoud, L., Tavani, A., La Vecchia, C., Jacobs, D. R., Negri, E., Levi, F., et al. (1998). Whole grain food intake and cancer risk. *Int. J. Cancer* 77, 24–28.

Connolly, M. L., Lovegrove, J. A., and Tuohy, K. M. (2010). *In vitro* evaluation of the microbiota modulation abilities of different sized whole oat grain flakes. *Anaerobe* 16, 483–488. doi: 10.1016/j.anaerobe.2010.07.001

Connolly, M. L., Tuohy, K. M., and Lovegrove, J. A. (2012). Wholegrain oat-based cereals have prebiotic potential and low glycaemic index. *Br. J. Nutr.* 108, 2198–2206. doi: 10.1017/S0007114512000281

Costabile, A., Klinder, A., Fava, F., Napolitano, A., Fogliano, V., Leonard, C., et al. (2008) Whole-grain wheat breakfast cereal has a prebiotic effect on the human gut microbiota: a double-blind, placebo-controlled, crossover study. *Br. J. Nutr.* 99, 110–120. doi: 10.1017/S0007114507793923

Daims, H., Stoecker, K., and Wagner, M. (2005). "Fluorescence in situ hybridization for the detection of prokaryotes" in *Molecular Microbial Ecology*, eds M. Osborn and C. Smith (New York, NY: Taylor & Francis), 192–201.

EFSA Panel on Dietetic Products, Nutrition and Allergies (NDA) (2010). Scientific Opinion on the substantiation of a health claim related to oat beta glucan and lowering blood cholesterol and reduced risk of (coronary) heart disease pursuant to Article 14 of Regulation (EC) No 1924/2006. *EFSA J.* 8:1885. doi: 10.2903/j.efsa.2010.1885

Ejtahed, H. S., Mohtadi-Nia, J., Homayouni-Rad, A., Niafar, M., Asghari-Jafarabadi, M., Mofid, V., et al. (2011). Effect of probiotic yogurt containing Lactobacillus acidophilus and *Bifidobacterium lactis* on lipid profile in individuals with type 2 diabetes mellitus. *J. Dairy Sci.* 94, 3288–3294. doi: 10.3168/jds.2010-4128

Franks, A. H., Harmsen, H. J., Raangs, G. C., Jansen, G. J., Schut, F., and Welling, G. W. (1998). Variations of bacterial populations in human feces measured by fluorescent *in situ* hybridization with group-specific 16S rRNA-targeted oligonucleotide probes. *Appl. Environ. Microbiol.* 64, 3336–3345.

Gunness, P., and Gidley, M. J. (2010). Mechanisms underlying the cholesterol-lowering properties of soluble dietary fiber polysaccharides. *Food Funct.* 1, 149–155. doi: 10.1039/c0fo00080a

Harmsen, H. J., Wildeboer-Veloo, A. C., Grijpstra, J., Knol, J., Degener, J. E., and Welling, G. W. (2000). Development of 16S rRNA-based probes for the *Coriobacterium* group and the *Atopobium* cluster and their application for enumeration of *Coriobacteriaceae* in human feces from volunteers of different age groups. *Appl. Environ. Microbiol.* 66, 4523–4527. doi: 10.1128/AEM.66.10.4523-4527.2000

He, M., van Dam, R. M., Rimm, E., Hu, F. B., and Qi, L. (2010). Whole-grain, cereal fiber, bran, and germ intake and the risks of all-cause and cardiovascular disease-specific mortality among women with type 2 diabetes mellitus. *Circulation* 121, 2162–2168. doi: 10.1161/CIRCULATIONAHA.109.907360

Hughes, S. A., Shewry, P. R., Gibson, G. R., McCleary, B. V., and Rastall, R. A. (2008). *In vitro* fermentation of oat and barley derived beta-glucans by human fecal microbiota. *FEMS Microbiol. Ecol.* 64, 482–493. doi: 10.1111/j.1574-6941.2008.00478.x

Jacobs, D. R., Meyer, K. A., Kushi, L. H., and Folsom, A. R. (1999). Is whole grain intake associated with reduced total and cause-specific death rates in older women? The Iowa Women's Health Study. *Am. J. Public Health.* 89, 322–329. doi: 10.2105/AJPH.89.3.322

Jones, M. L., Martoni, C. J., Parent, M., and Prakash, S. (2012a). Cholesterol-lowering efficacy of a microencapsulated bile salt hydrolase-active *Lactobacillus reuteri* NCIMB 30242 yoghurt formulation in hypercholesterolaemic adults. *Br. J. Nutr.* 107, 1505–1513. doi: 10.1017/S0007114511004703

Jones, M. L., Martoni, C. J., and Prakash, S. (2012b). Cholesterol lowering and inhibition of sterol absorption by *Lactobacillus reuteri* NCIMB 30242: a randomized controlled trial. *Eur. J. Clin. Nutr.* 66, 1234–1241. doi: 10.1038/ejcn.2012.126

Kim, G. B., Yi, S. H., and Lee, B. H. (2004). Purification and characterization of three different types of bile salt hydrolases from *Bifidobacterium* strains. *J. Dairy Sci.* 87, 258–266. doi: 10.3168/jds.S0022-0302(04)73164-1

Kim, H. J., and White, P. J. (2009). *In vitro* fermentation of oat flours from typical and high beta-glucan oat lines. *J. Agric. Food Chem.* 57, 7529–7536. doi: 10.1021/jf900788c

Langendijk, P. S., Schut, F., Jansen, G. J., Raangs, G. C., Kamphuis, G. R., Wilkinson, M. H., et al. (1995). Quantitative fluorescence *in situ* hybridization of *Bifidobacterium* spp. with genus-specific 16S rRNA-targeted probes and its application in fecal samples. *Appl. Environ. Microbiol.* 61, 3069–3075.

Levrat, M. A., Favier, M. L., Moundras, C., Rémésy, C., Demigné, C., and Morand, C. (1994). Role of dietary propionic acid and bile acid excretion in the hypocholesterolemic effects of oligosaccharides in rats. *J. Nutr.* 124, 531–538.

Manz, W., Amann, R., Ludwig, W., Vancanneyt, M., and Schleifer, K. H. (1996). Application of a suite of 16S rRNA-specific oligonucleotide probes designed to investigate bacteria of the phylum cytophaga-flavobacter-bacteroides in the natural environment. *Microbiology* 142, 1097–1106. doi: 10.1099/13500872-142-5-1097

Mellen, P. B., Walsh, T. F., and Herrington, D. M. (2008). Whole grain intake and cardiovascular disease: a meta-analysis. *Nutr. Metab. Cardiovasc. Dis.* 18, 283–290. doi: 10.1016/j.numecd.2006.12.008

Mensink, R. P., Zock, P. L., Kester, A. D., and Katan, M. B. (2003). Effects of dietary fatty acids and carbohydrates on the ratio of serum total to HDL cholesterol and on serum lipids and apolipoproteins: a meta-analysis of 60 controlled trials. *Am. J. Clin. Nutr.* 77, 1146–1155.

Montonen, J., Knekt, P., Järvinen, R., Aromaa, A., and Reunanen, A. (2003). Whole-grain and fiber intake and the incidence of type 2 diabetes. *Am. J. Clin. Nutr.* 77, 622–629.

Nilsson, U., Johansson, M., Nilsson, A., Björck, I., and Nyman, M. (2008). Dietary supplementation with beta-glucan enriched oat bran increases fecal concentration of carboxylic acids in healthy subjects. *Eur. J. Clin. Nutr.* 62, 978–984. doi: 10.1038/sj.ejcn.1602816

Nordlund, E., Aura, A. M., Mattila, I., Kössö, T., Rouau, X., and Poutanen, K. (2012). Formation of phenolic microbial metabolites and short-chain fatty acids from rye, wheat, and oat bran and their fractions in the metabolical *in vitro* colon model. *J. Agric. Food Chem.* 60, 8134–8145. doi: 10.1021/jf3008037

Ooi, L. G., Ahmad, R., Yuen, K. H., and Liong, M. T. (2010). *Lactobacillus gasseri* [corrected] CHO-220 and inulin reduced plasma total cholesterol and low-density lipoprotein cholesterol via alteration of lipid transporters. *J. Dairy Sci.* 93, 5048–5058. doi: 10.3168/jds.2010-3311

Park, O. J., Kang, N. E., Chang, M. J., and Kim, W. K. (2004). Resistant starch supplementation influences blood lipid concentrations and glucose control in overweight subjects. *J. Nutr. Sci. Vitaminol.* 50, 93–99. doi: 10.3177/jnsv.50.93

Queenan, K. M., Stewart, M. L., Smith, K. N., Thomas, W., Fulcher, R. G., and Slavin, J. L. (2007). Concentrated oat beta-glucan, a fermentable fiber, lowers serum cholesterol in hypercholesterolemic adults in a randomized controlled trial. *Nutr. J.* 6, 6–6. doi: 10.1186/1475-2891-6-6

Ripsin, C. M., Keenan, J. M., Jacobs, D. R., Elmer, P. J., Welch, R. R., Van Horn, L., et al. (1992). Oat products and lipid lowering. A meta-analysis. *JAMA* 267, 3317–3325. doi: 10.1001/jama.1992.03480240079039

Ryan, D., Kendall, M., and Robards, K. (2007). Bioactivity of oats as it relates to cardiovascular disease. *Nutr. Res. Rev.* 20, 147–162. doi: 10.1017/S0954422407782884

Rycroft, C. E., Jones, M. R., Gibson, G. R., and Rastall, R. A. (2001). A comparative *in vitro* evaluation of the fermentation properties of prebiotic oligosaccharides. *J. Appl. Microbiol.* 91, 878–887. doi: 10.1046/j.1365-2672.2001.01446.x

Shapiro, D. J., and Rodwell, V. W. (1971). Regulation of hepatic 3-hydroxy-3-methylglutaryl coenzyme A reductase and cholesterol synthesis. *J. Biol. Chem.* 246, 3210–3216.

Sheu, W. H.-H., Lee, I. T., Chen, W., and Chan, Y.-C. (2008). Effects of xylooligosaccharides in type 2 diabetes mellitus. *J. Nutr. Sci. Vitaminol.* 54, 396–401. doi: 10.3177/jnsv.54.396

Tanaka, H., Doesburg, K., Iwasaki, T., and Mierau, I. (1999). Screening of lactic acid bacteria for bile salt hydrolase activity. *J. Dairy Sci.* 82, 2530–2535. doi: 10.3168/jds.S0022-0302(99)75506-2

Tiwari, U., and Cummins, E. (2011). Meta-analysis of the effect of β-glucanăintake on blood cholesterol and glucose levels. *Nutrition* 27, 1008–1016. doi: 10.1016/j.nut.2010.11.006

Turunen, K., Tsouvelakidou, E., Nomikos, T., Mountzouris, K. C., Karamanolis, D., Triantafillidis, J., et al. (2011). Impact of beta-glucan on the fecal microbiota of polypectomized patients: a pilot study. *Anaerobe* 17, 403–406. doi: 10.1016/j.anaerobe.2011.03.025

Vulevic, J., Juric, A., Tzortzis, G., and Gibson, G. R. (2013). A mixture of trans-galactooligosaccharides reduces markers of metabolic syndrome and modulates the fecal microbiota and immune function of overweight adults. *J. Nutr.* 143, 324–331. doi: 10.3945/jn.112.166132

Wolever, T. M., Gibbs, A. L., Brand-Miller, J., Duncan, A. M., Hart, V., Lamarche, B., et al. (2011). Bioactive oat β-glucan reduces LDL cholesterol in Caucasians and non-Caucasians. *Nutr. J.* 10,130. doi: 10.1186/1475-2891-10-130

Ye, E. Q., Chacko, S. A., Chou, E. L., Kugizaki, M., and Liu, S. (2012). Greater whole-grain intake is associated with lower risk of type 2 diabetes, cardiovascular disease, and weight gain. *J. Nutr.* 142, 1304–1313. doi: 10.3945/jn.111.155325

Zhang, J., Li, L., Song, P., Wang, C., Man, Q., Meng, L., et al. (2012). Randomized controlled trial of oatmeal consumption versus noodle consumption on blood lipids of urban Chinese adults with hypercholesterolemia. *Nutr. J.* 11:54. doi: 10.1186/1475-2891-11-54


**Conflict of Interest Statement:** The authors declare that the research was conducted in the absence of any commercial or financial relationships that could be construed as a potential conflict of interest.

# 9

# Phylogenetic Analysis of the *Bifidobacterium* Genus using Glycolysis Enzyme Sequences

*Katelyn Brandt[1,2] and Rodolphe Barrangou[1,2]**

*[1] Functional Genomics Graduate Program, North Carolina State University, Raleigh, NC, USA, [2] Department of Food, Bioprocessing and Nutrition Sciences, North Carolina State University, Raleigh, NC, USA*

Bifidobacteria are important members of the human gastrointestinal tract that promote the establishment of a healthy microbial consortium in the gut of infants. Recent studies have established that the *Bifidobacterium* genus is a polymorphic phylogenetic clade, which encompasses a diversity of species and subspecies that encode a broad range of proteins implicated in complex and non-digestible carbohydrate uptake and catabolism, ranging from human breast milk oligosaccharides, to plant fibers. Recent genomic studies have created a need to properly place *Bifidobacterium* species in a phylogenetic tree. Current approaches, based on core-genome analyses come at the cost of intensive sequencing and demanding analytical processes. Here, we propose a typing method based on sequences of glycolysis genes and the proteins they encode, to provide insights into diversity, typing, and phylogeny in this complex and broad genus. We show that glycolysis genes occur broadly in these genomes, to encode the machinery necessary for the biochemical spine of the cell, and provide a robust phylogenetic marker. Furthermore, glycolytic sequences-based trees are congruent with both the classical 16S rRNA phylogeny, and core genome-based strain clustering. Furthermore, these glycolysis markers can also be used to provide insights into the adaptive evolution of this genus, especially with regards to trends toward a high GC content. This streamlined method may open new avenues for phylogenetic studies on a broad scale, given the widespread occurrence of the glycolysis pathway in bacteria, and the diversity of the sequences they encode.

**Edited by:**
*Francesca Turroni,*
*University College of Cork, Ireland*

**Reviewed by:**
*Abelardo Margolles,*
*Consejo Superior de Investigaciones Científicas, Spain*
*Gabriele Andrea Lugli,*
*University of Parma, Italy*

***Correspondence:**
*Rodolphe Barrangou*
*rbarran@ncsu.edu*

**Keywords: *Bifidobacterium*, glycolysis, phylogeny, probiotic, evolution**

## INTRODUCTION

*Bifidobacterium* species are an important component of the human gastrointestinal tract (GIT) microbiome, and exert critical functional roles, especially during the establishment of gut microbial composition early in life. Consequently, they are the subject of extensive microbiological and genetics studies, to investigate their probiotic phenotypes, and genotypes, respectively. Actually, many studies are investigating the genetic basis for their health-promoting functionalities, both in industry and academia. This genus is often found in the GIT of animals (Ventura et al., 2014), and is the predominant phylogenetic group early in human life (Turroni et al., 2012a). Indeed, a mounting body of evidence has established vertical transmission between the mother and infants (Milani et al., 2015), notably through the selective nurture of bifidobacteria through diverse non-digestible human-milk oligosaccharides (HMOs) that are a critical component of breast milk (Sela, 2011). These HMOs selectively drive the colonization of the infantile GIT by species that encode prebiotic transporters and hydrolases (Turroni et al., 2012b). Recently, a dichotomy has been established between healthy term babies with a normal gut microbiome, and preterm infants whom have not been colonized by *Bifidobacterium* species (Arboleya et al., 2015). Several studies have implicated the expansive carbohydrate uptake and catabolism gene repertoire of bifidobacteria as the key driver of adaptation of this genus to the infant diet (Milani et al., 2014). In fact, several species of bifidobacteria have shown unique genome composition adaptation trajectories in their carbohydrate utilization machinery, rendering them competitive in this environment (Pokusaeva et al., 2011; Ventura et al., 2012).

To better understand how these organisms have emerged as potent early-life colonizers, there has been a surge in genome sequencing in recent years. At the time of writing, 47 established species and subspecies have been sequenced (Milani et al., 2016), providing a wealth of genomic information, which serves as a valuable tool for understanding the species and strain diversity within this polymorphic genus, as well as unraveling the key elements that drive health-promoting and colonization phenotypes in humans. However, given the democratization of sequencing technologies in general, and genome and microbiome sequencing in particular, it is imperative that tools and methods be available to analyze this high-throughput data, and specifically allow experimentalists to parse out the complex phylogeny of this broad genus. Indeed, basic questions being addressed regarding the occurrence, diversity and functions of various *Bifidobacterium* species in the human GIT will require the ability to accurately and consistently assign phylogeny.

Fundamentally, as new sequences become available, it is important to know where to place strains on the phylogenetic tree of *Bifidobacterium*. Whereas the affordability, accessibility and ability to generate high-throughput data have become somewhat straightforward, a key challenge lies in the analysis of these sequences, regarding assembly, comparative analyses and phylogenetic assignments. Historically, 16S rRNA sequences have been used across the phylogenetics field for classification and sequence tree-based assignments, but there are growing concerns about the adequacy and sustainability of this method (Fox et al., 1992), notably with regards to the availability of proper references (Clarridge, 2004), and the actual levels of conservation of sequences targeted by "universal" primers (Baker et al., 2003). Because of this, new approaches have been suggested, ranging from multi-locus approaches, using housekeeping genes (Eisen, 1995), to core-genome analyses (Medini et al., 2005). For *Bifidobacterium*, efforts have been focused on creating a phylogeny based on whole and/or conserved genomic sequences, namely the pan-genome and the core-genome, respectively (Lukjancenko et al., 2011; Lugli et al., 2014). While the core-genome is arguably comprehensive, core-genome assembly is time consuming and computationally intense. Alternative methods need to be developed, to allow rapid and convenient phylogenetic screening of new and potentially unknown sequences. Preferably, such a method would provide high resolution, low-throughput, robust, accurate, and affordable information.

Notwithstanding phenotypic diversity between organisms that have specialized metabolic pathway combinations, and the corresponding genomic complement, there are core biochemical pathways and processes that are broadly distributed across the Tree of Life. Noteworthy, glycolysis is a fundamental process for most cells, and may be construed as the biochemical backbone of most, if not all, living organisms. Indeed, this process allows for the genesis of energy through the catabolism of simple carbohydrates. This pathway is, at least partially, present in all genomes (Fothergill-Gilmore and Michels, 1993) and consequently constitutes a promising biochemical, and thus genetic, marker for phylogenetic studies. Because these genes are important, they are typically members of the house-keeping genomic set, and are widely dispersed across the Tree of Life. However, they are likely subject to less selective pressure than other phylogenetic markers (i.e., ribosomal sequences), and thus afford a more diverse set of sequences to encompass a broad range of assorted sequences (Fothergill-Gilmore, 1986). Therefore, we set out to assess the potential of glycolytic genes, and the sequences of the proteins they encode, for bifidobacteria phylogenetic studies. In particular, we determined the occurrence and diversity of these glycolytic enzyme genes in the genomes of bifidobacteria, and compared and contrasted sequence alignment-based trees with one another, and to those derived from alternative sequences, notably the core-genome, and the 16S rRNA-based reference tree. Our results show how the glycolysis protein sequences can be used as suitable markers to create a phylogeny of *Bifidobacterium* that is as accurate as the core-genome based phylogeny, but much less computationally demanding. We also explore how basic features of the genetic sequences of glycolysis can reveal trends and patterns of evolution among the different *Bifidobacterium* species and the genus as a whole.

## MATERIALS AND METHODS

### Genetic Sequences Sampling and Reference Genomes

We used sequences derived from a total of 48 *Bifidobacterium* genomes from distinct species and subspecies, as listed in **Table 1**. *Bifidobacterium stercoris* was included in this analysis, as a separate species, but it was recently renamed as a strain of *Bifidobacterium adolescentis* (Killer et al., 2013). Our results (see below) show that *B. stercoris* is always a close neighbor of *B. adolescentis*, consistent with the newest findings. These genomes were mined for the presence of glycolytic enzymes using Geneious version 9.0.5(Kearse et al., 2012). We selectively elected to pursue a scheme based on canonical glycolysis genes, as to generate a broadly applicable method. Nevertheless, the classical glycolysis genes do not universally occur in bacterial genomes. Furthermore, some organisms do carry alternative pathways, such as the bifid shunt in *bifidobacterium*, which could prove valuable, but are not widely distributed. The nine canonical glycolysis enzymes from bifidobacteria (de Vries and

Stouthamer, 1967) were found in each genome. Four reference species (*Bifidobacterium longum* subsp. *longum*, *B. adolescentis*, *Bifidobacterium animalis sub. lactis*, and *Bifidobacterium breve*) were used to make a database of the nine genes. The Annotate from Database feature was used (with 40% nucleotide sequence similarity cut-off) to identify glycolytic orthologs in the other genomes. As all genomes had been previously annotated, we confirmed the original annotation to the database annotation

**TABLE | Species and genome list.**

| Genus | Species | Subspecies | Strain | Accession number | Naming convention | Locus tag |
|---|---|---|---|---|---|---|
| *Bifidobacterium* | *actinocoloniiforme* | | DSM 22766 | NZ_CP011786 | B_actinocoloniiforme | AB656 |
| *Bifidobacterium* | *adolescentis* | | ATCC 15703 | NC_008618 | B_adolescentis | BAD |
| *Bifidobacterium* | *angulatum* | | LMG 11039 | NZ_JGYL00000000 | B_angulatum | BIANG |
| *Bifidobacterium* | *animalis* | *animalis* | ATCC 22527 | NC_017834 | B_animalis_a | BANAN |
| *Bifidobacterium* | *animalis* | *lactis* | DSM 10140 | NC_012815 | B_animalis_l | BALAT |
| *Bifidobacterium* | *asteroides* | | PRL 2011 | NC_018720 | B_asteroides | BAST |
| *Bifidobacterium* | *biavatii* | | DSM 23969 | NZ_JDUU00000000 | B_biavatti | OU23 |
| *Bifidobacterium* | *bifidum* | | LMG 13200 | NZ_JSEB00000000 | B_bifidum | LMG13200 |
| *Bifidobacterium* | *bohemicum* | | DSM 22767 | NZ_JDUS00000000 | B_bohemicum | OU21 |
| *Bifidobacterium* | *bombi* | | DSM 19703 | NZ_JDTS00000000 | B_bombi | OT95 |
| *Bifidobacterium* | *boum* | | LMG 10736 | NZ_JGYQ00000000 | B_boum | BBOU |
| *Bifidobacterium* | *breve* | | UCC 2003 | NC_020517 | B_breve | Bbr |
| *Bifidobacterium* | *callitrichos* | | DSM 23973 | NZ_JGYS00000000 | B_callitrichos | BCAL |
| *Bifidobacterium* | *catenulatum* | | JCM 1194 | NZ_AP012325 | B_catenulatum | BBCT |
| *Bifidobacterium* | *choerinum* | | LMG 10510 | NZ_JGYU00000000 | B_choerinum | BCHO |
| *Bifidobacterium* | *coryneforme* | | LMG 18911 | NZ_CP007287 | B_coryneforme | BCOR |
| *Bifidobacterium* | *crudilactis* | | LMG 23609 | NZ_JHAL00000000 | B_crudilactis | DB51 |
| *Bifidobacterium* | *cuniculi* | | LMG 10738 | NZ_JGYV00000000 | B_cuniculi | BCUN |
| *Bifidobacterium* | *dentium* | | Bd1 | NC_013714 | B_dentium | BDP |
| *Bifidobacterium* | *gallicum* | | DSM 20093 | NZ_ABXB00000000 | B_gallicum | BIFGAL |
| *Bifidobacterium* | *gallinarum* | | LMG 11586 | NZ_JGYX00000000 | B_gallinarum | BIGA |
| *Bifidobacterium* | *indicum* | | LMG 11587 | NZ_CP006018 | B_indicum | BINDI |
| *Bifidobacterium* | *kashiwanohense* | | JCM 15439 | NZ_AP012327 | B_kashiwanohense | BBKW |
| *Bifidobacterium* | *longum* | *longum* | NCC 2705 | NC_004307 | B_longum | BL |
| *Bifidobacterium* | *longum* | *infantis* | ATCC 15697 | NC_011593 | B_longum_i | Blon |
| *Bifidobacterium* | *longum* | *suis* | LMG 21814 | NZ_JGZA00000000 | B_longum_s | BLSS |
| *Bifidobacterium* | *magnum* | | LMG 11591 | NZ_JGZB00000000 | B_magnum | BMAGN |
| *Bifidobacterium* | *merycicum* | | LMG 11341 | NZ_JGZC00000000 | B_merycicum | BMERY |
| *Bifidobacterium* | *minimum* | | LMG 11592 | NZ_JGZD00000000 | B_minimum | BMIN |
| *Bifidobacterium* | *mongoliense* | | DSM 21395 | NZ_JGZE00000000 | B_mongoliense | BMON |
| *Bifidobacterium* | *moukalabense* | | DSM 27321 | NZ_AZMV00000000 | B_moukalabense | BMOU |
| *Bifidobacterium* | *pseudocatenulatum* | | JCM 1200 | NZ_AP012330 | B_pseudocatenulatum | BBPC |
| *Bifidobacterium* | *pseudolongum* | *globosum* | LMG 11569 | NZ_JGZG00000000 | B_pseudolongum_g | BPSG |
| *Bifidobacterium* | *pseudolongum* | *pseudolongum* | LMG 11571 | NZ_JGZH00000000 | B_pseudolongum_p | BPSP |
| *Bifidobacterium* | *psychraerophilum* | | LMG 21775 | NZ_JGZI00000000 | B_psychraerophilum | BPSY |
| *Bifidobacterium* | *pullorum* | | LMG 21816 | NZ_JGZJ00000000 | B_pullorum | BPULL |
| *Bifidobacterium* | *reuteri* | | DSM 23975 | NZ_JGZK00000000 | B_reuteri | BREU |
| *Bifidobacterium* | *ruminantium* | | LMG 21811 | NZ_JGZL00000000 | B_ruminantium | BRUM |
| *Bifidobacterium* | *saeculare* | | LMG 14934 | NZ_JGZM00000000 | B_saeculare | BSAE |
| *Bifidobacterium* | *saguini* | | DSM 23967 | NZ_JGZN00000000 | B_saguini | BISA |
| *Bifidobacterium* | *scardovii* | | LMG 21589 | NZ_JGZO00000000 | B_scardovii | BSCA |
| *Bifidobacterium* | *stellenboschense* | | DSM 23968 | NZ_JGZP00000000 | B_stellenboschense | BSTEL |
| *Bifidobacterium* | *stercoris* | | DSM 24849 | NZ_JGZQ00000000 | B_stercoris | BSTER |
| *Bifidobacterium* | *subtile* | | LMG 11597 | NZ_JGZR00000000 | B_subtile | BISU |
| *Bifidobacterium* | *thermacidophilum* | *porcinum* | LMG 21689 | NZ_JGZS00000000 | B_thermacidophilum_p | BPORC |
| *Bifidobacterium* | *thermacidophilum* | *thermacidophilum* | LMG 21395 | NZ_JGZT00000000 | B_thermacidophilum_t | THER5 |
| *Bifidobacterium* | *thermophilum* | | JCM 7027 | – | B_thermophilum | BTHER |
| *Bifidobacterium* | *tsurumiense* | | JCM 13495 | NZ_JGZU00000000 | B_tsurumiense | BITS |

*List of the 48 species and subspecies used in this study. Accession numbers and naming conventions included.*

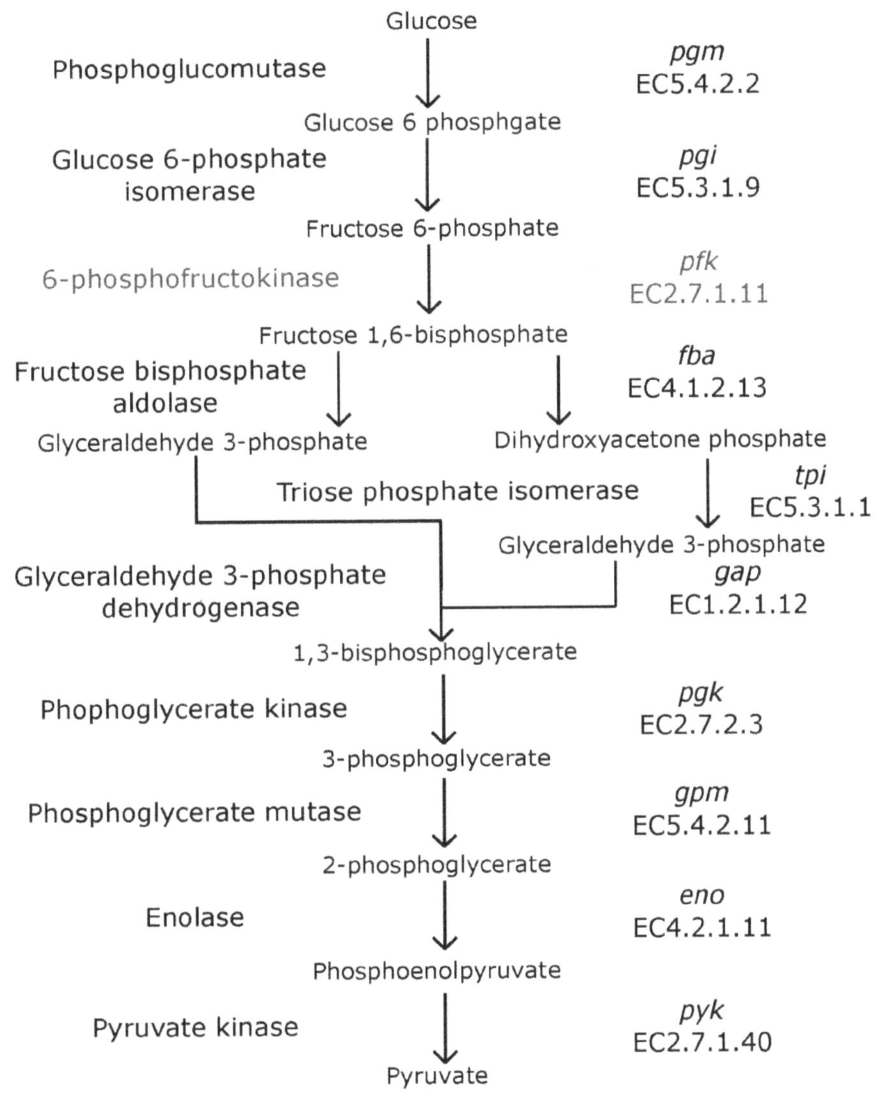


**FIGURE | Glycolysis pathway.** Traditional biochemical pathway of glycolysis. Enzyme names are listed to left of arrows, and gene names and EC numbers are shown on the right. 6-phosphofructokinase is faded to represent its absence in *Bifidobacterium*.

manually to validate this method of mining. In cases where multiple hits were obtained, BLAST (Altschul et al., 1990) analyses were carried out to select the correct homolog. Translated sequences were confirmed using ExPasy (Gasteiger et al., 2003). For the 16S rRNA analysis, the 16S rRNA sequences were extracted manually from each genome. In case of multiple hits, BLAST analyses were carried out to select the right sequences. For increased robustness, the glycolysis enzyme sequences were concatenated in order of occurrence in the glycolysis pathway (Lang et al., 2013).

## Genesis of Sequence Alignment-based Trees

Five different alignments were made for each tree using Geneious version 9.0.5. ClustalW (Larkin et al., 2007) was used, with the BLOSUM scoring matrix, and settings of gap creation at −10 cost, and gap extension at −0.1 cost per element. For the 16S rRNA alignment, ClustalW was set so that the cost matrix was IUB, with a gap opening penalty of 15, and gap extension cost of 6.66. MUSCLE (Edgar, 2004) was used with the setting of eight maximum number of iterations for the amino acid sequences and the 16S rRNA alignments. The Geneious Pairwise Alignment was set so that the alignment type was global alignment with free end gaps and the cost matrix was BLOSUM62 for the amino acid sequences. For the 16S rRNA gene analysis, the alignment type was global alignment with free end gaps and a cost matrix of 65% similarity (5.0/−4.0). MAFFT (Katoh et al., 2002) was used twice, for both the amino acid sequences and the 16S rRNA sequences. For the amino acid sequences the first alignment had an algorithm setting of auto, a scoring matrix of BLOSUM62, a gap open penalty of 1.53, and an offset value of 0.123. The second

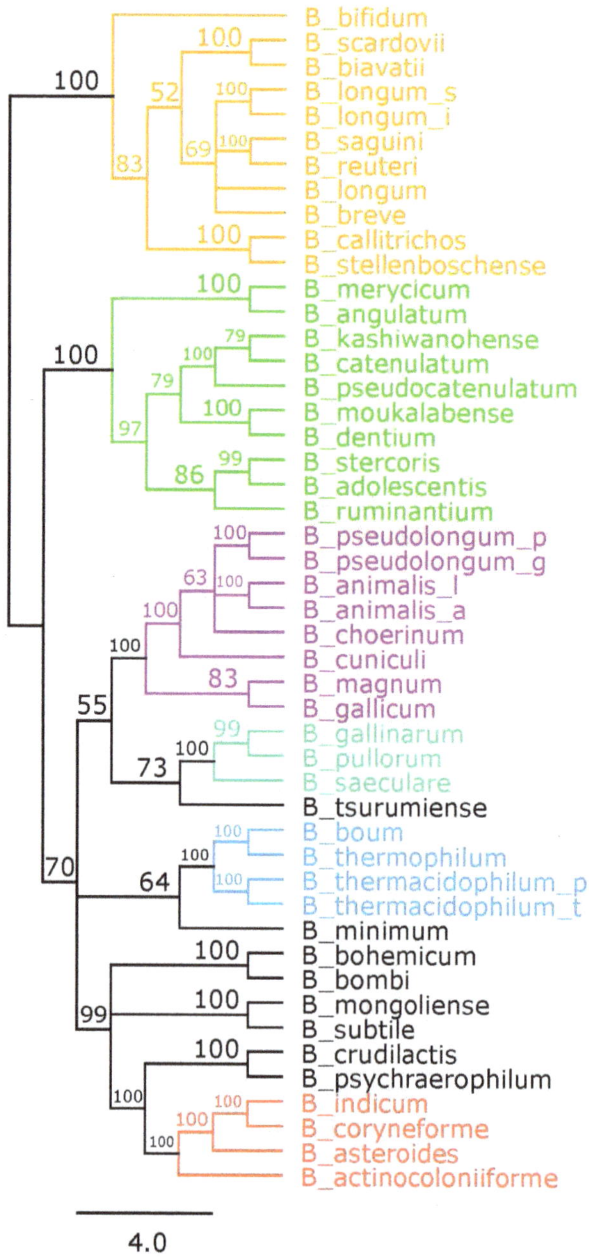


**FIGURE | Glycolytic proteins concatenated tree.** Consensus tree based on alignment of the concatenated amino acid sequences of the glycolysis pathway found in *Bifidobacterium*. Trees were made using RaxML. Bootstrap values are found on each node. Phylogenetic groups are colored as follows: *Bifidobacterium longum* is orange, *Bifidobacterium adolescentis* is green, *Bifidobacterium psdeudolongum* is purple, *Bifidobacterium pollorum* is blue-green, *Bifidobacterium boum* is blue, and *Bifidobacterium asteroides* is red. Species names follow the naming convention from **Table**

alignment had an algorithm setting of auto, a scoring matrix of BLOSUM80, a gap open penalty of 1.53, and an offset value of 0.123. For the first 16S rRNA alignment, the algorithm was set to auto, the scoring matrix was set to 100 PAM/k = 2, the gap open penalty was set to 1.53, and the offset value was set to 0.123. The second alignment for the 16S rRNA was set so that the algorithm was auto, the scoring matrix was 200 PAM/k = 2, the gap open penalty was 1.53, and the offset value was 0.123. trimAl (Capella-Gutiérrez et al., 2009) was used to select a consistent alignment between the five alignments. The parameters were compareset and automated1. Using Geneious, trees were made from the respective consistent alignments. The trees were generated using RaxML version 7.2.8 (Stamatakis, 2006b, 2014). For the protein based trees the parameters were set so that the model was CAT (Lartillot and Philippe, 2004) BLOSUM62, the algorithm was Bootstrap using rapid hill climbing with random seed 1, and the number of bootstrap replicates was 100 (Stamatakis, 2006a). For the 16S rRNA tree, the nucleotide model was GTR CAT, the algorithm was Bootstrap using rapid hill climbing with random seed 1, and the number of bootstrap replicates was 100. A consensus tree was then built using the consensus builder in Geneious, at a 50% support threshold. The consensus tree was used in all further analyses. The sums of branch lengths for each tree were found by adding the branch lengths together in Mega6 (Tamura et al., 2013).

**TABLE | Sum of branch lengths for each tree.**

| Gene | E. C. number | Sum |
|---|---|---|
| Phosphoglucomutase (pgm,1) | 5.4.2.2 | 125.03 |
| Glucose-6-phosphate isomerase (pgi,2) | 5.3.1.9 | 153.43 |
| Fructose bisphosphate aldolase (fba, 4) | 4.1.2.13 | 151.76 |
| Triose phosphate isomerase (tpi, 5) | 5.3.1.1 | 170.61 |
| Glyceraldehyde 3-phosphate dehydrogenase (gap, 6) | 1.2.1.12 | 103.07 |
| Phsophoglycerate kinase (pgk, 7) | 2.7.2.3 | 132.41 |
| Phosphoglycerate mutase (gpm, 8) | 5.4.2.11 | 174.7 |
| Enolase (eno, 9) | 4.2.1.11 | 145.06 |
| Pyruvate kinase (pyk, 10) | 2.7.1.40 | 107.56 |
| Concatenated | – | 99.56 |
| 16S rRNA | – | 204.99 |

*Sum of branch lengths for each tree. EC number for each enzyme is also listed.*

## Statistical Analyses

All statistical analyses were carried out using R version 3.2.2 (R Core Team, 2015). This software was also used to generate plots, graphs and display quantitative data throughout the manuscript.

# RESULTS

## Glycolytic Enzyme Sequence-based Phylogeny

Bifidobacteria contain nine of the 10 traditional enzymes (**Figure** ) commonly found in the glycolysis pathway (de Vries and Stouthamer, 1967). Phylogenetic analyses were carried out using the amino acid sequences of the proteins encoded by the aforementioned glycolysis genes. A comprehensive tree based on sequence alignment of the concatenated sequences of the glycolytic enzymes found in *Bifidobacterium* is shown in **Figure** . Six separate phylogenetic groups were identified, as previously established from the core-genome (Milani et al., 2016). These groups are: the *B. longum* group (orange), the *B. adolescentis* group (green), the *Bifidobacterium pseudolongum* group (purple), the *Bifidobacterium pollurom* group (blue-green), the *Bifidobacterium boum* group (blue), and the *Bifidobacterium asteroides* group (red; Bottacini et al., 2014). The number of individuals in each group varied between 3 and 11, with the *B. longum* group being the most diverse. *Bifidobacterium angulatum* and *Bifidobacterium merycicum* were moved to the *B. adolescentis* group due to a high bootstrap value in the concatenated tree. The concatenated tree has bootstrap values that range from 52 to 100. We observe a total of 34 bootstrap values of 70 and above (Supplementary Figure ). Trees based on sequence alignments of the individual enzymes of glycolysis can be found in Supplementary Figures . Interestingly, all of the individual trees resolved the phylogenetic groups found in the core-genome with only the Gap and Eno trees providing alternative locations for a few branches, notably *Bifidobacterium magnum*, *Bifidobacterium gallicum*, and *Bifidobacterium thermacidophilum sub. thermacidophilum*. **Table** shows the sum of branch lengths for each tree. The 16S rRNA tree has the largest sum at 204.99, while the concatenated tree had the smallest sum at 99.56. The consistent clustering into these six phylogenetic trees illustrates how robust and valuable the glycolytic sequences are with regards to phylogenetic information. It also shows that this method is congruent with the core-genome.

## 16S rRNA-based Reference Phylogeny

A reference phylogeny was generated using the 16S rRNA sequences of each of the 48 species and sub-species included in this study (**Figure** ). The six phylogenetic groups are identified and colored the same as in the concatenated tree. We elected to assign the *B. angulatum* and *B. merycicum* from the *B. longum* group to the *B. adolescentis* group, consistent with the concatenated tree. Noteworthy, the tree has bootstrap values that range from 51 to 100, with 17 nodes at values of 70 and above, which is half the amount found in the concatenated tree (Supplementary Figure ). With regards to size, we point out that the concatenated tree is based on overall sequences ranging between 3,205 amino acids and 3,479 amino acids, which quantitatively compares as approximately twice the amount to the 16S rRNA ~1,600 nt range, in terms of input-information amounts.

## Genome-Wide Analyses

The overall genome sizes in this study ranged from 1.73 Mb for *Bifidobacterium indicum* to 3.26 Mb for *Bifidobacterium biavatii*, with an average of 2.28 Mb and a median of 2.17 Mb. The GC content ranged from 52.8% for *Bifidobacterium tsurumiense* to 65.5% for *Bifidobacterium choerinum*, with an average of 60.4% and a median of 60.2%. This substantiates the perception that bifidobacteria are generally categorized as high-GC content organisms, at the genome-wide level (Ventura et al., 2007). However, a thorough analysis of GC content across the phylogenetic groups revealed that even among these high-GC organisms there are three distinct subsets of high, medium, and low-GC bifidobacteria (**Figure** ). Most of the species fall in the upper medium-GC range, with the low-GC range being the least populated. There are some noteworthy groupings between the phylogenetic groups, specifically the *B. pullorum* and the *B. boum* groups, for which the entire groups are packed tightly

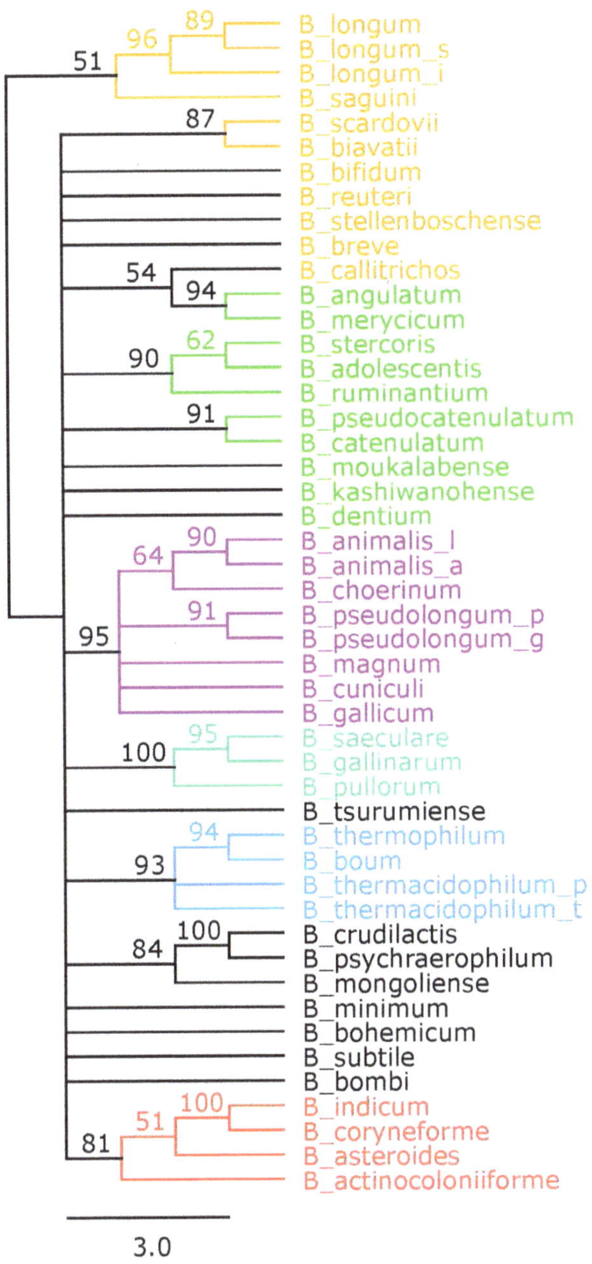


**FIGURE | 16S rRNA phylogenetic tree.** Consensus tree based on alignment of the 16S rRNA sequences. Trees were made using RaxML. Bootstrap values are found on each node. Phylogenetic groups are colored as follows: *B. longum* is orange, *B. adolescentis* is green, *B. psdeudolongum* is purple, *B. pollorum* is blue-green, *B. boum* is blue, and *B. asteroides* is red. Species names follow the naming convention from **Table** .

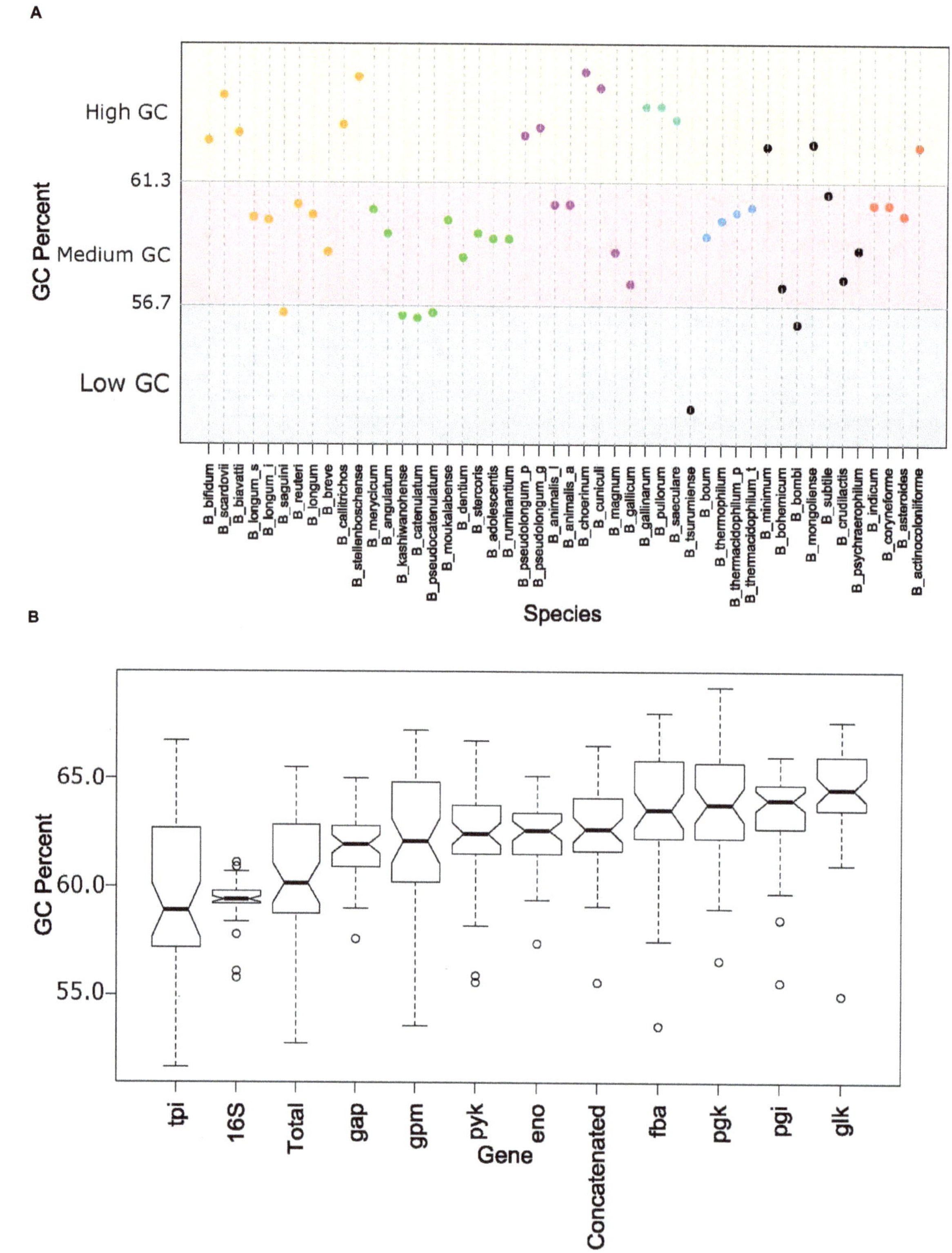


**FIGURE | GC content by species and glycolytic genes. (A)** Shows the total GC content of each species organized by the glycolytic concatenated tree. Spectrum is split into three groups: low GC from 0.52 to 0.567 (gray), Medium GC from 0.567 to 0.613 (pink), and High GC from 0.613 to 0.66 (yellow). Phylogenetic groups are colored as follows: *B. longum* is orange, *B. adolescentis* is green, *B. psdeudolongum* is purple, *B. pollorum* is blue-green, *B. boum* is blue, and *B. asteroides* red. Species names following the naming convention from **Table** . contains notched boxplots of the GC values of each gene and total GC. Boxes are ranked in order of median. Notches that do not overlap are indicative of strong evidence of difference between two medians.

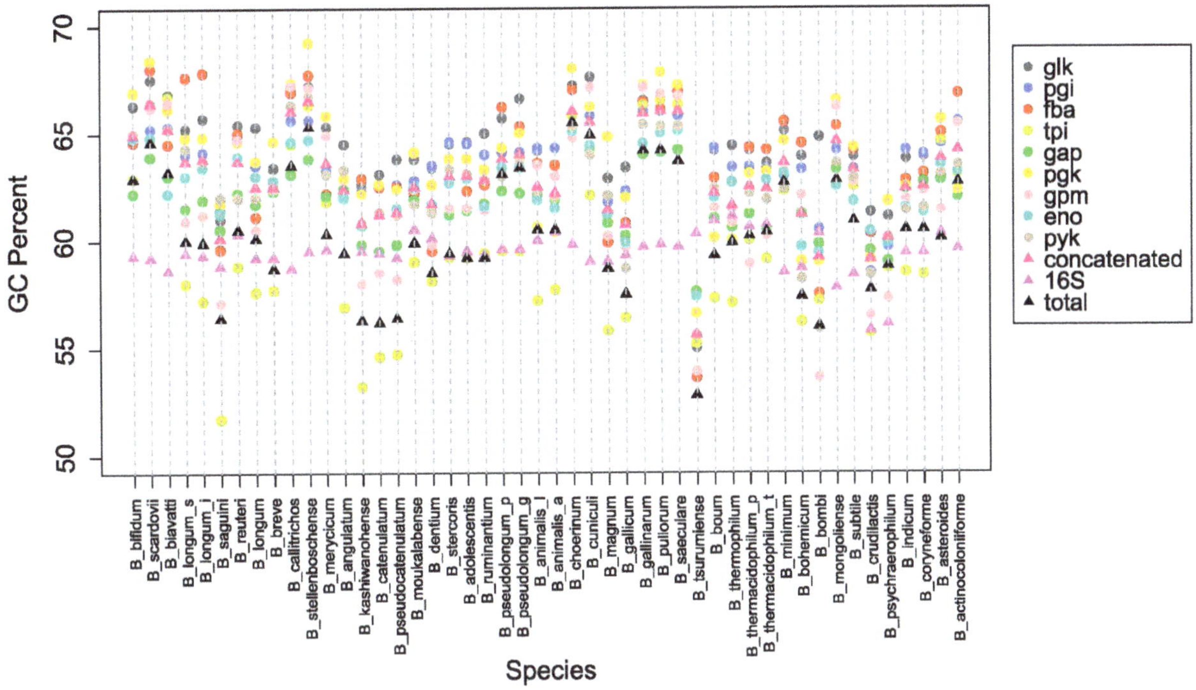


**FIGURE | Overall GC content patterns across species.** GC percent for each glycolysis gene, 16SrRNA and overall genome, for species listed in **Table** .

in the high GC region and the medium GC region, respectively. All of the other groups, except the *B. longum* group, span two of these subsets. For the *B. longum* group, *Bifidobacterium saguini* lies just at the border between the low and medium GC subsets. This group has the largest spread, consistent with being the most diverse in the concatenated and 16S rRNA trees.

Next we looked at how the GC content varied across the trees. **Figure B** shows boxplots of the GC content of each tree and the total GC content. Except for the 16S rRNA and *tpi* trees, all other trees had median GC values with strong evidence of being higher than the median total GC content (Chambers, 1983). Looking on an individual basis, over half of the genomes have 16S rRNA and *tpi* GC values below their total GC, while the other genes are either above or close to their total GC (**Figure** ). Again, the *B. pullorum* and *B. boum* groups are tightly packed in regards to their GC spread amongst their glycolysis genes, 16S rRNA, and total GC. In contrast, the *B. longum* group has the largest spread, a parallel to its higher diversity in the phylogenetic trees.

## DISCUSSION

*Bifidobacterium* is a diverse genus of human intestinal beneficial microbes that provide health-promoting functionalities, as illustrated by their broad use as probiotics in foods and dietary supplements (Turroni et al., 2014). Recently, extensive genomic analyses of diverse species, subspecies and phylogenetic groups have provided insights into their adaptation to the human gut, notably with regards to their ability to colonize the intestinal cavity in general, and utilize non-digestible carbohydrates in particular (Milani et al., 2016). Studies investigating the use of human breast milk oligosaccharides illustrate the important contribution of these probiotics in establishing the human gut microbiome at the early stages of life (Sela, 2011). Yet, these studies also reveal that there are many distinct and diverse *Bifidobacterium* species and phylogenetic groups that colonize the human GIT, perhaps with idiosyncratic genomic attributes, and their corresponding functionalities (Chaplin et al., 2015). These organisms have specifically adapted to their environment to competitively utilize available nutrients (Sánchez et al., 2013). In the human gut, these consist of non-digestible complex oligosaccharides that are not adsorbed, nor broken down in the upper GIT. Whereas, plant-based fibers are important in the adult diet, HMOs are important components of the infant diet. Furthermore, *Bifidobacterium* have even been successful in helping each other through cross-feeding (Turroni et al., 2015). Thus, we addressed the need to establish practical means to allocate phylogeny with minimalistic information based on sequences that encode glycolysis, the biochemical spine of most cells.

Here, we have shown that a multigene approach using glycolysis sequences can be used to uncover genomic trends and to make an accurate phylogenetic tree, based on a relatively small amount of information. The concatenated glycolysis tree in **Figure** is congruent with both the 16S rRNA tree and the established core-genome-based tree (Milani et al., 2016). The only notable exception is the placement of *B. merycicum* and *B. angulatum*. However, the relocation was between two neighboring phylogenetic groups in the concatenated and core-genome based trees. The glycolysis pathway is perhaps as, if not more, robust and accurate than the 16S rRNA tree. Compared to the 16S rRNA, the bootstrap values of the concatenated tree were higher on average. This leads to more confidence in the placement of species and the identification of phylogenetic groups, which in comparison, can appear arbitrarily located on the 16S rRNA. The concatenated tree is able to identify groups as well as the core-genome based tree. In fact, all of the phylogenetic groups from the core-genome were consistently found across the glycolytic pathway based trees. However, the glycolysis-based trees have the advantage of being much less labor intensive than the core-genome approach. This allows for accurate phylogenetic mapping of new strains or species, possibly encompassing unknown species, in less time and with less data than a core-genome. This approach is high resolution, low throughput, affordable, and accurate. Part of the success of this approach is the universality of glycolysis. Glycolysis is the biochemical backbone of the cell, and as such all organisms have at least some part of the glycolysis pathway represented (Fothergill-Gilmore and Michels, 1993). Even though these are slower-evolving genes, the changes that are made are enough to make an accurate phylogeny (Fothergill-Gilmore, 1986), evidenced from the congruence between our trees and the core-genome based tree. Even though the glycolysis enzymes are considered "slow evolvers," our data shows they are evolving at different rates amongst themselves. This can be explained by the fact that the glycolysis pathway is adapted by organisms to best fit their own unique environment and requirements (Bar-Even et al., 2012), as seen here in the *Bifidobacterium* and their bifid shunt (Sela et al., 2010). Some of the genes have specialized secondary functions, such as enolase acting as a cell surface receptor in *Bifidobacterium* (Candela et al., 2009). All of this makes the glycolysis pathway an excellent phylogenetic marker candidate. The various rates in evolution and moonlighting abilities also allow for further applications in recognizing adaptive trends.

The functional diversity of bifidobacteria is underpinned by multi-dimensional variety in their genomes, including overall content, organization, sequence diversity, and others. In extreme cases, even a two-fold difference in genome size can be observed. Despite being generally perceived as high GC organisms, they vary enough to have distinct relative classes of high, middle, and low-GC, amongst themselves (**Figure** ). Yet, there are non-random patterns and phenomena that drive these differences. The phylogenetic groups are clustered in specific regions of the GC continuum. Some groups are more tightly packed than others. A general trend that is observed across the genus is an evolutionary movement toward a high(er) GC content. The higher end of the spectrum is more densely populated then the lower end of the spectrum, indicative of an upward trend. This is reflected by the increased GC content in the individual glycolysis genes, when compared to the total GC content. Of the glycolysis genes, only one, *tpi*, does not show strong evidence for being different from the genome-wide (total) GC content. Critically, all of the other genes are above the total GC content. When we combine the overall genomic data with the GC-content groupings and trends discovered using glycolysis as phylogenetic markers, we posit the hypothesis that, over time, the GC content within the genomes of bifidobacteria increases, as to deviate further away from the 50% value, as the organisms adapt, and their genomes evolve accordingly.

Because of the broad occurrence of the glycolysis pathway in the Tree of Life, it is a suitable candidate marker to use in phylogenetic studies, likely beyond its application in bifidobacteria. In addition to being conserved genes that capture genetic diversity, glycolysis genes are consistently amongst the most highly expressed in not only *Bifidobacterium* (Turroni et al., 2015), but other organisms as well (Barrangou et al., 2006). This reflects both the importance of these sequences genetically (as illustrated by GC content drift), and functionally (as illustrated by their propensity for high levels of constitutive transcription). Because of this, it may be possible to correlate transcriptional data to phylogenetic studies on a broader scale. From here, it could be feasible to assign species and map data to known references using transcriptomic, genomic, or meta-data. Indeed, as the democratization of metagenomic technologies continues, and the need to assign phylogenetic information to partial genomic information increases, we propose that this method be used to provide insights into the phylogeny of un-assigned contigs. Overall, this approach allows for accurate phylogenetic mapping, congruent with a core-genome and more robust than the 16S rRNA phylogenetic approach, as well as inference on genomic adaptation, using either genomic, transcriptomic, or meta-data in a timely fashion and with minimal computation.

## AUTHOR CONTRIBUTIONS

KB and RB designed and carried out experiments, interpreted results, and wrote the manuscript.

## ACKNOWLEDGMENT

We would like to thank the Dr. Todd Klaenhammer lab and the CRISPR lab for providing insights and support during this project.

## FUNDING

This study was supported by startup funds from North Carolina State University. KB is a recipient of a NIEHS training grant.

## SUPPLEMENTARY MATERIAL

The Supplementary Material for this article can be found online at: http://journal.frontiersin.org/article/10.3389/fmicb.2016.00657

## REFERENCES


Altschul, S. F., Gish, W., Miller, W., Myers, E. W., and Lipman, D. J. (1990). Basic local alignment search tool. *J. Mol. Biol.* 215, 403–410. doi: 10.1016/s0022-2836(05)80360-2

Arboleya, S., Sánchez, B., Milani, C., Duranti, S., Solís, G., Fernández, N., et al. (2015). Intestinal microbiota development in preterm neonates and effect of perinatal antibiotics. *J. Pediatr.* 166, 538–544. doi: 10.1016/j.jpeds.2014.09.041

Baker, G. C., Smith, J. J., and Cowan, D. A. (2003). Review and re-analysis of domain-specific 16S primers. *J. Microbiol. Methods* 55, 541–555. doi: 10.1016/j.mimet.2003.08.009

Bar-Even, A., Flamholz, A., Noor, E., and Milo, R. (2012). Rethinking glycolysis: on the biochemical logic of metabolic pathways. *Nat. Chem. Biol.* 8, 509–517. doi: 10.1038/nchembio.971

Barrangou, R., Azcarate-Peril, M. A., Duong, T., Conners, S. B., Kelly, R. M., and Klaenhammer, T. R. (2006). Global analysis of carbohydrate utilization by *Lactobacillus acidophilus* using cDNA microarrays. *Proc. Natl. Acad. Sci. U.S.A.* 103, 3816–3821. doi: 10.1073/pnas.0511287103

Bottacini, F., Ventura, M., van Sinderen, D., and O'Connell Motherway, M. (2014). Diversity, ecology and intestinal function of bifidobacteria. *Microb. Cell Fact.* 13(Suppl. 1), S4–S4. doi: 10.1186/1475-2859-13-S1-S4

Candela, M., Biagi, E., Centanni, M., Turroni, S., Vici, M., Musiani, F., et al. (2009). Bifidobacterial enolase, a cell surface receptor for human plasminogen involved in the interaction with the host. *Microbiology* 155, 3294–3303. doi: 10.1099/mic.0.028795-0

Capella-Gutiérrez, S., Silla-Martínez, J. M., and Gabaldón, T. (2009). trimAl: a tool for automated alignment trimming in large-scale phylogenetic analyses. *Bioinformatics* 25, 1972–1973. doi: 10.1093/bioinformatics/btp348

Chambers, J. M. (1983). "Notched box plots," in *Graphical Methods for Data Analysis* (Belmont, CA: Wadsworth International Group), 60–63.

Chaplin, A. V., Efimov, B. A., Smeianov, V. V., Kafarskaia, L. I., Pikina, A. P., and Shkoporov, A. N. (2015). Intraspecies genomic diversity and long-term persistence of *Bifidobacterium longum*. *PLoS ONE* 10:e0135658. doi: 10.1371/journal.pone.0135658

Clarridge, J. E. (2004). Impact of 16S rRNA gene sequence analysis for identification of bacteria on clinical microbiology and infectious diseases. *Clin. Microbiol. Rev.* 17, 840–862. doi: 10.1128/CMR.17.4.840-862.2004

de Vries, W., and Stouthamer, A. H. (1967). Pathway of glucose fermentation in relation to the taxonomy of bifidobacteria. *J. Bacteriol.* 93, 574–576.

Edgar, R. C. (2004). MUSCLE: multiple sequence alignment with high accuracy and high throughput. *Nucleic Acids Res.* 32, 1792–1797. doi: 10.1093/nar/gkh340

Eisen, J. A. (1995). The RecA protein as a model molecule for molecular systematic studies of bacteria: comparison of trees of RecAs and 16S rRNAs from the same species. *J. Mol. Evol.* 41, 1105–1123. doi: 10.1007/BF00173192

Fothergill-Gilmore, L. A. (1986). The evolution of the glycolytic pathway. *Trends Biochem. Sci.* 11, 47–51. doi: 10.1016/0968-0004(86)90233-1

Fothergill-Gilmore, L. A., and Michels, P. A. M. (1993). Evolution of glycolysis. *Prog. Biophys. Mol. Biol.* 59, 105–135. doi: 10.1016/0079-6107(93)90001-Z

Fox, G. E., Wisotzkey, J. D., and Jurtshuk, P. (1992). How close is close: 16S rRNA sequence identity may not be sufficient to guarantee species identity. *Int. J. Syst. Evol. Microbiol.* 42, 166–170. doi: 10.1099/00207713-42-1-166

Gasteiger, E., Gattiker, A., Hoogland, C., Ivanyi, I., Appel, R. D., and Bairoch, A. (2003). ExPASy: the proteomics server for in-depth protein knowledge and analysis. *Nucleic Acids Res.* 31, 3784–3788. doi: 10.1093/nar/gkg563

Katoh, K., Misawa, K., Kuma, K. I, and Miyata, T. (2002). MAFFT: a novel method for rapid multiple sequence alignment based on fast Fourier transform. *Nucleic Acids Res.* 30, 3059–3066. doi: 10.1093/nar/gkf436

Kearse, M., Moir, R., Wilson, A., Stones-Havas, S., Cheung, M., Sturrock, S., et al. (2012). Geneious basic: an integrated and extendable desktop software platform for the organization and analysis of sequence data. *Bioinformatics* 28, 1647–1649. doi: 10.1093/bioinformatics/bts199

Killer, J., Sedláèek, I., Rada, V., Havlík, J., and Kopeèn*ı*, J. (2013). Reclassification of *Bifidobacterium* stercoris Kim et al. 2010 as a later heterotypic synonym of *Bifidobacterium adolescentis*. *Int. J. Syst. Evol. Microbiol.* 63, 4350–4353. doi: 10.1099/ijs.0.054957-0

Lang, J. M., Darling, A. E., and Eisen, J. A. (2013). Phylogeny of bacterial and archaeal genomes using conserved genes: supertrees and supermatrices. *PLoS ONE* 8:e62510. doi: 10.1371/journal.pone.0062510

Larkin, M. A., Blackshields, G., Brown, N. P., Chenna, R., McGettigan, P. A., McWilliam, H., et al. (2007). Clustal W and Clustal X version 2.0. *Bioinformatics* 23, 2947–2948. doi: 10.1093/bioinformatics/btm404

Lartillot, N., and Philippe, H. (2004). A Bayesian mixture model for across-site heterogeneities in the amino-acid replacement process. *Mol. Biol. Evol.* 21, 1095–1109. doi: 10.1093/molbev/msh112

Lugli, G. A., Milani, C., Turroni, F., Duranti, S., Ferrario, C., Viappiani, A., et al. (2014). Investigation of the evolutionary development of the genus *Bifidobacterium* by comparative genomics. *Appl. Environ. Microbiol.* 80, 6383–6394. doi: 10.1128/aem.02004-14

Lukjancenko, O., Ussery, D. W., and Wassenaar, T. M. (2011). Comparative genomics of *Bifidobacterium*, *Lactobacillus* and related probiotic genera. *Microb. Ecol.* 63, 651–673. doi: 10.1007/s00248-011-9948-y

Medini, D., Donati, C., Tettelin, H., Masignani, V., and Rappuoli, R. (2005). The microbial pan-genome. *Curr. Opin. Genet. Dev.* 15, 589–594. doi: 10.1016/j.gde.2005.09.006

Milani, C., Lugli, G. A., Duranti, S., Turroni, F., Bottacini, F., Mangifesta, M., et al. (2014). Genomic encyclopedia of type strains of the genus *Bifidobacterium*. *Appl. Environ. Microbiol.* 80, 6290–6302. doi: 10.1128/aem.02308-14

Milani, C., Mancabelli, L., Lugli, G. A., Duranti, S., Turroni, F., Ferrario, C., et al. (2015). Exploring vertical transmission of bifidobacteria from mother to child. *Appl. Environ. Microbiol.* 81, 7078–7087. doi: 10.1128/aem.02037-15

Milani, C., Turroni, F., Duranti, S., Lugli, G. A., Mancabelli, L., Ferrario, C., et al. (2016). Genomics of the genus *Bifidobacterium* reveals species-specific adaptation to the glycan-rich gut environment. *Appl. Environ. Microbiol.* 82, 980–991. doi: 10.1128/aem.03500-15

Pokusaeva, K., Fitzgerald, G. F., and Sinderen, D. (2011). Carbohydrate metabolism in bifidobacteria. *Genes Nutr.* 6, 285–306. doi: 10.1007/s12263-010-0206-6

R Core Team (2015). *R: A Language and Environment for Statistical Computing*. Vienna: R Foundation for Statistical Computing.

Sánchez, B., Ruiz, L., Gueimonde, M., Ruas-Madiedo, P., and Margolles, A. (2013). Adaptation of bifidobacteria to the gastrointestinal tract and functional consequences. *Pharmacol. Res.* 69, 127–136. doi: 10.1016/j.phrs.2012.11.004

Sela, D. A. (2011). Bifidobacterial utilization of human milk oligosaccharides. *Int. J. Food Microbiol.* 149, 58–64. doi: 10.1016/j.ijfoodmicro.2011.01.025

Sela, D. A., Price, N. P. J., and Mills, D. (2010). "Metabolism of bifidobacteria," in *Bifidobacteria: Genomics and Molecular Aspects*, eds B. Mayo and D. van Sinderen (Norwich: Caister Academic Press).

Stamatakis, A. (2006a). "Phylogenetic models of rate heterogeneity: a high performance computing perspective," in *Proceedings 20th IEEE International Parallel & Distributed Processing Symposium* (Rhodes Island: IEEE).

Stamatakis, A. (2006b). RAxML-VI-HPC: maximum likelihood-based phylogenetic analyses with thousands of taxa and mixed models. *Bioinformatics* 22, 2688–2690. doi: 10.1093/bioinformatics/btl446

Stamatakis, A. (2014). RAxML version 8: a tool for phylogenetic analysis and post-analysis of large phylogenies. *Bioinformatics* 30, 1312–1313. doi: 10.1093/bioinformatics/btu033

Tamura, K., Stecher, G., Peterson, D., Filipski, A., and Kumar, S. (2013). MEGA6: molecular evolutionary genetics analysis version 6.0. *Mol. Biol. Evol.* 30, 2725–2729. doi: 10.1093/molbev/mst197

Turroni, F., Duranti, S., Bottacini, F., Guglielmetti, S., Van Sinderen, D., and Ventura, M. (2014). *Bifidobacterium bifidum* as an example of a specialized human gut commensal. *Front. Microbiol.* 5:437. doi: 10.3389/fmicb.2014.00437

Turroni, F., Özcan, E., Milani, C., Mancabelli, L., Viappiani, A., van Sinderen, D., et al. (2015). Glycan cross-feeding activities between bifidobacteria under in vitro conditions. *Front. Microbiol.* 6:1030. doi: 10.3389/fmicb.2015.01030

Turroni, F., Peano, C., Pass, D. A., Foroni, E., Severgnini, M., Claesson, M. J., et al. (2012a). Diversity of bifidobacteria within the infant gut microbiota. *PLoS ONE* 7:e36957. doi: 10.1371/journal.pone.0036957

Turroni, F., Strati, F., Foroni, E., Serafini, F., Duranti, S., van Sinderen, D., et al. (2012b). Analysis of predicted carbohydrate transport systems encoded by *Bifidobacterium bifidum* PRL2010. *Appl. Environ. Microbiol.* 78, 5002–5012. doi: 10.1128/AEM.00629-12

Ventura, M., Canchaya, C., Tauch, A., Chandra, G., Fitzgerald, G. F., Chater, K. F., et al. (2007). Genomics of Actinobacteria: tracing the evolutionary history of an ancient phylum. *Microbiol. Mol. Biol. Rev.* 71, 495–548. doi: 10.1128/MMBR.00005-07

Ventura, M., Turroni, F., Lugli, G. A., and van Sinderen, D. (2014). Bifidobacteria and humans: our special friends, from ecological to genomics perspectives. *J. Sci. Food Agric.* 94, 163–168. doi: 10.1002/jsfa.6356

Ventura, M., Turroni, F., Motherway, M. O. C., MacSharry, J., and van Sinderen, D. (2012). Host-microbe interactions that facilitate gut colonization by commensal bifidobacteria. *Trends Microbiol.* 20, 467–476. doi: 10.1016/j.tim.2012.07.002

**Conflict of Interest Statement:** The authors declare that the research was conducted in the absence of any commercial or financial relationships that could be construed as a potential conflict of interest.

# 10

# Phylogenomic Analyses and Comparative Studies on Genomes of the *Bifidobacteriales*: Identification of Molecular Signatures Specific for the Order *Bifidobacteriales* and its Different Subclades

*Grace Zhang*[1]*, Beile Gao*[2]*, Mobolaji Adeolu*[1]*, Bijendra Khadka*[1] *and Radhey S. Gupta*[1*]

[1] *Department of Biochemistry and Biomedical Sciences, McMaster University, Hamilton, ON, Canada,* [2] *CAS Key Laboratory of Tropical Marine Bio-resources and Ecology, Guangdong Key Laboratory of Marine Materia Medica, South China Sea Institute of Oceanology, Chinese Academy of Sciences, Guangzhou, China*

**Edited by:**
Marco Ventura,
University of Parma, Italy

**Reviewed by:**
Gabriele Andrea Lugli,
University of Parma, Italy
Longxian Lv,
Zhejiang University, China

***Correspondence:**
Radhey S. Gupta
gupta@mcmaster.ca

The order *Bifidobacteriales* comprises a diverse variety of species found in the gastrointestinal tract of humans and other animals, some of which are opportunistic pathogens, whereas a number of others exhibit health-promoting effects. However, currently very few biochemical or molecular characteristics are known which are specific for the order *Bifidobacteriales*, or specific clades within this order, which distinguish them from other bacteria. This study reports the results of detailed comparative genomic and phylogenetic studies on 62 genome-sequenced species/strains from the order *Bifidobacteriales*. In a robust phylogenetic tree for the *Bifidobacteriales* constructed based on 614 core proteins, a number of well-resolved clades were observed including a clade separating the *Scarodvia*-related genera (*Scardovia* clade) from the genera *Bifidobacterium* and *Gardnerella*, as well as a number of previously reported clusters of *Bifidobacterium* spp. In parallel, our comparative analyses of protein sequences from the *Bifidobacteriales* genomes have identified numerous molecular markers that are specific for this group of bacteria. Of these markers, 32 conserved signature indels (CSIs) in widely distributed proteins and 10 signature proteins are distinctive characteristics of all sequenced *Bifidobacteriales* species and provide novel and highly specific means for distinguishing these bacteria. In addition, multiple other molecular signatures are specific for the following clades of *Bifidobacteriales*: (i) 5 CSIs specific for a clade comprising of the *Scardovia*-related genera; (ii) 3 CSIs and 2 CSPs specific for a clade consisting of the *Bifidobacterium* and *Gardnerella* spp.; (iii) multiple other signatures demarcating a number of clusters of the *B. asteroides*-and *B. longum*- related species. The described molecular markers provide novel and reliable means for distinguishing the *Bifidobacteriales* and a number of their clades in molecular terms and for the classification of these bacteria. The *Bifidobacteriales*-specific CSIs, found in important proteins, are predicted to play important roles in modifying the cellular functions of the affected proteins. Hence, biochemical studies on the cellular functions of these CSIs could lead to discovery of novel characteristics of either all *Bifidobacteriales*, or specific groups of bacteria within this order. Some of the functions affected/modified by these genetic changes could also be important for the probiotic/pathogenic activities of the bifidobacteria.

**Keywords: molecular signatures for bifidobacteria, phylogeny, taxonomy, conserved signature indels, conserved signature proteins, *Bifidobacterium asteroides*-clade, *Scardovia*-clade**

## INTRODUCTION

The order *Bifidobacteriales* contains a large collection of bacterial species, many of which are significant constituents of the gastrointestinal tract of humans, other mammals, birds and honey bees (Biavati et al., 2000; Biavati and Mattarelli, 2006; Turroni et al., 2009, 2011; Biavati, 2012; Milani et al., 2014). In addition to widely recognized health-promoting effects of bifidobacterial species (Leahy et al., 2005; Ventura et al., 2009a; Cronin et al., 2011), some members of the group found in human and animal oral cavities are implicated in the development of dental caries (Huys et al., 2007; Mantzourani et al., 2009; Ventura et al., 2009b). Additionally, *Gardnerella vaginalis* is indicated to play an important role in the pathogenesis of bacterial vaginosis and urinary tract infections (Smith et al., 1992; Bradshaw et al., 2006; Alves et al., 2014; Kenyon and Osbak, 2014). The order *Bifidobacteriales* is part of the phylum Actinobacteria (Ventura et al., 2007b; Zhi et al., 2009; Gao and Gupta, 2012) and it harbors a single family, *Bifidobacteriaceae,* containing >50 recognized species (Biavati, 2012; Lugli et al., 2014; Milani et al., 2014; Parte, 2014) that are grouped into eight genera: *Aeriscardovia, Alloscardovia, Bifidobacterium, Gardnerella, Pseudoscardovia, Neoscardovia, Parascardovia,* and *Scardovia* (Jian and Dong, 2002; Simpson et al., 2004; Huys et al., 2007; Biavati and Mattarelli, 2012; García-Aljaro et al., 2012; Storms and Vandamme, 2012; Killer et al., 2013). Of these, the genus *Bifidobacterium*, encompassing 39 species and 9 subspecies, forms the largest group and accounts for more than 75% of the described taxa within the order *Bifidobacteriales* (Biavati and Mattarelli, 2012; Milani et al., 2014).

Phylogenetic analyses based on 16S rRNA, as well as sequences for a number of housekeeping genes/proteins, are the main approaches used in the past to distinguish among the *Bifidobacteriales* species and genera (Miyake et al., 1998; Ventura and Zink, 2003; Ventura et al., 2004, 2006, 2007a; Biavati and Mattarelli, 2006; Yarza et al., 2008; Bottacini et al., 2010; Turroni et al., 2011; Mattarelli et al., 2014). In recent years, complete or draft genome sequences have become available for all currently recognized *Bifidobacterium* species and subspecies (Ventura et al., 2009b; Milani et al., 2014). Based on these sequences, a panel of multiplex PCR primers has been developed enabling rapid and specific identification of different *Bifidobacterium* species and subspecies (Ferrario et al., 2015). Based on genome sequences, two recent studies have also examined the evolutionary relationships among *Bifidobacterium* species employing large datasets of sequences comprising the core proteins of this genus (Lugli et al., 2014; Sun et al., 2015). The robust phylogenetic trees obtained in these studies provide important insights concerning the evolutionary relationships among the *Bifidobacterium* species and they strongly support the existence of 6-7 distinct clusters within this genus. These clusters are referred to as the *B. asteroides, B. pseudolongum, B. longum, B. bifidum, B. adolescentis, B. pullorum,* and *B. boum* groups (Lugli et al., 2014; Sun et al., 2015). Similar clusters are also observed in phylogenetic trees based on the 16S and 23S rRNA genes as well trees based on other gene/protein sequences. Comparative analyses of the *Bifidobacterales* genomes are also providing useful insights concerning species-specific characteristics that are suggested to play important roles in the adaptation of particular species to either human or insect gut environment (Ventura et al., 2009b; Bottacini et al., 2010, 2012; Turroni et al., 2010).

Due to the health-promoting effects of bifidobacteria, it is of much interest to identify genetic and biochemical characteristics that are specific for the *Bifidobacteriales* or particular groups/clusters within this order of bacteria. Currently, very few such characteristics are known. One important class of genome sequence-based molecular markers, which have proven very useful for evolutionary, taxonomic and functional studies are conserved signature insertions or deletions (CSIs) that are uniquely present in the genes/proteins homologs from a defined group of organisms (Gao and Gupta, 2005, 2012; Gupta, 2010, 2014). Conserved signature proteins (CSPs), which are genes/proteins that are uniquely found within a monophyletic group of organisms, provide another class of useful molecular makers for evolutionary and functional studies (Gao et al., 2006; Ventura et al., 2007a; Gao and Gupta, 2012; Gupta, 2016a,b). Both these types of markers constitute highly reliable characteristics of specific groups of organisms and they have been extensively utilized for the identification/demarcation of prokaryotic taxa of different ranks in molecular terms (Gao and Gupta, 2012; Gupta et al., 2013a,b, 2016).

In the present work, we report detailed phylogenetic and comparative analyses on protein sequences from the sequenced members of the order *Bifidobacteriales* in order to identify CSIs and CSPs that are specific for different groups within this order. These studies have led to identification of 32 CSIs in widely distributed proteins and 10 CSPs that are uniquely found in all or most of the genome sequenced *Bifidobacteriales* species providing novel molecular markers that distinguish this order from all other bacteria. In addition, our work has also identified multiple other CSIs and CSPs that distinguish a number of clades of *Bifidobacteriales,* including a clade consisting of the *Bifidobacterium* and *Gardnerella* species, another clade consisting of the *Scardovia*-related genera, and multiple signatures that are specific for different clusters of *B. asteroides* or *B. longum* related species. These signatures provide novel means for the identification and demarcation of the members of the described clades in molecular terms and for functional studies that could lead to discovery of novel biochemical and/or other novel properties of these bacteria.

## METHODS

### Phylogenetic Analysis

A phylogenetic tree for 62 genome-sequenced members from the order *Bifidobacteriales* was constructed based on concatenated sequences of 614 proteins. The protein families used in this phylogeny were identified using the UCLUST algorithm (Edgar, 2010) to identify proteins families present in at least 80% of the input genomes which shared at least 50% sequence identity and 50% sequence length. Each identified protein family was individually aligned using Clustal Omega (Sievers et al., 2011) and trimmed using Gblocks 0.91b (Castresana, 2000) with relaxed parameters (Talavera and Castresana, 2007). The concatenated dataset of the trimmed sequence alignments contained 197, 777 aligned amino acid residues. A maximum-likelihood tree based on this alignment was constructed using FastTree 2 (Price et al., 2010) employing the Whelan and Goldman model of protein sequence evolution (Whelan et al., 2001) and RAxML 8 (Stamatakis, 2014) using the Le and Gascuel model of protein sequence evolution (Le and Gascuel, 2008). SH-like statistical support values (Guindon et al., 2010) for each branch node in the final phylogenetic tree were calculated using RAxML 8 (Stamatakis, 2014). This process was completed using an internally developed software pipeline.

In parallel, a phylogenetic tree based on the 16S rRNA gene sequences of type strains covering all described species within the order *Bifidobacteriales* was also constructed. The 16S rRNA sequences were retrieved from Ribosomal Database Project (Cole et al., 2014) and aligned using the SINA aligner (Pruesse et al., 2012) to form a multiple sequence alignment that was 1604 aligned nucleotides long with common gaps removed. A maximum-likelihood phylogenetic tree based on this multiple sequence alignment was created using MEGA 6 employing the General Time-Reversible model of sequence evolution with branch support based on 1000 bootstrap replicates (Tamura et al., 2013).

### Identification of Conserved Signature Indels

Conserved signature indels (CSIs) were identified by the procedures described in detail recently (Gupta, 2014). Briefly, BLASTp (Altschul et al., 1997) searches were performed on each protein in the genome of *Bifidobacterium adolescentis* ATCC 15703 (Accession number AP009256.1) against all available sequences in the GenBank non-redundant database. Multiple sequence alignments were then created using ClustalX (Jeanmougin et al., 1998) for proteins that returned high scoring matches from *Bifidobacteriales* and other prokaryotes. The alignments were then visually inspected for the presence of insertions or deletions that were flanked on both sides by at least 5-6 conserved amino acid residues in the neighboring 30–40 amino acids. Detailed BLASTp searches were then carried out on short sequence segments containing the indel and the flanking conserved regions (60-100 amino acids long) to determine the specificity of the indels. SIG_CREATE and SIG_STYLE (available on Gleans.net) were then used to create Signature files for CSIs that were specific for the *Bifidobacteriales* order or its subgroups as described in earlier work (Gupta et al., 2013a; Gupta, 2014). Due to space limitations, sequence information for all *Bifidobacterium* species, particularly for different subspecies of *B. longum*, *B. animalis*, *B. pseudolongum*, and *B. thermacidophilum*, is not shown in the presented alignment files. However, unless otherwise noted, all of the described CSIs are specific for the indicated groups (i.e., similar CSIs were not present in the protein homologs from other bacteria in the top 500 Blast hits). It should be noted that significant blast hits for a number of CSIs and CSPs described here are also observed for one of the following three *Chlamydia trachomatis* strains (SwabB1, H1 IMS, and H17 IMS) deposited by the Sanger Institute. We suspect that these anomalous results are due to cross contamination of the sequenced cultures from the above *Chlamydia trachomatis* strains by a *Gardnerella* vaginalis strain. We have communicated our concern with the supporting evidence to the Sanger Institute.

### Identification of Conserved Signature Proteins

BLASTp searches were carried out to examine the specificity of some previously described conserved signature proteins (CSPs), which were indicated to be specific for the order *Bifidobacteriales* (Gao and Gupta, 2012). Additionally, limited work to identify CSPs for the *B. asteroides* group of species was carried out by conducting BLASTp searches on all proteins from the genomes of *Bifidobacterium asteroides* (Bottacini et al., 2012) as query sequences. BLASTp searches were performed against all available sequences in the GenBank non-redundant sequence database and the results of these searches were then manually inspected, as described in earlier work (Gao et al., 2006; Gao and Gupta, 2012), for proteins for which all significant hits were from the *B. asteroides* group of species.

### Homology Modeling of Elongation Factor Tu from *Bifidobacterium longum*

Homology models of EF-Tu homolog from *Bifidobacterium longum* were built using the solved crystallographic structure of EF-Tu from *Escherichia coli* (PDB ID: 3U6K) as the template. Initially, 200 models were generated using MODELER v9.14 (Sali and Blundell, 1993) and ranked/selected using assigned discrete optimized potential (DOPE) scores (Shen and Sali, 2006). The model with the highest DOPE score was then submitted to the ModRefiner program to obtain atomic-level energy minimization and to obtain a model with reliable stereochemistry quality (Xu and Zhang, 2011).

# RESULTS

## Phylogenetic Analysis of the Species from the Order *Bifidobacteriales*

Phylogenomic analyses of members of the genus *Bifidobacterium* have been previously reported based on core protein sequences from 45 and 48 described species from this genus (Lugli et al., 2014; Sun et al., 2015). However, these studies did not include the other members of the order *Bifidobacteriales* such as *Gardnerella, Scardavia, Alloscardovia,* and *Parascardovia*, as well as several unnamed *Bifidobacterium* spp. (viz. strains A11, 7101, AGR2158, MSTE12, 12.1.47BFAA) whose genomes have been sequenced. Additionally, the genome sequence of a recently described species *B. aesculapii* is also now available (Toh et al., 2015). To comprehensively examine the evolutionary relationships among different members of the order *Bifidobacteriales*, a phylogenetic tree was constructed for all 62 genome sequenced members of the family which included 54 *Bifidobacterium* species/strains, 5 species from *Scardovia* and related genera (viz. *Alloscardovia* and *Parascardovia*) and three strains of *Gardnerella vaginalis.* The tree was constructed based on the concatenated sequences of 614 universally or nearly universally present core proteins for which sequence information could be obtained from the 62 sequenced genomes. A maximum-likelihood tree based on these sequences, which represents the most comprehensive phylogenetic analysis of the order *Bifidobacteriales* to date, is presented in **Figure** .

In the tree shown, members of the order *Bifidobacteriales,* at the highest level, form two main clusters. One of these clusters referred to as the *Scardovia* cluster groups together the genera *Scardovia, Parascardovia,* and *Alloscardovia*, whereas the second cluster is comprised of members of the genus *Bifidobacterium* and *Gardnerella.* Importantly in this tree, as well as in an earlier study in a phylogenetic tree based on concatenated sequences for RpoB, RpoC, and GyrB proteins, different strains of *Gardnerella vaginalis* were found to branch in between the *Bifidobacterium* species (Gao and Gupta, 2012), making the genus *Bifidobacterium* polyphyletic. Earlier phylogenetic studies on members of the genus *Bifidobacterium* have identified a number of different clusters, which are referred to as the *B. asteroides, B. pseudolongum, B. longum, B. bifidum, B. adolescentis, B. pullorum,* and *B. boum* groups (Ventura et al., 2006; Turroni et al., 2011; Lugli et al., 2014; Sun et al., 2015). The existence of these groups/clusters is also confirmed and supported by the tree shown in **Figure** . Of these clusters, the species-related to *B. asteroides* cluster exhibited the deepest branching within the genus *Bifidobacterium*, as also observed in earlier work (Lugli et al., 2014; Sun et al., 2015). The *B. asteroides* clade is generally demarcated as being comprised of the *B. asteroides, B. indicum, B. coryneforme,* and *B. actinocoloniiforme* species (marked as cluster III in **Figure** ). However, as discussed later, a number of clusters, marked I, II, and IV, which are either part of the *B. asteroides* clade or are related to this clade are also distinguished in phylogenetic trees and by the CSIs identified in this work.

We have also created a phylogenetic tree based on 16S rRNA gene sequences of all named *Bifidobacteriales* species (Supplementary Figure ). The overall branching pattern in the 16S rRNA tree is similar to that observed in the concatenated protein tree with *Scardovia* and related genera forming the deepest branches in the tree and the genera *Scardovia, Alloscardovia,* and *Parascardovia* were part of one of the deepest branching clusters. The different clusters of the *Bifidobacterium* spp. that are observed in the concatenated protein tree were also supported by the 16S rRNA tree and *G. vaginalis* was found to branch in between these clusters. The polyphyletic nature of the genus *Bifidobacterium* in 16S rRNA gene based phylogenies is also observed in earlier work (Yilmaz et al., 2014).

## Identification of Molecular Markers That Are Specific for the Order *Bifidobacteriales*

The main focus of this work is the identification of molecular characteristics that are specific for the *Bifidobacteriales* species and could be used for their identification as well as functional studies. As noted earlier, conserved inserts and deletions (i.e., indels or CSIs) in genes/proteins and conserved signature proteins that are uniquely found in a phylogenetically coherent group of organisms provide very useful molecular markers for such purposes. The indels that provide useful molecular markers are of defined size and they are flanked on both sides by conserved regions to ensure that they are reliable characteristics (Gupta, 1998; Gupta and Griffiths, 2002; Ajawatanawong and Baldauf, 2013). These conserved indels in gene/protein sequences result from highly specific and rare genetic changes, hence when such an indel is uniquely found in a phylogenetically coherent group of species, its simplest explanation is that the genetic change responsible for it occurred once in a common ancestor of the indicated group and then the change was passed on to various descendants (Gupta, 1998, 2014; Rokas and Holland, 2000; Gao and Gupta, 2005). Based upon the presence or absence of a conserved indel in outgroup species, it is also possible to determine whether a given indel represents an insert or a deletion (Gupta, 1998; Gao and Gupta, 2012).

Comparative analyses of protein sequence alignments from bifidobacteia species carried out in this work have led to the identification of 32 CSIs in a broad range of highly conserved proteins, which are specifically found in different *Bifidobacteriales* taxa (see **Table** ). One example of a CSI that is specific for all members of the order *Bifidobacteriales* is shown in **Figure** . In this case, a 4 amino acid (aa) insertion is present in a highly conserved region of the protein synthesis elongation factor EF-Tu, which is commonly shared by all sequenced bifidobacteria species, but it is not found in any other bacteria in the top 500 BLAST hits. The protein EF-Tu is a highly conserved protein, which is universally present in all organisms (Harris et al., 2003) and the 4 aa CSI in this protein is a distinctive characteristic of homologs from all sequenced *Bifidobacteriales* species. Sequence information for 31 other CSIs, which are also specifically shared by members of the order *Bifidobacteriales*, and which are present in proteins involved in different other functions, is provided in Supplementary Figures and some of their

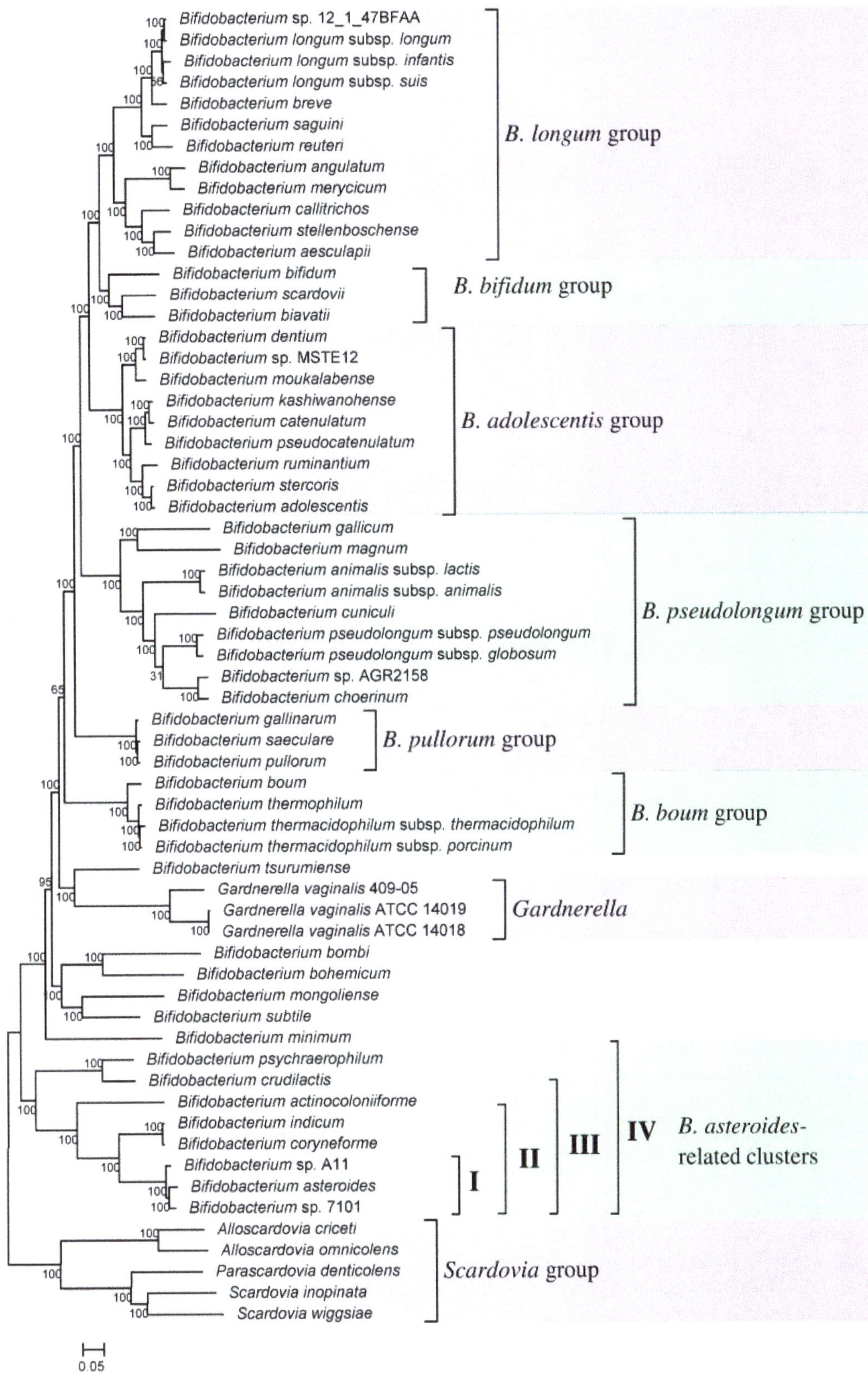


**FIGURE | A maximum-likelihood tree based on concatenated sequences of 614 core proteins from 62 sequenced genome-sequenced members of the order *Bifidobacteriales*.** The tree was rooted at the midpoint and SH-like support values are indicated at nodes. A number of different clades/clusters that are consistently observed in phylogenetic trees are marked.

**TABLE | Characteristics of conserved signature indels that are Specific for the order *Bifidobacteriales*.**

| Protein name | GI number | Figure no. | Indel size | Indel region[a] |
|---|---|---|---|---|
| Elongation factor Tu | 38606895 | **Figure 2** | 4 aa ins | 106–144 |
| DNA topoisomerase I | 489904111 | Supplementary Figure S2 | 1 aa del | 31–80 |
| DNA polymerase sliding clamp subunit | 408500301 | Supplementary Figure S3 | 1 aa ins | 79–118 |
| Beta-galactosidase | 504834401 | Supplementary Figure S4 | 1–2 aa ins | 371–423 |
| Ketol-acid reductoisomerase | 651881972 | Supplementary Figure S5 | 2 aa del | 242–284 |
| Serine-pyruvate aminotransferase | 489903803 | Supplementary Figure S6 | 2 aa ins | 74–119 |
| 50S ribosomal protein L21 | 489922190 | Supplementary Figure S7 | 1 aa ins | 42–82 |
| Methionine aminopeptidase | 547078960 | Supplementary Figure S8 | 1 aa ins | 34–70 |
| Bifunctional acetaldehyde-CoA/alcohol dehydrogenase | 500062906 | Supplementary Figure S9 | 1 aa ins | 534–574 |
| Bifunctional acetaldehyde-CoA/alcohol dehydrogenase | 500062906 | Supplementary Figure S10 | 1 aa ins | 809–845 |
| Formate acetyltransferase | 500063439 | Supplementary Figure S11 | 2 aa ins | 367–416 |
| ATP synthase F0 subunit A | 547078870 | Supplementary Figure S12 | 1 aa ins | 131–163 |
| Peptide chain release factor 1 | 489924412 | Supplementary Figure S13 | 2 aa ins | 197–237 |
| Arginine ABC transporter ATP-binding protein | 489905014 | Supplementary Figure S14 | 1 aa del | 224–280 |
| Transketolase | 489905793 | Supplementary Figure S15 | 4 aa ins | 338–388 |
| Histidine kinase | 547084095 | Supplementary Figure S16 | 1 aa ins | 362–405 |
| DNA repair ATPase | 489905284 | Supplementary Figure S17 | 3 aa ins | 353–394 |
| n-acetyl-gamma-glutamyl-phosphate reductase | 547072106 | Supplementary Figure S18 | 1 aa ins | 10–60 |
| Arginine biosynthesis bifunctional protein ArgJ | 547072098 | Supplementary Figure S19 | 1 aa ins | 1–42 |
| Excinuclease ABC subunit C | 494111998 | Supplementary Figure S20 | 1 aa ins | 103–150 |
| Cysteine desulfurase | 500063210 | Supplementary Figure S21 | 4 aa ins | 54–105 |
| 2-C-methyl-D-erythritol 2,4-cyclodiphosphate synthase | 489906135 | Supplementary Figure S22 | 1 aa ins | 58–81 |
| Argininosuccinate lyase | 547072080 | Supplementary Figure S23 | 5 aa ins | 405–454 |
| CarD family transcriptional regulator | 500063173 | Supplementary Figure S24 | 1 aa ins | 30–79 |
| Acetyltransferase GNAT family | 547074268 | Supplementary Figure S25 | 1 aa ins | 112–152 |
| Acetyltransferase GNAT family | 547074268 | Supplementary Figure S25 | 2 aa ins | 112–152 |
| Signal recognition particle protein | 489904236 | Supplementary Figure S26 | 1 aa ins | 70–110 |
| 50S ribosomal protein L13 | 489923970 | Supplementary Figure S27 | 1 aa del | 51–90 |
| DNA gyrase B subunit protein | 547082727 | Supplementary Figure S28 | 2 aa del | 637–686 |
| Hemolysin III | 489923478 | Supplementary Figure S29 | 1 aa del | 171–216 |
| Pseudouridine synthase | 547071034 | Supplementary Figure S30 | 1 aa ins | 56–95 |
| Guanylate kinase | 500063064 | Supplementary Figure S31 | 4 aa ins | 85–124 |
| D-alanine–D-alanine ligase | 493336643 | Supplementary Figure S32 | 2–7 aa ins | 202–244 |

[a] *The indel region indicates the region of the protein where the described CSI is present.*

characteristics are summarized in **Table** . Barring an isolated exception, all of the CSIs listed in **Table** are specifically found in different members of the order *Bifidobacteriales* and are not present in the protein homologs from other bacteria. Due to their specific presence in bifidobacteria species, the described CSIs provide novel molecular markers for distinguishing and demarcating members of the order *Bifidobacteriales* from all other bacteria. We have previously described 14 CSPs, whose homologs were specifically found in the 13 different sequenced bifidobacteria species that were available at the time (Gao and Gupta, 2012). Updated BLASTp searches on the sequences of these CSPs confirm that 10 of these CSPs, information for whom is provided in **Table** , are still distinctive characteristics of members of the order *Bifidobacteriales* and they provide additional molecular markers for identification and functional studies on bifidobacteria.

## Molecular Signatures for Some of the Subclades of *Bifidobacteriales*

In the phylogenetic tree based on concatenated protein sequences, *Bifidobacteriales* species form a number of different clusters. At the deepest level, of the two main clusters that are observed, one consists of the genus *Scardovia* and related genera, whereas the other is comprised of species from the genera *Bifidobacterium* and *Gardnerella*. In our analyses, we have also identified a number of CSIs and CSPs which distinguish these two clades of the *Bifidobacteriales*. **Figure** shows one example of a CSI consisting of a 1 aa insertion in the DNA polymerase IV protein that is specifically found in different *Bifidobacterium* species and *Gardnerella*, but which is not found in any of the sequenced *Scardovia*-related genera of the *Bifidobacteriales*. Two other CSIs in the ribosomal RNA small subunit methyltransferase E protein and GTP-binding

| Group | Organism | Accession | 106 | | 144 |
|---|---|---|---|---|---|
| *Bifidobacteriales* | Bifidobacterium adolescentis | 38606895 | DCPVIHVSAYGALHDDA | PDHE | KWVEQIKKLMDAVDDYIP |
| | Bifidobacterium angulatum | 254777842 | ------T---------- | ---- | ---QSV-D---------- |
| | Bifidobacterium animalis | 221163935 | ----V-T---------- | ---D | ---ATV-E---D------ |
| | Bifidobacterium bifidum | 221163938 | -----XT-------X-- | ---X | ---QXV-D------X--- |
| | Bifidobacterium boum | 221163940 | -----RT---------- | ---- | ---QT--D------E--- |
| | Bifidobacterium breve | 221163942 | ------T---------- | ---- | ---QSV-D---------- |
| | Bifidobacterium catenulatum | 38606893 | Y-----T---------- | ---- | ----SV-E--K---E--- |
| | Bifidobacterium choerinum | 639202362 | ----V-T---------- | ---- | ---AT--E---D------ |
| | Bifidobacterium dentium | 489933530 | ------T---------- | ---D | ----SV-E--K---E--- |
| | Bifidobacterium gallicum | 493338026 | ----V-T---------- | ---D | ----TV-E--KD--E--- |
| | Bifidobacterium gallinarum | 221163945 | ----------------- | ---- | ----AV-E--K---E--- |
| | Bifidobacterium indicum | 221163949 | -----RT---------- | ---D | ---QTV-D---------- |
| | Bifidobacterium longum | 170516895 | ------T---------- | ---- | ---QSV-D---------- |
| | Bifidobacterium magnum | 551239258 | ----V-T---------- | ---- | ----SV-E--NY--E--- |
| | Bifidobacterium minimum | 221163959 | -----RT---------- | ---D | ---QT--D---------- |
| | Bifidobacterium pseudocatenulatum | 254777852 | ------T---------- | ---- | ----SV-E--K---E--- |
| | Bifidobacterium pseudolongum | 221163947 | ----V-T---------- | ---- | ----TV-E--ND------ |
| | Bifidobacterium ruminantium | 254777834 | ----------------- | ---- | --------------E--- |
| | Bifidobacterium subtile | 221163961 | ------T---------- | ---- | ---QSV-D---------- |
| | Bifidobacterium thermacidophilum | 657872572 | -----RT---------- | ---- | ---QTV-D---------- |
| | Bifidobacterium tsurumiense | 651882426 | ----VRT---------- | ---D | ----TV-E--KD--E--- |
| | Bifidobacterium aesculapii | 943595915 | ----------------- | ---- | --------------E--- |
| | Bifidobacterium actinocoloniiforme | 705419702 | -----RT---------- | ---D | ---QT--D---------- |
| | Bifidobacterium stellenboschense | 736510954 | ----------------- | ---- | --------------E--- |
| | Bifidobacterium callitrichos | 759445153 | ----------------- | ---- | H-------------E--- |
| | Bifidobacterium saguini | 727802107 | ---I-R----------- | ---- | -----V--------E--- |
| | Bifidobacterium reuteri | 763217958 | -----R--------N-- | ---- | H-------------E--- |
| | Bifidobacterium biavatii | 705394035 | -----RL---------- | ---- | H----V--------E--- |
| | Bifidobacterium saeculare | 705430964 | ----------------- | ---- | ----AV-E--K---E--- |
| | Bifidobacterium pullorum | 705440708 | ------L---------- | ---- | ----AV-E--N---E--- |
| | Bifidobacterium kashiwanohense | 705412687 | ------T---------- | ---- | ----SV-E--K---E--- |
| | Bifidobacterium merycicum | 705457021 | ------T---------- | ---- | ----SV-E--K---E--- |
| | Bifidobacterium cuniculi | 705444485 | ----V-T---------- | ---- | ----T--E--ND--E--- |
| | Bifidobacterium mongoliense | 705436553 | -----RT---------- | ---- | ---QS--D------N--- |
| | Bifidobacterium thermophilum | 763423717 | -----RT---------- | ---- | ---QTV-D------E--- |
| | Bifidobacterium bohemicum | 705455516 | -----RT---------- | ---D | ---QTV-D---------- |
| | Bifidobacterium asteroides | 799166623 | -----RT---------- | ---D | ---QT--D--N------- |
| | Bifidobacterium bombi | 763214630 | -----RT---------- | ---D | ---QT--D------E--- |
| | Bifidobacterium crudilactis | 736120398 | -----RT---------- | ---- | ---KTVED--K------- |
| | Bifidobacterium coryneforme | 705388555 | -----RT---------- | ---D | ---QTV-D---------- |
| | Bifidobacterium psychraerophilum | 705400794 | -----RT---------- | ---D | ---KTVQD--E------- |
| | Bifidobacterium moukalabense | 489933530 | ------T---------- | ---D | ----SV-E--K---E--- |
| | Bifidobacterium scardovii | 705449472 | -----RT---------- | ---D | ---QTV-D------E--- |
| | Bifidobacterium stercoris | 673001648 | ----------------- | ---- | ------------------ |
| | Alloscardovia omnicolens | 545373117 | -----RT---------- | ---- | --AQTV-D--N---E--- |
| | Gardnerella vaginalis | 523608067 | -----RT---------- | ---D | ----TV-E--K---E--- |
| | Alloscardovia criceti | 516877532 | -----RT---------- | ---- | --AQTVRD--N---E--- |
| | Parascardovia denticolens | 493331502 | -A---RT---------- | ---- | ----TV-Q---T--E--- |
| | Scardovia inopinata | 493335588 | -A-I-RT---------- | ---- | ----SV-----T--E--- |
| | Scardovia wiggsiae | 494250700 | -A-I-RT---------- | ---D | --AKTV-E---T--E--- |
| | Pseudoscardovia suis | 459945644 | -----RT---------- | ---- | H---T--Q--------- |
| Other Bacteria | Cesiribacter andamanensis | 496485643 | -I---KG--L-G-NG-- | | Q--AK-EE--N---E--- |
| | Acidobacteriaceae bacterium | 522212729 | -L--VR---L---NG-E | | --EKA-DE--EK--E--- |
| | Actinomyces urogenitalis | 566245879 | NA---R---FQ--QG-E | | --TAS--E------E--- |
| | Alicyclobacillus acidocaldarius | 495615214 | -V---RG--LK--EG-P | | Q--AK-EE------E--- |
| | Collinsella tanakaei | 496433145 | -I---RG--L---NGEE | | --MDA-RE--NS--E--- |
| | Fibrella aestuarina | 505143096 | NI---QG--L-G-NG-- | | ---KT-EE---S--SF-- |
| | Granulicoccus phenolivorans | 652534224 | -A--VQ----Q--QG-E | | --GKSVLD------E--- |
| | Hymenobacter ocellatus | 1169497 | NI--VQG--L-G-NG-- | | ---GT-EQ---S--NW-- |
| | Ilyobacter polytropus | 503153272 | -I---AG--L---NGE- | | Q--AK-EE------S--- |
| | Leptotrichia buccalis | 502496882 | -V-I-AG--L---NGE- | | ----K-ME------E--- |
| | Mobiluncus curtisii | 490107593 | ---I-----LK--EG-P | | E-TKK-EE--E---T--- |
| | Pedobacter agri | 498286935 | -I---QG--L-G-NG-- | | ---GK-ME------S--- |
| | Propionibacterium thoenii | 653569667 | N---VR---FQ--QG-P | | --TQS-LD------E--- |
| | Pseudoclavibacter soli | 654791955 | -A--VR---LK--EG-E | | --AATVAD---K--EFV- |
| | Spirosoma panaciterrae | 522086969 | NI---QG--L-G-NG-- | | ---KT-EE---N--SW-- |
| | Tsukamurella paurometabola | 502893068 | -A--VR--G-Q--QG-- | | ----S-VE--N---ES-- |

**FIGURE | Partial sequence alignment of the protein synthesis elongation factor-Tu showing a 4 aa insertion in a conserved region that is specific for members of the order *Bifidobacteriales*.** The dashes in this alignment as well as all other alignments show identity with the amino acid on the top line. The

*(Continued)*

**FIGURE | Continued**
Genebank Identification numbers of the protein sequences are shown, and the topmost numbers indicate the position of this sequence in the species shown on the top line. Due to space constraints, sequence information for different subspecies is not shown. However, unless otherwise indicated, these CSIs are present in the sequenced subspecies of *B. longum*, *B. animalis*, *B. pseudolongum*, and *B. thermacidophilum*. Information for large numbers of other CSIs, which are also specific for the order *Bifidobacteriales* is presented in **Table** and Supplementary Figures .

**TABLE | Conserved signature proteins that are uniquely found in the *Bifidobacteriales*.**

| Accession no. | Length | Function | Species specificity |
|---|---|---|---|
| ZP_02917512 | 73 | Unknown, hypothetical | *Bifidobacteriales* |
| ZP_02917322 | 275 | Unknown, hypothetical | *Bifidobacteriales* |
| ZP_02917261 | 336 | Unknown, hypothetical | *Bifidobacteriales* |
| ZP_02917147 | 228 | Unknown, hypothetical | *Bifidobacteriales* |
| ZP_02917106 | 399 | Unknown, hypothetical | *Bifidobacteriales* |
| ZP_02919152 | 201 | Unknown, hypothetical | *Bifidobacteriales* |
| ZP_02918813 | 121 | Unknown, hypothetical | *Bifidobacteriales* |
| ZP_02916931 | 84 | Unknown, hypothetical | *Bifidobacteriales* |
| ZP_02917770 | 76 | Unknown, hypothetical | *Bifidobacteriales* |
| ZP_02918933 | 321 | Unknown, hypothetical | *Bifidobacteriales* |
| ZP_02917048 | 222 | Unknown, hypothetical | *Bifidobacterium* and *Gardnerella* |
| ZP_02919141 | 299 | Unknown, hypothetical | *Bifidobacterium* and *Gardnerella* |
| ZP_02919088 | 260 | Unknown, hypothetical | *Bifidobacterium* |
| ZP_02918031 | 283 | Unknown, hypothetical | *Bifidobacterium* |
| ZP_02919040 | 189 | Unknown, hypothetical | *Bifidobacterium* |
| WP_015021123.1 | 152 | Unknown, hypothetical | *B. asteroides* cluster I |
| WP_033511744.1 | 116 | Unknown, hypothetical | *B. asteroides* cluster II |
| WP_015021403.1 | 283 | Unknown, hypothetical | *B. asteroides* cluster III |
| WP_015022574.1 | 190 | Unknown, hypothetical | *B. asteroides* cluster III |
| WP_015022150.1 | 300 | Unknown, hypothetical | *B. asteroides* cluster III |

*The species that are part of the B. asteroides clusters I, II, and III are indicated in* ***Figure*** .

protein YchF are also specifically shared by members of the genera *Bifidobacterium* and *Gardnerella.* Sequence information for these CSIs is presented in Supplementary Figures , S34 and some of their characteristics are summarized in **Table** . Additionally, we have also confirmed that the homologs of 5 of the 6 previously described CSPs (Gao and Gupta, 2012), information for which is summarized in **Table** , are also present in only members of these two genera.

We have also identified a number of CSIs that are commonly and specifically shared by members of the genus *Scardovia* and related genera for which sequence information is available. One example of a CSI which is specifically found in members of the genera *Scardovia, Parascardovia* and *Alloscardovia*, consisting of 1 aa insertion in the triosephosphate isomerase protein, is presented in **Figure** . Four other CSIs in four different proteins (viz. FHA domain protein, Glycosyl transferase, PAC2 family protein and Phosphate-ABC- transporter substrate-binding protein) are also largely specific for these genera of *Bifidobacteriales.* Sequence information for these CSIs is provided in Supplementary Figures S36–S39 and their characteristics are summarized in **Table** . Interestingly, the CSIs in the Glycosyl transferase and PAC2 family proteins are also commonly shared by *G. vaginalis.*

A number of distinct clusters within the genus *Bifidobacterium* are consistently observed in different phylogenetic studies including in the phylogenetic trees constructed in this work (**Figure** ). A number of CSIs identified in our work serve to distinguish some of the *Bifidobacterium* clusters. Three of the identified CSIs are specific for the *B. longum* group and sequence information for one of these CSIs, consisting of a 1 aa insertion in the phosphogluconate dehydrogenase, is shown in **Figure** . Sequence information for the other 2 CSIs that are also specific for a subgroup of species from the *B. longum* clade are presented in Supplementary Figures S40, S41 and their characteristics are summarized in **Table** . One additional CSI consisting of a 1 aa insertion in transketolase protein is specifically shared by members of the *B. longum, B. bifidum,* and *B. adolescentis* clades. Members of these clusters group together in phylogenetic trees and the shared presence of this CSI supports the view that that the members of these taxa are more closely and specifically related to each other.

The members of the *B. asteroides* cluster forms the deepest branching group within the genus *Bifidobacterium.* A number of CSIs identified in this study are specific for group of species, which are either part of the *B. asteroides* clade or related to this clade. The *B. asteroides* clade is demarcated as being made up of the species *B. asteroides, B. indicum, B. coryneforme,* and *B. actinocoloniiforme* species (marked cluster III in **Figure** ) (Lugli et al., 2014; Sun et al., 2015). Surprisingly, in our work no CSI was identified that was commonly shared by all of the species from this clade. However, our work identified four CSIs for a cluster (cluster II) comprising of all of other species from the *B. asteroides* clade, except *B. actinocoliniiforme*, which shows the deepest branching within this clade. One example of a CSI specific for members of the *B. asteroides* cluster II consisting of 1 aa insertion in the purine biosynthesis protein purH is shown in **Figure A.** Sequence information for three other CSIs that are also specific for the *B. asteroides* group is presented in Supplementary Figures . In our phylogenetic trees as well as in different identified signatures, two *Bifidobacterium* spp. strains A11 and 7101, isolated from honey bee guts (Anderson et al., 2013), also consistently group with the *B. asteroides.* Two CSIs identified in our work are specifically shared by *B. asteroides* and the *Bifidobacterium* sp. A11 and *Bifidobacterium* sp.7101 (referred to as *B. asteroides* cluster I) providing additional evidence of the close relationship of these *Bifidobacterium* strains to the *B. asteroides.* Sequence information for these CSIs is presented in Supplementary Figure . Lastly, one additional CSI identified in this work, consisting of a 3 aa insertion in a conserved region of the protein 5'-methylthioadenosine nucleosidase, is commonly shared by all the members of the

| Group | Species | ID | 88 | | 125 |
|---|---|---|---|---|---|
| *Bifidobacterium-Gardnerella* clade | Bifidobacterium adolescentis | 489904486 | FLPVDIHYYREMSRRI | F | TEVFSRVTDRIEQVSVDECYM |
| | Bifidobacterium angulatum | 489923981 | ------R---KV-QS- | M | RD--LT----F-------G-- |
| | Bifidobacterium animalis | 490389093 | -----H---SAV-HE- | - | HT--RE--------------- |
| | Bifidobacterium asteroides | 504835263 | -----MT---SI---- | - | DT----I-NQ----------- |
| | Bifidobacterium bifidum | 489911054 | -----M----M----- | V | E----Q----F-------A-- |
| | Bifidobacterium boum | 651390579 | -----MR---M----- | - | ED----I---F-------G-- |
| | Bifidobacterium breve | 489926514 | -----MA---M--H-- | M | H---LTI--QF-------G-- |
| | Bifidobacterium catenulatum | 489932716 | -----MS---A--H-- | - | K--L--I-G------------ |
| | Bifidobacterium choerinum | 639202583 | -----H---SQI--D- | - | GTI------Q----------- |
| | Bifidobacterium dentium | 489934088 | -----MQ---A----- | - | ------I-------------- |
| | Bifidobacterium gallicum | 493338339 | -----GR----I--Q- | - | QT--HQ------P-------- |
| | Bifidobacterium longum | 494112237 | -----MA---M--H-- | M | H---LT----F-------G-- |
| | Bifidobacterium magnum | 651390243 | ------A---TV-HQ- | - | -DI--K---Q----------- |
| | Bifidobacterium minimum | 551240467 | -----MR----T---- | - | SD-LG------------A-- |
| | Bifidobacterium moukalabense | 575769369 | -----MQ---A----- | - | ------I-------------- |
| | Bifidobacterium pseudocatenulatum | 490330617 | -----MS---A--H-- | - | K--L--I-G------------ |
| | Bifidobacterium pseudolongum | 651883820 | -----GA---RV--Q- | - | -DI--Q---A----------- |
| | Bifidobacterium ruminantium | 651886874 | ------R--------- | - | -----Q--------------- |
| | Bifidobacterium subtile | 639065366 | -----M----SI-H-- | V | S---GA------------A-- |
| | Bifidobacterium thermacidophilum | 657871235 | -----MS---M----- | - | D----QI--Q--------G-- |
| | Bifidobacterium thermophilum | 505263881 | -----MS---M----- | - | D----QI--Q--------G-- |
| | Bifidobacterium tsurumiense | 651881823 | -----MA---A--H-- | - | S----QI--H----------- |
| | Bifidobacterium pullorum | 705444526 | -----MR---M----- | - | D-----I-------------- |
| | Bifidobacterium biavatii | 705395711 | -----MT---M----- | - | R----Q--------------- |
| | Bifidobacterium scardovii | 705445973 | -----MA---M--H-- | - | S-----I-------------- |
| | Bifidobacterium saeculare | 705428972 | -----MR---M--H-- | - | S----QI-------------- |
| | Bifidobacterium crudilactis | 736119830 | -----MR--------- | - | S---A-I--N--------G-- |
| | Bifidobacterium mongoliense | 705435860 | ------D---R----- | - | ----G----A--------G-- |
| | Bifidobacterium psychraerophilum | 705397424 | -----M----SV---- | - | N---AS------------G-- |
| | Bifidobacterium kashiwanohense | 746131641 | ---I-MS---A--H-- | - | K-----I-G------------ |
| | Bifidobacterium actinocoloniiforme | 705420974 | -----MS---AV--S- | - | SQ--A----QV---------- |
| | Bifidobacterium bohemicum | 917511722 | -V---MT---T--K-- | - | ----QQ---Q--------G-- |
| | Bifidobacterium stellenboschense | 736511145 | -----MR---AV---- | M | ----HA----V-------G-- |
| | Bifidobacterium cuniculi | 705443047 | -----HT---QV-HT- | - | R-I--T--------------- |
| | Bifidobacterium indicum | 705389070 | -----MA--HYI-HE- | - | KT----I------------F- |
| | Bifidobacterium coryneforme | 799123986 | -----MA--HYV-HE- | - | KT----I------------F- |
| | Bifidobacterium bombi | 763215354 | YVS--MA---R---C- | - | -Q---Q------------G-- |
| | Bifidobacterium aesculapii | 943597392 | -----MR---SV--Q- | M | -D--HA----V-------G-- |
| | Bifidobacterium saguini | 672997532 | -----MR---M--H-- | M | H---LT----F-------G-- |
| | Bifidobacterium reuteri | 763216660 | -----MR---M--H-- | M | H---LT----F-------G-- |
| | Bifidobacterium callitrichos | 759442531 | -----MR---SV--E- | M | -T--HA---QV-------G-- |
| | Bifidobacterium merycicum | 705458204 | -----MA--CGV--A- | M | HD--HA----V-------G-- |
| | Bifidobacterium gallinarum | 672964903 | -----MR---M--H-- | - | S----QI-------------- |
| | Bifidobacterium stercoris | 673001421 | ---------------- | - | --------------------- |
| | Gardnerella vaginalis | 490207126 | ------P--SSV-KS- | - | ENI-LHI-NQV-K-------- |
| *Scardovia*-related genera | Alloscardovia omnicolens | 545376757 | -----MA--KSL-AQV | | FAI-EE---QV-H-----G-- |
| | Alloscardovia criceti | 516877857 | -----MK---DI-AQV | | F-L--Q---Q--R-------- |
| | Parascardovia denticolens | 493331646 | Y----MA---S---KV | | FAI-HQ---Q--------G-- |
| | Scardovia inopinata | 493335694 | -----MA---S--HQV | | F-I-HQI--Q--P-------- |
| | Scardovia wiggsiae | 494250503 | -----MQ---S--SQV | | FGI-A-I-----K-------- |
| Other Bacteria | Acaricomes phytoseiuli | 516860991 | -PRH-L--QV-AQ- | | M-I-RS---QV--L-I--AFL |
| | Alkaliphilus metalliredigens | 501012359 | Y---RHER---V--QV | | FDILYE---LV-PL-I--AFL |
| | Caldithrix abyssi | 493987017 | Y--PNGAL-QQV-QN- | | FQI-YEF--LV-P--I--AFL |
| | Clostridium arbusti | 497919451 | ----RHER---I-NKV | | FNI-NEI-PL--PL-I--A-L |
| | Eikenella corrodens | 565108605 | -V-PRF-L---V-QQ- | | HA--RQH--LV-PL-L--A-L |
| | Enterococcus raffinosus | 498440001 | -K-G-HAK-SAV-KQV | | R-I-H-Y--I--P--I--A-L |
| | Eubacterium ramulus | 545613317 | MV-P-HT------QKL | | MHHLRSIC-E-----I----- |
| | Facklamia ignava | 493752464 | -I-PN-D---QI-QS- | | HHI-RQY--L--P--L--A-L |
| | Gottschalkia acidurici | 504780063 | ----RMER-K-V--QV | | FGILYEIA-L--P--I--A-L |
| | Haemophilus parasuis | 491998641 | L---NMSL-KAV-EQ- | | HCI-R-Y--I--PI-L--A-L |
| | Lactobacillus fructivorans | 497707820 | -KAPEFKL--DV-HQ- | | HDI-HEY-----T-AL--A-L |
| | Leuconostoc lactis | 497687618 | --TP-FAK-KAI-KQ- | | HAI-RTI-PK--P-AY--A-L |
| | Longispora albida | 517162097 | Y--P-FTE-SAV-KLV | | MAM-RD--PLV--L----AFL |
| | Methanoculleus bourgensis | 504680316 | Y---NRPL-SRV-DE- | | MAIL-ERA-------I--A-L |
| | Pasteurella multocida | 492120851 | L---NMPL-KQV-QQ- | | HQI-R-Y--V--PL-L--A-L |
| | Prevotella bryantii | 493325432 | IAASHHS--AKI-HQ- | | H-I-QDY--K--PI-I--AFL |
| | Sporomusa ovata | 544739083 | ---S-HEK-SRV-QE- | | LHI-ADFSPQV-PL-L--AFL |
| | Streptococcus peroris | 493116418 | -I-GNYEK-TAV-QQ- | | RAI-K-Y--K--PM-I--A-L |

**FIGURE | Partial sequence alignment of DNA polymerase IV showing a 1 aa insertion that is specific for the *Bifidobacterium* and *Gardnerella* species, but not found in any other *Bifidobacteriales*.** Information for other CSIs specific for this clade is presented in **Table** and Supplementary Figures S33–S35.

**TABLE | Characteristics of Conserved Signature Indels Distinguishing a number of subgroups within the order *Bifidobacteriales*.**

| Protein name | GI number | Figure number | Indel size | Indel position | Specificity |
|---|---|---|---|---|---|
| DNA polymerase IV | 489904486 | **Figure 3** | 1 aa ins | 88–125 | *Bifidobacterium-Gardnerella* |
| Ribosomal RNA small subunit methyltransferase E | 547081721 | Supplementary Figure S33 | 3 aa del | 118–160 | *Bifidobacterium-Gardnerella* |
| GTP-binding protein YchF | 547055080 | Supplementary Figure S34 | 1 aa ins | 309–354 | *Bifidobacterium-Gardnerella* |
| Cytochrome C | 500062679 | Supplementary Figure S35 | 3 aa del | 730–765 | *Bifidobacterium* |
| Triosephosphate isomerase | 651360171 | **Figure 4** | 1 aa ins | 251–286 | Scardovia clade |
| FHA domain protein | 493335662 | Supplementary Figure S36 | 1 aa ins | 37–67 | Scardovia clade |
| Glycosyl transferase | 648490110 | Supplementary Figure S37 | 2 aa ins | 23–67 | Scardovia clade |
| PAC2 family protein | 294458767 | Supplementary Figure S38 | 2 aa ins | 32–77 | Scardovia clade |
| Phosphate ABC transporter substrate-binding protein | 493336671 | Supplementary Figure S39 | 2 aa ins | 167–206 | Scardovia clade |
| Phosphogluconate dehydrogenase | 497766884 | **Figure 5** | 1 aa ins | 360–401 | *B. longum* cluster |
| PhoU family transcriptional regulator | 489926631 | Supplementary Figure S40 | 2 aa del | 159–190 | *B. longum* cluster |
| Cystathionine gamma-synthase | 494112910 | Supplementary Figure S41 | 2 aa ins | 262–302 | *B. longum* cluster |
| Transketolase | 489905793 | Supplementary Figure S42 | 1 aa ins | 234–274 | *B. longum, B. bifidum* and *B. adolescentis* clade |
| Purine biosynthesis protein purH | 658453400 | **Figure 6A** | 1 aa ins | 247–278 | *B. asteroides* cluster II # |
| Shikimate dehydrogenase | 658453363 | Supplementary Figure S43 | 1 aa ins | 264–301 | *B. asteroides* cluster II # |
| 5-methyltetrahydropteroyltriglutamate–homocysteine methyltransferase | 504834759 | Supplementary Figure S44 | 1 aa ins | 336–369 | *B. asteroides* cluster II # |
| ABC transporter substrate-binding protein | 504835116 | Supplementary Figure S45 | 1 aa del | 253–286 | *B. asteroides* cluster II # |
| 5′-methylthioadenosine nucleosidase | 504835309 | **Figure 6B** | 3 aa ins | 1–33 | *B. asteroides*-related cluster IV # |
| Peptide ABC transporter ATP-binding protein | 504834913 | Supplementary Figure S46A | 20 aa ins | 76–127 | *B. asteroides* cluster I# |
| N-acetyl-gamma-glutamyl-phosphate reductase | 504834965 | Supplementary Figure S46B | 1 aa ins | 34–74 | *B. asteroides* cluster I# |

# *The B. asteroides-related cluster I, II, and IV are demarcated in* ***Figure*** .

*B. asteroides* as well as by *B. crudilactis* and *B. psychaerophilum*. The latter two species form a deeper branching cluster that appears to be specifically related to the *B. asteroides* clade in the tree based on concatenated protein sequences (marked as *B. asteroides* cluster IV in **Figure** ). The shared presence of this CSI by the *B. asteroides* clade and *B. crudilactis* and *B. psychaerophilum* support the inference that these species are specifically related to the *B. asteroides* clade.

In addition to the described CSIs, BLAST searches on the protein sequences of *B. asteroides* have also identified 5 CSPs, whose homologs are specifically present in the members of *B. asteroides* group of species. Information for these CSPs is also presented in **Table** . Of these CSPs, three CSPs are specific for the commonly described *B. asteroides* clade (Cluster III in **Figure** ), whereas the remaining two are specific for the clusters I and II.

## DISCUSSION

Members of the order *Bifidobacteriales* are one of the main groups within bacteria where several members exhibit health-promoting probiotic effects on humans (Biavati et al., 2000; Biavati and Mattarelli, 2006; Ventura et al., 2007b, 2009a; Cronin et al., 2011; Turroni et al., 2011). Other *Bifidobacteriales* species are also responsible for implicated in the development of dental caries as well as bacterial vaginosis and urinary tract infections (Bradshaw et al., 2006; Mantzourani et al., 2009; Ventura et al., 2009b; Kenyon and Osbak, 2014). However, very little is known at present concerning the genetic or biochemical characteristics of these bacteria that mediate their beneficial or pathogenic effects. In the present work, we have carried out detailed phylogenetic and comparative analyses of protein sequences from the genomes of *Bifidobacteriales* species to examine in depth their evolutionary relationships and also to identify molecular markers that are unique to these bacteria at multiple phylogenetic levels. Based on a robust and comprehensive phylogenetic tree for the *Bifidobacteriales* species based on 614 core proteins from the sequenced genomes, the following inferences regarding the evolutionary relationships among the *Bifidobacteriales* species could be made. (i) The sequenced *Bifidobacteriales* species appear to form two main clusters, a deeper clade consisting of the *Scardovia*-related genera (viz. *Scardovia, Parascardovia* and *Alloscardovia*) and another cluster grouping together *Bifidobacterium* and *Gardnerella* genera. (ii) *Gardnerella vaginalis* rather than branching separately is found to consistently branch in between different *Bifidobacterium* species.

| Group | Species | ID | 251 | | 286 |
|---|---|---|---|---|---|
| *Scardovia*-related genera | Alloscardovia omnicolens | 651360171 | YGGSVTAQNATLIMNEA | R | DMDGFLIGGASLNPEEFS |
| | Metascardovia criceti | 648490205 | -------N----L---- | - | ------V-----DAD--A |
| | Parascardovia denticolens | 493334606 | -----SSK---DL--Q- | H | -V----V-----D----A |
| | Scardovia inopinata | 535476484 | ------S---MDL--Q- | H | -V----V-----D----- |
| | Scardovia wiggsiae | 515714829 | ------S---MDL-DQ- | H | -V----V-----D----- |
| Other *Bifidobacteriales* | Bifidobacterium angulatum | 489922926 | ------SK--AELIA-P | | -V----V-----DAN--A |
| | Bifidobacterium asteroides | 504834734 | ------SK--SQMITQP | | -V----V-----DV---- |
| | Bifidobacterium bifidum | 489909858 | -----SSK---HLIG-P | | -V----------D-D-LA |
| | Bifidobacterium boum | 651391054 | ------SK--DQLIRQP | | -V----V-----HA--LA |
| | Bifidobacterium breve | 585130875 | ------SK--AELISQP | | -V--------A-DVD-L- |
| | Bifidobacterium longum | 547239381 | ------SK--AELISQP | | -V--------A-DV--LA |
| | Bifidobacterium magnum | 651390093 | ------SK--VELIGQE | | -V--------A-HVD-LT |
| | Bifidobacterium minimum | 651887732 | ------SRDVAQLIEQP | | -I----V-----DVD--T |
| | Bifidobacterium sp. 7101 | 658453366 | ------SK--SQMITQP | | -V----V-----DV---V |
| | Bifidobacterium sp. A11 | 639450006 | ------SK--SQMITQP | | -V----V-----DV---A |
| | Bifidobacterium subtile | 639063916 | ------SG--AELIE-- | | -V----------DV--LA |
| | Bifidobacterium thermacidophilum | 657872129 | ------SK--SQLITQP | | -V----V-----H---LA |
| | Bifidobacterium thermophilum | 505263497 | ------SK--SQLITQP | | -V----V-----HA--LA |
| | Bifidobacterium tsurumiense | 651882159 | ------SK--SQLIS-P | | -V----V-----DVD-LA |
| | Bifidobacterium psychraerophilum | 705400252 | -----SS---AQLIQ-- | | -V----V-----EV---- |
| | Bifidobacterium bohemicum | 705453892 | -----SSR--VELIS-P | | -V----------D---LA |
| | Bifidobacterium gallinarum | 705423798 | ------SH--VQLITQP | | -V----------D---LA |
| | Bifidobacterium crudilactis | 736120758 | -----SSR--AQLIQ-R | | -V----V-----EVD--- |
| | Bifidobacterium coryneforme | 799123296 | ------SK--SQ-IQ-H | | -V----V-----DVQ--A |
| | Bifidobacterium pullorum | 705442364 | ------SN--VQLITQP | | -V----------D---LA |
| | Bifidobacterium mongoliense | 705436912 | ------SK-C-Q-IA-P | | -V----------DV--LA |
| | Bifidobacterium reuteri | 763215760 | ------SK--AELISQP | | -V-----------A--LT |
| | Bifidobacterium merycicum | 705458800 | ------SK--AELIE-P | | -V----V-----DAN--A |
| | Bifidobacterium saguini | 727803668 | ------SK--AELISQP | | -V----------DA--LA |
| | Bifidobacterium bombi | 763213935 | -----SSK--VETLDQN | | -V----V---A-DAD--- |
| | Bifidobacterium kashiwanohense | 705413968 | -----SSK--VELIG-P | | -V--------A--VD-LT |
| | Bifidobacterium callitrichos | 672961096 | -----SSK--VELIG-P | | -V--------A-DAD-LA |
| | Bifidobacterium pseudocatenulatum | 524473885 | -----SSK--EELIG-P | | -V--------A--VD-LT |
| | Bifidobacterium sp. AGR2158 | 651885675 | ------SK--VELIS-P | | -V--------A-HVD-LT |
| | Bifidobacterium choerinum | 639202923 | ------SK--VELIS-P | | -V--------A-HVD-LT |
| | Bifidobacterium gallicum | 270277799 | ------SK--VELIG-P | | -V--------A-DVN-LT |
| | Bifidobacterium adolescentis | 489905779 | -----SSK--VELI--P | | -V-------SA-KVD-LT |
| | Bifidobacterium actinocoloniiforme | 705421242 | ------SK-VQELIGQP | | -V--------A-DVD--A |
| | Bifidobacterium cuniculi | 705445566 | ------SR--IELIE-P | | -V--------A-HVD-LT |
| | Bifidobacterium biavatii | 705395331 | -----SSK--VELIG-P | | -V--------A-DAD-LA |
| | Bifidobacterium catenulatum | 489931487 | -----SSK--VELIG-P | | -V--------A-HVD-LT |
| | Bifidobacterium stellenboschense | 736510025 | -----SSK--VELIG-P | | -V--------A-DVD-LT |
| | Bifidobacterium moukalabense | 736875813 | -----SSK--VELIG-P | | -V--------A-KVD-LT |
| | Bifidobacterium ruminantium | 651886380 | -----SSK--VELIR-P | | -V--------A-KVD-LT |
| | Bifidobacterium dentium | 171278564 | -----SSK--VELIG-P | | -V--------A-KVD-LT |
| | Bifidobacterium aesculapii | 943597283 | -----SSR--VELIG-P | | -V--------A-DVD-L |
| | Bifidobacterium scardovii | 672995889 | -----SSK--VELIG-P | | -V--------A-DVD-LA |
| | Bifidobacterium sp. MSTE12 | 570842516 | -----SSK--VELIG-P | | -V--------A-KVD-LT |
| | Bifidobacterium stercoris | 673003375 | -----SSK--VELI--P | | -V-------SA-KVD-LT |
| | Bifidobacterium saeculare | 672990889 | ------SH--VQLITQP | | -V----------D---LA |
| | Bifidobacterium indicum | 655535248 | ------SK--SQMIQ-H | | -V----V-----DVQ--A |
| | Bifidobacterium animalis | 490328467 | -----SSK-CVELIE-P | | -V--------A-DVD-LT |
| | Bifidobacterium pseudolongum | 746105858 | -----SSK--VELIT-P | | -V--------A-DVD-LT |
| | Bifidobacterium longum | 524120693 | ------SK--AELISQP | | -V--------A-DV--LA |
| | Gardnerella vaginalis | 490207959 | ------SS--SSLIA-P | | -V----------DAQ-LA |
| Other Bacteria | Buchnera aphidicola | 499991861 | ------KN--KELIYQK | | -I----V-----QL---- |
| | Cellulomonas fimi | 503536946 | -----KSS-VAS--AKP | | -V--A-V-----D----A |
| | Acidimicrobium ferrooxidans | 506278511 | -----DVG--VAFLAQ- | | -I--L-V-----D--R-A |
| | Persephonella marina | 501943652 | -----NEK--RDLIK-P | | NV----V-T---D--R-Y |
| | Verrucomicrobia bacterium | 640165320 | ------P---RELLAQP | | -V--L-V-----EA-S-T |
| | Waddlia chondrophila | 502947249 | -----KPDS-RVM-EQS | | -V--V-V-----D--T-A |
| | Xylanimonas cellulosilytica | 502641898 | -----KSS-VAD--AKP | | -V--A-V-----D-A--A |

**FIGURE | Example of 1 aa conserved signature indel in the protein triosephosphate isomerase that is specific for the *Scardovia* clade comprising of the genera *Scardovia, Parascardovia, Metascardovia,* and *Alloscardovia*.** Information for other CSIs specific for this clade is presented in **Table** and Supplementary Figures S36–S39.

| Clade | Species | Accession | 360 | | 401 |
|---|---|---|---|---|---|
| *B. longum* clade | Bifidobacterium longum subsp. longum | 227213792 | LSEVARIWRGGCIIRAQF | L | LDRITEAFRGDNPPASLLFDPYF |
| | Bifidobacterium longum subsp. infantis | 213522589 | ------------------ | - | --------K-------------- |
| | Bifidobacterium longum subsp. suis | 672977536 | ------------------ | - | --------K---L---------- |
| | Bifidobacterium breve | 489924740 | ------------------ | - | --------K-------------- |
| | Bifidobacterium sp. 12_1_47BFAA | 316915838 | ------------------ | - | ----------------------- |
| | Bifidobacterium aesculapii | 943596306 | ------------------ | - | --------K-------------- |
| | Bifidobacterium saguini | 727803447 | ------------------ | - | --------K-------------- |
| | Bifidobacterium callitrichos | 759445384 | ------------------ | - | --------K-------------- |
| | Bifidobacterium stellenboschense | 736514481 | ------------------ | - | --------K-------------- |
| | Bifidobacterium reuteri | 763216847 | ------------------ | - | --------K-------------- |
| Other *Bifidobacteriales* | Bifidobacterium adolescentis | 547078766 | -AA-------------K- | | -N--S---ESGEANV----A--- |
| | Bifidobacterium angulatum | 489923209 | -AA-------------K- | | -N--S---ESGEANV----A--- |
| | Bifidobacterium animalis[1] | 490328228 | MAHL-----A-------- | | ----S---ESGQADV----A--- |
| | Bifidobacterium asteroides | 504834148 | MGAI------------K- | | -N--SD-YASGRE-G------F- |
| | Bifidobacterium bifidum | 489911235 | -AA-------------K- | | -N--SD--ENGEANV----A--- |
| | Bifidobacterium boum | 651391314 | -AA--------------- | | -N-VA---ESGEADI----A--- |
| | Bifidobacterium catenulatum | 489931996 | -AA-------------K- | | -N--S---ESGEANV----A--- |
| | Bifidobacterium choerinum | 639202818 | MAHL-----A------K- | | ----S---EDGAADV----A--- |
| | Bifidobacterium dentium | 489937347 | -AA-------------K- | | -N--S---ESGEANV----A--- |
| | Bifidobacterium gallicum | 493336974 | MAQL-----A------K- | | -N--SD--ESGAADV----A--- |
| | Bifidobacterium magnum | 551238766 | -AA------A------K- | | -N--SD--ESGAADV----A--- |
| | Bifidobacterium minimum | 551241056 | -AA--------------- | | -NVVSD--ESGDADV----A--- |
| | Bifidobacterium moukalabense | 575769825 | -AA-------------K- | | -N--S---ESGEANV----A--- |
| | Bifidobacterium pseudocatenulatum | 490332077 | -AA-------------K- | | -N--S---ESGEANV----A--- |
| | Bifidobacterium pseudolongum | 651883513 | -AA------A------E- | | ----S-S-ETGQADV----A--- |
| | Bifidobacterium ruminantium | 651886121 | -AA-------------K- | | -N--S---ESGEANV----A--- |
| | Bifidobacterium sp. 7101 | 658452142 | MGAI------------K- | | -N--SD-YASGRE-G------F- |
| | Bifidobacterium sp. A11 | 639449736 | MGAI------------K- | | -N--SD-YASGRE-G------F- |
| | Bifidobacterium sp. AGR2158 | 651885376 | MAHL-----A------K- | | ----S---EDGAADV----A--- |
| | Bifidobacterium sp. Hma3N | 563426038 | MGAI------------K- | | -N--SD-YASGRE-G------F- |
| | Bifidobacterium subtile | 639064830 | -AA--------------- | | -NVVSD--ESGEANV----A--- |
| | Bifidobacterium thermacidophilum[1] | 657871497 | -AA--------------- | | -N-VA---ESGEANV----AS-- |
| | Bifidobacterium thermophilum | 505262608 | -AA--------------- | | -N-VS---ESGEADI----A--- |
| | Bifidobacterium kashiwanohense | 746132357 | -AS-------------K- | | -N--S---ESGEANV----A--- |
| | Bifidobacterium merycicum | 705460238 | -AA-------------K- | | -N--S---ESGEANV----A--- |
| | Bifidobacterium biavatii | 705398390 | -AA-------------K- | | -N--SD--ESGEADV----A--- |
| | Bifidobacterium scardovii | 705447194 | -AA-------------K- | | -N--SD--ESGEANV----A--- |
| | Bifidobacterium psychraerophilum | 705399550 | -GA-------------K- | | -N--SD-YDSSETFG------F- |
| | Bifidobacterium crudilactis | 736121669 | -GA-------------K- | | -N--SD-YDSSETFG------F- |
| | Bifidobacterium bohemicum | 705456353 | MAA--------------- | | -NVVSD--QSGEADV----A--- |
| | Bifidobacterium indicum | 705391329 | MGAI------------K- | | -N--SD-YASGDEFG------F- |
| | Bifidobacterium coryneforme | 705404673 | MGAI------------K- | | -N--SD-YASGDEFG------F- |
| | Bifidobacterium actinocoloniiforme | 705420678 | -GAI------------K- | | -N--SD-YGSGEQFG------F- |
| | Bifidobacterium bombi | 763215194 | -AA--------------- | | -NKVSD--ESGEAEV-M--AS-- |
| | Bifidobacterium mongoliense | 705438389 | -AA--------------- | | -NVVSD--ESGEADV-M--A--- |
| | Bifidobacterium cuniculi | 705443828 | MAG------A------R- | | --T-SD--ETGTADV----A--- |
| | Bifidobacterium tsurumiense | 673010562 | -AA-------------K- | | -N-VSD--ESGEANI----AR-- |
| | Gardnerella vaginalis | 523593112 | -AA--------------- | | -N-VS---ESGQASV----A--- |
| | Alloscardovia omnicolens | 545370500 | -GA-------------K- | | -N--S---DSGEKFE---Q-A-- |
| | Scardovia inopinata | 493335449 | -GA-------------K- | | -N--SD-YDSGRTFS-----D-- |
| | Scardovia wiggsiae | 656313997 | -GA-------------K- | | -N--SD-YGSGRTFT-----E-- |

**FIGURE | Partial sequence alignment of phosphogluconate dehydrogenase showing a 1 aa insertion that is specific for the *B. longum* cluster.**

(iii) Within *Bifidobacterium* species, a number of distinct clusters, referred to as the *B. asteroides, B. pseudolongum, B. longum, B. bifidum, B. adolescentis, B. pullorum,* and *B. boum* groups, are observed as described in earlier work (Lugli et al., 2014; Sun et al., 2015). Of these clusters, the *B. asteroides* group forms the deepest branching lineage within the *Bifidobacterium* (Bottacini et al., 2012; Lugli et al., 2014; Sun et al., 2015).

The present work also identified large number of novel molecular signatures in the forms of CSIs and CSPs, which are specific characteristics of the members of the order *Bifidobacteriales* at multiple phylogenetic levels. Of these signatures, 32 CSIs and 10 CSPs are specific for the entire order *Bifidobacteriales*. The identified *Bifidobacteriales*-specific CSIs are present in assorted widely distributed proteins carrying out wide variety of cellular functions. All of the 10 *Bifidobacteriales*-specific CSPs are proteins of unknown functions. Given the specificity of these CSIs and CSPs for the *Bifidobacteriales*, the genetic changes leading to these molecular characteristics have likely occurred in a common ancestor of the *Bifidobacteriales* (Gao and Gupta, 2005, 2012). Additionally, our analyses have also identified many other molecular signatures (CSIs and CSPs), which independently support the existence of a number of clades of bifidobacteria that are consistently observed in phylogenetic trees. The clades identified by these molecular signatures include, (i) a clade encompassing the genera *Scardovia, Parascardovia* and *Alloscardovia*, (ii) signatures that are commonly shared by

**A**

| Group | Species | ID | 247 … 278 |
|---|---|---|---|
| *B. asteroides* Cluster II | Bifidobacterium sp. 7101 | 658453400 | GFAAAEHLGG A KEMSYNNYVDADSAWRSVWDF |
| | Bifidobacterium indicum | 705387921 | ---SANQ--- - -A-------------T---- |
| | Bifidobacterium coryneforme | 799123273 | ---SANQ--- - -A-------------T---- |
| | Bifidobacterium sp. A11 | 639449677 | ---------- - ------------A-------- |
| | Bifidobacterium asteroides | 504834707 | ---------- - ------------A-------- |
| Other *Bifidobacteriales* | Bifidobacterium adolescentis | 489905846 | ---H--Q--- -P----------A---A---- |
| | Bifidobacterium animalis | 490328508 | ---D--Q--- -P----------A---A---- |
| | Bifidobacterium bifidum | 547761202 | ---H--Q--- -P----------A---A---M |
| | Bifidobacterium boum | 651391078 | ---H--Q--- -P----------A---A---- |
| | Bifidobacterium breve | 489926028 | ---H--Q--- -P----------A---T---M |
| | Bifidobacterium catenulatum | 489931437 | ---H--Q--- -P----------A---A---- |
| | Bifidobacterium choerinum | 639201990 | ---D--Q--- -P----------A---A---- |
| | Bifidobacterium dentium | 489936809 | ---H--Q--- -P----------A---A---- |
| | Bifidobacterium gallicum | 493337638 | ---H--Q--- -P----------A---A---- |
| | Bifidobacterium longum | 494112083 | ---H--Q--- -P----------A---T---M |
| | Bifidobacterium magnum | 651390100 | --H-RQ--- -P----------A---A---- |
| | Bifidobacterium minimum | 551240262 | ---Q-DQ--- -P----------A---A---M |
| | Bifidobacterium moukalabense | 575770122 | ---H--Q--- -P----------A---A---- |
| | Bifidobacterium pseudocatenulatum | 490332749 | ---H--Q--- -P----------A---A---- |
| | Bifidobacterium pseudolongum | 651883334 | ---N--Q--- -P----------A---A---- |
| | Bifidobacterium ruminantium | 651886301 | ---H--Q--- -P----------A---A---- |
| | Bifidobacterium sp. 12_1_47BFAA | 496059019 | ---H--Q--- -P----------A---T---M |
| | Bifidobacterium sp. AGR2158 | 651885634 | ---D--Q--- -P----------A---A---- |
| | Bifidobacterium subtile | 639065922 | ---H-VQ--- -P----------A---T---- |
| | Bifidobacterium thermacidophilum | 657872150 | ---H-DQ--- -P----------A---A---- |
| | Bifidobacterium thermophilum | 505263470 | ---H--Q--- -P----------A---A--Y- |
| | Bifidobacterium tsurumiense | 651882128 | ---H-QQ--- -P----------A---A---- |
| | Gardnerella vaginalis | 657897086 | -L-Q--L--- -P----------A---A---- |
| | Alloscardovia omnicolens | 551235973 | --H--Q--- -P------------------ |
| | Parascardovia denticolens | 493332087 | D--H-QQ--- -P-----F------------- |
| | Scardovia inopinata | 493336115 | D--H-RQ--- -P------------------ |
| | Scardovia wiggsiae | 494249089 | --H-RQ-N- -P------------------ |
| | Metascardovia criceti | 516878262 | --H--Q--- -P------------------ |
| Other Bacteria | Actinomyces neuii | 490943764 | -I-N-LT--- -P------S---A-L-AAY-H |
| | Micromonospora chokoriensis | 663715182 | -L-Q--Q-H- ------------A---AAN-- |
| | Nocardioidaceae bacterium | 495639466 | -L----Q-H- ----------T-A-R-AAY-Q |
| | Pseudonocardia asaccharolytica | 655577930 | -L-G--Q-H- ------------A---AAH-H |
| | Salinispora tropica | 656097905 | -L-Q--Q-H- ------------A---AAN- |

**B**

| Group | Species | ID | 1 … 33 |
|---|---|---|---|
| *B. asteroides* Cluster IV | Bifidobacterium asteroides | 504835309 | MDEEVALIGQGLKQA DQD SHAGDAGLKVISGTL |
| | Bifidobacterium coryneforme | 705402914 | --------AK--EDP IH- ERS-Q-S-D-VR-S- |
| | Bifidobacetrium indicum | 705389850 | --------AK--EDP IH- ERSAQ-S-D-VR-S- |
| | Bifidobacterium actinocoloniiforme | 705421108 | -------VAQ--DGV VH- PASRS-G-D-VR-R- |
| | Bifidobacterium sp. 7101 | 658452650 | --------------- --- ---S-------C--- |
| | Bifidobacterium sp. A11 | 658450567 | ---------R----- -H- P----------C--- |
| | Bifidobacterium crudilactis | 917260499 | LE---N--ARS-HDV EHM EASRR---D-V---- |
| | Bifidobacterium psychraerophilum | 917315014 | L----E--ARS-DD- RHL ERSKQ-S-D-VA-N- |
| Other *Bifidobacteriales* | Bifidobacterium bifidum | 503129232 | L-D-----AAS-NDV T-DRQ---D-TR--- |
| | Bifidobacterium breve | 504297181 | L-------AKS-TRV T-TAKGS-DIV---V |
| | Bifidobacterium longum | 548569142 | L-------AKS-THV T-TAKGS-DIV---V |
| | Bifidobacterium boum | 651390606 | L-------AAS-DHV E-RRE-S-D-AH--- |
| | Bifidobacterium thermacidophilum | 657871210 | L-------AAS-DHV E-HRE-S-D-AH--- |
| | Bifidobacterium thermophilum | 505263911 | L-------AAS-DHV E-HRE-S-D-AH--- |
| | Gardnerella vaginalis | 490235908 | L-----H-ASA-ENV --DRS-S-N-SC--- |
| Other Bacteria | Glaciecola polaris | 494165993 | ----IT-LK-SIR-L QEHKH-H-TLYT-Q- |
| | Paenibacillus assamensis | 655149029 | ----I--LL-AMG-V EEHEQ--NRYYR--- |
| | Oceanobacillus picturae | 595760994 | -------LL-NVSNQ NEETI--C-FVK-QI |

**FIGURE | Conserved signature indels that are specific for the *B. asteroides*-related clades of the *Bifidobacteriales*. (A)** Partial sequence alignment of the purine biosynthesis protein purH showing a 1 aa insertion which is specific for the *B. asteroides* cluster II species in the protein tree (**Figure** ); **(B)** Excerpt from sequence alignment of the protein 5′′-methylthioadenosine nucleosidase showing a 3 aa insertion that is specific for the *B. asteroides*-related cluster IV in the protein tree.

*Bifidobacterium* and *Gardnerella* species to the exclusion of other bifidobacteria, and (iii) signatures demarcating specific clusters of *B. asteroides*- or *B. longum*- related species.

The order *Bifidobacteriales* presently contains a single family, *Bifidobacteriaceae*. Based upon the results of phylogenomic studies and identified molecular signatures, it appears that the members of this order could be divided into two family-level groups, one comprising of the *Scardovia*-related genera (viz. *Scarodivia, Parascardovia,* and *Alloscardovia*) and the other consisting of the genera *Bifidobacterium* and *Gardnerella*. However, genome sequence information for members of several newly described *Scardovia*-related genera (viz. *Aeriscardovia, Neoscardovia,* and *Pseudoscardovia*), is lacking at present (Simpson et al., 2004; García-Aljaro et al., 2012; Killer et al., 2013). In future studies, depending upon whether the species from these genera branch with the *Scardovia*-clade and their sharing of the molecular signatures specific for this clade, the possibility of dividing the order *Bifidobacteriales* into two or more families could be considered.

The genus *Bifidobacterium*, which is comprised of 49 species and subspecies, contains most of the recognized taxa within the order *Bifidobacteriales*. Although earlier phylogenetic studies have consistently observed 6–7 distinct clusters of *Bifidobacterium* species (Ventura et al., 2006, 2007b; Turroni et al., 2011; Lugli et al., 2014; Sun et al., 2015), due to lack of any other distinguishing characteristics, no attempt has been made to formally recognize any of these clusters. In our work, we have identified a number of molecular signatures that are either completely or largely specific for the members of two of these clusters (viz. the *B. asteroides* and *B. longum* groups). Of these clusters, the distinctness of the *B. asteroides* group (comprising of the species *B. asteroides, B. indicum, B. coryneforme, B. actinocoloniiforme, B.* sp. A11, and *B.* sp. 7101) which forms the deepest branching lineage within the *Bifidobacterium*, is supported by 2 CSIs and 4 CSPs that are uniquely shared by most of the members of this clade. Further, most of the species which are part of the *B. asteroides* clade have been isolated from the gastrointestinal tract of honey bees, and unlike other bifidobacteria, they are also capable of carrying out respiratory metabolism (Killer et al., 2010, 2011; Bottacini et al., 2012; Lugli et al., 2014; Sun et al., 2015). All of these characteristics indicate that the members of the *B. asteroides* clade are a good candidate for recognition as a distinct genus level taxon within the order *Bifidobacteriales*.

The molecular markers for the order *Bifidobacteriales* and some of its clades, in addition to their utility for taxonomic and diagnostic studies (Ahmod et al., 2011; Gupta, 2014; Wong et al., 2014), also provide important new tools for genetic and biochemical studies. Earlier work on a number of CSIs in the Hsp60 and Hsp70 proteins has established that both large and small CSIs in conserved proteins are essential for the group of organisms in which they are found (Singh and Gupta, 2009; Gupta, 2016b). Removal of these CSIs, or any significant change in them, was shown to be incompatible with the cellular growth of the CSI-containing organisms. Thus, the identified CSIs are predicted to play essential role in the organisms in which they are found. Structural studies on several studied CSIs show that the sequences corresponding to them are present in the surface loops of the proteins (Singh and Gupta, 2009; Gupta and Khadka, 2016). Limited structural work on some of the *Bifidobacteriales*-specific CSIs that we have carried out also shows that these CSIs are located in the surface loops of the proteins. One example of the structural location of a *Bifidobacteriales*-specific CSIs is illustrated in **Figure** . In this case, a homology model of protein synthesis elongation factor Tu from *B. longum* was created to determine the location of the 4 aa *Bifidobacteriales*-specific CSI found in this protein. A structural comparison of the EF-Tu from *B. longum* and *E. coli* shown in **Figure** reveals that the CSI in the *B. longum* homolog is present in the protein surface loop within the GTPase domain of EF-Tu. The surface loops in proteins play important role in mediating protein-protein or protein-ligand interactions and it is expected that the identified CSIs are involved in mediating novel interactions that are specific and essential for the CSI-containing organisms (Akiva et al., 2008; Hashimoto and Panchenko, 2010). Similar to the CSIs in the EF-Tu protein, our work has identified numerous other CSIs in different essential proteins, which are specific for the *Bifidobacteriales* species. Functional studies on proteins harboring these CSIs provide an important means for discovering novel biochemical characteristics that are unique to either all *Bifidobacteriales* or specific clades of these bacteria, and which could possibly also provide useful insights into the growth-promoting as well as pathogenic effects of some of these bacteria.

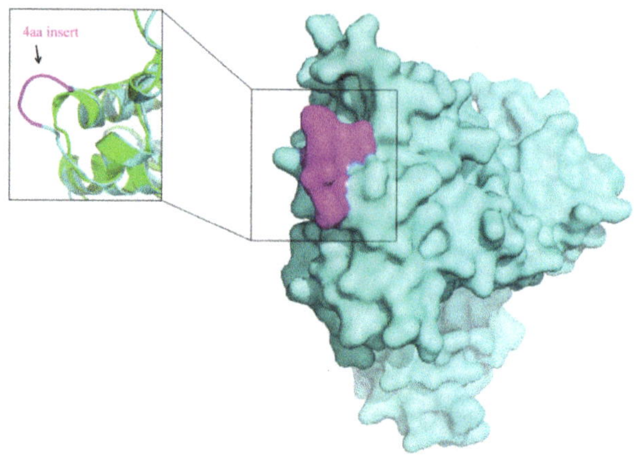


**FIGURE | Surface representation of the homology model of Elongation factor Tu (EF-Tu) from *B. longum* (Cyan).** The conserved 4 aa insert which is located on the surface of the EF-Tu is shown in magenta. A superposition of the homology model of the *B. longum* homolog of EF-Tu (Cyan) with the *E. coli* homolog of EF-Tu of (PDB ID: 3U6K) (Green) shows that the conserved 4 aa insert forms a surface loop on the protein.

## AUTHOR CONTRIBUTIONS

GZ, BG, MA, BK carried out comparative analyses of the bifidobacteriales genomes to identify signatures reported here. ZG and MA constructed phylogenetic trees and BK carried out homology modeling of the protein sequences. BG, MA, and RG were responsible for the writing and editing of the manuscript. All of the work was carried out under the direction of RG.

## ACKNOWLEDGMENTS

The work from RSG' lab was supported by research grant No. 249924 from the Natural Science and Engineering Research Council of Canada. Work by BG was supported in part by research grants from the National Science Foundation of China (31570011) and Natural Science Foundation of Guangdong Province (2015A030306039). BG is also a scholar of the "100 Talents Project" of the Chinese Academy of Sciences.

## SUPPLEMENTARY MATERIAL

The Supplementary Material for this article can be found online at: http://journal.frontiersin.org/article/10.3389/fmicb.2016.00978

## REFERENCES


Ahmod, N. Z., Gupta, R. S., and Shah, H. N. (2011). Identification of a *Bacillus anthracis* specific indel in the yeaC gene and development of a rapid pyrosequencing assay for distinguishing *B. anthracis* from the *B. cereus* group. *J. Microbiol. Methods* 87, 278–285. doi: 10.1016/j.mimet.2011.08.015

Ajawatanawong, P., and Baldauf, S. L. (2013). Evolution of protein indels in plants, animals and fungi. *BMC Evol. Biol.* 13:140. doi: 10.1186/1471-2148-13-140

Akiva, E., Itzhaki, Z., and Margalit, H. (2008). Built-in loops allow versatility in domain-domain interactions: lessons from self-interacting domains. *Proc. Natl. Acad. Sci. U.S.A.* 105, 13292–13297. doi: 10.1073/pnas.0801207105

Altschul, S. F., Madden, T. L., Schaffer, A. A., Zhang, J., Zhang, Z., Miller, W., et al. (1997). Gapped BLAST and PSI-BLAST: a new generation of protein databases search programs. *Nucleic Acids Res.* 25, 3389–3402. doi: 10.1093/nar/25.17.3389

Alves, P., Castro, J., Sousa, C., Cereija, T. B., and Cerca, N. (2014). *Gardnerella vaginalis* outcompetes 29 other bacterial species isolated from patients with bacterial vaginosis, using in an *in vitro* biofilm formation model. *J. Infect. Dis.* 210, 593–596. doi: 10.1093/infdis/jiu131

Anderson, K. E., Johansson, A., Sheehan, T. H., Mott, B. M., Corby-Harris, V., Johnstone, L., et al. (2013). Draft genome sequences of two *Bifidobacterium* sp. from the honey bee (*Apis mellifera*). *Gut Pathog.* 5:42. doi: 10.1186/1757-4749-5-42

Biavati, B. (2012). "Family I *Bifidobacteriaceae* Stackebrandt, Rainey and Ward-Rainey 1997, 487$^{VP}$," in *Bergey's Manual of Systematic Bacteriology, Vol. 5, The Actinobacteria*, eds W. Whitman, M. Goodfellow, P. Kampfer, H. J. Busse, M. E. Trujillo, W. Ludwig, et al. (New York, NY: Springer), 171.

Biavati, B., and Mattarelli, P. (2012). "Genus I *Bifidobacterium* Orla-Jensen 1924, 472$^{AL}$," in *Bergey's Manual of Systematic Bacteriology,* Vol. 5, *The Actinobacteria*, eds W. Whitman, M. Goodfellow, P. Kampfer, H. J. Busse, M. E. Trujillo, W. Ludwig, et al. (New York, NY: Springer), 171–206.

Biavati, B., and Mattarelli, P. (2006). "The family *Bifidobacteriaceae*," in *The Prokaryotes: An Evolving Electronic Resource for the Microbiological Community*, eds M. Dworkin, S. Falkow, E. Rosenberg, K. H. Schleifer, E. Stackebrandt (New York, NY: Springer-Verlag), 322–382. doi: 10.1007/0-387-30743-5_17

Biavati, B., Vescovo, M., Torriani, S., and Bottazzi, V. (2000). Bifidobacteria: histroy, ecology, physiology and applications. *Ann. Microbiol.* 50, 117–131.

Bottacini, F., Medini, D., Pavesi, A., Turroni, F., Foroni, E., Riley, D., et al. (2010). Comparative genomics of the genus *Bifidobacterium*. *Microbiology* 156, 3243–3254. doi: 10.1099/mic.0.039545-0

Bottacini, F., Milani, C., Turroni, F., Sanchez, B., Foroni, E., Duranti, S., et al. (2012). *Bifidobacterium* asteroides PRL2011 genome analysis reveals clues for colonization of the insect gut. *PLoS ONE* 7:e44229. doi: 10.1371/journal.pone.0044229

Bradshaw, C. S., Tabrizi, S. N., Fairley, C. K., Morton, A. N., Rudland, E., and Garland, S. M (2006). The association of *Atopobium vaginae* and *Gardnerella vaginalis* with bacterial vaginosis and recurrence after oral metronidazole therapy. *J. Infect. Dis.* 194, 828–836. doi: 10.1086/506621

Castresana, J. (2000). Selection of conserved blocks from multiple alignments for their use in phylogenetic analysis. *Mol. Biol. Evol.* 17, 540–552. doi: 10.1093/oxfordjournals.molbev.a026334

Cole, J. R., Wang, Q., Fish, J. A., Chai, B., McGarrell, D. M., Sun, Y., et al. (2014). Ribosomal Database Project: data and tools for high throughput rRNA analysis. *Nucleic Acids Res.* 42, D633–D642. doi: 10.1093/nar/gkt1244

Cronin, M., Ventura, M., Fitzgerald, G. F., and van Sinderen, D. (2011). Progress in genomics, metabolism and biotechnology of bifidobacteria. *Int. J. Food Microbiol.* 149, 4–18. doi: 10.1016/j.ijfoodmicro.2011.01.019

Edgar, R. C. (2010). Search and clustering orders of magnitude faster than BLAST. *Bioinformatics* 26, 2460–2461. doi: 10.1093/bioinformatics/btq461

Ferrario, C., Milani, C., Mancabelli, L., Lugli, G. A., Turroni, F., Duranti, S., et al. (2015). A genome-based identification approach for members of the genus *Bifidobacterium*. *FEMS Microbiol. Ecol.* 91:fiv009. doi: 10.1093/femsec/fiv009

Gao, B., and Gupta, R. S. (2005). Conserved indels in protein sequences that are characteristic of the phylum *Actinobacteria*. *Int. J. Syst. Evol. Microbiol.* 55, 2401–2412. doi: 10.1099/ijs.0.63785-0

Gao, B., and Gupta, R. S. (2012). Phylogenetic framework and molecular signatures for the main clades of the phylum Actinobacteria. *Microbiol. Mol. Biol. Rev.* 76, 66–112. doi: 10.1128/MMBR.05011-11

Gao, B., Parmanathan, R., and Gupta, R. S. (2006). Signature proteins that are distinctive characteristics of *Actinobacteria* and their subgroups. *Antonie Van Leeuwenhoek* 90, 69–91. doi: 10.1007/s10482-006-9061-2

García-Aljaro, C., Ballesté, E., Rosselló-Móra, R., Cifuentes, A., Richter, M., and Blanch, A. R. (2012). *Neoscardovia arbecensis* gen. nov., sp. nov., isolated from porcine slurries. *Syst. Appl. Microbiol.* 35, 374–379. doi: 10.1016/j.syapm.2012.06.007

Guindon, S., Dufayard, J. F., Lefort, V., Anisimova, M., Hordijk, W., and Gascuel, O. (2010). New algorithms and methods to estimate maximum-likelihood phylogenies: assessing the performance of PhyML 3.0. *Syst. Biol.* 59, 307–321. doi: 10.1093/sysbio/syq010

Gupta, R. S. (1998). Protein phylogenies and signature sequences: a reappraisal of evolutionary relationships among archaebacteria, eubacteria, and eukaryotes. *Microbiol. Mol. Biol. Rev.* 62, 1435–1491.

Gupta, R. S. (2010). "Applications of conserved indels for understanding microbial phylogeny," in *Molecular Phylogeny of Microorganisms*, eds A. Oren, R. T. Papke (Norfolk: Caister Academic Press), 135–150.

Gupta, R. S. (2014). "Identification of conserved indels that are useful for classification and evolutionary studies," in *Bacterial Taxonomy, Methods in Microbiology, Vol. 41*, eds M. Goodfellow, I. C. Sutcliffe, J. Chun (London: Elsevier), 153–182. doi: 10.1016/bs.mim.2014.05.003

Gupta, R. S. (2016a). Editorial: applications of genome sequences for discovering characteristics that are unique to different groups of organisms and provide insights into evolutionary relationships. *Front. Genet.* 7:27. doi: 10.3389/fgene.2016.00027

Gupta, R. S. (2016b). Impact of genomics on the understanding of microbial evolution and classification: the importance of Darwin's views on classification. *FEMS Microbiol. Rev.* doi: 10.1093/femsre/fuw011. [Epub ahead of print].

Gupta, R. S., Chander, P., and George, S. (2013a). Phylogenetic framework and molecular signatures for the class Chloroflexi and its different clades; proposal for division of the class *Chloroflexia* class. nov. [corrected] into the suborder *Chloroflexineae* subord. nov., consisting of the emended family Oscillochloridaceae and the family *Chloroflexaceae* fam. nov., and the suborder *Roseiflexineae subord.* nov., containing the family Roseiflexaceae fam. nov. Antonie van Leeuwenhoek 103, 99–119. doi: 10.1007/s10482-012-9790-3

Gupta, R. S., Chen, W. J., Adeolu, M., and Chai, Y. (2013b). Molecular signatures for the class *Coriobacteriia* and its different clades; Proposal for division of the class *Coriobacteriia* into the emended order *Coriobacteriales*, containing the emended family *Coriobacteriaceae* and *Atopobiaceae* fam. nov., and

*Eggerthellales* ord. nov., containing the family *Eggerthellaceae* fam. nov. *Int. J. Syst. Evol. Microbiol.* 63, 3379–3397. doi: 10.1099/ijs.0.048371-0

Gupta, R. S., and Griffiths, E. (2002). Critical issues in bacterial phylogeny. *Theor. Popul. Biol.* 61, 423–434. doi: 10.1006/tpbi.2002.1589

Gupta, R. S., and Khadka, B. (2016). Evidence for the presence of key chlorophyll-biosynthesis-related proteins in the genus Rubrobacter (Phylum Actinobacteria) and its implications for the evolution and origin of photosynthesis. *Photosyn. Res.* 127, 201–218. doi: 10.1007/s11120-015-0177-y

Gupta, R. S., Naushad, S., Fabros, R., and Adeolu, M. (2016). A phylogenomic reappraisal of family-level divisions within the class *Halobacteria*: proposal to divide the order Halobacteriales into the families *Halobacteriaceae, Haloarculaceae* fam. nov., and *Halococcaceae* fam. nov., and the order Haloferacales into the families, *Haloferacaceae* and *Halorubraceae* fam nov. *Antonie van Leeuwenhoek* 109, 565–587. doi: 10.1007/s10482-016-0660-2

Harris, J. K., Kelley, S. T., Spiegelman, G. B., and Pace, N. R. (2003). The genetic core of the universal ancestor. *Genome Res.* 13, 407–412. doi: 10.1101/gr.652803

Hashimoto, K., and Panchenko, A. R. (2010). Mechanisms of protein oligomerization, the critical role of insertions and deletions in maintaining different oligomeric states. *Proc. Natl. Acad. Sci. U.S.A.* 107, 20352–20357. doi: 10.1073/pnas.1012999107

Huys, G., Vancanneyt, M., D'Haene, K., Falsen, E., Wauters, G., and Vandamme, P. (2007). *Alloscardovia omnicolens* gen. nov., sp nov., from human clinical samples. *Int. J. Syst. Evol. Microbiol.* 57, 1442–1446. doi: 10.1099/ijs.0.64812-0

Jeanmougin, F., Thompson, J. D., Gouy, M., Higgins, D. G., and Gibson, T. J. (1998). Multiple sequence alignment with Clustal x. *Trends Biochem. Sci.* 23, 403–405. doi: 10.1016/S0968-0004(98)01285-7

Jian, W. Y., and Dong, X. Z. (2002). Transfer of *Bifidobacterium* incipinatum and *Bifidobacterium* denticolens to *Scardovia inopinata* gen. nov., comb. nov., and Parascardovia denticolens gen. nov., comb. nov., respectively. *Int. J. Syst. Evol. Microbiol.* 52, 809–812. doi: 10.1099/00207713-52-3-809

Kenyon, C. R., and Osbak, K. (2014). Recent progress in understanding the epidemiology of bacterial vaginosis. *Curr. Opin. Obstet. Gynecol.* 26, 448–454. doi: 10.1097/GCO.0000000000000112

Killer, J., Kopecný, J., Mrázek, J., Koppová, I., Havlík, J., Benada, O., et al. (2011). *Bifidobacterium actinocoloniiforme* sp. nov. and *Bifidobacterium bohemicum* sp nov., from the bumblebee digestive tract. *Int. J. Syst. Evol. Microbiol.* 61, 1315–1321. doi: 10.1099/ijs.0.022525-0

Killer, J., Kopecný, J., Mrázek, J., Rada, V., Dubná, S., and Marounek, M. (2010). Bifidobacteria in the digestive tract of bumblebees. *Anaerobe* 16, 165–170. doi: 10.1016/j.anaerobe.2009.07.007

Killer, J., Mrazek, J., Bunesova, V., Havlik, J., Koppova, I., Benada, O., et al. (2013). *Pseudoscardovia suis* gen. nov., sp. nov., a new member of the family *Bifidobacteriaceae* isolated from the digestive tract of wild pigs *(Sus scrofa)*. *Syst. Appl. Microbiol.* 36, 11–16. doi: 10.1016/j.syapm.2012.09.001

Le, S. Q., and Gascuel, O. (2008). An improved general amino acid replacement matrix. *Mol. Biol. Evol.* 25, 1307–1320. doi: 10.1093/molbev/msn067

Leahy, S. C., Higgins, D. G., Fitzgerald, G. F., and van Sinderen, D. (2005). Getting better with bifidobacteria. *J. Appl. Microbiol.* 98, 1303–1315. doi: 10.1111/j.1365-2672.2005.02600.x

Lugli, G. A., Milani, C., Turroni, F., Duranti, S., Ferrario, C., Viappiani, A., et al. (2014). Investigation of the evolutionary development of the genus *Bifidobacterium* by comparative genomics. *Appl. Environ. Microbiol.* 80, 6383–6394. doi: 10.1128/AEM.02004-14

Mantzourani, M., Fenlon, M., and Beighton, D. (2009). Association between *Bifidobacteriaceae* and the clinical severity of root caries lesions. *Oral Microbiol. Immunol.* 24, 32–37. doi: 10.1111/j.1399-302X.2008.00470.x

Mattarelli, P., Holzapfel, W., Franz, C. M. A. P., Endo, A., Felis, G. E., Hammes, W., et al. (2014). Recommended minimal standards for description of new taxa of the genera *Bifidobacterium*, Lactobacillus and related genera. *Int. J. Syst. Evol. Microbiol.* 64, 1434–1451. doi: 10.1099/ijs.0.060046-0

Milani, C., Lugli, G. A., Duranti, S., Turroni, F., Bottacini, F., Mangifesta, M., et al. (2014). Genomic encyclopedia of type strains of the genus *Bifidobacterium*. *Appl. Environ. Microbiol.* 80, 6290–6302. doi: 10.1128/AEM.02308-14

Miyake, T., Watanabe, K., Watanabe, T., and Oyaizu, H. (1998). Phylogenetic analysis of the genus *Bifidobacterium* and related genera based on 16S rDNA sequences. *Microbiol. Immunol.* 42, 661-667. doi: 10.1111/j.1348-0421.1998.tb02337.x

Parte, A. C. (2014). LPSN-list of prokaryotic names with standing in nomenclature. *Nucleic Acids Res.* 42, D613–D616. doi: 10.1093/nar/gkt1111

Price, M. N., Dehal, P. S., and Arkin, A. P. (2010). FastTree 2-approximately maximum-likelihood trees for large alignments. *PLoS ONE* 5:e9490. doi: 10.1371/journal.pone.0009490

Pruesse, E., Peplies, J., and Glöckner, F. O. (2012). SINA: accurate high-throughput multiple sequence alignment of ribosomal RNA genes. *Bioinformatics* 28, 1823–1829. doi: 10.1093/bioinformatics/bts252

Rokas, A., and Holland, P. W. (2000). Rare genomic changes as a tool for phylogenetics. *Trends Ecol. Evol. (Amst).* 15, 454–459. doi: 10.1016/S0169-5347(00)01967-4

Sali, A., and Blundell, T. L. (1993). Comparative protein modelling by satisfaction of spatial restraints. *J. Mol. Biol.* 234, 779–815. doi: 10.1006/jmbi.1993.1626

Shen, M. Y., and Sali, A. (2006). Statistical potential for assessment and prediction of protein structures. *Protein Sci.* 15, 2507–2524. doi: 10.1110/ps.0624 16606

Sievers, F., Wilm, A., Dineen, D., Gibson, T. J., Karplus, K., Li, W., et al. (2011). Fast, scalable generation of high-quality protein multiple sequence alignments using Clustal Omega. *Mol. Syst. Biol.* 7, 539. doi: 10.1038/msb.2011.75

Simpson, P. J., Ross, R. P., Fitzgerald, G. F., and Stanton, C. (2004). *Bifidobacterium psychraerophilum* sp. nov. and *Aeriscardovia aeriphila* gen. nov., sp. nov., isolated from a porcine caecum. *Int. J. Syst. Evol. Microbiol.* 54, 401–406. doi: 10.1099/ijs.0.02667-0

Singh, B., and Gupta, R. S. (2009). Conserved inserts in the Hsp60 (GroEL) and Hsp70 (DnaK) proteins are essential for cellular growth. *Mol. Genet. Genomics* 281, 361–373. doi: 10.1007/s00438-008-0417-3

Smith, S. M., Ogbara, T., and Eng, R. H. K. (1992). Involvement of *Gardnerella-vaginalis* in urinary-tract infections in men. *J. Clin. Microbiol.* 30, 1575–1577.

Stamatakis, A. (2014). RAxML version 8: a tool for phylogenetic analysis and post-analysis of large phylogenies. *Bioinformatics* 30, 1312–1313. doi: 10.1093/bioinformatics/btu033

Storms, V., and Vandamme, P. (2012). "Genus IV gardnerella greenwood and pickett 1980, 176$^{VP}$," in *Bergey's Manual of Systematic Bacteriology, Vol. 5, The Actinobacteria*, eds W. Whitman, M. Goodfellow, P. Kampfer, H. J. Busse, M. E. Trujillo, W. Ludwig, et al. (New York, NY: Springer), 208–211.

Sun, Z. H., Zhang, W. Y., Guo, C. Y., Yang, X. W., Liu, W. J., Wu, Y., et al. (2015). Comparative genomic analysis of 45 type strains of the genus *Bifidobacterium*: a snapshot of its genetic diversity and evolution. *PLoS ONE* 10:e117912. doi: 10.1371/journal.pone.0117912

Talavera, G., and Castresana, J. (2007). Improvement of phylogenies after removing divergent and ambiguously aligned blocks from protein sequence alignments. *Syst. Biol.* 56, 564–577. doi: 10.1080/10635150701472164

Tamura, K., Stecher, G., Peterson, D., Filipski, A., and Kumar, S. (2013). MEGA6: molecular evolutionary genetics analysis version 6.0. *Mol. Biol. Evol.* 30, 2725–2729. doi: 10.1093/molbev/mst197

Toh, H., Yamazaki, Y., Tashiro, K., Kawarai, S., Oshima, K., Nakano, A., et al. (2015). Draft genome sequence of *Bifidobacterium* aesculapii DSM 26737T, isolated from feces of baby common marmoset. *Genome Announc.* 3:e01463-15. doi: 10.1128/genomeA.01463-15

Turroni, F., Bottacini, F., Foroni, E., Mulder, I., Kim, J. H., Zomer, A., et al. (2010). Genome analysis of *Bifidobacterium bifidum* PRL2010 reveals metabolic pathways for host-derived glycan foraging. *Proc. Natl. Acad. Sci. U.S.A.* 107, 19514–19519. doi: 10.1073/pnas.1011100107

Turroni, F., van Sinderen, D., and Ventura, M. (2009). *Bifidobacteria*: from ecology to genomics. *Front. Biosci.* 14, 4673–4684. doi: 10.2741/3559

Turroni, F., van Sinderen, D., and Ventura, M. (2011). Genomics and ecological overview of the genus *Bifidobacterium*. *Int. J. Food Microbiol.* 149, 37–44. doi: 10.1016/j.ijfoodmicro.2010.12.010

Ventura, M., Canchaya, C., Del Casale, A., Dellaglio, F., Neviani, E., Fitzgerald, G. F., et al. (2006). Analysis of bifidobacterial evolution using a multilocus approach. *Int. J. Syst. Evol. Microbiol.* 56, 2783–2792. doi: 10.1099/ijs.0.64233-0

Ventura, M., Canchaya, C., Fitzgerald, G. F., Gupta, R. S., and van Sinderen, D. (2007a). Genomics as a means to understand bacterial phylogeny and ecological adaptation: the case of bifidobacteria. *Antonie Van Leeuwenhoek* 91, 351–372. doi: 10.1007/s10482-006-9122-6

Ventura, M., Canchaya, C., Tauch, A., Chandra, G., Fitzgerald, G. F., Chater, K. F., et al. (2007b). Genomics of *Actinobacteria*: tracing the evolutionary history of an ancient phylum. *Microbiol. Mol. Biol. Rev.* 71, 495–548. doi: 10.1128/MMBR.00005-07

Ventura, M., Canchaya, C., Zink, R., Fitzgerald, G. F., and van Sinderen, D. (2004). Characterization of the groEL and groES loci in *Bifidobacterium breve* UCC 2003: genetic, transcriptional, and phylogenetic analyses. *Appl. Environ. Microbiol.* 70, 6197–6209. doi: 10.1128/AEM.70.10.6197-6209.2004

Ventura, M., O'Flaherty, S., Claesson, M. J., Turroni, F., Klaenhammer, T. R., van Sinderen, D., et al. (2009a). Genome-scale analyses of health-promoting bacteria: probiogenomics. *Nat. Rev. Microbiol.* 7, 61–71. doi: 10.1038/nrmicro2047

Ventura, M., Turroni, F., Zomer, A., Foroni, E., Giubellini, V., Bottacini, F., et al. (2009b). The *Bifidobacterium dentium* Bd1 genome sequence reflects its genetic adaptation to the human oral cavity. *PLoS Genet.* 5:e1000785. doi: 10.1371/journal.pgen.1000785

Ventura, M., and Zink, R. (2003). Comparative sequence analysis of the tuf and recA genes and restriction fragment length polymorphism of the internal transcribed spacer region sequences supply additional tools for discriminating *Bifidobacterium lactis* from *Bifidobacterium animalis*. *Appl. Environ. Microbiol.* 69, 7517–7522. doi: 10.1128/AEM.69.12.7517-7522.2003

Whelan, S., Liò, P., and Goldman, N. (2001). Molecular phylogenetics: state-of-the-art methods for looking into the past. *Trends Genet.* 17, 262–272. doi: 10.1016/S0168-9525(01)02272-7

Wong, S. Y., Paschos, A., Gupta, R. S., and Schellhorn, H. E. (2014). Insertion/deletion-based approach for the detection of *Escherichia coli* O157:H7 in freshwater environments. *Environ. Sci. Technol.* 48, 11462–11470. doi: 10.1021/es502794h

Xu, D., and Zhang, Y. (2011). Improving the physical realism and structural accuracy of protein models by a two-step atomic-level energy minimization. *Biophys. J.* 101, 2525–2534. doi: 10.1016/j.bpj.2011.10.024

Yarza, P., Richter, M., Peplies, J., Euzeby, J., Amann, R., Schleifer, K. H., et al. (2008). The All-Species Living Tree project: a 16S rRNA-based phylogenetic tree of all sequenced type strains. *Syst. Appl. Microbiol.* 31, 241–250. doi: 10.1016/j.syapm.2008.07.001

Yilmaz, P., Parfrey, L. W., Yarza, P., Gerken, J., Pruesse, E., Quast, C., et al. (2014). The SILVA and "All-species Living Tree Project (LTP)" taxonomic frameworks. *Nucleic Acids Res.* 42, D643–D648. doi: 10.1093/nar/gkt1209

Zhi, X. Y., Li, W. J., and Stackebrandt, E. (2009). An update of the structure and 16S rRNA gene sequence-based definition of higher ranks of the class *Actinobacteria*, with the proposal of two new suborders and four new families and emended descriptions of the existing higher taxa. *Int. J. Syst. Evol. Microbiol.* 59, 589–608. doi: 10.1099/ijs.0.65780-0

**Conflict of Interest Statement:** The authors declare that the research was conducted in the absence of any commercial or financial relationships that could be construed as a potential conflict of interest.

# 11

# Microencapsulation in Alginate and Chitosan Microgels to Enhance Viability of *Bifidobacterium longum* for Oral Delivery

*Timothy W. Yeung*[1], *Elif F. Üçok*[1], *Kendra A. Tiani*[1,2], *David J. McClements*[1,3] and *David A. Sela*[1,3,4]*

[1] Department of Food Science, University of Massachusetts, Amherst, MA, USA, [2] Commonwealth Honors College, University of Massachusetts, Amherst, MA, USA, [3] Center for Bioactive Delivery, Institute of Applied Life Science, University of Massachusetts, Amherst, MA, USA, [4] Center for Microbiome Research, University of Massachusetts Medical School, Worcester, MA, USA

**Edited by:**
David Berry,
University of Vienna, Austria

**Reviewed by:**
Borja Sanchez,
Instituto de Productos Lácteos de Asturias – Consejo Superior de Investigaciones Científicas, Spain
Chiara Ferrario,
University of Parma, Italy

***Correspondence:**
David A. Sela
davidsela@umass.edu

Probiotic microorganisms are incorporated into a wide variety of foods, supplements, and pharmaceuticals to promote human health and wellness. However, maintaining bacterial cell viability during storage and gastrointestinal transit remains a challenge. Encapsulation of bifidobacteria within food-grade hydrogel particles potentially mitigates their sensitivity to environmental stresses. In this study, *Bifidobacterium longum* subspecies and strains were encapsulated in core-shell microgels consisting of an alginate core and a microgel shell. Encapsulated obligate anaerobes *Bifidobacterium longum* subsp. *infantis* and *Bifidobacterium longum* subsp. *longum* exhibited differences in viability in a strain-dependent manner, without a discernable relationship to subspecies lineage. This includes viability under aerobic storage conditions and modeled gastrointestinal tract conditions. Coating alginate microgels with chitosan did not improve viability compared to cells encapsulated in alginate microgels alone, suggesting that modifying the surface charge alone does not enhance delivery. Thus hydrogel beads have great potential for improving the stability and efficacy of bifidobacterial probiotics in various nutritional interventions.

**Keywords: microencapsulation, bifidobacteria, probiotics, simulated digestion, oral delivery**

## INTRODUCTION

Beneficial bacteria are often incorporated into functional foods and nutritional interventions to be ingested orally as probiotics. This includes humans and livestock animals that receive direct-fed microbials to enhance health outcomes and reduce pathogen load (Braat et al., 2006; Puccio et al., 2007; Neal-McKinney et al., 2012; Watson and Preedy, 2015). *Bifidobacterium longum* colonizes the human gastrointestinal tract (GIT) and is one of the 48 recognized taxa that are encompassed within the genus *Bifidobacterium* (Milani et al., 2014; Sun et al., 2015). This obligate anaerobe is one of the earliest colonizers of the infant GIT, and is present in lower concentrations in the adult gut (Schell et al., 2002; Sela et al., 2008). The bifidobacterial taxa *longum, infantis,* and *suis* were previously classified as discrete species, but more recently they were reclassified as subspecies of *B. longum* (Sakata et al., 2002). Their unification as a single species is based primarily on genomic and phenotypic similarities shared between these groups. While *Bifidobacterium animalis* subsp. *lactis* is often used for probiotic applications, *B. longum* strains are of particular interest due to its likely co-evolution with humans. This is evident in *B. longum* utilization of human milk oligosaccharides and establishes a protective gut microbiome in infants through adulthood (Sela and Mills, 2010).

*Bifidobacterium longum* is deployed in several probiotic applications using a variety of delivery formats (Adhikari et al., 2000; Fortin et al., 2011; Amine et al., 2014; Lewis et al., 2015). A relatively large dose of probiotics is recommended to impart health benefits, typically $10^6$–$10^7$ CFU/g per day (Krasaekoopt et al., 2003; Roy, 2005). However, the direct incorporation of free probiotic cells into food products and supplements results in a significant decrease in cell viability throughout storage and gastrointestinal transit (Sultana et al., 2000; de Vos et al., 2010). Therefore, prolonged storage and the process of ingesting these probiotics may reduce their viability below recommended levels to achieve health benefits. Microencapsulating probiotic cells within hydrogel matrices protects them against extrinsic environmental factors thereby enhancing bacterial survival during processing, storage, and digestion (de Vos et al., 2010; Fareez et al., 2015; Yeung et al., 2016). Encapsulation may also dictate the controlled release of the probiotic at the precise anatomical site of activity within the GIT, thereby enhancing the efficacy of the probiotic through specific targeting after oral delivery (de Barros et al., 2015; Zhang et al., 2015b).

Several biopolymer materials are available to encapsulate microbes in hydrogel matrices, depending on the desired physicochemical properties of the delivery vehicle. The most commonly used food-grade biopolymers are proteins (e.g., whey proteins and caseins) and carbohydrates (e.g., starch and gums; Bagchi et al., 2010; Gaonkar et al., 2014; Etchepare et al., 2015). For many food applications, it is advantageous to encapsulate probiotics within hydrogel beads that trap bacteria within small particles containing cross-linked biopolymer molecules. These microgels must be engineered to encapsulate high concentrations of probiotics and protect them from environmental stresses, such as acidic pH, bile salts, and digestive enzymes (Zhang et al., 2015a). Alginate has been widely used as a biopolymer suitable for food applications as it is relatively inexpensive, easy to gel, biodegradable, and compatible with many food systems (Gombotz and Wee, 2012; Lee and Mooney, 2012). Indeed, recently studies have shown that lactococcal-based probiotics can be encapsulated within alginate microgels to improve their stability (Yeung et al., 2016).

There are appreciable differences between probiotic strain tolerance toward environmental and gastrointestinal stresses. Consequently, it is possible to identify particular strains that are more resistant to these stresses than others, which are therefore more suitable for commercial application (Godward et al., 2000; Krasaekoopt et al., 2004; Capela et al., 2006). As an anaerobe, bifidobacterial species including *B. longum* differ in their sensitivity to oxygen exposure and other environmental stresses during the preparative phase prior to probiotic deployment (Kawasaki et al., 2006; Ruiz et al., 2012). Therefore, bifidobacterial probiotics may be encapsulated to restrict oxidative damage during preparation and storage and to limit exposure to degradative processes within the GIT.

The aim of this study was to design, fabricate, and characterize a food-grade encapsulation system to protect *B. longum* cells during simulated storage and gastrointestinal passage. Previously, we demonstrated that encapsulation of probiotics within alginate microgels could improve their viability during storage (Yeung et al., 2016). In the current study, we encapsulated *B. longum* cells within alginate beads to determine if their viability could be enhanced in storage and gastrointestinal transit. Moreover, the impact of coating these alginate beads with a layer of chitosan was investigated as well. Chitosan coated alginate beads have previously been used to enhance the mucoadhesive properties of probiotic bacteria (Chen et al., 2013).

## MATERIALS AND METHODS

### Preparation of Bacterial Cultures

Four strains of both *Bifidobacterium longum* subsp. *longum* (*B. longum*) and *Bifidobacterium longum* subsp. *infantis* (*B. infantis*) were studied (**Table** ). All strains were originally isolated from infant feces. Stock solutions were maintained by storing bacteria at −80°C in deMann, Regosa, Sharpe (MRS) media with 0.05% L-cysteine in 25% glycerol. Bacteria were propagated in MRS with L-cysteine at 37°C for 24 h, checked for purity, and maintained on MRS agar anaerobically. Anaerobic conditions were maintained in a double chamber anaerobic hood with an airlock (88% $N_2$, 10% $CO_2$, and 2% $H_2$) from Coy Laboratory Products (Grass Lake, Mississippi, USA).

Isolated colonies were routinely propagated in MRS broth (50 mL) for 40 h at 37° C. Cells were harvested by centrifugation at 4000 × *g* for 10 min, washed twice with 0.85% NaCl (physiological saline) solution (25 mL), and suspended in 0.85% NaCl (2 mL). The resulting cell suspensions were used either directly for assessing survival of free cells (i.e., no encapsulation) or subjected to encapsulation as described in section "Microencapsulation of Bifidobacterial Cells." Free cell suspensions (2 mL) were stored in 0.85% NaCl solution (50 mL) at 2–5°C for up to 5 weeks to model long-term storage conditions.

**TABLE** | ***Bifidobacterium longum* strains selected for encapsulation.**

| Subspecies | Strain designation |
|---|---|
| *infantis* | UMA 298 |
| | UMA 299 |
| | UMA 300 |
| | UMA 305 |
| *longum* | UMA 306 |
| | UMA 318 |
| | UMA 401 |
| | UMA 402 |

## General Chemicals Used in Encapsulation and Modeled Digestion

For bacterial culture preparation, MRS broth was obtained from Becton Dickinson and Company (Sparks, MD, USA). Agar, L-cysteine hydrochloric acid, and sodium chloride (NaCl) were purchased from Sigma–Aldrich (St. Louis, MO, USA). Glycerol and sodium citrate dihydrate was purchased from Fisher Scientific (Fair Lawn, NJ, USA).

For encapsulation experiments, sodium alginate (TICA-algin HG 400 powder) was donated by TIC Gums (White Marsh, ML, USA). Calcium chloride hexahydrate, chitosan (medium molecular weight) was obtained from Sigma-Aldrich. Glacial acetic acid was purchased from Fisher Scientific.

For simulated digestion, ammonium nitrate, bile extract porcine, lipase from porcine pancreas type II, pepsin from porcine gastric mucosa, porcine gastric mucin type II, potassium chloride, potassium citrate, potassium phosphate, sodium DL-lactate, sodium hydroxide (NaOH), and uric acid sodium salt were also purchased from Sigma-Aldrich. Hydrochloric acid (HCl), phosphate buffer saline (PBS), and urea were purchased from Fisher Scientific.

## Microencapsulation of Bifidobacterial Cells

Bifidobacteria were encapsulated within alginate microgels using an injection–gelation method (Whelehan and Marison, 2011; Seiffert, 2013). Briefly, 1% (w/v) sodium alginate solution was prepared, autoclaved, and then cooled to ambient temperature. The sterile alginate solution (198 mL) was mixed with 2 mL of $\sim 10^9$ CFU/mL probiotic organisms suspended in physiological saline. The polymeric solution was agitated to uniformly distribute cells throughout the mixture. The alginate beads were prepared aseptically using an encapsulator (Büchi B-390®, Büchi Labortechnik AG, Flawil, Switzerland) with a nozzle size of 120 μm, using the manufacturer's standard operating conditions (amplitude 3, frequency 800 Hz, electrode 800 V, pressure 250–300 mbar). Aliquots of probiotic/alginate solution were injected into 0.1 M calcium chloride solution (350 mL). After 1-h gelation under agitation, the resulting calcium alginate beads were collected by filtration, rinsed with sterile deionized water (200 mL), and re-filtered. Microbeads ($\sim$30 mL) were stored in physiological saline solution (50 mL) at 4°C for up to 4 weeks to model long-term storage conditions. This process was repeated for all eight strains of bifidobacteria.

Unfilled alginate beads were prepared identically but without the addition of bacterial strains to the alginate solution. 1% alginate solution (200 mL) was extruded into of 0.1 M $CaCl_2$ (350 mL) solution under continuous agitation. The working parameters (nozzle diameter, frequency, charge, and pressure), filtering steps and storage conditions used were the same as those for the preparation of filled alginate beads.

An aqueous chitosan solution (0.4% w/v) was prepared as described previously by Zhou et al. (1998). Briefly, chitosan (0.4 g) was dissolved in distilled water (90 mL) and glacial acetic acid (0.8 mL). The pH was adjusted to 5.0–5.1 with NaOH, and the total volume was adjusted to 100 mL. The solution was autoclaved and filtered to remove undissolved solids. Subsequently, the alginate beads were submerged in the chitosan solution to provide a secondary coating by electrostatic attraction of the cationic chitosan molecules to the surfaces of the anionic alginate beads. The mixture was agitated for 1 h before filtering and rinsing beads with sterile distilled water. Chitosan-coated alginate beads were then stored and analyzed as described in section "Alginate and Chitosan-coated Alginate Microbead Characterization."

## Alginate and Chitosan-Coated Alginate Microbead Characterization

### Particle Size Distribution

The particle size distribution was determined by static light scattering (Mastersizer S, Malvern Instruments, Worcestershire, UK). Each sample (1–2 mL) was suspended in distilled water (10 mL) and vortexed to avoid multiple scattering effects and to ensure homogeneity prior to analysis. Volume-weighted (D [4,3]) and surface-weighted (D [3,2]) mean particle diameters were obtained for all samples.

### Optical Microscopy Characterization

The overall appearance of alginate and chitosan-coated alginate beads was characterized with an optical microscope (C1 Digital Eclipse, Nikon, Tokyo, Japan). Microgel suspensions (1–2 mL) were immersed in physiological saline (10 mL) and vortexed to separate individual beads. Optical images were obtained using a digital camera and further analyzed using the instrument software (EZ CSI version 3.8, Nikon).

### Scanning Electronic Microscopy (SEM)

The bead microstructure was characterized using a bench-top scanning electron microscope (JCM-6000 NeoScope, JEOL, Tokyo, Japan). To prepare the samples prior to analysis, alginate, and chitosan-coated alginate beads were freeze-dried and sputter-coated with gold (10 nm) before loading onto the microscope. Images of the microgels were documented in representative fields.

### Electrical Properties

The surface potential (ζ-potential) of alginate and chitosan-coated alginate microgels was evaluated by electrophoretic light scattering (Zetasizer Nano ZS, Malvern Instruments, Worcestershire, UK). For each sample, refrigerated microgels (1–2 mL) were suspended in distilled water (10 mL) and vortexed to separate the beads. Samples were then loaded into the measurement cells and analyzed.

## Modeled Long-term Storage Conditions of Encapsulated Bifidobacteria

Total cell counts of free and encapsulated bifidobacteria were determined by a modified drop plate method as previously described (Herigstad et al., 2001). Briefly, 10 drops (10 μL) of a dilution within a series ($10^0$–$10^7$) were deposited on MRS agar plates and counted after incubation under anaerobic conditions at 37°C.

To determine viable counts of the encapsulated bacteria, beads (1 mL) were re-suspended in 10% sodium citrate dihydrate solution (9 mL) followed by vortexing. The number of released cells was determined by plate count using MRS agar, dilutions of dissolved beads ($10^{-1}$–$10^{-7}$) were plated in duplicate and incubated at 37°C anaerobically for 40 h. For lower viability samples later, beads (2 mL) were re-suspended in 10% sodium citrate dihydrate solution (2 mL) instead, and dilutions ($10^{0}$–$10^{-3}$) were plated as before. Samples were taken over a 4-week period on days 0 (initial), 1, 3, 5, 7, 10, 14, 21, and 28. Day 24 was also plated for free cell samples.

## *In Vitro* Simulated Digestion of Alginate and Chitosan-Coated Alginate Microbeads

Fluids used in *in vitro* modeling of digestion were prepared based on the method described by Li et al. (2011). One liter of modeled saliva stock solution was prepared with ammonium nitrate (0.328 g), potassium chloride (0.202 g), potassium citrate (0.308 g), potassium phosphate (0.636 g), sodium chloride (1.594 g), sodium DL-lactate (0.146 g), urea (0.198 g), and uric acid sodium salt (0.021 g) in distilled water. The stock solution was then filter-sterilized. The day before digestion experiments were carried out, the salivary phase was prepared by adding porcine gastric mucin type II (2.4 g) to saliva stock solution (80 mL). The solution was stirred overnight at room temperature to completely dissolve the powder.

One liter of simulated gastric stock solution was prepared by adding sodium chloride (2 g) and 6 M hydrochloric acid (7 mL) to distilled water and filter sterilizing. The simulated intestinal stock solution (500 ml) was prepared by adding calcium chloride hexahydrate (27.38 g) and sodium chloride (109.685 g) to distilled water and autoclaved. Pepsin extracted from porcine gastric mucosa (0.32 g) was then added to gastric stock solution (100 mL).

The day before digestion experiments were carried out, porcine bile extract (0.75 g) was added to PBS solution (14 mL) for the modeled intestinal phase. The solution was stirred overnight at room temperature to completely dissolve the powder. Lipase from porcine pancreas type II (0.24 g) was dissolved in PBS solution (10 mL); the solution (5 mL) was then added with bile salt solution (7 mL) and intestinal stock solution (33 mL).

Free and encapsulated bifidobacteria cells were separately added to simulated saliva fluids (22 mL, pH adjusted to 6.7–6.8), simulated gastric fluids (45 mL, pH adjusted to 2.5–2.6) or simulated intestinal fluids (45 mL, pH adjusted to 7.0–7.2) and stored in a shaking incubator (MaxQ 6000, Thermo Scientific, Waltham, MA, USA) set at 37°C with a shaking speed of 110 rpm (Supplementary Figure ). Dilutions ($10^{0}$–$10^{5}$) were plated on MRS agar for initial, 5, 10, 15, or 30 min exposure and incubated anaerobically for at least 48 h.

## Statistical Analysis

The mean of two or three individual determinations was used to calculate particle size, ζ-potential. The mean of 10 replicate drops was used to calculate cell counts. Analysis of variance (ANOVA) followed by Tukey honest significant difference test was use to analyze all data and compare individual means. This was performed using statistical software (GraphPad Prism 6, GraphPad Software, La Jolla, CA, USA).

# RESULTS

## Particle Size Analysis of Alginate and Chitosan-Coated Alginate Microgels

Light scattering was used to determine the mean particle diameter of the different microgel samples (**Table** ). The mean particle sizes of alginate beads containing similar strains were similar, ranging from 135 to 185 μm (D [3,2]) for encapsulated *B. infantis* strains and 149–216 μm (D [3,2]) for encapsulated *B. longum* strains. The chitosan-coated alginate beads were significantly larger compared than the alginate beads, ranging from 191 to 292 μm (D [3,2]). This increase in particle size may have been because of the additional coating formed by the alginate molecules, or because of some aggregation of the microgels. Microgel aggregation may have occurred due to bridging flocculation, which is the ability of the chitosan cation to adsorb to the surfaces of two or more anionic alginate beads. Additional information regarding the structural configuration of the microgels was therefore obtained through microscopy.

## Optical Microscopy of Alginate and Chitosan-Coated Alginate Microbeads

The structures of samples containing free cells or bacterial-loaded microgels were determined using optical microscopy

**TABLE | Volume-based (D [4,3]) and surface-based (D [3,2]) mean particle diameters measured by static light scattering alginate and chitosan-coated alginate beads with strains of bifidobacteria.**

| Beads | | μm | |
|---|---|---|---|
| | | D [4,3] | D [3,2] |
| **Alg.** | | | |
| Subsp. *infantis* | UMA 298 | 233 ± 4$^{ab}$ | 167 ± 6$^{abc}$ |
| | UMA 299 | 230 ± 3$^{ab}$ | 162 ± 3$^{b}$ |
| | UMA 300 | 251 ± 6$^{a}$ | 185 ± 12$^{cd}$ |
| | UMA 305 | 211 ± 4$^{b}$ | 135 ± 2$^{e}$ |
| Subsp. *longum* | UMA 306 | 247 ± 13$^{ac}$ | 164 ± 4$^{ab}$ |
| | UMA 318 | 228 ± 11$^{ab}$ | 149 ± 13$^{abe}$ |
| | UMA 401 | 277 ± 4$^{cd}$ | 216 ± 3$^{f}$ |
| | UMA 402 | 287 ± 13$^{de}$ | 195 ± 3$^{dfg}$ |
| **Chit.-alg.** | | | |
| Subsp. *infantis* | UMA 299 | 327 ± 2$^{fg}$ | 292 ± 3$^{h}$ |
| | UMA 300 | 344 ± 14$^{f}$ | 237 ± 3$^{i}$ |
| Subsp. *longum* | UMA 401 | 310 ± 20$^{eg}$ | 213 ± 4$^{f}$ |
| | UMA 402 | 315 ± 26$^{efg}$ | 191 ± 11$^{g}$ |

*Values are shown as average particle size ± standard deviation. Values followed by the same letters in the same column are not significantly different (p > 0.05) from each other.*

immediately after encapsulation (**Figure** ). Free cells appeared rod-shaped as expected for bifidobacteria (**Figures A,D**). The unfilled alginate and chitosan-coated alginate microgels were similar in morphology, although the individual coated alginate beads did appear larger than the uncoated ones, which is consistent with the particle size analysis (**Figures C,F**). Encapsulated bifidobacteria were clearly visualized within the microgels for both alginate and chitosan-coated alginate microgels (**Figures E,F**). The bifidobacterial-loaded alginate and chitosan-coated alginate beads had a similar external appearance as the equivalent unloaded beads. The microgels were generally spherical with diameters around 100–300 μm for all samples.

## Scanning Electron Microscopy

Scanning electron microscopy (SEM) was used to inspect the structure of the alginate and chitosan-coated alginate beads (**Figure 2**). Freeze-dried microgels were uniform in size and shape. However, the surfaces of the microgels observed by SEM appeared wrinkled, whereas they presented as smooth when observed by optical microscopy. This is likely due to sublimation of water originally trapped within the hydrogel matrix, as has been described previously (Yeung et al., 2016). The chitosan-coated alginate beads appeared to be more irregular in shape compared to alginate beads. Qualitatively, the alginate beads had smoother wrinkles and microstructures, whereas the chitosan-coated beads exhibited sharp jagged edges. This observation suggests that the chitosan layer has been successfully deposited onto the external surfaces of the alginate microgels.

## Zeta Potential Analysis of Microencapsulated Bifidobacteria

Electrophoretic light scattering was used to evaluate the electrical characteristics of the microgels (**Table** ). The ζ-potentials of all the alginate beads were negative, ranging from −4.2 to −9.4 mV for *B. infantis* and −2.6 to −4.4 mV for *B. longum* as predicted with this coating. In contrast, all chitosan-coated alginate bead samples had positive surface potentials ranging from +9.9 to +14.9 mV for *B. infantis* and +0.8 to +9.0 mV for *B. longum*. These results indicate that the cationic chitosan molecules formed a secondary shell around the anionic calcium alginate beads.

## Survival of Bifidobacterial Strains during Long-term Storage

### Non-encapsulated Bifidobacterial Cells

The viability of four *B. longum* and four *B. infantis* strains that were not encapsulated was determined during 5 weeks of storage

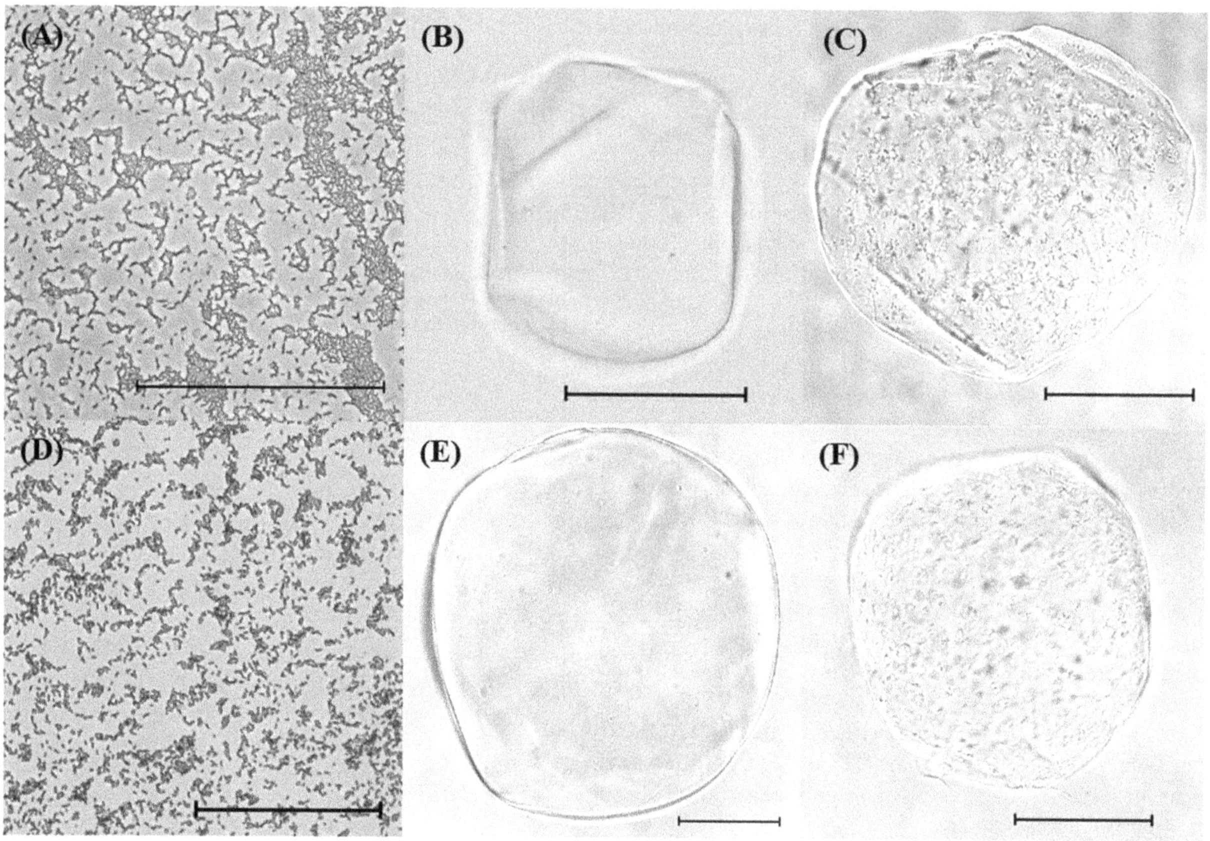


**FIGURE | Optical microscope images of (A) *Bifidobacterium longum* subsp. *infantis* UMA 300 (20×), (B) unfilled alginate bead (20×), (C) filled alginate bead with *B. longum* subsp. *longum* UMA 306 (20×), (D) *B. longum* subsp. *longum* UMA 318 (20×), (E) unfilled chitosan-coated alginate bead (20×), (F) filled chitosan-coated alginate bead with *B. longum* subsp. *infantis* UMA 299 (20×).** Scale bars represent 100 μm.

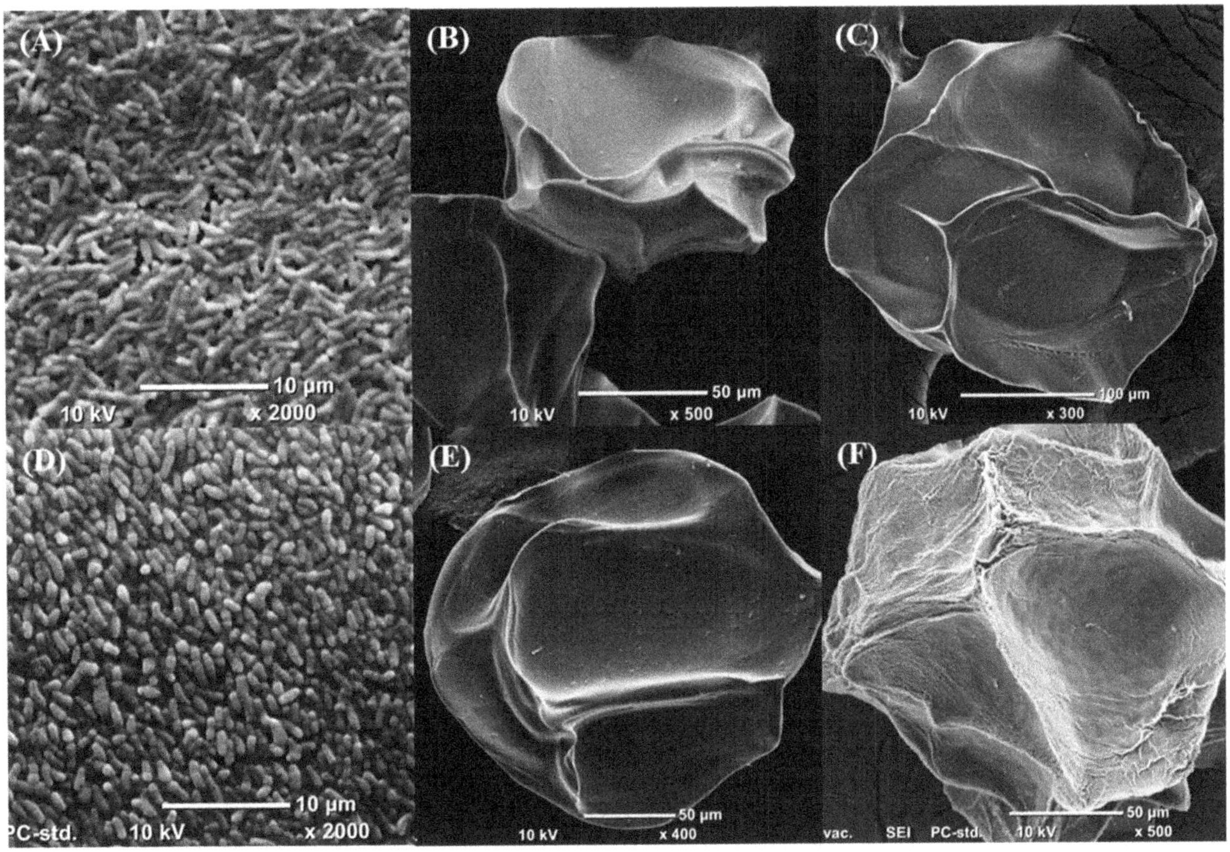


**FIGURE | Scanning electron micrographs of (A) *B. longum* subsp. *infantis* UMA299, (B) unfilled alginate bead, (C) unfilled chitosan-coated alginate bead, (D) *B. longum* subsp. *longum* UMA 306, (E) alginate bead containing *B. longum* subsp. *longum* UMA 401, (F) chitosan-coated alginate bead containing *B. longum* subsp. *longum* UMA 300.** Samples were dried before sputter-coating with gold. SEM was set at high-vacuum, 10 kV.

(**Figure A**; Supplementary ). As expected, there was a decrease in the viability of the bifidobacteria evaluated, but the rate of the decrease was strain dependent. A sharp decrease in viability was observed for *B. infantis* UMA318 and *B. longum* UMA401, diminishing by 9–10 log CFU over the course of a week under aerobic conditions. *B. infantis* UMA 300 and *B. infantis* UMA 305 remained viable for slightly longer, with a 10-log reduction observed within 10 days. Whereas, *B. infantis* UMA 298 and *B. infantis* 306 exhibited a 9–10 log decrease over 2 weeks of storage. Interestingly, *B. infantis* UMA 299 and *B. longum* UMA 402 survived the longest, as viable cell counts diminished by 7–8 logs over 3 weeks before decreasing to undetectable levels.

## Encapsulated Bifidobacterial Cells

Viability following encapsulation was determined for all eight bifidobacterial strains (**Figure B**; Supplementary Table S2). There were distinct differences between the effects of encapsulation depending on strain type. The cell viability of *B. infantis* UMA 298, *B. infantis* UMA 305, and *B. longum* UMA 318 stains rapidly decreased and were undetectable after 3 days. Unexpectedly, *B. infantis* UMA 298 and *B. infantis* UMA 305 in alginate were inactivated faster than the corresponding free cells, being undetectable after 24 and 10 days, respectively. Viability of the encapsulated *B. longum* UMA 318 was identical to that of free cells. *B. infantis* UMA 300, *B. longum* UMA 306, *B. longum* 401,

**TABLE | Zeta potential of alginate and chitosan-coated alginate beads with strains of bifidobacteria.**

| Beads | | mV |
|---|---|---|
| **Alg.** | | |
| Subsp. *infantis* | UMA 298 | $-5.23 \pm 2.06^{ab}$ |
| | UMA 299 | $-9.42 \pm 2.54^{a}$ |
| | UMA 300 | $-8.73 \pm 4.88^{a}$ |
| | UMA 305 | $-4.15 \pm 1.17^{ab}$ |
| Subsp. *longum* | UMA 306 | $-3.14 \pm 2.24^{ab}$ |
| | UMA 318 | $-2.60 \pm 0.04^{ab}$ |
| | UMA 401 | $-4.38 \pm 0.64^{ab}$ |
| | UMA 402 | $-4.28 \pm 1.12^{ab}$ |
| **Chit.-alg.** | | |
| Subsp. *infantis* | UMA 299 | $9.92 \pm 3.92^{c}$ |
| | UMA 300 | $14.87 \pm 4.26^{c}$ |
| Subsp. *longum* | UMA 401 | $0.79 \pm 2.53^{bd}$ |
| | UMA 402 | $9.03 \pm 4.90^{cd}$ |

*Values are shown as average ζ-potential ± standard deviation. Values followed by the same letters are not significantly different ($p > 0.05$) from each other.*

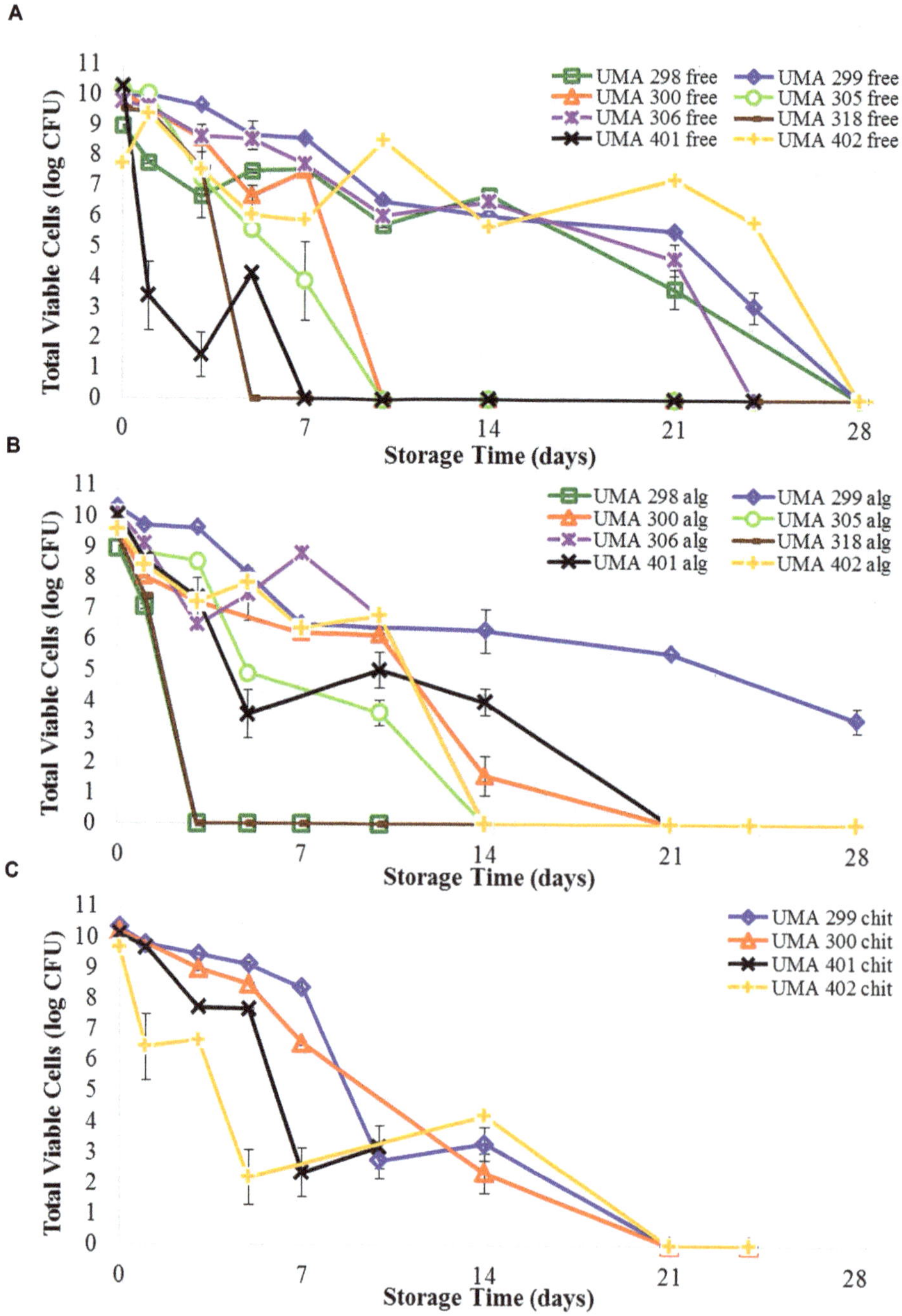


**FIGURE | Survival of (A) free *B. longum* cells, (B) *B. longum* cells in calcium alginate microbeads, (C) *B. longum* cells in chitosan-coated alginate microbeads in wet refrigerated storage over time.** Counts based on samples drop-plated on MRS agar and incubated at 37°C anaerobically. Error bars indicate the standard error of replicate counts ($n = 10$).

and *B. longum* 402 maintained viable populations that decreased by 3–4 log CFU after 10 days of storage before rapidly decreasing to zero. Encapsulated *B. infantis* UMA 300 survived 3 days longer than corresponding free cells; encapsulated *B. longum* UMA 401 lasted a week longer than free cells. Encapsulated *B. longum* UMA 306 survived similarly to free cells, and encapsulated *B. longum* UMA 402 survived over a week shorter than corresponding free cells. Interestingly, *B. infantis* UMA 299 viability was enhanced as it experienced a 5 log CFU reduction in 3 weeks compared to an 8 log CFU reduction during this time for the non-encapsulated cells. Thus, encapsulating with alginate extended cell viability of *B. infantis* UMA 299 and 300 by a few days, and extended viability of *B. longum* UMA 401 cells by a week. Encapsulating other *B. infantis* and *B. longum* strains did not appear to extend viability over the storage conditions used.

Two strains each of *B. infantis* (UMA 299 and 300) and *B. longum* (UMA 401 and 402) were encapsulated in a secondary coating of chitosan applied to the alginate bead core and submitted to testing over time (**Figure C**; Supplementary Table S3). *B. longum* UMA 401 and 402 both decreased 3–4 log within 3 days, and fell to undetectable levels by 2 weeks. Viability of *B. infantis* UMA 299 and *B. infantis* UMA 300 decreased only 2 logs in 5 days, before falling to undetectable levels after 2 weeks. Encapsulating *B. infantis* UMA 299, *B. infantis* UMA 300, and *B. longum* UMA 401 cells in chitosan-coated alginate beads did not appear to extend viability compared with uncoated alginate microbeads. *B. longum* UMA 402 cells in chitosan-coated alginate decreased 2.4 log CFU, whereas *B. longum* UMA 402 cells in alginate alone decreased 7.2 log CFU, between day 3 and day 14. Hence, encapsulating *B. longum* UMA 402 in chitosan-coated alginate extended detectible viability 4 days more than encapsulating *B. longum* UMA 402 in alginate alone.

## Survival of Encapsulated Cells during Simulated Digestion

Free and bifidobacterial cells encapsulated in chitosan-coated alginate beads were subjected to simulated digestion in a GIT model as previously described (Li et al., 2011). Free *B. infantis* UMA 299 and cells encapsulated chitosan-coated alginate were immersed separately in simulated salivary, gastric, and intestinal phases and assessed over time for cell viability (**Table** ). The strain was selected based on its high viability during storage in free and encapsulated forms (**Table** ; Supplementary Tables ). The bacteria appeared to be relatively stable within simulated saliva fluids, as less than a 1 log CFU reduction was experienced in 30 min of exposure regardless of encapsulation. The model salivary juice did not greatly inhibit cell viability in general, as less than one log CFU reduction was experienced in 30 min of exposure regardless of encapsulation. However, microencapsulation provided enhanced protection for UMA299 by shielding the strain from the low pH of the gastric phase. Encapsulated cells decreased by 1.4 logs CFU, whereas untreated cells decreased by 2.7 logs following exposure to pH 2.5 conditions (5 min). This indicates a significant, albeit fleeting protection afforded to the encapsulated cells as viability was abrogated after 10 min of exposure to the gastric phase. Similarly, UMA299 cell viability was not detectible after 5 min of exposure to the intestinal phase. *B. longum* UMA 402 encapsulated in chitosan-coated alginate was also subjected to simulated digestion in preliminary tests (data not shown). As with *B. infantis*, cell viability remained stable in the modeled salivary phase, but underwent a 6-log reduction after only a few minutes exposure to gastric phase (pH 2.5).

## DISCUSSION

Initially, chitosan-coating of alginate beads was postulated to enhance the viability of encapsulated probiotics by reducing their exposure to environmental stresses during storage and within the GIT (Kamalian et al., 2014). Accordingly, the influence of encapsulation on a panel of *B. longum* strains to assess differential viability was systematically studied. The calcium alginate beads formed using an injection–gelation method were roughly spherical in shape, negatively charged, and had dimensions around 130–220 μm. Coating the alginate beads with chitosan caused a small increase in their size and changed their charge from negative to positive. Optical microscopy (**Figures C,F**) confirmed that the bifidobacteria were immobilized within the hydrogel beads, which is consistent with previous encapsulation studies (Hansen et al., 2002; Fareez et al., 2015; Yeung et al., 2016).

Interestingly, encapsulation of bifidobacteria in chitosan-coated alginate beads led to decreased improvement in their storage or gastrointestinal stability compared with cells in alginate beads. One possible explanation for this observation is that the alginate hydrogel used had relatively large pores, and so small molecules, such as oxygen, acids, bile salts, or digestive enzymes, could easily diffuse into the microgels and inactivate the encapsulated bacteria (McClements, 2015). These results suggest that a simple secondary layer of chitosan alone

**TABLE | Simulated digestion of free and encapsulated *B. longum* subsp. *infantis* UMA 299 in three separate stages.**

| Time (minutes) | log CFU | | | | | |
|---|---|---|---|---|---|---|
| | Free | | | Encapsulated | | |
| | Saliva | Gastric | Intestinal | Saliva | Gastric | Intestinal |
| pH | 6.74 | 2.53 | 7.04 | 6.78 | 2.57 | 7.12 |
| 0 | $9.63 \pm 0.07^{aA}$ | $9.63 \pm 0.07^{aA}$ | $9.63 \pm 0.07^{aA}$ | $8.40 \pm 0.84^{abB}$ | $8.40 \pm 0.84^{aB}$ | $8.40 \pm 0.84^{aB}$ |
| 5 | $9.17 \pm 0.06^{a}$ | $6.99 \pm 0.03^{b}$ | ND | $8.10 \pm 0.05^{a}$ | $6.90 \pm 0.04^{a}$ | ND |
| 10 | $9.28 \pm 0.09^{a}$ | ND | ND | $7.89 \pm 0.07^{a}$ | ND | ND |
| 15 | $9.14 \pm 0.06^{a}$ | ND | ND | $6.42 \pm 1.07^{b}$ | ND | ND |
| 30 | $9.01 \pm 0.06^{a}$ | ND | ND | $8.14 \pm 0.040^{a}$ | ND | ND |

*Counts based on samples drop-plated on MRS agar and incubated at 37°C anaerobically. Values are shown as mean cell number ± standard error of replicate counts (n = 10). Means within each column followed by the same lowercase letters are not significantly different (p > 0.05) from each other. Means within each row followed by the same uppercase letters are not significantly different (p > 0.05) from each other.*

will not fully protect encapsulated bifidobacteria, and that further optimization is required to engineer more effective delivery systems. Previous studies have shown that alginate has a prebiotic effect on bifidobacteria, which might account for its ability to enhanced viability, potentially through a non-encapsulation mechanism (Wang et al., 2006; Ramnani et al., 2012). In future studies, it may be useful to examine the influence of different biopolymer materials and methods on the ability of microgels to enhance probiotic viability. As an example, the hydrogel pore size may be decreased to limit molecular diffusion, with the addition of anti-oxidants to limit oxidation reactions and prebiotics to stimulate probiotic growth in the colon. Since bifidobacteria ferment oligosaccharides within the gut, a synbiotic approach that integrates prebiotic substrates including plant or milk oligosaccharides may advance bifidobacterial-based delivery (Sela, 2011). Alternatively, judicious selection of strain selection that are resistant to acids, bile salts, or digestive enzymes may enhance the delivery scheme. However, previous studies indicate that most bifidobacteria strains typically exhibit a significant decrease in survival around pH 4 which would necessitate shielding from gastric conditions (Sun and Griffiths, 2000).

Bifidobacteria have been exposed to simulated digestive fluids in previously conducted studies (O'Riordan et al., 2001; Hansen et al., 2002; Kamalian et al., 2014). Although specific strains tested and experimental schemes vary between studies. Hansen et al. (2002) encapsulated several bifidobacterial strains in microgels formed by an emulsion-templating method, and then exposed them to simulated gastric and small intestinal fluids. In this study, *B. infantis* and *B. longum* strains showed a 4–6 log CFU/mL decrease between exposure to gastric fluids set at pH 6.0 and pH 2.0 for 2 h, and 3–5 log CFU/mL reduction between exposure to intestinal fluid containing 0 and 1% bile for 24 h. Hansen et al. (2002) also encapsulated *B. longum* experienced a 5-log CFU/mL reduction after 30 min exposure to gastric juice (pH 2.0). In the study herein, *B. infantis* UMA 299 encapsulated in chitosan-coated alginate underwent an 8-log reduction in a 10-min exposure to gastric fluid (pH 2.6), and an 8-log reduction in 5 min exposure to intestinal fluid (**Figure 3C**; Supplementary Table ). This study included 0.75% bile extract, pepsin, and lipases in the gastric and intestinal fluids for the purpose of simulating the harsh conditions of the human GIT. The bifidobacterial general stress response has been studied (Sánchez et al., 2007; Zomer et al., 2009). As with the phylogenetically dissimilar lactic acid bacteria, bifidobacteria employ ATPases to pump protons from the cell when exposed to acidic conditions (Matsumoto et al., 2004; Ventura et al., 2004). In addition, when exposed to bile during gastrointestinal transit, certain bifidobacterial strains deploy bile salt hydrolase to promote cell survival in the small intestine (Ruiz et al., 2014).

In an additional study, an emulsion encapsulation method was performed on *B. pseudocatenulatum* G4 in chitosan-coated alginate and exposed to gastric conditions (pH 1.5) for 2 h followed by intestinal phase for 5 h (Kamalian et al., 2014). The encapsulated *B. pseudocatenulatum* experienced a 4-log reduction when encapsulated in alginate and a 2-log reduction in chitosan-coated alginate, relative to the 5-log reduction in the control. However, this was accomplished in the absence of digestive enzymes or bile salts in simulated gastric and intestinal fluids that would present additional hurdles to the bifidobacterial cells. O'Riordan et al. (2001) studied spray-dried *Bifidobacterium* spp. PL1 in starch and subjected the resultant granules to simulated digestion. After 3 h of exposure to buffer with pH 2.8, they were unable to detect viable cells as well as other sampling points in between 0 and 3 h. This is consistent with the results presented in this study.

In summary, bifidobacterial viability following encapsulation varied between subspecies as well as strains. This suggests that there is a range of genotypic and phenotypic factors contributing to stress responses that promote enhanced viability. Further functional genomic analysis of encapsulated probiotic organisms can aid in matching strains with the particular encapsulation process to optimize cell integrity during storage. Moreover, similar approaches may be used in selecting ideal delivery vehicles to shield bifidobacteria during GIT transit to arrive intact and metabolically poised to exert beneficial activities in the distal colon. Subsequent formulations may optimize delivery vehicles by incorporating antioxidants and cryoprotectants within the encapsulation gel matrix to preserve bifidobacterial cell viability.

## AUTHOR CONTRIBUTIONS

TY, DM, and DS conceived the experimental plan. TY conducted laboratory experiments, data analysis, and drafted the manuscript. EÜ and KT assisted with experiments, analyses, and contributed to the manuscript. DS supervised execution of the experimental plan, analyzed data, and critically reviewed the final manuscript. All authors read and approved the manuscript prior to submission.

## ACKNOWLEDGMENTS

The authors thank TIC Gums for donating sodium alginate for the microencapsulation experiments. We thank David Prodanas for assistance with freeze-drying samples prior to SEM and Ezgi Özcan for assistance with statistical analysis. We also thank Jean Alamed and various individuals at the University of Massachusetts for helpful advice. EÜ thanks the Scientific and Technological Research Council of Turkey (TUBITAK) for financial support under Grant BIDEB 2214/A.

## SUPPLEMENTARY MATERIAL

The Supplementary Material for this article can be found online at: http://journal.frontiersin.org/article/10.3389/fmicb.2016.00494

## REFERENCES


Adhikari, K., Mustapha, A., Grün, I. U., and Fernando, L. (2000). Viability of microencapsulated bifidobacteria in set yogurt during refrigerated storage. *J. Dairy Sci.* 83, 1946–1951. doi: 10.3168/jds.S0022-0302(00)75070-3

Amine, K. M., Champagne, C. P., Raymond, Y., St-Gelais, D., Britten, M., Fustier, P., et al. (2014). Survival of microencapsulated *Bifidobacterium longum* in cheddar cheese during production and storage. *Food Control* 37, 193–199. doi: 10.1016/j.foodcont.2013.09.030

Bagchi, D., Lau, F. C., and Ghosh, D. K. (2010). *Biotechnology in Functional Foods and Nutraceuticals*. Boca Raton, FL: CRC Press.

Braat, H., Rottiers, P., Hommes, D. W., Huyghebaert, N., Remaut, E., Remon, J.-P., et al. (2006). A phase I trial with transgenic bacteria expressing interleukin-10 in Crohn's disease. *Clin. Gastroenterol. Hepatol.* 4, 754–759. doi: 10.1016/j.cgh.2006.03.028

Capela, P., Hay, T. K. C., and Shah, N. P. (2006). Effect of cryoprotectants, prebiotics and microencapsulation on survival of probiotic organisms in yoghurt and freeze-dried yoghurt. *Food Res. Int.* 39, 203–211. doi: 10.1016/j.foodres.2005.07.007

Chen, S., Cao, Y., Ferguson, L. R., Shu, Q., and Garg, S. (2013). Evaluation of mucoadhesive coatings of chitosan and thiolated chitosan for the colonic delivery of microencapsulated probiotic bacteria. *J. Microencapsul.* 30, 103–115. doi: 10.3109/02652048.2012.700959

de Barros, J. M., Lechner, T., Charalampopoulos, D., Khutoryanskiy, V. V., and Edwards, A. D. (2015). Enteric coated spheres produced by extrusion/spheronization provide effective gastric protection and efficient release of live therapeutic bacteria. *Int. J. Pharm.* 493, 483–494. doi: 10.1016/j.ijpharm.2015.06.051

de Vos, P., Faas, M. M., Spasojevic, M., and Sikkema, J. (2010). Encapsulation for preservation of functionality and targeted delivery of bioactive food components. *Int. Dairy J.* 20, 292–302. doi: 10.1016/j.idairyj.2009.11.008

Etchepare, M. D. A., Barin, J. S., Cichoski, A. J., Jacob-Lopes, E., Wagner, R., Fries, L. L. M., et al. (2015). Microencapsulation of probiotics using sodium alginate. *Ciênc. Rural* 45, 1319–1326.

Fareez, I. M., Lim, S. M., Mishra, R. K., and Ramasamy, K. (2015). Chitosan coated alginate-xanthan gum bead enhanced pH and thermotolerance of *Lactobacillus plantarum* LAB12. *Int. J. Biol. Macromol.* 72, 1419–1428. doi: 10.1016/j.ijbiomac.2014.10.054

Fortin, M.-H., Champagne, C., St-Gelais, D., Britten, M., Fustier, P., and Lacroix, M. (2011). Viability of *Bifidobacterium longum* in cheddar cheese curd during manufacture and storage: effect of microencapsulation and point of inoculation. *Dairy Sci. Technol.* 91, 599–614. doi: 10.1007/s13594-011-0034-5

Gaonkar, A. G., Vasisht, N., Khare, A. R., and Sobel, R. (2014). *Microencapsulation in the Food Industry: A Practical Implementation Guide*. Amsterdam: Elsevier.

Godward, G., Sultana, K., Kailasapathy, K., Peiris, P., Arumugaswamy, R., and Reynolds, N. (2000). The importance of strain selection on the viavility and survival of probiotic bacteria in dairy foods. *Michwissenschaft* 55, 441–445.

Gombotz, W. R., and Wee, S. F. (2012). Protein release from alginate matrices. *Adv. Drug Deliv. Rev.* 64, 194–205. doi: 10.1016/j.addr.2012.09.007

Hansen, L. T., Allan-Wojtas, P., Jin, Y. L., and Paulson, A. T. (2002). Survival of Ca-alginate microencapsulated *Bifidobacterium* spp. in milk and simulated gastrointestinal conditions. *Food Microbiol.* 19, 35–45. doi: 10.1006/fmic.2001.0452

Herigstad, B., Hamilton, M., and Heersink, J. (2001). How to optimize the drop plate method for enumerating bacteria. *J. Microbiol. Methods* 44, 121–129. doi: 10.1016/S0167-7012(00)00241-4

Kamalian, N., Mirhosseini, H., Mustafa, S., and Manap, M. Y. A. (2014). Effect of alginate and chitosan on viability and release behavior of *Bifidobacterium pseudocatenulatum* G4 in simulated gastrointestinal fluid. *Carbohydr. Polym.* 111, 700–706. doi: 10.1016/j.carbpol.2014.05.014

Kawasaki, S., Mimura, T., Satoh, T., Takeda, K., and Niimura, Y. (2006). Response of the microaerophilic *Bifidobacterium* species, *B. boum* and *B. thermophilum*, to oxygen. *Appl. Environ. Microbiol.* 72, 6854–6858. doi: 10.1128/AEM.01216-06

Krasaekoopt, W., Bhandari, B., and Deeth, H. (2003). Evaluation of encapsulation techniques of probiotics for yoghurt. *Int. Dairy J.* 13, 3–13. doi: 10.1016/S0958-6946(02)00155-3

Krasaekoopt, W., Bhandari, B., and Deeth, H. (2004). The influence of coating materials on some properties of alginate beads and survivability of microencapsulated probiotic bacteria. *Int. Dairy J.* 14, 737–743. doi: 10.1016/j.idairyj.2004.01.004

Lee, K. Y., and Mooney, D. J. (2012). Alginate: properties and biomedical applications. *Prog. Polym. Sci.* 37, 106–126. doi: 10.1016/j.progpolymsci.2011.06.003

Lewis, Z. T., Shani, G., Masarweh, C. F., Popovic, M., Frese, S. A., Sela, D. A., et al. (2015). Validating bifidobacterial species and subspecies identity in commercial probiotic products. *Pediatr. Res.* doi: 10.1038/pr.2015.244 [Epub ahead of print].

Li, Y., Hu, M., and McClements, D. J. (2011). Factors affecting lipase digestibility of emulsified lipids using an in vitro digestion model: proposal for a standardised pH-stat method. *Food Chem.* 126, 498–505. doi: 10.1016/j.foodchem.2010.11.027

Matsumoto, M., Ohishi, H., and Benno, Y. (2004). H+-ATPase activity in *Bifidobacterium* with special reference to acid tolerance. *Int. J. Food Microbiol.* 93, 109–113. doi: 10.1016/j.ijfoodmicro.2003.10.009

McClements, D. J. (2015). Encapsulation, protection, and release of hydrophilic active components: potential and limitations of colloidal delivery systems. *Adv. Colloid Interface Sci.* 219, 27–53. doi: 10.1016/j.cis.2015.02.002

Milani, C., Lugli, G. A., Duranti, S., Turroni, F., Bottacini, F., Mangifesta, M., et al. (2014). Genomic encyclopedia of type strains of the genus *Bifidobacterium*. *Appl. Environ. Microbiol.* 80, 6290–6302. doi: 10.1128/AEM.02308-14

Neal-McKinney, J. M., Lu, X., Duong, T., Larson, C. L., Call, D. R., Shah, D. H., et al. (2012). Production of organic acids by probiotic lactobacilli can be used to reduce pathogen load in poultry. *PLoS ONE* 7:e43928. doi: 10.1371/journal.pone.0043928

O'Riordan, K., Andrews, D., Buckle, K., and Conway, P. (2001). Evaluation of microencapsulation of a *Bifidobacterium* strain with starch as an approach to prolonging viavility during storage. *J. Appl. Microbiol.* 91, 1059–1066. doi: 10.1046/j.1365-2672.2001.01472.x

Puccio, G., Cajozzo, C., Meli, F., Rochat, F., Grathwohl, D., and Steenhout, P. (2007). Clinical evaluation of a new starter formula for infants containing live *Bifidobacterium longum* BL999 and prebiotics. *Nutrition* 23, 1–8. doi: 10.1016/j.nut.2006.09.007

Ramnani, P., Chitarrari, R., Tuohy, K., Grant, J., Hotchkiss, S., Philp, K., et al. (2012). In vitro fermentation and prebiotic potential of novel low molecular weight polysaccharides derived from agar and alginate seaweeds. *Anaerobe* 18, 1–6. doi: 10.1016/j.anaerobe.2011.08.003

Roy, D. (2005). Technological aspects related to the use of bifidobacteria in dairy products. *Lait* 85, 39–56. doi: 10.1051/lait:2004026

Ruiz, L., Gueimonde, M., Ruas-Madiedo, P., Ribbera, A., Clara, G., Ventura, M., et al. (2012). Molecular clues to understand the aerotolerance phenotype of *Bifidobacterium animalis* subsp. lactis. *Appl. Environ. Microbiol.* 78, 644–650. doi: 10.1128/AEM.05455-11

Ruiz, L., Margolles, A., and Sánchez, B. (2014). "Bile resistance mechanisms in *Lactobacillus* and *Bifidobacterium*," in *Microbial Mechanisms of Tolerance to Weak Acids: An Overview*, eds N. P. Mira and M. C. Teixeira (Lausanne: Frontiers E-books), 65.

Sakata, S., Kitahara, M., Sakamoto, M., Hayashi, H., Fukuyama, M., and Benno, Y. (2002). Unification of *Bifidobacterium infantis* and *Bifidobacterium suis* as *Bifidobacterium longum*. *Int. J. Syst. Evol. Microbiol.* 52, 1945–1951. doi: 10.1099/00207713-52-6-1945

Sánchez, B., Champomier-Vergès, M.-C., Stuer-Lauridsen, B., Ruas-Madiedo, P., Anglade, P., Baraige, F., et al. (2007). Adaptation and response of *Bifidobacterium animalis* subsp. lactis to bile: a proteomic and physiological approach. *Appl. Environ. Microbiol.* 73, 6757–6767. doi: 10.1128/AEM.00637-07

Schell, M. A., Karmirantzou, M., Snel, B., Vilanova, D., Berger, B., Pessi, G., et al. (2002). The genome sequence of *Bifidobacterium longum* reflects its adaptation to the human gastrointestinal tract. *Proc. Natl. Acad. Sci. U.S.A.* 99, 14422–14427. doi: 10.1073/pnas.212527599

Seiffert, S. (2013). Microgel capsules tailored by droplet-based microfluidics. *ChemPhysChem* 14, 295–304. doi: 10.1002/cphc.201200749

Sela, D., Chapman, J., Adeuya, A., Kim, J., Chen, F., Whitehead, T., et al. (2008). The genome sequence of *Bifidobacterium longum* subsp. infantis reveals adaptations for milk utilization within the infant microbiome. *Proc. Natl. Acad. Sci. U.S.A.* 105, 18964–18969. doi: 10.1073/pnas.0809584105

Sela, D. A. (2011). "Bifidobacterial utilization of human milk oligosaccharides," in *Proceedings of the 3rd International Symposium on Propionibacteria and Bifidobacteria: Dairy and Probiotic applications, 1-4 June 2010*, Vol. 149, Oviedo, 58–64.

Sela, D. A., and Mills, D. A. (2010). Nursing our microbiota: molecular linkages between bifidobacteria and milk oligosaccharides. *Trends Microbiol.* 18, 298–307. doi: 10.1016/j.tim.2010.03.008

Sultana, K., Godward, G., Reynolds, N., Arumugaswamy, R., Peiris, P., and Kailasapathy, K. (2000). Encapsulation of probiotic bacteria with alginate-starch and evaluation of survival in simulated gastrointestinal conditions and in yoghurt. *Int. J. Food Microbiol.* 62, 47–55. doi: 10.1016/S0168-1605(00)00380-9

Sun, W., and Griffiths, M. W. (2000). Survival of bifidobacteria in yogurt and simulated gastric juice following immobilization in gellan–xanthan beads. *Int. J. Food Microbiol.* 61, 17–25. doi: 10.1016/S0168-1605(00)00327-5

Sun, Z., Zhang, W., Guo, C., Yang, X., Liu, W., Wu, Y., et al. (2015). Comparative genomic analysis of 45 type strains of the genus *Bifidobacterium*: a snapshot of its genetic diversity and evolution. *PLoS ONE* 10:e0117912. doi: 10.1371/journal.pone.0117912

Ventura, M., Canchaya, C., van Sinderen, D., Fitzgerald, G. F., and Zink, R. (2004). *Bifidobacterium lactis* DSM 10140: identification of the atp (atpBEFHAGDC) operon and analysis of its genetic structure, characteristics, and phylogeny. *Appl. Environ. Microbiol.* 70, 3110–3121. doi: 10.1128/AEM.70.5.3110-3121.2004

Wang, Y., Han, F., Hu, B., Li, J., and Yu, W. (2006). In vivo prebiotic properties of alginate oligosaccharides prepared through enzymatic hydrolysis of alginate. *Nutr. Res.* 26, 597–603. doi: 10.1016/j.nutres.2006.09.015

Watson, R. R., and Preedy, V. R. (2015). *Probiotics, Prebiotics, and Synbiotics: Bioactive Foods in Health Promotion*. Boston, MA: Academic Press.

Whelehan, M., and Marison, I. W. (2011). Microencapsulation using vibrating technology. *J. Microencapsul.* 28, 669–688. doi: 10.3109/02652048.2011.586068

Yeung, T. W., Arroyo-Maya, I. J., McClements, D. J., and Sela, D. A. (2016). Microencapsulation of probiotics in hydrogel particles: enhancing *Lactococcus lactis* subsp. cremoris LM0230 viability using calcium alginate beads. *Food Funct.* doi: 10.1039/C5FO00801H [Epub ahead of print].

Zhang, Z., Zhang, R., Chen, L., Tong, Q., and McClements, D. J. (2015a). Designing hydrogel particles for controlled or targeted release of lipophilic bioactive agents in the gastrointestinal tract. *Eur. Polym. J.* 72, 698–716. doi: 10.1016/j.eurpolymj.2015.01.013

Zhang, Z., Zhang, R., Decker, E. A., and McClements, D. J. (2015b). Development of food-grade filled hydrogels for oral delivery of lipophilic active ingredients: pH-triggered release. *Food Hydrocoll.* 44, 345–352. doi: 10.1016/j.foodhyd.2014.10.002

Zhou, Y., Martins, E., Groboillot, A., Champagne, C., and Neufeld, R. (1998). Spectrophotometric quantification of lactic bacteria in alginate and control of cell release with chitosan coating. *J. Appl. Microbiol.* 84, 342–348. doi: 10.1046/j.1365-2672.1998.00348.x

Zomer, A., Fernandez, M., Kearney, B., Fitzgerald, G. F., Ventura, M., and van Sinderen, D. (2009). An interactive regulatory network controls stress response in *Bifidobacterium breve* UCC2003. *J. Bacteriol.* 191, 7039–7049. doi: 10.1128/JB.00897-09

**Conflict of Interest Statement:** The authors declare that the research was conducted in the absence of any commercial or financial relationships that could be construed as a potential conflict of interest.

# 12

# Modulation of the Bifidobacterial Communities of the Dog Microbiota by Zeolite

*Alberto Sabbioni[1†], Chiara Ferrario[2†], Christian Milani[2], Leonardo Mancabelli[2], Enzo Riccardi[1], Francesco Di Ianni[1], Valentino Beretti[1], Paola Superchi[1] and Maria C. Ossiprandi[1]**

[1] *Department of Veterinary Medical Science, University of Parma, Parma, Italy,* [2] *Laboratory of Probiogenomics, Department of Life Sciences, University of Parma, Parma, Italy*

During last decades canine health and well being is becoming an important issue for human owners. In dogs, several factors including diet, pathogenic bacterial and stress conditions can affect the composition of the gut microbiota. In this study, we evaluated the effect of dietary chabazitic zeolitite (CZ) supplementation on the contribution of bifidobacteria to the fecal microbiota in training hunting dogs. Fecal microbiota cataloging based on 16S rRNA microbial profiling analyses highlighted an increase of *Lactobacillus* and *Bifidobacterium* in animals treated with CZ, with a simultaneous decrease of pathogens associated with dog gastrointestinal infections, such as *Klebsiella* and *Enterobacter*. A detailed profiling of the bifidobacterial population of dogs receiving CZ based on the ITS-based sequencing approach, revealed an enhancement bifidobacterial of species typical of animals such as *Bifidobacterium animalis* and *B. pseudolongum*. Moreover, these analyses identified the occurrence of putative new bifidobacterial taxa in both treated and untreated samples.

**Keywords: dogs, gut microbiota, chabazitic zeolitite, *Bifidobacterium*, adsorptive capacity**

**Edited by:**
*Francesca Turroni,*
*University College Cork, Ireland*

**Reviewed by:**
*Claudio Hidalgo Cantabrana,*
*Dairy Research Institute of Asturias,*
*IPLA-CSIC, Spain*
*Francesca Bottacini,*
*University College Cork, Ireland*

***Correspondence:**
*Maria C. Ossiprandi*
*mariacristina.ossiprandi@unipr.it*

† *These authors have contributed equally to this work.*

## INTRODUCTION

Pet population is increasing in western countries, and dogs are the major human companions. Mutual interest has evolved into companion animals being a stable part of human life and therefore, the health and wellbeing of pets have increasingly raised interest during last decades. During history, the dog diet has changed, starting from a carnivorous behavior and a high protein diet (Clauss et al., 2010) to a carbohydrate rich diet and an urban life-style.

Despite the long span history of human-dog co-evolution, the knowledge of canine intestinal microbiota composition is much less complete than for humans. The dog gastro-intestinal tract (GIT) represents a rich ecosystem, composed of a wide range of metabolically active microorganisms (Simpson et al., 2002; Suchodolski et al., 2008; Kerr et al., 2013a). The predominant bacterial phyla in the colon and faeces of dogs are represented by *Firmicutes* (40–60%), *Bacteroidetes* (5–10%), *Proteobacteria* (15–20%), and *Fusobacteria* (5%) (Kerr et al., 2013b; Deng and Swanson, 2015), representing approximately 99% of the gut microbiota in dogs. However, very little is known about the occurrence of healthy promoting microorganisms such as bifidobacteria in the gut especially using metagenomics based approaches (Gavini et al., 2006; Jia et al., 2010).

Bifidobacteria are Gram positive bacteria that colonize different ecological niches, but represents one of the dominant colonizers of mammals at the very early first stages of life (Milani et al., 2015). The analyses of the gut microbiota of different mammals indicate that some bifidobacterial species, usually detected in the human GIT, were also identified in many other animals (Lamendella et al., 2008). For example, *Bifidobacterium bifidum*, *B. adolescentis*, *B. catenulatum*, and *B. dentium* are human-type bifidobacteria (Duranti et al., 2015, 2016), but these taxa displayed a cosmopolitan ecological behavior among different mammals (Lamendella et al., 2008).

In hunting dogs, emotional stress to which they are submitted during the training, can alter the habitat of the GIT (Rutgers et al., 1996). Therefore, to keep a suitable function of the GIT through appropriate feeding strategies is interesting, to avoid the intestinal colonization by enteropathogens (e.g., *Escherichia coli*, *Salmonella* ssp., *Clostridium perfringens*, *C. difficile*) (McKenzie et al., 2010; Kerr et al., 2013a).

To avoid antibiotic therapies, alternative products are under investigation. Zeolitites are aluminosilicates characterized by an open structure, which can accommodate a wide variety of ions. The particle size, crystallite size, and the degree of aggregation of the zeolitic material, as well as the porosity of individual particles, determine the access of ingesta fluids to the zeolitic surface during the passage across the GIT, and strongly affect its ion exchange, adsorption and catalytic properties (Papaioannou et al., 2005). The mechanism of action of zeolite is likely to be multifunctional. Different health and performance promoting properties were highlighted for zeolite in animal diet. These include ammonia binding effect, fecal elimination of *p*-cresol, retarding effect on digesta transit, enhanced pancreatic ezymes activity, and aflatoxin sequestering effect (Papaioannou et al., 2005). Moreover, recently it was reported the application of zeolite in reducing pathogens counts in broiler chicken (Prasai et al., 2016). Among zeolitites, the chabazitic zeolitite (CZ) has a high cation-exchange capacity and bulk density (Pabalan and Bertetti, 2001). Dietary inclusion of zeolitites has been effective in animals (e.g., pigs, calves) and humans suffering from gastrointestinal disturbances (RodriguezFuentes et al., 1997; Papaioannou et al., 2005). To date, no data exist about the evaluation of the effects of zeolitites on dog intestinal microbiota. The aim of the present study was to assess the effect of dietary CZ supplementation on the fecal microbiota with particular emphasis on bifidobacterial populations in training hunting dogs through culture-dependent methods and 16S rRNA/ITS (internal transcribed spacer) microbial profiling approach.

## MATERIALS AND METHODS

### Ethics Statement

This study was carried out in accordance with the recommendations of the ethical committee of the University of Parma. The protocol was approved by the "Comitato di Etica Università degli Studi di Parma", Italy. All animal procedures were performed according to national guidelines (Decreto legislativo 26/2014) on the protection of animals used for scientific purposes.

### Animals and Experimental Procedure

Twenty adult English Setter dogs, reared in the same kennel, were selected to be homogeneous with reference to age (mean age ± SD: 3.50 ± 1.9 years), body weight (mean weight ± SD: 18.83 ± 2.96 kg) and gender (10 males, 10 not pregnant females). Based on age, weight, and sex animals were equally divided into two groups (10 dogs group$^{-1}$), individually penned with a rest area inside (2.70 m× 1.40 m) and a paddock outside (4.50 m× 1.40 m). Animals were free of any clinical symptoms indicating gastrointestinal disease and they did not receive medications that are expected to alter the gut microbiota such as antibiotics. Dogs were wormed one month before the start of study. The characteristics of the groups are reported in **Table** . During a period of 28 days, both groups received a diet, based on raw poultry meat (25% crude protein, 24% ether extract, 5% ash, 2% crude fiber, and 18.4 MJ kg$^{-1}$ ME, on dry matter). The individual ration, administered at about 25 g dry matter kg$^{-1}$ of body weight$^{0.75}$, once a day, was supplemented (group Tr) or not (group NTr) with CZ powder at the dose of 5 g day$^{-1}$. For each dog, zeolitite was weighed and added to the ration at each meal. Free access to water was provided. During the study, all dogs were daily subjected to an aerobic physical activity characterized by gallop for 20 min, according to the trainer's practices. Training was performed in two outdoor next areas, at a mean temperature and relative humidity of 24 ± 3°C and of 67 ± 10%, respectively. Inside each group, five pairs of dogs were identified and each of them assigned alternatively to one or to the other of the training areas.

### Chabazitic Zeolitite Source

The powder of CZ, was obtained after sterilization at 200°C for 20 min (Chabasite 70® Verdi S.p.A, Italy). The total zeolitic content was 70 ± 5%, of which 65 ± 3% due to chabazite (Na0.14K1.03Ca1.00Mg0.17) [Al3.46Si8.53O24] × 9.7$H_2O$ and 5 ± 3% to phillipsite (Na0.9Ca0.5K0.6) [Si5.2A12.8O16] × 6$H_2O$. No traces of clinoptilolite were found. The composition of zeolitic powder was determined by Rietveld-RIR method (Gualtieri, 2000). The cation-exchange capacity and bulk density in relation to particles size were 2.2 ± 0.1 mEq g$^{-1}$ and 0.70–0.90 g (cm$^3$)$^{-1}$, respectively (Gualtieri, 2000; Cresswell and Hamilton, 2002). Water retention in relation to particles size was about 30–40% (w/w). The granulometry of the powder was less than 100 μm.

**TABLE | Characteristics of the experimental groups (mean ± SD).**

| Parameter | Groups* | |
|---|---|---|
| | NTr | Tr |
| Animals (No.) | 10 | 10 |
| Age (years) | 3.41 ± 1.59 | 3.50 ± 1.60 |
| Body weight (kg) | 19.59 ± 2.85 | 18.08 ± 2.74 |

**NTr, untreated group; Tr, treated group.*

## Collection of Fecal Samples

Feces consistency was scored using a scale of 1 (hard) to 5 (watery) (Grieshop et al., 2002) at days 0 (Time point 0, T0), 16 (Time point 1, T1) from the beginning of the dietary treatment, and at the end of experimental period (day 29, Time point 2, T2). During the same days, individual fecal samples were collected directly from the rectum, using a sterile glove lubricated with water. The feces were placed in sterile polyethylene bags, immediately transported to the laboratory on ice packs and frozen at −20°C until analysis.

## 16S rRNA/ITS Microbial Profiling

Upon arrival at the laboratory, individual fecal samples were aliquoted and combined with other individual samples from the same treatment to form pooled samples. In fact, in animal health it has been shown recently that pooling stool samples allows a rapid assessment of infection intensity and drug efficacy (Mekonnen et al., 2013). Each individual dog sample was equally represented in the respective pooled sample. DNA was extracted from pooled fecal samples using the QIAamp DNA Stool Mini kit following the manufacturer's instructions (Qiagen Ltd., Strasse, Germany).

Partial 16S rRNA gene sequences were amplified from extracted DNA using primer pair Probio_Uni and /Probio_Rev, which target the V3 region of the 16S rRNA gene sequence, as previously reported (Milani et al., 2013). Partial ITS sequences were amplified from extracted DNA using the primer pair Probio-bif_Uni/Probiobif_Rev as described by Milani et al. (2014b). The PCR conditions used were 5 min at 95°C and 35 cycles of 30 s at 94°C, 30 s at 55°C, and 90 s at 72°C, followed by 10 min at 72°C. Amplification was carried out using a Veriti Thermocycler (Applied Bio-systems).

16S rRNA gene and ITS sequencing were performed using a MiSeq (Illumina) according to the protocols previously published (Milani et al., 2013, 2014b).

## 16S rRNA Gene-Based Microbiota Analysis

The achieved individual sequence reads were filtered by the Illumina software to remove low quality and polyclonal sequences. All Illumina quality-approved, trimmed, and filtered data were exported as.fastq files. The.fastq files were processed using a custom script based on the QIIME software suite (Caporaso et al., 2010). Paired-end reads pairs were assembled to reconstruct the complete Probio_Uni/Probio_Rev amplicons. Quality control retained sequences with a length between 140 and 400 bp and mean sequence quality score >20 while sequences with homopolymers >7 bp and mismatched primers were omitted. In order to calculate downstream diversity measures (alpha diversity indices, Unifrac analysis), 16S rRNA Operational Taxonomic Units (OTUs) were defined at ≥97 % sequence homology using uclust (Edgar, 2010) and OTUs with less than 10 sequences were filtered. All reads were classified to the lowest possible taxonomic rank using QIIME (Caporaso et al., 2010) and a reference dataset from the SILVA database (Quast et al., 2013). Biodiversity of the samples (alpha-diversity) were calculated with Chao1 index.

## ITS-Based Microbiota Analysis

For ITS-based microbiota analysis Fastq files obtained by sequencing of the ITS amplicons were analyzed using a custom script, named bif_ITS_analysis.sh script[1]. This script requires QIIME (Caporaso et al., 2010) to be installed (or works in a QIIME virtual machine) and accepts.bam or.fastq input files containing sequencing reads. Input data were processed as previously described (Milani et al., 2014b).

## Bacterial Counts

The homogenates fecal specimens were serially diluted with both half-strength Wilkins-Chalgren Anaerobe Broth (WCAB) and Buffered Peptone Water (ThermoScientific-Oxoid, UK). Dilutions in duplicate were plated on MacConkey agar (Merck, Germany) for Enterobacteriaceae, Perfringens agar Base (OPSP) (Oxoid, UK) for *C. perfringens*, vancomycin and bromocresol green (LAMVAB) agar (Hartemink and Rombouts, 1999) for lactobacilli, and Azide maltose agar (Biolife, Italy) for enterococci counts. MacConkey agar and Azide maltose agar plates were incubated aerobically at 37°C for 24 and 48 h, respectively. Other media were incubated anaerobically at 37°C for 48–72 h. The taxonomy of colonies isolated random on selective media were determined at genus and species level by API System (Bio-Merieux, Italy) to verify the reliability of the media utilized.

## *In vitro* CZ Adsorptive Capacity

The ability of CZ to bind to enteropathogens bacteria was evaluated in pooled feces using two reference strains, i.e., *E. coli* ATCC 35218 and *C. perfringens* ATCC 13124. Strains were grown in Mueller-Hinton Broth (Difco, MI, USA) at 37°C for 24 h, then transferred to 10 ml of broth and grown for another 8 h to reach the final exponential phase.

Adsorptive capacity of CZ was evaluated, measuring spectrophotometrically the OD of the samples (An and Friedman, 1997). Twenty-five grams of pooled feces obtained by NTr groups and collected on days 0, 16, and 29 were placed, in triplicate, into flasks containing 225 ml of Buffered Sodium Chloride-Peptone Solution pH 7.0 (Oxoid, UK). CZ was added in different quantities (0, 0.25, 0.5, 1 g). Lastly, *C. perfringens* ATCC 13124 or *E. coli* ATCC 35218 strains were added to medium and incubated at 37°C. At 0, 2, 4, 6, and 24 h, 150 μl of the suspension were transferred into a microtiter plate in four replicates and the absorbance was immediately evaluated (VICTOR3, 1420 multilabel counter, PerkinElmer, Italy) at 620 nm.

## Statistical Analysis

Data for fecal score and fecal bacteria counts were checked for normality and then analyzed by ANOVA using the GLM procedure in SAS (Version 9.4, SAS Institute Inc., USA). The mixed model included the fixed effects of group (two levels), of sampling time (three levels), the interaction between group and

[1]http://probiogenomics.unipr.it/sw/bif_ITS_analysis.zip

sampling time and the random effect of animal. Values of colony forming units (CFU) have been expressed as $\log_{10}$ $g^{-1}$ of feces.

Statistical significance was reached for $P \leq 0.05$ as a $P$-value $>0.05$ and $\leq 0.10$ was considered as a trend.

## Data Deposition

Raw sequences of 16S rRNA gene profiling are accessible through SRA study accession number SRP075756. Raw sequences of ITS profiling are accessible through SRA study accession number SRP080281.

# RESULTS

## 16S rRNA Profiling of CZ Treated Dog

Pooled fecal samples from CZ treated (Tr) and no-treated (NTr) dogs were obtained in order to assess the microbiota composition based on 16S rRNA-sequencing analysis as described previously (Milani et al., 2013). The sequencing produced a total of 589784 reads with an average of 98297 reads per sample (Supplementary Table ).

Assessment of rarefaction curves, based on the Chao1 biodiversity indexes calculated for 10 subsampling of sequenced read pools, indicated that both curves tend to reach a plateau. Therefore, in all cases the obtained sequencing data was deemed adequate to cover the vast majority of the biodiversity contained within the samples (**Figure 1A**). Moreover, the two curves did not show relevant differences, thus indicating that the analyzed samples have similar biodiversity.

## Gut Microbiota Composition of CZ Treated Dogs

During the study, diarrhea events were not observed in the CZ treated dogs. CZ did not affect the palatability of the feed, which was eaten completely within 30 min after dosing. Fecal scores were not affected by the factors in the statistical model ($P > 0.05$; **Table** ). Fecal microbiota differences were observed in relation to group and sampling time ($P < 0.05$).

Inspection of predicted taxonomic profiles at phylum level for all NTr samples (T0, NTrT1, NTrT2) highlighted that *Firmicutes* (average 51.15% ± 11.46%) represented the dominant phylum of the cecal community in dogs, outnumbering the *Proteobacteria* (average 27.06% ± 15.75%), the *Fusobacteria* (average 8.54% ± 3.46%) and the *Bacteroidetes* (average 5.49% ± 2.62%) phyla (**Figures B,C**).

The comparison of the average relative abundance of NTr and Tr samples at time point T1 revealed a decrease of members of the *Enterobacteriaceae* family (−66.99 %), such as *Escherichia* (−67.16%), *Klebsiella* (−94.75%), and *Hafnia* (−74.87%), in Tr samples (**Figure D**) and an increase of *Lactobacillus* (205.16%) and *Bifidobacterium* (75.35%) in CZ treated animals (**Figure 1D**). At time point T2 in CZ treated animals (**Figure 1D**), the decrease in *Enterobacteriaceae* (−15.34%), includes a reduction of the genera *Hafnia* (-67.85 %), *Klebsiella* (−77.18%), and *Enterobacter* (−84.69%), along with an increase in relative abundance of *Lactobacillus* (861.64%) and *Bifidobacterium* (157.73%) (**Figure D**).

Notably, data achieved with culture-dependent approaches largely confirmed results obtained with 16S rRNA microbial profiling. In fact, *Lactobacillus* ssp. and *Enterococcus* ssp. counts were higher, while *Enterobacteriaceae* counts were lower in Tr than in NTr group ($P < 0.05$). *Lactobacillus* ssp. counts tended to be higher in Tr than in NTr group on day 16 (T1; 7.43 vs. 7.24; $P < 0.10$) and were higher on day 29 (T2; 8.18 vs. 7.25; $P < 0.05$). An increase of *Enterococcus* ssp. concentration (8.10 vs. 7.27) and a decrease of *Enterobacteriaceae* counts (6.24 vs. 7.14) were found in Tr compared to NTr group on day 29 (T2; $P < 0.05$). Besides, no change on the fecal *C. perfringens* counts was reported in relation both to the sampling time and to the treatment ($P > 0.05$).

## Bifidobacterial Community Modulation by CZ

Focusing on the contribution of bifidobacteria to the overall dog microbiota, it is worth noticing that at day 0 (T0) and in NTr animals at days 16 and 29 (T1 and T2, respectively), this genus represents 2.32% ± 1.88% of the gut microbiota of hunting dogs. In treated animals (Tr) the presence of the *Bifidobacterium* genus showed an increase of about 157.73% compared with Tr animals at T2, after 29 days of CZ diet (**Figure D**).

In order to precisely catalog the effects on the bifidobacterial population of dogs after CZ treatment, we performed an ITS profiling of bifidobacterial communities in stool samples of Tr and NTr dogs.

Quality filtering of the sequenced ITS amplicons produced an average of 52468 high-quality and full-length reads per sample (Supplementary Table S2) that were taxonomically attributed reaching the minimal taxonomical rank of species.

The composition of bifidobacterial populations of dogs included in the analysis showed the presence of peculiar species, such as *B. pseudolongum* (average of 60.70% ± 24.61% for T0 and NTr samples) and *B. animalis* (average of 7.84% ± 7.50% in T0 and NTr dogs) (**Figure A**), which have been previously described to be typical of the animal GIT (Milani et al., 2014a) and especially of the dog GIT (Gavini et al., 2006). Notably, other bifidobacterial species previously described to be typical of the human gut such *B. catenulatum* and *B. bifidum* were detected at a lower extend (**Figure B**).

Furthermore, in untreated animal samples (NTr1 and NTr2), ITS analysis revealed the occurrence of *B. longum*, *B. gallinarum*, and *B. pseudocatenulatum* species, typical human bifidobacterial taxa (Milani et al., 2014a). One possible explanation of the presence of these species in the canine gut microbiota could be a bacterial transmission between animals and trainers as previously reported in literature (Song et al., 2013). However, further investigations will be needed. Notably, a large proportion of the OTUs defined as 'unclassified' in T0 dog samples (**Figures A,B**) clusters separately from any current known bifidobacterial taxon, thus putatively representing novel *Bifidobacterium* taxa. These putative new unclassified bifidobacterial species represents the second most present

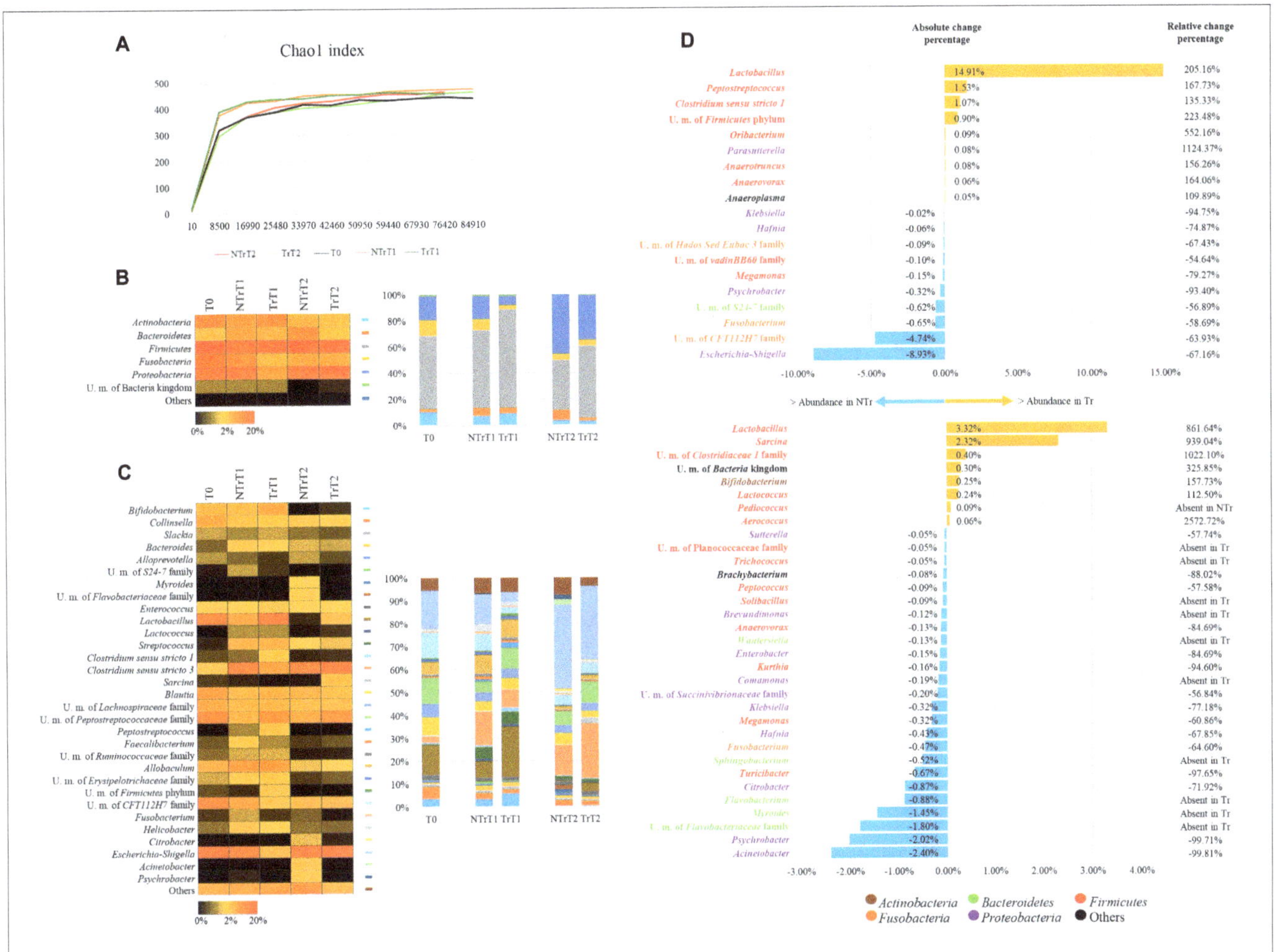


**FIGURE | Exploration of the taxonomic profile of NTr and Tr groups. (A)** Shows the rarefaction curves representing variation of the Chao1 and the Shannon diversity indexes at increasing sequencing depth of NTr and Tr fecal samples. **(B)** Displays bar plots and heat map of the identified bacterial phyla in the pooled CZ treated or untreated samples. **(C)** Represents bar plots and heat map of the identified bacterial genera in the pooled Tr or NTr samples. **(D)** exhibits the variation of taxa at time point T1 (upward) and T2 (below). We reported the bacterial genera with absolute change percentage >0.05% and showing increase >100% or decrease <−50% of relative change percentage in Tr data sets as compared to those obtained from NTr samples. In all panels the term unclassified member is abbreviated to U. m..

**TABLE | Effects of chabazitic zeolitite (CZ) supplementation on fecal score and fecal microbial concentration (least squares means of $log_{10}$ CFU $g^{-1}$ of feces).**

| Parameter | Groups* | | Sampling time | | | SEM† | *P*-values | | |
|---|---|---|---|---|---|---|---|---|---|
| | NT | T | 0 | T1 | T2 | | G‡ | St‡ | GxSt‡ |
| Faecal score§ | 3.15 | 3.35 | 3.07 | 3.33 | 3.37 | 1.18 | NS | NS | NS |
| *Lactobacillus* ssp. | 7.22 | 7.59 | 7.18 | 7.33 | 7.72 | 0.08 | <0.001 | <0.001 | <0.001 |
| *Enterococcus* ssp. | 7.19 | 7.51 | 7.15 | 7.22 | 7.68 | 0.05 | <0.001 | <0.001 | <0.001 |
| *Enterobacteriaceae* | 7.18 | 6.85 | 7.17 | 7.19 | 6.69 | 0.06 | <0.001 | <0.001 | <0.001 |
| *Clostridium perfringens* | 7.36 | 6.99 | 8.18 | 6.64 | 6.71 | 1.27 | NS | NS | NS |

**NTr, untreated group; Tr, treated group. ‡SEM, standard error of the difference of means. ‡G, group effect; St, sampling time effect; G × St, interaction. § On a scale of 1 (hard) to 5 (watery).*

bifidobacterial taxa in the dog microbiota, in both Tr and NTr animals (**Figures A,B**).

As reported above, at time point T2 in CZ treated animals, there was an increase in relative abundance of the genus *Bifidobacterium* (**Figure D**). ITS profiling experiments revealed an increase of 803.14 and 54.76% of *B. animalis* species and *B. pseudolongum* species, respectively, after the addition of CZ. Moreover, a slight increase was detected also for the here

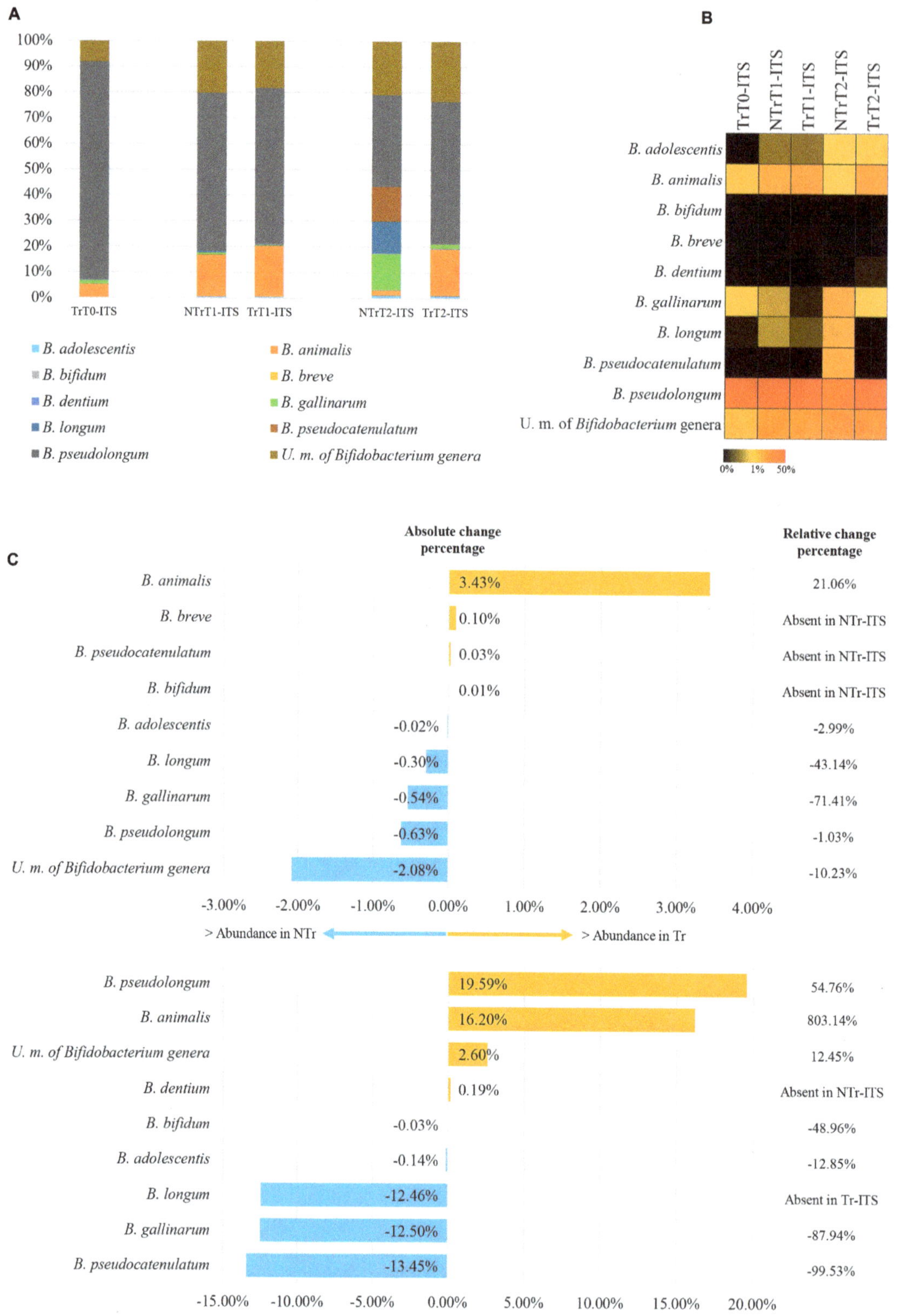


**FIGURE | Exploration of the bifidobacterial population of NTr and Tr groups. (A)** Represents the bar plots of the identified bifidobacteria in the pooled CZ treated or untreated samples through the ITS analysis. **(B)** Shows heat map of the identified bifidobacteria in the pooled Tr-ITS or NTr-ITS samples. **(C)** Displays the variation of the bifidobacterial population at time point T1 (upward) and T2 (below). We reported the *Bifidobacterium* species with absolute change percentage >0.05 % and showing increase >100% or decrease <−50% of relative change percentage in Tr-ITS data sets as compared to those obtained from NTr-ITS samples. In all panels the term unclassified member is abbreviated to U. m..

identified putative new bifidobacterial taxa in TrT2 (12.45 %) compared to NTrT2 (**Figure C**).

## *In Vitro* Bacterial Adsorptive Test

CZ showed an adsorptive capacity toward *E. coli* (**Figure 3A**) and *C. perfringens* (**Figure 3B**) strains in a dose- and time-dependent trial. Differences among CZ levels were registered for both strains after 2, 4, 6, and 24 h of incubation ($P < 0.05$). In particular, higher adsorptive capability against *E. coli* strain, was observed when CZ was added to the medium at a dose of 0.5 and 1 g rather than of 0 and 0.25 g ($P < 0.05$). When CZ was added at a dose of 1 g, negative values of OD starting from 0 h of incubation was observed for *E. coli*. During the first six hours of incubation, the adsorptive effect of CZ on *C. perfringens* strain was higher for levels of 0.5 and 1 g, than of 0 and 0.25 g ($P < 0.05$; **Figure 3**).

## DISCUSSION

In hunting dogs, emotional factors, such as those to which they are submitted during the training, can affect the GIT permeability, motility, secretion and mucin production. Thus, ultimately altering the habitat of resident gut bacteria and promoting changes in the gut microbiota composition (Gagne et al., 2013). Therefore, various feeding strategies have been developed in order to keep a suitable function of the GIT tract. Zeolite and in particular CZ have shown efficacy in animals (such as pigs, calves) and humans suffering from gastrointestinal disturbances (RodriguezFuentes et al., 1997; Papaioannou et al., 2005).

In this study, 20 adult English Setter dogs were trained and fed with a diet supplemented with CZ to evaluate how the microbiota and in particular bifidobacterial population as well as specific gut pathogens, could be modulated.

The results obtained after 29 days of CZ diet, showed that CZ affects the fecal microbial concentration but not the fecal score, which remained in a desirable range (well-formed, soft stools) for healthy dogs (Gagne et al., 2013). Notably, we observed an increase in relative abundance of *Lactobacillus* ssp. as well as *Bifidobacterium* ssp. phylotypes, accompanied by a decrease in phylotypes belonging to *Enterobacteriaceae* family in CZ fecal samples. This could be supported by the adsorptive capacity exploited by CZ toward *E. coli* and *C. perfringens*. Furthermore, *E. coli* and *Enterobacter* are common causes of extra-intestinal opportunistic infections in dogs (Ogeer-Gyles et al., 2006), while *C. perfringens* is strongly related to hemorrhagic gastroenteritis (Schlegel et al., 2012).

Moreover, the major presence of lactobacilli and bifidobacteria could be very interesting since these bacterial taxa are considered to exploit beneficial roles on the health of their hosts (Gibson et al., 2005). In this context, various members of *Lactobacillus* and *Bifidobacterium* species are the most exploited probiotic bacteria utilized for pet (Kelley et al., 2010; Strompfova et al., 2014) and some of them have been suggested to improve the health and brain function of dogs (Biagi et al., 2007; Bravo et al., 2011). Increased concentrations of these microorganisms have been associated with decreased fecal concentrations of potentially pathogenic bacteria and decreased levels of carcinogenic and putrefactive compounds in digesta (Grieshop et al., 2002).

This is the first study where the bifidobacterial community of healthy dog was explored through a Next Generation Sequencing approach involving bifidobacterial ITS profiling. The obtained results allowed the identification of a bifidobacterial profile in English setter hunting dogs and revealed the presence of typical animal bifidobacteria such as *B. animalis* and *B. pseudolongum* and many putative new taxa. CZ treatment led to an increase of the abundance of *B. animalis* and *B. pseudolongum* species, which are characterized by the presence of genes encoding for exopolysaccharides structures that could lead to a special cell protection (Ferrario et al., 2016; Hidalgo-Cantabrana et al., 2016). Increase of the bifidobacterial strains coupled with the adsorptive capacity of CZ could bring to a reduction of species belonging to the *Enterobacteriaceae* family, such as *Klebsiella* and *Enterobacter*, typical dog pathogens (Gibson et al., 2008). Combined CZ treatment with probiotic supplementation, such as bifidobacterial strains, might enhance the reduction of canine pathogens as well as strength the beneficial effects on the animal health.

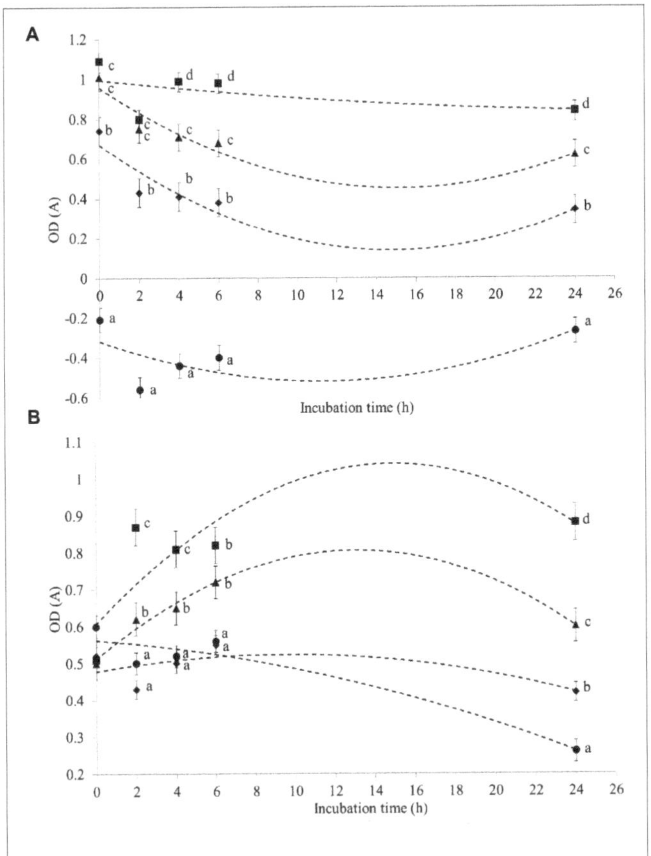


**FIGURE | *In vitro* adsorptive capacity of chabazitic zeolitite (CZ) toward *Escherichia coli* (A) and *Clostridium perfringens* (B).** CZ levels: (■) 0 g, (▲) 0.25 g, (◆) 0.5 g, (●) 1 g. Error bars indicate standard errors; a, b, c, d, $P < 0.05$ (differences among CZ levels).

## CONCLUSION

Dietary CZ supplementation can help to maintain a balanced intestinal microbial ecosystem and to prevent stress-related GIT upsets in healthy dogs, with a decrease of gut pathogens and a remarkable increase of bifidobacteria. This is particularly relevant in training hunting dogs where the mental and physical stress, to which they are subjected during training periods, can affect GI permeability and motility. Further studies are needed to confirm the beneficially effect by CZ also in diseased dogs.

## AUTHOR CONTRIBUTIONS

AS, PS, VB, and MO designed and performed experiments. MO, LM, AS, PS, and CF wrote the manuscript. LM and CM performed bioinformatic analyses. AS, CF, PS, MO, and VB performed experiments. CM, LM, ER, and FDI commented the manuscript. PS, AS, and MO conceived the study, revised and approved the manuscript. All authors reviewed the manuscript.

## FUNDING

LM is supported by Fondazione Cariparma, Parma, Italy.

## ACKNOWLEDGMENTS

The authors express sincere thanks to Verdi S.p.A, Italy for generously providing the Chabasite 70® product, Della Bassana kennel, Pegognaga (MN), Italy for providing necessary facilities and dogs and Ms Mirella Masini for her operating contribution in the present study.

## SUPPLEMENTARY MATERIAL

The Supplementary Material for this article can be found online at: http://journal.frontiersin.org/article/10.3389/fmicb.2016.01491

## REFERENCES

An, Y. H., and Friedman, R. J. (1997). Laboratory methods for studies of bacterial adhesion. *J. Microbiol. Methods* 30, 141–152. doi: 10.1016/S0167-7012(97)00058-4

Biagi, G., Cipollini, I., Pompei, A., Zaghini, G., and Matteuzzi, D. (2007). Effect of a *Lactobacillus animalis* strain on composition and metabolism of the intestinal microflora in adult dogs. *Vet. Microbiol.* 124, 160–165. doi: 10.1016/j.vetmic.2007.03.013

Bravo, J. A., Forsythe, P., Chew, M. V., Escaravage, E., Savignac, H. M., Dinan, T. G., et al. (2011). Ingestion of *Lactobacillus* strain regulates emotional behavior and central GABA receptor expression in a mouse via the vagus nerve. *Proc. Natl. Acad. Sci. U.S.A.* 108, 16050–16055. doi: 10.1073/pnas.1102999108

Caporaso, J. G., Kuczynski, J., Stombaugh, J., Bittinger, K., Bushman, F. D., Costello, E. K., et al. (2010). QIIME allows analysis of high-throughput community sequencing data. *Nat. Methods* 7, 335–336. doi: 10.1038/nmeth.f.303

Clauss, M., Kleffner, H., and Kienzle, E. (2010). Carnivorous mammals: nutrient digestibility and energy evaluation. *Zoo Biol.* 29, 687–704. doi: 10.1002/zoo.20302

Cresswell, H. P., and Hamilton, G. J. (2002). "Particle size analysis," in *Soil Physical Measurement and Interpretation for Land Evaluation*, eds N. J. McKenzie, H. P. Cresswell, and K. J. Coughlan (Collingwood, VIC: CSIRO Publishing), 224–239.

Deng, P., and Swanson, K. S. (2015). Gut microbiota of humans, dogs and cats: current knowledge and future opportunities and challenges. *Br. J. Nutr.* 113, S6–S17. doi: 10.1017/S0007114514002943

Duranti, S., Milani, C., Lugli, G. A., Mancabelli, L., Turroni, F., Ferrario, C., et al. (2016). Evaluation of genetic diversity among strains of the human gut commensal *Bifidobacterium adolescentis*. *Sci. Rep.* 6, 23971. doi: 10.1038/srep23971

Duranti, S., Milani, C., Lugli, G. A., Turroni, F., Mancabelli, L., Sanchez, B., et al. (2015). Insights from genomes of representatives of the human gut commensal *Bifidobacterium bifidum*. *Environ. Microbiol.* 17, 2515–2531. doi: 10.1111/1462-2920.12743

Edgar, R. C. (2010). Search and clustering orders of magnitude faster than BLAST. *Bioinformatics* 26, 2460–2461. doi: 10.1093/bioinformatics/btq461

Ferrario, C., Milani, C., Mancabelli, L., Lugli, G. A., Duranti, S., Mangifesta, M., et al. (2016). Modulation of the eps-ome transcription of bifidobacteria through simulation of human intestinal environment. *FEMS Microbiol. Ecol.* 92, fiw056. doi: 10.1093/femsec/fiw056

Gagne, J. W., Wakshlag, J. J., Simpson, K. W., Dowd, S. E., Latchman, S., Brown, D. A., et al. (2013). Effects of a synbiotic on fecal quality, short-chain fatty acid concentrations, and the microbiome of healthy sled dogs. *BMC Vet. Res.* 9:246. doi: 10.1186/1746-6148-9-246

Gavini, F., Delcenserie, V., Kopeinig, K., Pollinger, S., Beerens, H., Bonaparte, C., et al. (2006). *Bifidobacterium* species isolated from animal feces and from beef and pork meat. *J. Food Prot.* 69, 871–877.

Gibson, G. R., McCartney, A. L., and Rastall, R. A. (2005). Prebiotics and resistance to gastrointestinal infections. *Br. J. Nutr.* 93, S31–S34. doi: 10.1079/Bjn20041343

Gibson, J. S., Morton, J. M., Cobbold, R. N., Sidjabat, H. E., Filippich, L. J., and Trott, D. J. (2008). Multidrug-resistant *E. coli* and enterobacter extraintestinal infection in 37 dogs. *J. Vet. Intern. Med.* 22, 844–850. doi: 10.1111/j.1939-1676.2008.00124.x

Grieshop, C. M., Flickinger, E. A., and Fahey, G. C. Jr. (2002). Oral administration of arabinogalactan affects immune status and fecal microbial populations in dogs. *J. Nutr.* 132, 478–482.

Gualtieri, A. F. (2000). Accuracy of XRPD QPA using the combined Rietveld-RIR method. *J. Appl. Crystallogr.* 33, 267–278. doi: 10.1107/S002188989901643x

Hartemink, R., and Rombouts, F. M. (1999). Comparison of media for the detection of bifidobacteria, lactobacilli and total anaerobes from faecal samples. *J. Microbiol. Methods* 36, 181–192. doi: 10.1016/S0167-7012(99)00031-7

Hidalgo-Cantabrana, C., Algieri, F., Rodriguez-Nogales, A., Vezza, T., Martinez-Camblor, P., Margolles, A., et al. (2016). Effect of a Ropy Exopolysaccharide-Producing *Bifidobacterium animalis* subsp. lactis strain orally administered on DSS-induced colitis mice model. *Front. Microbiol.* 7:868. doi: 10.3389/fmicb.2016.00868

Jia, J., Frantz, N., Khoo, C., Gibson, G. R., Rastall, R. A., and McCartney, A. L. (2010). Investigation of the faecal microbiota associated with canine chronic diarrhoea. *FEMS Microbiol. Ecol.* 71, 304–312. doi: 10.1111/j.1574-6941.2009.00812.x

Kelley, R. L., Park, J. S., O'Mahony, L., Minikhiem, D., and Fix, A. (2010). Safety and tolerance of dietary supplementation with a canine-derived probiotic (*Bifidobacterium animalis* strain AHC7) fed to growing dogs. *Vet. Ther.* 11, E1–E14.

Kerr, K. R., Beloshapka, A. N., and Swanson, K. S. (2013a). 2011 and 2012 early careers achievement awards: use of genomic biology to study companion animal intestinal microbiota. *J. Anim. Sci.* 91, 2504–2511. doi: 10.2527/jas.2012-6225

Kerr, K. R., Forster, G., Dowd, S. E., Ryan, E. P., and Swanson, K. S. (2013b). Effects of dietary cooked navy bean on the fecal microbiome of healthy companion dogs. *PLoS ONE* 8:e74998. doi: 10.1371/journal.pone.0074998

Lamendella, R., Santo Domingo, J. W., Kelty, C., and Oerther, D. B. (2008). Bifidobacteria in feces and environmental waters. *Appl. Environ. Microbiol.* 74, 575–584. doi: 10.1128/AEM.01221-07

McKenzie, E., Riehl, J., Banse, H., Kass, P. H., Nelson, S., and Marks, S. L. (2010). Prevalence of diarrhea and enteropathogens in racing sled dogs. *J. Vet. Intern. Med.* 24, 97–103. doi: 10.1111/j.1939-1676.2009.0418.x

Mekonnen, Z., Meka, S., Ayana, M., Bogers, J., Vercruysse, J., and Levecke, B. (2013). Comparison of individual and pooled stool samples for the assessment of soil-transmitted helminth infection intensity and drug efficacy. *PLoS Negl. Trop. Dis.* 7:e2189. doi: 10.1371/journal.pntd.0002189

Milani, C., Hevia, A., Foroni, E., Duranti, S., Turroni, F., Lugli, G. A., et al. (2013). Assessing the fecal microbiota: an optimized ion torrent 16S rRNA gene-based analysis protocol. *PLoS ONE* 8:e68739. doi: 10.1371/journal.pone.0068739

Milani, C., Lugli, G. A., Duranti, S., Turroni, F., Bottacini, F., Mangifesta, M., et al. (2014a). Genomic encyclopedia of type strains of the genus *Bifidobacterium*. *Appl. Environ. Microbiol.* 80, 6290–6302. doi: 10.1128/AEM.02308-14

Milani, C., Lugli, G. A., Turroni, F., Mancabelli, L., Duranti, S., Viappiani, A., et al. (2014b). Evaluation of bifidobacterial community composition in the human gut by means of a targeted amplicon sequencing (ITS) protocol. *FEMS Microbiol. Ecol.* 90, 493–503. doi: 10.1111/1574-6941.12410

Milani, C., Mancabelli, L., Lugli, G. A., Duranti, S., Turroni, F., Ferrario, C., et al. (2015). Exploring vertical transmission of Bifidobacteria from mother to child. *Appl. Environ. Microbiol.* 81, 7078–7087. doi: 10.1128/AEM.02037-15

Ogeer-Gyles, J., Mathews, K., Weese, J. S., Prescott, J. F., and Boerlin, P. (2006). Evaluation of catheter-associated urinary tract infections and multi-drug-resistant *Escherichia coli* isolates from the urine of dogs with indwelling urinary catheters. *J. Am. Vet. Med. Assoc.* 229, 1584–1590. doi: 10.2460/javma.229.10.1584

Pabalan, R. T., and Bertetti, F. P. (2001). Cation-exchange properties of natural zeolites. *Nat. Zeolites* 45, 453–518. doi: 10.2138/rmg.2001.45.14

Papaioannou, D., Katsoulos, P. D., Panousis, N., and Karatzias, H. (2005). The role of natural and synthetic zeolites as feed additives on the prevention and/or the treatment of certain farm animal diseases: a review. *Microporous Mesoporous Mater.* 84, 161–170. doi: 10.1016/j.micromeso.2005.05.030

Prasai, T. P., Walsh, K. B., Bhattarai, S. P., Midmore, D. J., Van, T. T., Moore, R. J., et al. (2016). Biochar, bentonite and zeolite supplemented feeding of layer chickens alters intestinal microbiota and reduces campylobacter load. *PLoS ONE* 11: e0154061. doi: 10.1371/journal.pone.0154061

Quast, C., Pruesse, E., Yilmaz, P., Gerken, J., Schweer, T., Yarza, P., et al. (2013). The SILVA ribosomal RNA gene database project: improved data processing and web-based tools. *Nucleic Acids Res.* 41, D590–D596. doi: 10.1093/nar/gks1219

RodriguezFuentes, G., Barrios, M. A., Iraizoz, A., Perdomo, I., and Cedre, B. (1997). Enterex: anti-diarrheic drug based on purified natural clinoptilolite. *Zeolites* 19, 441–448. doi: 10.1016/S0144-2449(97)00087-0

Rutgers, H. C., Batt, R. M., Proud, F. J., Sorensen, S. H., Elwood, C. M., Petrie, G., et al. (1996). Intestinal permeability and function in dogs with small intestinal bacterial overgrowth. *J. Small Anim. Pract.* 37, 428–434. doi: 10.1111/j.1748-5827.1996.tb02443.x

Schlegel, B. J., Van Dreumel, T., Slavic, D., and Prescott, J. F. (2012). Clostridium perfringens type A fatal acute hemorrhagic gastroenteritis in a dog. *Can. Vet. J.* 53, 555–557.

Simpson, J. M., Martineau, B., Jones, W. E., Ballam, J. M., and Mackie, R. I. (2002). Characterization of fecal bacterial populations in canines: effects of age, breed and dietary fiber. *Microb. Ecol.* 44, 186–197. doi: 10.1007/s00248-002-0001-z

Song, S. J., Lauber, C., Costello, E. K., Lozupone, C. A., Humphrey, G., Berg-Lyons, D., et al. (2013). Cohabiting family members share microbiota with one another and with their dogs. *Elife* 2, e00458. doi: 10.7554/eLife.00458

Strompfova, V., Pogany Simonova, M., Gancarcikova, S., Mudronova, D., Farbakova, J., Mad'ari, A., et al. (2014). Effect of *Bifidobacterium* animalis B/12 administration in healthy dogs. *Anaerobe* 28, 37–43. doi: 10.1016/j.anaerobe.2014.05.001

Suchodolski, J. S., Camacho, J., and Steiner, J. M. (2008). Analysis of bacterial diversity in the canine duodenum, jejunum, ileum, and colon by comparative 16S rRNA gene analysis. *FEMS Microbiol. Ecol.* 66, 567–578. doi: 10.1111/j.1574-6941.2008.00521.x

**Conflict of Interest Statement:** The authors declare that the research was conducted in the absence of any commercial or financial relationships that could be construed as a potential conflict of interest.

The reviewer FB and handling Editor declared their shared affiliation, and the handling Editor states that the process nevertheless met the standards of a fair and objective review.

# 13

# Proteomic Profiling of *Bifidobacterium bifidum* S17 Cultivated under *In Vitro* Conditions

*Xiao Wei[1], Simiao Wang[1], Xiangna Zhao[1], Xuesong Wang[1], Huan Li[1], Weishi Lin[1], Jing Lu[1], Daria Zhurina[2], Boxing Li[1], Christian U. Riedel[2]*, Yansong Sun[1]* and Jing Yuan[1]**

[1] *Institute of Disease Control and Prevention, Academy of Military Medical Sciences, Beijing, China,* [2] *Institute of Microbiology and Biotechnology, University of Ulm, Ulm, Germany*

**Edited by:**
*Francesca Turroni,*
*University College Cork, Ireland*

**Reviewed by:**
*Michael C. Nelson,*
*University of Connecticut, USA*
*Marco Ventura,*
*University of Parma, Italy*
*Reetta Satokari,*
*University of Helsinki, Finland*

***Correspondence:**
*Jing Yuan*
*yuanjing6216@163.com;*
*Yansong Sun*
*sunys@qq.com;*
*Christian U. Riedel*
*christian.riedel@uni-ulm.de*

Bifidobacteria are frequently used in probiotic food and dairy products. *Bifidobacterium bifidum* S17 is a promising probiotic candidate strain that displays strong adhesion to intestinal epithelial cells and elicits potent anti-inflammatory capacity both *in vitro* and in murine models of colitis. The recently sequenced genome of *B. bifidum* S17 has a size of about 2.2 Mb and encodes 1,782 predicted protein-coding genes. In the present study, a comprehensive proteomic profiling was carried out to identify and characterize proteins expressed by *B. bifidum* S17. A total of 1148 proteins entries were identified by liquid chromatography coupled to tandem mass spectrometry (LC-MS/MS), representing 64.4% of the predicted proteome. 719 proteins could be assigned to functional categories according to cluster of orthologous groups of proteins (COGs). The COG distribution of the detected proteins highly correlates with that of the complete predicted proteome suggesting a good coverage and representation of the genomic content of *B. bifidum* S17 by the proteome. COGs that were highly present in the proteome of *B. bifidum* S17 were Translation, Amino Acid Transport and Metabolism, and Carbohydrate Transport and Metabolism. Complete sets of enzymes for both the bifidus shunt and the Embden-Meyerh of pathway were identified. Further bioinformatic analysis yielded 28 proteins with a predicted extracellular localization including 14 proteins with an LPxTG-motif for cell wall anchoring and two proteins (elongation factor Tu and enolase) with a potential moonlighting function in adhesion. Amongst the predicted extracellular proteins were five of six pilin proteins encoded in the *B. bifidum* S17 genome as well as several other proteins with a potential role in interaction with host structures. The presented results are the first compilation of a proteomic reference profile for a *B. bifidum* strain and will facilitate analysis of the molecular mechanisms of physiology, host-interactions and beneficial effects of a potential probiotic strain.

**Keywords: proteomic profiling, metabolic pathways, pilin proteins, LC-MS/MS, *Bifidobacterium bifidum***

---

**Abbreviations:** DTT, DL-dithiothreitol; ECM, extracellular matrix; EDTA, ethylene diamine tetraacetie acid; HEPES, 4-(2-hydroxyethyl)-1-piperazineethanesulfonic acid; IECs, intestinal epithelial cells; KEGG, Kyoto Encyclopedia of Genes and Genomes; LC-MS/MS, liquid chromatography coupled to tandem mass spectrometry; LPXTG, Leu-Pro-X-Thr-Gly; MRS, Man-Rogosa-Sharpe; ORF(s), open reading frame(s); PMSF, phenylmethanesulfonyl fluoride; ThDP, thiamine diphosphate; TMDs, transmembrane domains; UHPLC, ultra-high-performance liquid chromatography.

## INTRODUCTION

Bifidobacteria represent an important group of the human intestinal microbiota (Bottacini et al., 2014). Due to their reported ability to reduce cholesterol levels, exclude intestinal pathogens, strengthen the intestinal barrier, alleviate symptoms of constipation, and/or modulate the immune response they are frequently used as active ingredients in food and dairy-based products (Leahy et al., 2005; Gareau et al., 2010). In order to reveal the molecular mechanisms responsible for these beneficial effects, several bifidobacterial strains have recently been sequenced (Ventura et al., 2009). Analysis of bifidobacterial genomes has led to the identification of various structures involved in host colonization and probiotic properties. For example, *Bifidobacterium breve* UCC2003 was shown to encode genes for production of exopolysaccharides (Fanning et al., 2012). These exopolysaccharides support persistence in the murine gastrointestinal tract and are required for the protective effect of *B. breve* UCC2003 against infections with the murine pathogen *Citrobacter rodentium*. *B. breve* UCC2003 also possesses type IV tight adherence pili and these pili were shown to support long term colonization of mice (O'Connell Motherway et al., 2011). Similarly, other strains and species of bifidobacteria were shown to contain gene clusters for Tad and/or sortase-dependent pili for some of which interaction with host structures has been shown (Foroni et al., 2011; Turroni et al., 2013, 2014b).

*Bifidobacterium bifidum* strains belong to the infant-type bifidobacteria and show a remarkable adaptation to their ecological niche in the intestinal tract of human neonates. This includes a large number of adhesive structures and a specific ability to utilize host-derived glycans (Turroni et al., 2014a). *B. bifidum* S17 was isolated from feces of a breast-fed infant and displays unusually strong adhesion to IECs (Riedel et al., 2006a; Preising et al., 2010; Gleinser et al., 2012). Additionally, the strain elicits a promising anti-inflammatory capacity both *in vitro* (Riedel et al., 2006b; Preising et al., 2010) and *in vivo* in three different murine models of colitis (Preising et al., 2010; Philippe et al., 2011; Grimm et al., 2015). The genome of *B. bifidum* S17 was sequenced and annotated and contains a predicted 1,782 protein-coding ORFs (Zhurina et al., 2011) including one Tad and three sortase-dependent pili gene clusters as well as several other genes suspected or shown to play a role in adhesion to host structures or host colonization (Gleinser et al., 2012; Westermann et al., 2012).

In recent years, proteomic analysis has become an indispensible tool to analyze the biology of microorganisms, their response to changes in the environmental conditions and their interaction with the host (Otto et al., 2014). One of first the reports on a proteomic analysis of a *Bifidobacterium* sp. strain was a proteomic reference map obtained by 2D electrophoresis and MALDI TOF-TOF mass spectrometry (Yuan et al., 2006). Since then, this technique has been used to study adaptation of various bifidobacteria to bile and oxidative stress (Sánchez et al., 2007; Xiao et al., 2011) and different carbon sources (Liu et al., 2011, 2015), host-induced proteome changes of *B. longum* NCC2705 (Yuan et al., 2008) or its interaction with IECs (Wei et al., 2014). Turroni et al. (2010) performed a proteomic analysis of *B. bifidum* PRL2010 during growth on different sugars and following contact with cultured epithelial cells (Turroni et al., 2010).

However, proteomic approaches that include 2D electrophoresis have limitations in the detection of alkaline and low-abundance proteins (Otto et al., 2014). These limitations can be overcome by 1D gel-based LC-MS/MS (Wickramasekara et al., 2011; Otto et al., 2014). Moreover, LC-MS/MS is more efficient and accurate when analyzing differential global protein expression quantitatively and was used for proteomics of *Burkholderia vietnamiensis* (Wickramasekara et al., 2011). Another example is the analysis of secretion profiles of *B. pseudomallei* MSHR668 (Burtnick et al., 2014). LC-MS/MS has also been employed for comparative proteomics of two *Lactobacillus rhamnosus* strains (Savijoki et al., 2011). For bifidobacteria, LC-MS/MS was used for differential proteomics of two *B. longum* strains (Guillaume et al., 2009) and a proteomic profiling of *B. longum* subsp. *infantis* during growth on lactose, glucose, and galacto-, fructo-, and human milk oligosaccharides (Kim et al., 2013).

In the present study, the proteome of *B. bifidum* S17 grown *in vitro* was analyzed by 1D gel-based ultra-high performance LC-MS/MS representing the first comprehensive proteomic profile for the species *B. bifidum*.

## MATERIALS AND METHODS

### Bacterial Strains and Growth Conditions

*Bifidobacterium bifidum* S17 was cultured in sealed jars anaerobically at 37°C in *Lactobacillus* MRS medium (Difco) supplemented with 0.05% L-cysteine. Anaerobic conditions were achieved and maintained using AnaeroGen sachets (Thermo Scientific). Bacteria were harvested for proteomic analysis in mid-exponential growth phase at an optical density at 600 nm of 0.9 corresponding to approximately $1.5 \times 10^8$ colony forming units/ml.

### Preparation of Whole Cell Protein Extracts

Bacteria were harvested by centrifugation, washed twice in phosphate-buffered saline (PBS), and pellets (about 0.30 g) were resuspended in 5 mL of lysis buffer (8 M urea, 30 mM HEPES, 1 mM phenylmethylsulfonyl fluoride, 2 mM ethylenediaminetetraacetic acid, 10 mM dithiothreitol) containing one dissolved tablet of complete protease inhibitor (Roche Diagnostics, Mannheim, Germany). Bacteria were then sonicated for 10 min on ice using a Sonifier 750 (Branson Ultrasonics Corp., Danbury, CT, USA) with the following conditions: 2 s of sonication with a 3 s interval set at 25% duty cycle. The cell lysate suspension was centrifuged for 30 min at $20,000 \times g$ to collect the supernatant. The proteins were reduced with 10 mM dithiothreitol at 56 °C for 1 h, and alkylated with 55 mM iodoacetamide at room temperature for 1 h in the dark. The treated proteins were precipitated in acetone at −20°C for 3 h. After centrifugation at $20,000 \times g$ for 30 min, the protein pellet was resuspended and ultrasonicated in pre-chilled 50% tetraethylammonium bromide buffer with 0.1% sodium dodecyl sulfate (SDS). The proteins were regained after centrifugation at 20,000 × *g* and the protein concentrations were measured by Bradford assay.

## 1D SDS Polyacrylamid Gelelectrophersis (SDS-PAGE) and in-Gel Digestion

The proteome was analyzed by 1D gel-based LC-MS/MS as described previously (Albrethsen et al., 2010). In brief, the proteins were separated via SDS polyacrylamid gelelectrophersis (SDS-PAGE) using precast 4–12 % gradient gels containing Bis-Tris buffer (NuPAGE MES system, Invitrogen). Protein gels were run by using 1x running buffer [259 mM Tris base, 2 M glycine, 1% (w/v) SDS in $ddH_2O$] at a constant voltage of 85 V for 20 min followed by 150 V for 40 min (until the dye runs off the gel). The gels were then stained with Coomassie R-250. A single lane of stained gel was cut into 10 pieces of approximately the same size and transferred into 1.5 mL Eppendorf tubes to perform in-gel tryptic digestion. Gel bands were destained and tryptically digested as described previously (Yuan et al., 2006). The digested peptides were extracted with 50% acetonitrile and 2% formic acid solution. After extraction, the peptides were transferred into 500 μL Eppendorf tubes and concentrated using a Speed-Vac Concentrator (Savant) and a volume of 10 μL containing 15 μg of protein was loaded into individual high-performance liquid chromatography (HPLC) autosampler vials and analyzed by liquid chromatography-tandem mass spectrometry (LC-MS/MS).

## LC-MS/MS Analysis

Peptide samples were analyzed by ultra-high performance LC-MS/MS on a quadrupole-Orbitrap mass spectrometer (Q-Exactive; Thermo Fisher, Germany) equipped with 15 cm (length) by 75 μm (inside diameter) column packed with 5 μm C18 medium (Thermo Fisher) which was kept at 21°C throughout the analysis. Mobile phase A was MilliQ water with 0.1% (v/v) formic acid. Mobile phase B was 99.9% (v/v) acetonitrile, 0.1% acetic acid. Gradient was run from 0% B to 30% B over 40 min and then to 80% B for 15 min. The LC was interfaced to a quadrupole-Orbitrap mass spectrometer (Q-Exactive; Thermo Fisher) via nano-electrospray ionization. An electrospray voltage of 1.8 kV was applied. The mass spectrometer was programmed to acquire, by data-dependent acquisition, tandem mass spectra from the top 20 ions in the full scan from 350 to 2,000 m/z. Dynamic exclusion was set to 15 s, singly charged ions were excluded, the isolation width was set to 2 m/z, the full MS resolution was set to 70,000, and the MS/MS resolution was set to 17,500. Normalized collision energy was set to 28, automatic gain control to 1e6, maximum fill MS to 20 ms, maximum fill MS/MS to 60 ms, and the under fill ratio to 0.1%.

## Protein Identification and Bioinformatic Analysis

Peptide identification were performed using Mascot v2.3.01 (Matrix Science Ltd.)[1] licensed in-house[2] (Albrethsen et al., 2010). Peptide mass finger printing searches were performed using Mascot v2.2.06 (Matrix Science Ltd.)[1] licensed in-house[2]. Monoisotopic peptide masses were used to search the databases, allowing a peptide mass accuracy of 30 ppm and fragment ion tolerance of 0.2 Dalton. Both methionine oxidation and cysteine carboxyamidomethylation were considered in the process. For protein identification by peptide mass fingerprinting, peptide masses were searched against the publically available database for *B. bifidum* S17 (NCBI Reference Sequence: NC_014616.1). For unambiguous identification of proteins, more than five peptides must be matched and the sequence coverage must be greater than 15%. The complete proteomic dataset is deposited and publically accessible on the iProX database[3] under project number IPX00067300.

Calculation of protein molecular weights and isoelectric points were carried out by the protparam software from the Expasy toolbox[4]. Functional classification of identified proteins was performed by BLASTPGP (Altschul et al., 1997) searching against the databases of Cluster of Orthologous Groups (COGs[5]) (Tatusov et al., 2000). The cellular localizations of all identified proteins were predicted by PSORTb version 3.0 (Yu et al., 2010). Prediction of signal peptides was carried out with SignalP Version 4.1 (Petersen et al., 2011), TMHMM server 2.0 (Sonnhammer et al., 1998) was used to predict transmembrane helices, and LocateP database (Zhou et al., 2008) was used for prediction of cell wall and lipid anchor motifs.

# RESULTS

The 2.2-Mb genome of *B. bifidum* S17 contains 1,782 predicted protein-coding ORFs. To obtain an overview on the proteins expressed by *B. bifidum* S17 under standard laboratory conditions, whole protein extracts of bacteria grown in MRS medium were subjected to LC-MS/MS and protein identification. A total of 1,148 proteins were successfully matched unambiguously to one of the 1,782 predicted proteins of *B. bifidum* S17, i.e., a coverage of 64.4% of the complete predicted proteome. All detected proteins the peptide data matches the protein sequence as predicted in the genome annotation (data not shown). Moreover, the proteome contained a total of 235 (conserved) hypothetical proteins demonstrating that these proteins are actually expressed by *B. bifidum* S17 at least under *in vitro* conditions, which changes their status from "hypothetical" to proteins with unknown function.

**Supplementary Table** summarizes these proteins along with information on predicted pI, molecular mass, function, signal peptides, and subcellular location. 76.7% of the *B. bifidum* S17 proteins detected by LC-MS/MS are acidic with pI between 3 and 7, 12.5% have a pI between 7 and 9, whereas 10.8% of the proteins have a pI greater than 9 (**Figure A**). Classification of the 1,148 detected proteins into clusters of

[1]http://www.matrixscience.com

[2]http://www.proteomics.cn

[3]www.iprox.org

[4]www.expasy.ch/tools/protparam.html

[5]ftp://ftp.ncbi.nih.gov/pub/COG/COG2014/static/lists/listCOGs.html

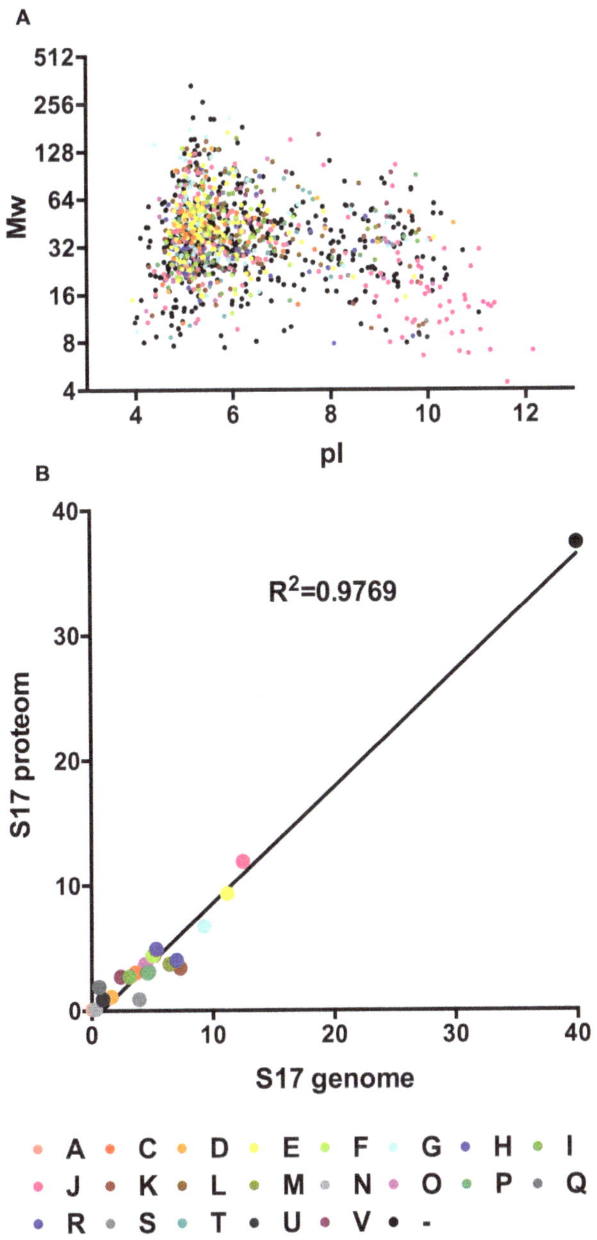


**FIGURE** | **(A)** *In silico* 2D gel electrophoresis of the identified proteins of the *B. bifidum* S17 proteome. For each identified protein, calculated molecular weight was plotted against the calculated pI. **(B)** Correlation analysis of the COG distributions of the detected proteins in the proteome and all proteins encoded on the genome of *B. bifidum* S17.

orthologous groups of proteins (COGs) assigned 719 proteins into 22 functional categories ( ). The remaining 429 unclassified proteins (37,4%) were denoted as COG-absent proteins. For none of the COGs an obvious clustering according to pI and Mw could be observed (**Figure** A). Comparison of the proteomes and genomes of *B. bifidum* S17 that the COG distribution of the *B. bifidum* S17 proteome is highly correlated to the distribution of all 1,782 predicted protein-coding ORFs of the genome (**Figure** B; coefficient of determination $R^2 = 0.9769$). Also, none of the COGs is skewed markedly off the regression in the proteome/genome comparison. Thus, none of the COGs seems to be over- or underrepresented in the proteome compared to the genome. Collectively, this indicates that the proteome captured by LC-MS/MS is comprehensive and covers all functional categories to a similar extend as encoded by the genomic information of *B. bifidum* S17.

The functional categories with highest representation amongst the detected proteins were Translation (functional category J; 137 proteins, 11.9% of all detected proteins), Amino acid Transport and Metabolism (functional category E; 107 proteins, 9.3%), and Carbohydrate Transport and Metabolism (functional category G; 77 proteins, 6.7%). Collectively, 26% of all detected proteins of the *B. bifidum* S17 proteome are assigned to functional categories J, A, K, and L. Thus, proteins possessing biological functions related to information storage, DNA replication, recombination and repair, RNA processing, transcription, and translation were highly prevalent in the proteome of *B. bifidum* S17. Of note, elongation factors Tu (EF-Tu; BBIF_1251) was amongst these proteins.

A total of 9.3% of the proteome have a (predicted) role in amino acid transport, biosynthesis, urea cycle, and metabolism of amino groups including 14 aminotransferases (**Table** ) and 6.7%

**TABLE** | **Relative abundance of functional categories according to clusters of orthologous groups of proteins (COGs) in the *Bifidobacterium bifidum* S17 proteome.**

| COG functional category | Abundance |
|---|---|
| A: RNA processing and modification | 0.1% |
| C: Energy production and conversion | 3.0% |
| D: Cell cycle control, mitosis and meiosis | 1.1% |
| E: Amino acid transport and metabolism | 9.3% |
| F: Nucleotide transport and metabolism | 4.4% |
| G: Carbohydrate transport and metabolism | 6.7% |
| H: Coenzyme transport and metabolism | 4.9% |
| I: Lipid transport and metabolism | 2.7% |
| J: Translation | 11.9% |
| K: Transcription | 3.4% |
| L: Replication, recombination and repair | 4.4% |
| M: Cell wall/membrane biogenesis | 3.7% |
| N: Cell motility | 0.1% |
| O: Posttranslational modification, protein turnover, chaperones | 3.7% |
| P: Inorganic ion transport and metabolism | 3.0% |
| Q: Secondary metabolites biosynthesis, transport and catabolism | 1.9% |
| R: General function prediction only | 4.0% |
| S: Function unknown | 0.9% |
| T: Signal transduction mechanisms | 3.1% |
| U: Intracellular trafficking and secretion | 0.9% |
| V: Defense mechanisms | 2.7% |
| X: Transposons | 0.2% |
| –: Not in COGs | 37.4% |

of all identified proteins are related to carbohydrate transport and metabolism. The major and characteristic metabolic pathway of bifidobacteria for fermentation of hexose sugars is the so-called fructose-6-phosphate or "bifid" shunt (Pokusaeva et al., 2011). We successfully detected a complete set of ten enzymes of the bifidus shunt (**Figure A**) including the key enzyme of the pathway xylulose-5-phosphate/fructose-6-phosphate phosphoketolase (BBIF_0798). Moreover, 11 enzymes for a complete Embden-Meyerhof pathway were identified (**Figure B**).

Subcellular localization of all 1,148 proteins detected by LC-MS/MS was predicted by PSORTb Version 3.0. Extracellular and cell wall proteins play critical roles in establishing and maintaining interactions between a microbe and its environment. Thus, proteins which, according to the PSORTB prediction, are located in the cell wall or extracellular were further analyzed for transmembrane helices and signal peptides by SignalP v4.1 and TMHMM Server v2.0 and searched or LPxTG cell wall anchor motifs using the LocateP database to confirm their cellular localization (**Table** ). Interestingly, these analysis suggest a different localization for several of these proteins as predicted by PSORTB. For example, three of the 28 proteins (BBIF_0312, BBIF_0337, BBIF_1026) contained none of these domains/sequence motifs and are thus probably cytoplasmatic. Following, corrections the PSORTB correction of the proteins in **Table** , the proteome of *B. bifidum* S17 consists of 743 (64.7%) cytoplasmic, 231 (21.1%) membrane, 14 (1.2%) cell wall, 3 (0.3%) extracellular proteins and 146 (12.7%) proteins with unknown cellular localization

**TABLE | Aminotransferases identified in the proteome of *B. bifidum* S17.**

| Locus_tag | Gene name | Description | Length [aa][a] | MW [kDa][b] |
|---|---|---|---|---|
| BBIF_0278 | *aspC1* | Aminotransferase | 401 | 42.8 |
| BBIF_0311 | *yhdR* | Aspartate aminotransferase | 396 | 43.2 |
| BBIF_0342 | *bbif_0342* | Multiple substrate aminotransferase (MsaT) containing domain of GntR family (transcriptional regulator) | 509 | 55.6 |
| BBIF_0469 | *bbif_0469* | Aspartate/tyrosine/aromatic aminotransferase | 398 | 43.7 |
| BBIF_0550 | *hisC* | Histidinol-phosphate aminotransferase | 391 | 43.2 |
| BBIF_0701 | *bbif_0701* | Aminotransferase | 370 | 40 |
| BBIF_0741 | *ilvE1* | Branched-chain amino acid aminotransferase | 262 | 29.1 |
| BBIF_0863 | *ilvE2* | Branched-chain amino acid aminotransferase | 375 | 41.5 |
| BBIF_0870 | *bbif_0870* | Aminotransferase | 522 | 58.1 |
| BBIF_1100 | *argD* | Acetylornithine aminotransferase | 429 | 45.4 |
| BBIF_1175 | *bbif_1175* | Aspartate aminotransferase | 394 | 43.5 |
| BBIF_1428 | *serC* | Phosphoserine aminotransferase | 380 | 40.6 |
| BBIF_1519 | *bbif_1519* | N-succinyldiaminopimelate aminotransferase | 392 | 41.5 |
| BBIF_1610 | *aspC2* | Aspartate aminotransferase | 409 | 44.5 |

[a]*aa, amino acids.* [b]*kDa, kilo Dalton.*

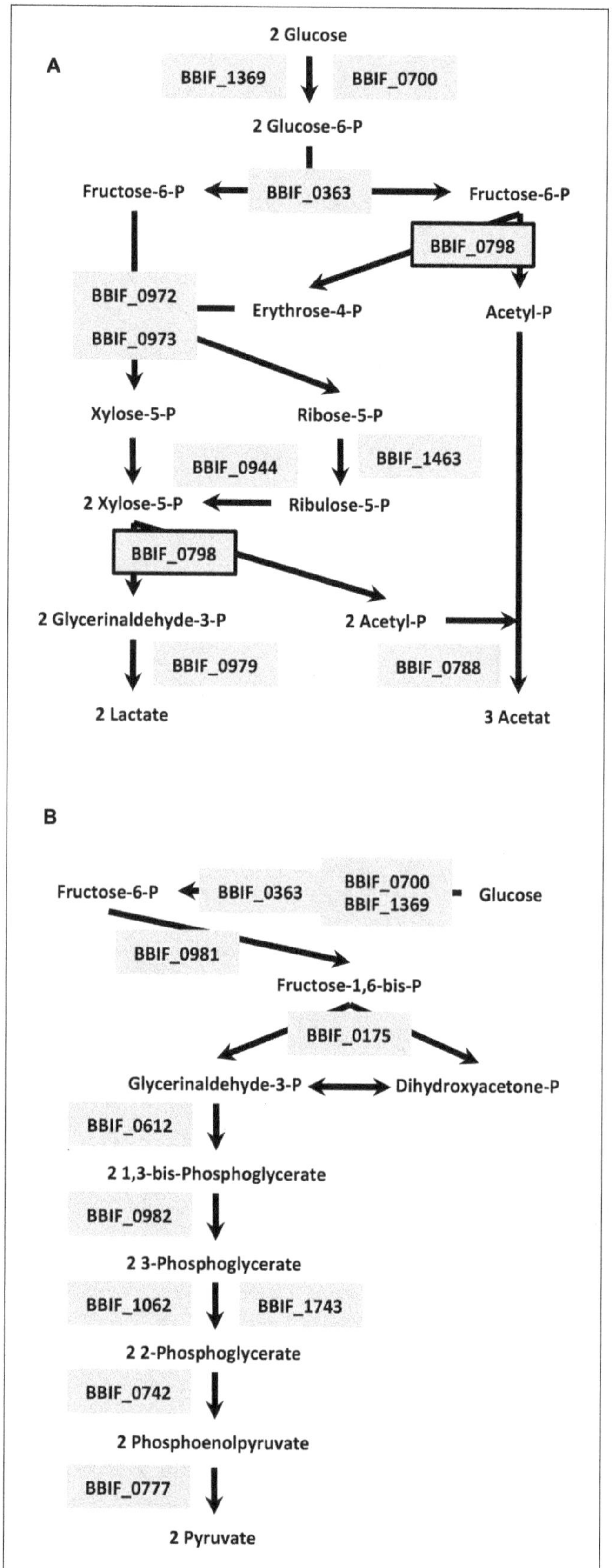


**FIGURE | Continued**

**FIGURE | Continued**

**Enzymes of the bifidus shunt (A) and Embden-Meyerhof pathway (B) detected in the proteome of *B. bifidum* S17: BBIF_1369: glucokinase; BBIF_0700: alternative glucokinase; BBIF_0798: xylulose-5-phosphate/ fructose-6-phosphate phosphoketolase; BBIF_0363: glucose-6-phosphate isomerase; BBIF_0973: transaldolase; BBIF_0972: transketolase, BBIF_1463: probable ribose-5-phosphate isomerase; BBIF_0944: ribulosephosphate 3-epimerase; BBIF_0788: acetate kinase; BBIF_0979: L-lactate dehydrogenase; BBIF_0175: fructose-bisphosphate aldolase; BBIF_0981: triosephosphate isomerase; BBIF_0612: glyceraldehyde 3-phosphate dehydrogenase; BBIF_0982: phosphoglycerate kinase; BBIF_1062: phosphoglycerate mutase; BBIF_1743: phosphoglycerate mutase family protein; BBIF_0742: enolase; BBIF_0777: pyruvate kinase.**

(**Figure** ). A number of proteins of the *B. bifidum* S17 proteome with extracellular or cell wall localization have been associated with bacterial adhesion to host structures, colonization and/or immunomodulation. For example, five of the six pilin proteins for sortase-dependent pili including major pilins BBIF_0301 and BBIF_1761 and the minor pilins BBIF_0302; BBIF_1648, and BBIF_1761 were detected in the proteome. Further proteins identified in the proteome of *B. bifidum* S17 that may have a role in host colonization are the potential adhesin BBIF_0636 (BopA), the s subtilisin family peptidase BBIF_1681, and BBIF_1317 and BBIF_1734, two glycoside hydrolases of the fucosidase and sialidase family.

**TABLE | Proteins of the *B. bifidum* S17 proteome with predicted cell wall or extracellular localization.**

| Locus_tag | Description | SP[a] | TMH[b] | Lipid anchor[c] | CW anchor[c] | Final prediction[d] |
|---|---|---|---|---|---|---|
| BBIF_0022 | Alpha-L-arabinofuranosidase | 1–36 | 13–35, 1138–1160 | – | 1133–1138 (LSHTG) | CW |
| BBIF_0285 | Conserved hypothetical protein containing multiple sugar recognition domains | 1–27 | 5–27, 1904–1926 | – | 1899–1903 (ISKTG) | CW |
| BBIF_0301 | Conserved hypothetical protein containing von Willebrand factor type A domain | 1–26 | 5–27, 1128–1150 | – | 1118–1123 (LPMTG) | CW |
| BBIF_0302 | Conserved hypothetical protein with Cna B-type domain | 1–29 | 7–29, 501–523 | - | 496–501 (LPKTG) | CW |
| BBIF_0507 | Beta-galactosidase BbgIII | 1–32 | – | – | 1903–1907 (LSKTG) | CW |
| BBIF_1317 | Alpha-L-fucosidase | 1–37 | 13–35, 1469–1491 | – | 1466–1470 (IAKTG) | CW |
| BBIF_1382 | Conserved hypothetical protein containing bacterial Ig-like domain (group 2) | 1–34 | 13–35, 1087–1106 | – | 1079–1083 (LSATG) | CW |
| BBIF_1461 | Beta-N-acetylglucosaminidase | 1–29 | 1935–1954 | – | 1926–1930 (ISKTG) | CW |
| BBIF_1576 | Beta-N-acetylglucosaminidase | 1–34 | 12–34, 1117–1139 | – | 1112–1116 (LSNTG) | CW |
| BBIF_1648 | Conserved hypothetical protein containing CnaB domain and LPXTG-anchor | 1–29 | 523–545 | – | 518–523 (LPLTG) | CW |
| BBIF_1681 | Subtilisin family peptidase (lactocepin) | 1–28 | 9–31,1325–1347 | – | 1320–1324 (VAKTG) | CW |
| BBIF_1734 | Sialidase | 1–35 | 13–35, 808–830 | – | 803–807 (LSKTG) | CW |
| BBIF_1761 | Cell surface protein with gram positive anchor and Cna protein B-type domains | 1–31 | 7–29, 505–527 | – | 500–505 (LPGTG) | CW |
| BBIF_1762 | Cell surface protein with LPXTG anchor | – | 2520–2542 | | 2514–2519 (LPDTG) | CW |
| BBIF_0048 | 1,4-beta-N-acetylmuramidase | 1–30 | – | – | – | E |
| BBIF_0483 | Conserved protein with the pectin lyase fold domain | 1–30 | – | – | – | E |
| BBIF_0522 | Conserved hypothetical protein with CHAP domain | 1–36 | 9–31 | – | – | E |
| BBIF_0158 | Trypsin-like serine protease | – | 203–225 | – | – | M |
| BBIF_0246 | Peptidylprolylisomerase, FKBP-type | 1–36 | 13–32 | 24–30 (VTLAACG) | – | M |
| BBIF_0313 | Hypothetical protein BBIF_0313 | – | 236–258 | – | – | M |
| BBIF_0592 | Peptide/nickel transport system, substrate-binding protein | 1–28 | 7–26 | 18–24 (ASLTACG) | – | M |
| BBIF_0636 | Peptide/nickel transport system, extracellular solute-binding protein (BopA) | 1–34 | – | 21–27 (LALGACG) | – | M |
| BBIF_1309 | Peptide/nickel transport system, substrate-binding protein | – | 55–77 | – | – | M |
| BBIF_1426 | Conserved hypothetical protein with NlpC/P60 domain | 1–25 | 16–38 | – | – | M |
| BBIF_1605 | ABC transporter solute-binding protein | – | 13–35 | – | – | M |
| BBIF_0312 | Conserved hypothetical protein | – | – | – | – | CP |
| BBIF_0337 | Hsp20-family heat shock chaperone | – | – | – | – | CP |
| BBIF_1026 | DNA polymerase III, delta subunit | – | – | – | – | CP |

[a]*SP, signal peptide predicted using SignalP v4.1.*
[b]*TMH, transmembrane helices, predicted usingTMHMM Server v2.0.*
[c]*Lipid anchor and cell wall (CW) anchor motifs obtained from the LocateP database (http://www.cmbi.ru.nl/locatep-db/cgi-bin/locatepdb.py).*
[d]*final prediction for cellular localization cell wall (Cw), extracellular (E), membrane (M), orcytoplasma (Cp).*

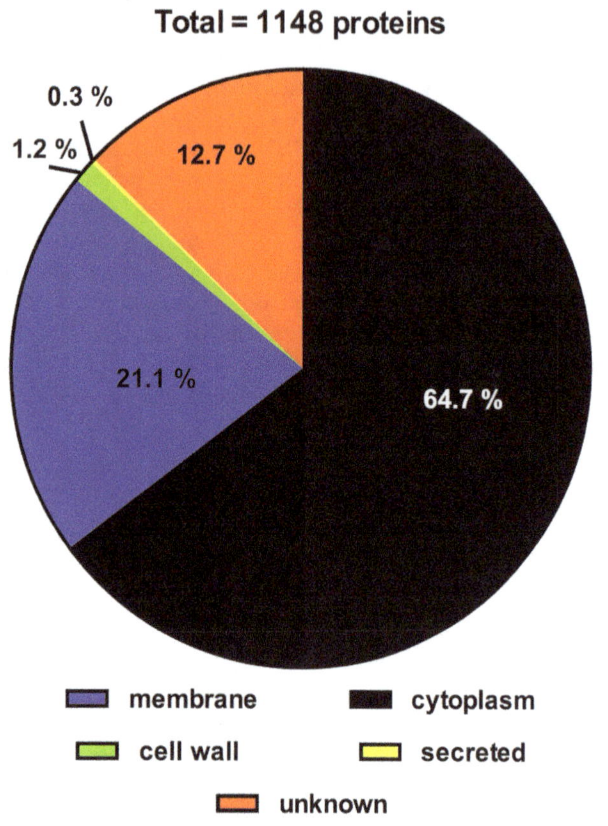


**FIGURE | Distribution of proteins of the *B. bifidum* S17 proteome according to subcellular location as predicted by PSORTb version 3.0 following correction for wrongly assigned localization of cell wall and extracellular proteins according to Table 3.**

## DISCUSSION

*Bifidobacterium bifidum* strains belong to the infant-type bifidobacteria and some of these strains possess interesting properties related to host-health (Turroni et al., 2014a). In an attempt to compile the first proteomic reference profile for the species, *B. bifidum*, we determined the complete proteome of *B. bifidum* S17 grown in MRS, i.e., the standard medium for routine culture of bifidobacteria, by LC-MS/MS. In total 1,148 proteins were detected, i.e., a coverage 64.4% of all predicted proteins of *B. bifidum* S17. This is comparable to the coverage obtained by proteome analysis of two *L. rhamnosus* strains (Savijoki et al., 2011). However, it may still be an underestimation of the complete proteome of *B. bifidum* S17 expressed under the conditions tested. The experimental approach utilized the bacterial pellet of cultures grown in MRS. Thus, most proteins that are secreted into the cell culture supernatant are probably not captured. In line with the proteomes of other bacteria (Yuan et al., 2006; Liu et al., 2012), proteins with a function in DNA replication, recombination and repair, RNA processing, transcription, and translation to were highly prevalent in the proteome of *B. bifidum* S17. One of these proteins is elongation factors Tu (EF-Tu; BBIF_1251), which was previously shown to posses moonlighting function as an adhesin that mediates binding of *B. longum* NCC2705 and *Lactobacillus* sp. mucus, IECs, and/or ECM components of the host (Granato et al., 2004; Ramiah et al., 2008; Dhanani and Bagchi, 2013; Wei et al., 2014).

Detection of large numbers of enzymes of the bifidus shunt, glycolysis, and amino acid metabolism in the proteome of *B. bifidum* S17 confirms previous findings on the proteome of *B. longum* NCC2705 (Yuan et al., 2006). *B. bifidum* S17 was grown in MRS and *B. longum* NCC2705 in modified Garches medium for proteome analysis. Both media contain high levels of complex components such as yeast extract, peptone and/or beef extract and glucose as additional carbon source (De Man et al., 1960; Krzewinski et al., 1997). This may explain the high numbers of enzymes detected in the two proteomes involved in the uptake, degradation and fermentation of these substrates. Similar to EF-Tu, one of the proteins of the Embden-Meyerhof pathway is a cytoplasmic enzyme that has a moonlighting function. Enolase (BBIF_0742) was shown to mediate adhesion of *B. bifidum* and other bifidobacteria to IECs and components of the ECM (Candela et al., 2009; Wei et al., 2014). An interesting finding is the detection of 14 aminotransferases in the proteome (**Table** ). Transamination reactions of amino acid converting pathways have recently attracted attention because they are the first step for the synthesis of important aroma compounds (Christensen et al., 1999) which may affect flavor of probiotic preparations.

A number of proteins of the *B. bifidum* S17 proteome contain domains and signal sequences for extracellular or cell wall location. Some of these proteins have been associated with bacterial adhesion to host structures, colonization and/or immunomodulation. Five of the six pilin proteins for sortase-dependent pili were identified in the proteome including major pilins BBIF_0301 and BBIF_1761 and the minor pilins BBIF_0302; BBIF_1648, and BBIF_1761. This suggests that at least two of the three sortase-dependent pili encoded on the genome of *B. bifidum* S17 are expressed under laboratory conditions. This is in line with the observation that sortase-dependent pili gene clusters are expressed by *B. bifidum* PRL2010 both *in vitro* and in the mouse GIT (Turroni et al., 2013) and pilus-like structures are detectable by atomic force microscopy in the same strain *in vitro* (Foroni et al., 2011). Expression of pili may have an important function in interaction of *B. bifidum* strains with the host. Heterologous expression of one of these gene clusters in *L. lactis* led to increased adhesion to ECM proteins and cultured IECs (Turroni et al., 2013). Moreover, the *L. lactis* strain expressing these pili elicits altered cytokine profiles in the murine gastrointestinal mucosa compared to the control. Another protein with potential role in adhesion to host structures is BBIF_0636. This protein is an extracellular solute-binding protein of a peptide/nickel transport system, which was also termed bifidobacterial outer protein A (BopA). Previous data obtained by us and others suggests that BopA mediates binding to cultured human IECs (Guglielmetti et al., 2008; Gleinser et al., 2012). However, these findings have been challenged recently (Kainulainen et al., 2013).

Three further proteins of the *B. bifidum* S17 proteome with a potential role in interaction with the host are BBIF_1317, BBIF_1681, and BBIF_1734. BBIF_1681 is a peptidase of the subtilisin family. The subtilisin family peptidase lactocepin of *L. casei* was shown to degrade pro-inflammatory cytokines contributing to the immunomodulatory effect of this probiotic bacterium (von Schillde et al., 2012). BBIF_1317 and BBIF_1734 are a glycoside hydrolases of the fucosidase and sialidase family, respectively. Sialidases are known as important virulence factors of bacterial pathogens that mediate attachment and degradation of host-derived mucus and were also shown to be involved in host colonization by commensal bacteria (Lewis and Lewis, 2012). Genes for mucin degradation pathways including sialidases are conserved amongst *B. bifidum* strains and most *B. bifidum* strains are able to grow on mucin as sole carbon source (Turroni et al., 2010). Similarly, fucosidases are involved in degradation of host-derived glucans such as human milk oligosaccharides and mucus (Turroni et al., 2010). Utilization of host-derived glucans is considered as nutritional adaptation to the intestinal tract of the (human) host (Turroni et al., 2010; Bottacini et al., 2014; Grimm et al., 2014). Thus, detection of proteins for degradation of host derived glycans under *in vitro* is somewhat surprising since these substrates are no present in standard growth medium, i.e., MRS. However, LC-MS/MS is a highly sensitive method that allows detection of even small amounts of a given protein (Otto et al., 2014). Moreover, pilin proteins that are only required *in vivo* were also present in the proteome of *B. bifidum* S17 and expression of pili genes and proteins *in vitro* has been observed previously (Foroni et al., 2011; Westermann et al., 2012). This suggests that regulation of host-interacting proteins may not strictly regulated in bifidobacteria.

In summary a total of 1,148 proteins of the predicted proteome were detected including important metabolic pathways, proteins known or suspected to be involved in adhesion and colonization as well as a large number of (previously) hypothetical proteins. This represents the first complete proteome analysis of the species *B. bifidum* and confirms previous findings on the proteome level.

## AUTHOR CONTRIBUTIONS

JY, CR, and XW designed research; XW, DZ, HL, and WL performed research; XW, SW, BL, and XZ contributed new reagents or analytic tools; XW, JL, and CR analyzed data; XW, CR, and YS wrote the paper.

## FUNDING

This work was supported by a grant from the National Natural Science Foundation of China (31370093) to JY and (81400592) to XW, Mega-projects of Science and Technology Research of China (Grant 2011ZX10004-001), the German Academic Exchange Service/Federal Ministry of Education and Research to CR (Grant D/09/04778), and a grant from the National High Technology Research and Development Program of China (863 Program; grant no. SS2014AA022210).

## SUPPLEMENTARY MATERIAL

The Supplementary Material for this article can be found online at: http://journal.frontiersin.org/article/10.3389/fmicb.2016.00097

**TABLE S1 | Characteristics of the 1,148 identified proteins.**

## REFERENCES


Albrethsen, J., Knol, J. C., Piersma, S. R., Pham, T. V., de Wit, M., Mongera, S., et al. (2010). Subnuclear proteomics in colorectal cancer: identification of proteins enriched in the nuclear matrix fraction and regulation in adenoma to carcinoma progression. *Mol. Cell. Proteomics* 9, 988–1005. doi: 10.1074/mcp.M900546-MCP200

Altschul, S. F., Madden, T. L., Schäffer, A. A., Zhang, J., Zhang, Z., Miller, W., et al. (1997). Gapped BLAST and PSI-BLAST: a new generation of protein database search programs. *Nucleic Acids Res.* 25, 3389–3402. doi: 10.1093/nar/25.17.3389

Bottacini, F., Ventura, M., van Sinderen, D., and O'Connell Motherway, M. (2014). Diversity, ecology and intestinal function of bifidobacteria. *Microb. Cell Fact.* 13(Suppl. 1), S4. doi: 10.1186/1475-2859-13-S1-S4

Burtnick, M. N., Brett, P. J., and DeShazer, D. (2014). Proteomic analysis of the Burkholderia pseudomallei type II secretome reveals hydrolytic enzymes, novel proteins, and the deubiquitinase TssM. *Infect. Immun.* 82, 3214–3226. doi: 10.1128/IAI.01739-14

Candela, M., Biagi, E., Centanni, M., Turroni, S., Vici, M., Musiani, F., et al. (2009). Bifidobacterial enolase, a cell surface receptor for human plasminogen involved in the interaction with the host. *Microbiology* 155, 3294–3303. doi: 10.1099/mic.0.028795-0

Christensen, J. E., Dudley, E. G., Pederson, J. A., and Steele, J. L. (1999). Peptidases and amino acid catabolism in lactic acid bacteria. *Antonie Van Leeuwenhoek* 76, 217–246. doi: 10.1023/A:1002001919720

De Man, J. C., Rogosa, M., and Sharpe, M. E. (1960). A medium for the cultivation of lactobacilli. *J. Appl. Microbiol.* 23, 130–135. doi: 10.1111/j.1365-2672.1960.tb00188.x

Dhanani, A. S., and Bagchi, T. (2013). The expression of adhesin EF-Tu in response to mucin and its role in *Lactobacillus* adhesion and competitive inhibition of enteropathogens to mucin. *J. Appl. Microbiol.* 115, 546–554. doi: 10.1111/jam.12249

Fanning, S., Hall, L. J., Cronin, M., Zomer, A., Macsharry, J., Goulding, D., et al. (2012). Bifidobacterial surface-exopolysaccharide facilitates commensal-host interaction through immune modulation and pathogen protection. *Proc. Natl. Acad. Sci. U.S.A.* 109, 2108–2113. doi: 10.1073/pnas.1115621109

Foroni, E., Serafini, F., Amidani, D., Turroni, F., He, F., Bottacini, F., et al. (2011). Genetic analysis and morphological identification of pilus-like structures in members of the genus *Bifidobacterium*. *Microb. Cell Fact.* 10(Suppl. 1), S16. doi: 10.1186/1475-2859-10-S1-S16

Gareau, M. G., Sherman, P. M., and Walker, W. A. (2010). Probiotics and the gut microbiota in intestinal health and disease. *Nat. Rev. Gastroenterol. Hepatol.* 7, 503–514. doi: 10.1038/nrgastro.2010.117

Gleinser, M., Grimm, V., Zhurina, D., Yuan, J., and Riedel, C. U. (2012). Improved adhesive properties of recombinant bifidobacteria expressing the *Bifidobacterium bifidum*-specific lipoprotein BopA. *Microb. Cell Fact.* 11, 80. doi: 10.1186/1475-2859-11-80

Granato, D., Bergonzelli, G. E., Pridmore, R. D., Marvin, L., Rouvet, M., and Corthésy-Theulaz, I. E. (2004). Cell surface-associated elongation factor Tu mediates the attachment of *Lactobacillus johnsonii* NCC533 (La1) to human intestinal cells and mucins. *Infect. Immun.* 72, 2160–2169. doi: 10.1128/IAI.72.4.2160-2169.2004

Grimm, V., Radulovic, K., and Riedel, C. U. (2015). Colonization of C57BL/6 mice by a potential probiotic *Bifidobacterium bifidum* strain under germ-free and specific pathogen-free conditions and during experimental colitis. *PLoS ONE* 10:e0139935. doi: 10.1371/journal.pone.0139935

Grimm, V., Westermann, C., and Riedel, C. U. (2014). Bifidobacteria-host interactions-an update on colonisation factors. *Biomed Res. Int.* 2014, 960826. doi: 10.1155/2014/960826

Guglielmetti, S., Tamagnini, I., Mora, D., Minuzzo, M., Scarafoni, A., Arioli, S., et al. (2008). Implication of an outer surface lipoprotein in adhesion of *Bifidobacterium bifidum* to Caco-2 cells. *Appl. Environ. Microbiol.* 74, 4695–4702. doi: 10.1128/AEM.00124-08

Guillaume, E., Berger, B., Affolter, M., and Kussmann, M. (2009). Label-free quantitative proteomics of two *Bifidobacterium longum* strains. *J. Proteomics* 72, 771–784. doi: 10.1016/j.jprot.2009.03.004

Kainulainen, V., Reunanen, J., Hiippala, K., Guglielmetti, S., Vesterlund, S., Palva, A., et al. (2013). BopA has no major role in the adhesion of *Bifidobacterium bifidum* to intestinal epithelial cells, extracellular matrix proteins and mucus. *Appl. Environ. Microbiol.* 79, 6989–6997. doi: 10.1128/AEM.01993-13

Kim, J.-H., An, H. J., Garrido, D., German, J. B., Lebrilla, C. B., and Mills, D. A. (2013). Proteomic analysis of *Bifidobacterium longum* subsp. infantis reveals the metabolic insight on consumption of prebiotics and host glycans. *PLoS ONE* 8:e57535. doi: 10.1371/journal.pone.0057535

Krzewinski, F., Brassart, C., Gavini, F., and Bouquelet, S. (1997). Glucose and galactose transport in *Bifidobacterium bifidum* DSM 20082. *Curr. Microbiol.* 35, 175–179. doi: 10.1007/s002849900234

Leahy, S. C., Higgins, D. G., Fitzgerald, G. F., and van Sinderen, D. (2005). Getting better with bifidobacteria. *J. Appl. Microbiol.* 98, 1303–1315. doi: 10.1111/j.1365-2672.2005.02600.x

Lewis, A. L., and Lewis, W. G. (2012). Host sialoglycans and bacterial sialidases: a mucosal perspective. *Cell. Microbiol.* 14, 1174–1182. doi: 10.1111/j.1462-5822.2012.01807.x

Liu, D., Wang, S., Xu, B., Guo, Y., Zhao, J., Liu, W., et al. (2011). Proteomics analysis of *Bifidobacterium longum* NCC2705 growing on glucose, fructose, mannose, xylose, ribose, and galactose. *Proteomics* 11, 2628–2638. doi: 10.1002/pmic.201100035

Liu, S., Ren, F., Zhao, L., Jiang, L., Hao, Y., Jin, J., et al. (2015). Starch and starch hydrolysates are favorable carbon sources for Bifidobacteria in the human gut. *BMC Microbiol.* 15:54. doi: 10.1186/s12866-015-0362-3

Liu, Y.-C., Lin, I.-H., Chung, W.-J., Hu, W. S., Ng, W. V., Lu, C.-Y., et al. (2012). Proteomics characterization of cytoplasmic and lipid-associated membrane proteins of human pathogen *Mycoplasma fermentans* M64. *PLoS ONE* 7:e35304. doi: 10.1371/journal.pone.0035304

O'Connell Motherway, M., Zomer, A., Leahy, S. C., Reunanen, J., Bottacini, F., Claesson, M. J., et al. (2011). Functional genome analysis of *Bifidobacterium breve* UCC2003 reveals type IVb tight adherence (Tad) pili as an essential and conserved host-colonization factor. *Proc. Natl. Acad. Sci. U.S.A.* 108, 11217–11222. doi: 10.1073/pnas.1105380108

Otto, A., Becher, D., and Schmidt, F. (2014). Quantitative proteomics in the field of microbiology. *Proteomics* 14, 547–565. doi: 10.1002/pmic.201300403

Petersen, T. N., Brunak, S., von Heijne, G., and Nielsen, H. (2011). SignalP 4.0: discriminating signal peptides from transmembrane regions. *Nat. Methods* 8, 785–786. doi: 10.1038/nmeth.1701

Philippe, D., Heupel, E., Blum-Sperisen, S., and Riedel, C. U. (2011). Treatment with *Bifidobacterium bifidum* 17 partially protects mice from Th1-driven inflammation in a chemically induced model of colitis. *Int. J. Food Microbiol.* 149, 45–49. doi: 10.1016/j.ijfoodmicro.2010.12.020

Pokusaeva, K., Fitzgerald, G. F., and van Sinderen, D. (2011). Carbohydrate metabolism in Bifidobacteria. *Genes Nutr.* 6, 285–306. doi: 10.1007/s12263-010-0206-6

Preising, J., Philippe, D., Gleinser, M., Wei, H., Blum, S., Eikmanns, B. J., et al. (2010). Selection of bifidobacteria based on adhesion and anti-inflammatory capacity in vitro for amelioration of murine colitis. *Appl. Environ. Microbiol.* 76, 3048–3051. doi: 10.1128/AEM.03127-09

Ramiah, K., van Reenen, C. A., and Dicks, L. M. T. (2008). Surface-bound proteins of *Lactobacillus plantarum* 423 that contribute to adhesion of Caco-2 cells and their role in competitive exclusion and displacement of *Clostridium sporogenes* and *Enterococcus faecalis*. *Res. Microbiol.* 159, 470–475. doi: 10.1016/j.resmic.2008.06.002

Riedel, C. U., Foata, F., Goldstein, D. R., Blum, S., and Eikmanns, B. J. (2006a). Interaction of bifidobacteria with Caco-2 cells-adhesion and impact on expression profiles. *Int. J. Food Microbiol.* 110, 62–68. doi: 10.1016/j.ijfoodmicro.2006.01.040

Riedel, C. U., Foata, F., Philippe, D., Adolfsson, O., Eikmanns, B. J., and Blum, S. (2006b). Anti-inflammatory effects of bifidobacteria by inhibition of LPS-induced NF-kappaB activation. *World J. Gastroenterol.* 12, 3729–3735.

Sánchez, B., Champomier-Vergès, M.-C., Collado, M., del, C., Anglade, P., Baraige, F., et al. (2007). Low-pH adaptation and the acid tolerance response of *Bifidobacterium longum* biotype longum. *Appl. Environ. Microbiol.* 73, 6450–6459. doi: 10.1128/AEM.00886-07

Savijoki, K., Lietzeìn, N., Kankainen, M., Alatossava, T., Koskenniemi, K., Varmanen, P., et al. (2011). Comparative proteome cataloging of *Lactobacillus rhamnosus* strains GG and Lc705. *J. Proteome Res.* 10, 3460–3473. doi: 10.1021/pr2000896

Sonnhammer, E. L., von Heijne, G., and Krogh, A. (1998). A hidden Markov model for predicting transmembrane helices in protein sequences. *Proc. Int. Conf. Intell. Syst. Mol. Biol.* 6, 175–182.

Tatusov, R. L., Galperin, M. Y., Natale, D. A., and Koonin, E. V. (2000). The COG database: a tool for genome-scale analysis of protein functions and evolution. *Nucleic Acids Res.* 28, 33–36. doi: 10.1093/nar/28.1.33

Turroni, F., Bottacini, F., Foroni, E., Mulder, I., Kim, J.-H., Zomer, A., et al. (2010). Genome analysis of *Bifidobacterium bifidum* PRL2010 reveals metabolic pathways for host-derived glycan foraging. *Proc. Natl. Acad. Sci. U.S.A.* 107, 19514–19519. doi: 10.1073/pnas.1011100107

Turroni, F., Duranti, S., Bottacini, F., Guglielmetti, S., Van Sinderen, D., and Ventura, M. (2014a). *Bifidobacterium bifidum* as an example of a specialized human gut commensal. *Front. Microbiol.* 5:437. doi: 10.3389/fmicb.2014.00437

Turroni, F., Serafini, F., Mangifesta, M., Arioli, S., Mora, D., van Sinderen, D., et al. (2014b). Expression of sortase-dependent pili of *Bifidobacterium bifidum* PRL2010 in response to environmental gut conditions. *FEMS Microbiol. Lett.* 357, 23–33. doi: 10.1111/1574-6968.12509

Turroni, F., Serafini, F., Foroni, E., Duranti, S., O'Connell Motherway, M., Taverniti, V., et al. (2013). Role of sortase-dependent pili of *Bifidobacterium bifidum* PRL2010 in modulating bacterium-host interactions. *Proc. Natl. Acad. Sci. U.S.A.* 110, 11151–11156. doi: 10.1073/pnas.1303897110

Ventura, M., O'Flaherty, S., Claesson, M. J., Turroni, F., Klaenhammer, T. R., van Sinderen, D., et al. (2009). Genome-scale analyses of health-promoting bacteria: probiogenomics. *Nat. Rev. Microbiol.* 7, 61–71. doi: 10.1038/nrmicro2047

von Schillde, M.-A., Hörmannsperger, G., Weiher, M., Alpert, C.-A., Hahne, H., Bäuerl, C., et al. (2012). Lactocepin secreted by *Lactobacillus* exerts anti-inflammatory effects by selectively degrading proinflammatory chemokines. *Cell Host Microbe* 11, 387–396. doi: 10.1016/j.chom.2012.02.006

Wei, X., Yan, X., Chen, X., Yang, Z., Li, H., Zou, D., et al. (2014). Proteomic analysis of the interaction of *Bifidobacterium longum* NCC2705 with the intestine cells Caco-2 and identification of plasminogen receptors. *J. Proteomics* 108, 89–98. doi: 10.1016/j.jprot.2014.04.038

Westermann, C., Zhurina, D. S., Baur, A., Shang, W., Yuan, J., and Riedel, C. U. (2012). Exploring the genome sequence of *Bifidobacterium bifidum* S17 for potential players in host-microbe interactions. *Symbiosis* 58, 191–200. doi: 10.1007/s13199-012-0205-z

Wickramasekara, S., Neilson, J., Patel, N., Breci, L., Hilderbrand, A., Maier, R. M., et al. (2011). Proteomics analyses of the opportunistic pathogen *Burkholderia vietnamiensis* using protein fractionations and mass spectrometry. *J. Biomed. Biotechnol.* 2011, 701928. doi: 10.1155/2011/701928

Xiao, M., Xu, P., Zhao, J., Wang, Z., Zuo, F., Zhang, J., et al. (2011). Oxidative stress-related responses of *Bifidobacterium longum* subsp. longum BBMN68 at the proteomic level after exposure to oxygen. *Microbiology* 157, 1573–1588. doi: 10.1099/mic.0.044297-0

Yu, N. Y., Wagner, J. R., Laird, M. R., Melli, G., Rey, S., Lo, R., et al. (2010). PSORTb 3.0: improved protein subcellular localization prediction with refined localization subcategories and predictive capabilities for all prokaryotes. *Bioinformatics* 26, 1608–1615. doi: 10.1093/bioinformatics/btq249

Yuan, J., Wang, B., Sun, Z., Bo, X., Yuan, X., He, X., et al. (2008). Analysis of host-inducing proteome changes in *Bifidobacterium longum* NCC2705 grown in vivo. *J. Proteome Res.* 7, 375–385. doi: 10.1021/pr0704940

Yuan, J., Zhu, L., Liu, X., Li, T., Zhang, Y., Ying, T., et al. (2006). A proteome reference map and proteomic analysis of *Bifidobacterium longum* NCC2705. *Mol. Cell. Proteomics* 5, 1105–1118. doi: 10.1074/mcp.M500410-MCP200

Zhou, M., Boekhorst, J., Francke, C., and Siezen, R. J. (2008). LocateP: genome-scale subcellular-location predictor for bacterial proteins. *BMC Bioinformatics* 9:173. doi: 10.1186/1471-2105-9-173

Zhurina, D., Zomer, A., Gleinser, M., Brancaccio, V. F., Auchter, M., Waidmann, M. S., et al. (2011). Complete genome sequence of *Bifidobacterium bifidum* S17. *J. Bacteriol.* 193, 301–302. doi: 10.1128/JB.01180-10

**Conflict of Interest Statement:** The authors declare that the research was conducted in the absence of any commercial or financial relationships that could be construed as a potential conflict of interest.

# 14

# A Critical Evaluation of Bifidobacterial Adhesion to the Host Tissue

*Christina Westermann[1†], Marita Gleinser[1†], Sinéad C. Corr[2] and Christian U. Riedel[1*]*

[1] *Institute of Microbiology and Biotechnology, University of Ulm, Ulm, Germany,* [2] *Department of Microbiology, Moyne Institute of Preventative Medicine, School of Genetics and Microbiology, Trinity College Dublin, Dublin, Ireland*

Bifidobacteria are common inhabitants of the human gastrointestinal tract that, despite a long history of research, have not shown any pathogenic potential whatsoever. By contrast, some bifidobacteria are associated with a number of health-related benefits for the host. The reported beneficial effects of bifidobacteria include competitive exclusion of pathogens, alleviation of symptoms of irritable bowel syndrome and inflammatory bowel disease, and modulation of intestinal and systemic immune responses. Based on these effects, bifidobacteria are widely used as probiotics by pharmaceutical and dairy industries. In order to exert a beneficial effect bifidobacteria have to, at least transiently, colonize the host in a sufficient population size. Besides other criteria such as resistance to manufacturing processes and intestinal transit, potential probiotic bacteria are tested for adhesion to the host structures including intestinal epithelial cells, mucus, and extracellular matrix components. In the present review article, we summarize the current knowledge on bifidobacterial structures that mediate adhesion to host tissue and compare these to similar structures of pathogenic bacteria. This reveals that most of the adhesive structures and mechanisms involved in adhesion of bifidobacteria to host tissue are similar or even identical to those employed by pathogens to cause disease. It is thus reasonable to assume that these structures and mechanisms are equally important for commensal or probiotic bacteria and play a similar role in the beneficial effects exerted by bifidobacteria.

***Edited by:***
*Francesca Turroni,*
*University College Cork, Ireland*

***Reviewed by:***
*Marco Ventura,*
*University of Parma, Italy*
*Simone Rampelli,*
*University of Bologna, Italy*

***Correspondence:***
*Christian U. Riedel*
*christian.riedel@uni-ulm.de*

*† These authors have contributed equally to this work.*

**Keywords: *Bifidobacterium*, probiotic, adhesion, host interaction**

## INTRODUCTION

The mammalian GIT is home to an extremely complex and diverse microbial ecosystem consisting primarily of prokaryotes. This microbial community is collectively referred to as the gut microbiota and exerts a number of profound effects on host health (Marchesi et al., 2016). During the first period of life when newborns are exclusively breast-fed, members of the genus *Bifidobacterium* are one of the predominant bacterial groups of the microbiota in the lower GIT (Yatsunenko et al., 2012; Bäckhed et al., 2015; Walker et al., 2015). The major source for bifidobacteria is the intestinal microbiota of the mother to which the newborn is exposed during (vaginal) delivery (Grönlund et al., 2011; Matamoros et al., 2013). More recently, breast milk has also been shown to contain viable bifidobacteria (Fernández et al., 2013; Jost et al., 2015). Although their relative proportion

**Abbreviations:** ECM, extracellular matrix; GIT, gastrointestinal tract; IECs, intestinal epithelial cells.

decreases over time, bifidobacteria are still a subdominant group amongst intestinal bacteria of adult humans (Arumugam et al., 2011).

With the exception of *Bifidobacterium dentium*, which has been associated with dental caries (Ventura et al., 2009), bifidobacteria have to date not shown any pathogenic potential. By contrast, a number of health promoting effects have been attributed to the presence of bifidobacteria in the GIT including improvement of symptoms of irritable bowel syndrome, inflammatory bowel disease and infectious diarrhea, modulation of intestinal and systemic immune responses, and resistance against colonization by pathogens (Gareau et al., 2010; Buffie and Pamer, 2013). Of note, a very recent study links bifidobacteria in the gut microbiota to enhanced anti-tumor immune responses and support of checkpoint-inhibition cancer therapy using a monoclonal antibody (Sivan et al., 2015). Based on these findings bifidobacteria are widely used as probiotics, i.e., live microorganisms which when administered in adequate amounts confer a health benefit to the host (Holmes et al., 2012; Foligné et al., 2013).

Besides the health promoting effects, several criteria are applied during selection of a suitable probiotic candidate strain including stability during manufacturing processes, viability during gastrointestinal transit and functionality at the desired target site (Foligné et al., 2013). One of the classical selection criteria for potential probiotic bacteria is adhesion to mucus and/or IECs (Klaenhammer and Kullen, 1999; Tuomola et al., 2001; Papadimitriou et al., 2015).

It may be argued that adhesion is not important for probiotic functionality since probiotic bacteria do not have access to host tissue due to the thick mucus layer covering the (healthy) gut epithelium. However, a number of bifidobacteria were shown to adhere to mucus (He et al., 2001; Izquierdo et al., 2008) and utilize host-derived mucins as a substrate for growth (Tailford et al., 2015). Also, bifidobacteria are discussed as potential treatment options for conditions with an impaired mucus layer (Whelan and Quigley, 2013; Johansson, 2014) facilitating direct access of (bifido)bacteria to the epithelium. Moreover, various bacterial pathogens must overcome the mucosal barriers and gain access to the epithelial layer to cause disease. For example, pathogenic *Escherichia coli* strains and related organisms use pili, fimbriae, and/or intimin with its translocated intimin receptor for adhesion to epithelial cells (Niemann et al., 2004). Another example is the interaction of InlA of *Listeria monocytogenes* with E-cadherin on host epithelial cells which is crucially required for infection (Stavru et al., 2011). Only once adhesion of these pathogens to the epithelium has been achieved despite the presence of an intact mucus layer, progression to later stages of infection and disease are possible (Bhavsar et al., 2007).

On the other hand, a number of probiotic traits may be directly linked to adhesion to host structures. One of the proposed health benefits of bifidobacteria is resistance against colonization or infection by pathogens. This may involve a variety of adhesion-independent mechanisms such as competition for nutrients or production of antimicrobial compounds (Buffie and Pamer, 2013; Lawley and Walker, 2013). Nevertheless, adhesion to IECs, mucus and ECM components by commensal and probiotic bacteria may also directly block access of pathogens to these structures (Bernet et al., 1993; Collado et al., 2005; Candela et al., 2008; Serafini et al., 2013) either by competition for attachment sites or steric hinderance. Also, there are numerous reports of immunomodulatory effects of bifidobacteria *in vitro* and in animal models (Bermudez-Brito et al., 2012). All these effects crucially depend on interaction with (and thus adhesion to) epithelial cells, dendritic cells, monocytes, macrophages and or other immune cells. Finally, even if not directly implicated mechanistically, adhesion might contribute to beneficial effects by allowing initial colonization or prolonging persistence of (probiotic) bifidobacteria in the GIT.

## FACTORS FOR ADHESION OF BIFIDOBACTERIA TO HOST STRUCTURES

A number of factors and structures involved in adhesion to IECs, ECM components, and mucus have been identified in bifidobacteria (**Table** ). These studies have been performed almost exclusively in *in vitro* model systems.

Adhesion to mucus is mostly analyzed using microtiter plate assays with immobilized mucus with quantification of adherent bacteria after metabolic labeling using radioisotopes or fluorescent dyes (He et al., 2001; Izquierdo et al., 2008; González-Rodríguez et al., 2012; Kainulainen et al., 2013). Similar assays are performed to analyze adhesion to immobilized ECM proteins (Kainulainen et al., 2013) or detection of ECM proteins bound to bacterial cells or protein extracts by specific antibodies (Candela et al., 2007, 2009, 2010).

The methods and cell lines used to determine adhesion to IECs differ largely between studies and groups. The most widely used cell lines are Caco-2, HT-29, and T84 (Guglielmetti et al., 2009; Preising et al., 2010; Gleinser et al., 2012; González-Rodríguez et al., 2012; Kainulainen et al., 2013; Grimm et al., 2014). In studies that employ more than one cell line, absolute adhesion of different strains may vary between cell lines but relative differences between strains are usually conserved (Riedel et al., 2006; Preising et al., 2010; Gleinser et al., 2012). One observation is that, although there is considerable strain-to-strain variability, strains of the species *B. bifidum* generally tend to adhere better to IECs than strains of other species (Guglielmetti et al., 2008; Gleinser et al., 2012). Detection of adherent bacteria is performed by metabolic labeling using radionucleotides (Riedel et al., 2006; Kainulainen et al., 2013), enumeration of colony forming units of adherent bacteria (Gleinser et al., 2012; González-Rodríguez et al., 2012), microscopic imaging and calculation of adhesion indices, i.e., the ratio of adherent bacteria and cells (Guglielmetti et al., 2008, 2009, 2010), or expression of fluorescent proteins (Grimm et al., 2014). However, the method of quantification does not seem to impact on adhesion itself as comparable results are obtained using radioactive or fluorescent labeling and plate counting (Riedel et al., 2006; Gleinser et al., 2012; Grimm et al., 2014).

In the following sections, the current knowledge on bifidobacterial adhesion to host structures will be summarized

**TABLE Adhesive structures identified in bifidobacteria.**

| Structure/protein/property | Species | Evidence and role | Reference |
|---|---|---|---|
| Type IVb (Tad) pili | *B. bifidum, B. breve, B. longum* subsp. *longum, B. adolescentis* | Genes expressed *in vitro* and *in vivo*, for *B. breve* UCC2003: pili detected *ex vivo* by transmission electron microscopy, required for efficient colonization of mice | O'Connell Motherway et al., 2011; Westermann et al., 2012; Zhurina et al., 2013; Bottacini et al., 2014; Duranti et al., 2014, 2015 |
| Type IVa pili | *B. adolescentis* | Genes expressed *in vitro*, regulation by carbon source | Duranti et al., 2014 |
| Sortase-dependent pili | *B. adolescentis, B. animalis* subsp. *lactis, B. bifidum, B. breve, B. dentium, B. longum* subsp. *longum, B. longum* subsp. *infantis* | Genes expressed *in vitro* and *in vivo*, regulated by carbon source and GIT-related stress, enhanced *in vivo*, pili detected on different strains by atomic force microscopy, heterologous expression of *pil2* and *pil3* genes of *B. bifidum* PRL2010 in *L. lactis* enhance binding to ECM proteins, expression of *pil2* increases adhesion to IECs | Foroni et al., 2011; O'Connell Motherway et al., 2011; Westermann et al., 2012; Turroni et al., 2013, 2014; Bottacini et al., 2014; Duranti et al., 2014, 2015; Wei et al., 2016 |
| BopA | *B. bifidum* | Purified BopA inhibits and homologous and heterologous expression increases adhesion to IECs, role in adhesion recently challenged as a BopA antibody did not inhibit adhesion | Guglielmetti et al., 2008; Gleinser et al., 2012; Kainulainen et al., 2013 |
| Transaldolase | *B. bifidum, B. longum* subsp. *Longum* | Binds to mucus, protein present on the surface of *B. bifidum* strains, protein level in *B. longum* proteome increased *in vivo*, differential expression of different isoforms in the presence of IECs | Yuan et al., 2008; González-Rodríguez et al., 2012; Wei et al., 2014 |
| DnaK | *B. animalis* subsp. *Lactis* | Binds to plasminogen | Candela et al., 2010 |
| Enolase | *B. bifidum, B. animalis* subsp. *lactis, B. longum* subsp. *longum* | Binds to plasminogen, in *B. longum* subsp. *longum* increased expression *in vivo* and in the presence of IECs, plasminogen binding site identified | Candela et al., 2007, 2009; Wei et al., 2014 |
| Hydrophobicity | *Bifidobacterium* sp. | Surface hydrophobicity correlates positively with autoaggregation and adhesion to IECs | Pérez et al., 1998; Del Re et al., 2000; Pan et al., 2006 |

and the involved factors will be compared to adhesion factors of pathogens.

## Pili

A wide range of Gram-positive and -negative bacteria possess proteinaceous surface appendages termed fimbriae or pili (Proft and Baker, 2009). In general, pili are adhesive structures that are involved in biofilm formation, conjugation, motility, and adhesion to biotic and abiotic surfaces (Maier and Wong, 2015). These hair-like structures extend to some distance (up to 3 μm) from the bacterial cell surface (Proft and Baker, 2009). It is hypothesized that they are able to bridge the repulsive forces between microbial cells and biological substrates, which under physiological conditions are both negatively charged (Proft and Baker, 2009). Pili are well-known for their role as virulence factors of Gram-positive and -negative pathogens and are important for initial attachment to host tissues (Telford et al., 2006; Proft and Baker, 2009).

There is increasing evidence that bifidobacteria also encode and express pilus-like structures on their cell surface. The first report of pili in bifidobacteria was the presence of genes encoding type IVb tight adherence (Tad) pili in *B. breve* UCC2003 (O'Connell Motherway et al., 2011). Since then, genes for Tad, Type VIa, and/or sortase-dependent pili were found in basically all sequenced genomes of bifidobacteria (**Table** ). Interestingly, in most cases bifidobacteria posses more than one pilus-coding locus and *B. dentium* harbours as much as seven gene clusters for sortase-dependent pili (Foroni et al., 2011).

Transcriptional analysis revealed that at least some of the genes are expressed under *in vitro* conditions and are regulated in response to substrate, presence of other bacteria, growth phase, or stress conditions related to the GIT (Foroni et al., 2011; Westermann et al., 2012; Duranti et al., 2014; Turroni et al., 2014). Moreover, pilin proteins are present in the *in vitro* proteome of *B. bifidum* S17 (Wei et al., 2016) and pilus-like structures were observed on several bifidobacteria by electron and atomic force microscopy (Foroni et al., 2011; O'Connell Motherway et al., 2011; Duranti et al., 2014). Collectively, this suggests that bifidobacteria possess functional pili. There is also evidence that bifidobacterial pili have a role in colonization of the host and attachment to epithelial cells. In *B. breve* UCC2003, expression of the *tad* locus was up-regulated in the GIT of mice and is required for efficient colonization in the presence of a competing microbiota (O'Connell Motherway et al., 2011). In a *B. adolescentis* strain, expression of pilus gene clusterss and presence of pili were enhanced when bacteria were isolated from the murine GIT or grown on starch, cellobiose or maltodextrin, i.e., substrates abundantly present in the GIT (Duranti et al., 2014). Similarly, expression of two of the three sortase-dependent pili clusters of *B. bifidum* PRL2010 is enhanced in the murine GIT and in the presence of human IECs *in vitro* (Turroni et al., 2013). Although a direct role for colonization and adhesion by inactivation of the corresponding genes is missing (probably due to the lack of appropriate genetic tools for the species *B. bifidum*), heterologous expression of the two gene clusters in *Lactococcus lactis* led to presence of pilus-like structures. Moreover, the recombinant *L. lactis* strains displayed increased adhesion to cultured IECs (*pil2* cluster) and ECM proteins laminin, fibronectin, fibrinogen, and plasminogen (*pil2* and *pil3* clusters; Turroni et al., 2013). Adhesion to fibronectin seems to be mediated by sugar-binding domains of the pili since enzymatic deglycosylation of fibronectin markedly reduced adhesion of the recombinant *L. lactis* strains expressing the *pil2* and *pil3* gene clusters of *B. bifidum* PRL2010 (Turroni et al., 2013).

## Moonlighting Proteins

A rather obscure group of proteins involved in adhesion of bacteria to host tissues are so-called moonlighting proteins (Huberts and van der Klei, 2010). These proteins are multifunctional and usually have an enzymatic role in bacterial metabolism or other cellular processes but at the same time are involved in totally unrelated biological functions (Huberts and van der Klei, 2010). In more than 90 pathogenic bacteria, proteins with a moonlighting function in virulence have been identified (Henderson, 2014). Interestingly, a large number of moonlighting proteins are cytoplasmic enzymes of the central metabolism that lack secretion signals raising the question if these proteins are actively exported to mediate virulence related functions. The best characterized examples are adhesins of pathogenic bacteria that are involved in primary attachment to host tissue and are important for later stages of infection (Henderson and Martin, 2011).

Enzymes of glycolysis with a moonlighting function in adhesion of pathogens include aldolase (or transaldolase), enolase, and glyceraldehyde-3-phosphate dehydrogenase (Henderson and Martin, 2011). These proteins were detected in proteomes of different bifidobacteria (Yuan et al., 2006, 2008; Ruiz et al., 2009; Gilad et al., 2011; Liu et al., 2011; Wei et al., 2016; Zhu et al., 2016). Transaldolase, a cytoplasmatic key enzyme of the bifidus shunt, was found to be present on the surface of several *B. bifidum* strains (González-Rodríguez et al., 2012). Using an *in vitro* binding assay the transaldolase could be identified as a mucin-binding protein and the specificity of this interaction was confirmed by increased mucus binding of recombinant *L. lactis* strains expressing transaldolase (González-Rodríguez et al., 2012). Enolase of different *B. longum*, *B. bifidum*, *B. animalis* subsp. *lactis* and *B. breve* strains was shown to interact with plasminogen (Candela et al., 2007, 2009; Wei et al., 2014). Moreover, the plasminogen binding site in the *B. lactis* enolase was shown to be homologous to that of *Streptococcus pneumoniae* and specific amino acid residues crucial for plasminogen binding have been identified (Candela et al., 2009). Another moonlighting protein that serves as an adhesin for bifidobacteria is DnaK, which has a primary function as a chaperone (Henderson and Martin, 2011). For *B. animalis* subsp. *lactis* BI07, DnaK was shown to bind plasminogen (Candela et al., 2007, 2010). Further potential moonlighting proteins of *B. animalis* subsp. *lactis* BI07 with plasminogen-binding activity are glutamine synthetase, bilesalt hydrolase, and phosphoglycerate mutase (Candela et al., 2007).

For *B. longum* NCC2705, transaldolase was detected at higher levels incubated *in vivo* in a rabbit intestinal loop compared to *in vitro* growth (Yuan et al., 2008) and enolase and transaldolase were more abundant in the proteome following co-cultivation with IECs (Wei et al., 2014). Also, expression of DnaK and enolase is upregulated in several bifidobacteria in response to bile (Savijoki et al., 2005; Candela et al., 2010). This indicates that bifidobacteria might be able to sense the conditions of the intestinal environment and presence of IECs (or receptors on IECs) and respond by enhancing expression of adhesive molecules.

## Other Adhesion Factors

A rather general, and non-specific property of bacteria that has been associated occasionally with adhesion of pathogens to host tissue is surface hydrophobicity (Hirt et al., 2000; Kouidhi et al., 2010). Several studies have tested different strains and species of bifidobacteria for hydrophobicity, autoaggregation and adhesion to IECs (Pérez et al., 1998; Del Re et al., 2000; Pan et al., 2006). Overall, the results suggest that (i) strains with higher surface hydrophobicity show higher autoaggregation and adhesion to IECs and (ii) *B. bifidum* strains tend to be more hydrophobic than strains of other *Bifidobacterium* sp. This is in line with other studies showing that *B. bifidum* strains adhere better to IECs than strains of other species (Preising et al., 2010; Gleinser et al., 2012).

Other non-proteinaceous component of the bacterial envelope that have been associated with adhesion to host tissue of Gram positive pathogens are glycoconjugates including exopolysaccharides, lipoteichoic, and wall teichoic acids (Weidenmaier and Peschel, 2008; Tytgat and Lebeer, 2014; Tan et al., 2015). Despite the presence of genes (potentially) involved in biosynthesis of exopolysaccharides and teichoic acids in most of the sequenced genomes of bifidobacteria (Hidalgo-Cantabrana et al., 2014; Colagiorgi et al., 2015), a contribution to adhesion have not been demonstrated conclusively so far. However, one study links exopolysaccharide production of bifidobacteria with adhesion to mucus by showing that purified exopolysaccharides of two bifidobacteria reduced adhesion of intact bacterial cells of these strains (Ruas-Madiedo et al., 2006).

Besides the abovementioned pili and moonlighting proteins no specific adhesins such as intimin, internalins, lectins, fibronectin-binding proteins as described for a number of pathogens (Niemann et al., 2004; Kline et al., 2009) have been characterized for bifidobacteria. A bioinformatic screen of the genome of *B. bifidum* S17 yielded a number of proteins with domains such as fibronectin type III domain, concanavalin A-like lectin, and collagen triple helix repeat domains, suggesting that bifidobacteria might have similar adhesins (Westermann et al., 2012). A definite role of the corresponding proteins in adhesion to host structures has yet to be demonstrated.

One specific protein that has been suspected to play a role in adhesion of bifidobacteria to IECs is BopA, lipoprotein of the cell envelope specifically found in *B. bifidum* strains (Guglielmetti et al., 2008; Gleinser et al., 2012). However, BopA contains the characteristic domains of a solute-binding protein and is part of an operon that encodes a putative oligopeptide ABC-transporter (Gleinser et al., 2012). Moreover, a recent study has challenged the idea that BopA serves a function in adhesion by showing that blocking BopA using a specific antibody does not affect adhesion of *B. bifidum* MIMBb75 to IECs (Kainulainen et al., 2013). Thus, BopA might be another example for a moonlighting protein but whether it has a role in adhesion of *B. bifidum* strains to intestinal tissue in humans needs to be elucidated in further studies.

## CONCLUSION

A large number of *Bifidobacterium* sp. strains were shown to adhere to IECs, mucus, and/or ECM proteins. For some bifidobacteria, adhesive structures have been characterized and include pili and different moonlighting proteins. Lactobacilli, another group of potential probiotic, Gram-positive microorganisms use exactly the same structures to adhere to the same target sites on host tissues (Vélez et al., 2007; van Tassell and Miller, 2011). Pathogenic microorganisms employ similar or even identical structures to adhere to host structures. The genus *Bacteroides* contains highly abundant commensal species as well as opportunistic pathogens that even may cause cancer (Wexler, 2007; Sears et al., 2014). Both commensal and pathogenic strains were shown to adhere to IECs, ECM, or mucus (Brook and Myhal, 1991; Ferreira et al., 2002; Macfarlane et al., 2005; de O Ferreira et al., 2006; Huang et al., 2011; Ferreira Ede et al., 2013) and pili, specific ECM-binding proteins, EPS etc. (Brook and Myhal, 1991; de O Ferreira et al., 2006; Pumbwe et al., 2006; Ferreira Ede et al., 2013) are involved in the process. Collectively, this illustrates that both pathogenic and commensal, in some cases even beneficial, bacteria employ the same strategies to attach to host structures. There is no doubt that adhesion of pathogens to host tissue is required or helps to promote infection. Bifidobacteria are generally regarded as safe microorganisms, which despite intensive studies of the past decades have not shown any pathogenic potential whatsoever. Instead, there are a number of health-related benefits associated with bifidobacteria. Although definitive proof is missing in most cases, it is reasonable to assume that adhesion to host tissue by beneficial bacteria are also required for or support their health-promoting effects. Moreover, the impressive number of different adhesion factors encoded by individual strains of bifidobacteria suggests that adhesion to host tissue is important for bifidobacteria to colonize and strive in the highly competitive ecosystem of the GIT.

## AUTHOR CONTRIBUTIONS

All authors listed, have made substantial, direct and intellectual contribution to the work, and approved it for publication.

## FUNDING

This study was partially supported by the German Academic Exchange Service/Federal Ministry of Education and Research to CUR (Grant D/09/04778). CW was supported by a by Ph.D. fellowship of the "Landesgraduiertenförderung Baden-Württemberg." The funders had no role in the decision to publish or the content of the manuscript.

## REFERENCES

Arumugam, M., Raes, J., Pelletier, E., Le Paslier, D., Yamada, T., Mende, D. R., et al. (2011). Enterotypes of the human gut microbiome. *Nature* 473, 174–180. doi: 10.1038/nature09944

Bäckhed, F., Roswall, J., Peng, Y., Feng, Q., Jia, H., Kovatcheva-Datchary, P., et al. (2015). Dynamics and stabilization of the human gut microbiome during the first year of life. *Cell Host Microbe* 17, 690–703. doi: 10.1016/j.chom.2015.04.004

Bermudez-Brito, M., Plaza-Díaz, J., Muñoz-Quezada, S., Gómez-Llorente, C., and Gil, A. (2012). Probiotic mechanisms of action. *Ann. Nutr. Metab.* 61, 160–174. doi: 10.1159/000342079

Bernet, M. F., Brassart, D., Neeser, J. R., and Servin, A. L. (1993). Adhesion of human bifidobacterial strains to cultured human intestinal epithelial cells and inhibition of enteropathogen-cell interactions. *Appl. Environ. Microbiol.* 59, 4121–4128.

Bhavsar, A. P., Guttman, J. A., and Finlay, B. B. (2007). Manipulation of host-cell pathways by bacterial pathogens. *Nature* 449, 827–834. doi: 10.1038/nature06247

Bottacini, F., O Connell Motherway, M., Kuczynski, J., O Connell, K. J., Serafini, F., Duranti, S., et al. (2014). Comparative genomics of the *Bifidobacterium breve* taxon. *BMC Genomics* 15:170. doi: 10.1186/1471-2164-15-170

Brook, I., and Myhal, M. L. (1991). Adherence of *Bacteroides fragilis* group species. *Infect. Immun.* 59, 742–744.

Buffie, C. G., and Pamer, E. G. (2013). Microbiota-mediated colonization resistance against intestinal pathogens. *Nat. Rev. Immunol.* 13, 790–801. doi: 10.1038/nri3535

Candela, M., Bergmann, S., Vici, M., Vitali, B., Turroni, S., Eikmanns, B. J., et al. (2007). Binding of human plasminogen to *Bifidobacterium*. *J. Bacteriol.* 189, 5929–5936. doi: 10.1128/JB.00159-07

Candela, M., Biagi, E., Centanni, M., Turroni, S., Vici, M., Musiani, F., et al. (2009). Bifidobacterial enolase, a cell surface receptor for human plasminogen involved in the interaction with the host. *Microbiology* 155, 3294–3303. doi: 10.1099/mic.0.028795-0

Candela, M., Centanni, M., Fiori, J., Biagi, E., Turroni, S., Orrico, C., et al. (2010). DnaK from *Bifidobacterium animalis* subsp. lactis is a surface-exposed human plasminogen receptor upregulated in response to bile salts. *Microbiology* 156, 1609–1618. doi: 10.1099/mic.0.038307-0

Candela, M., Perna, F., Carnevali, P., Vitali, B., Ciati, R., Gionchetti, P., et al. (2008). Interaction of probiotic *Lactobacillus* and *Bifidobacterium* strains with human intestinal epithelial cells: adhesion properties, competition against enteropathogens and modulation of IL-8 production. *Int. J. Food Microbiol.* 125, 286–292. doi: 10.1016/j.ijfoodmicro.2008.04.012

Colagiorgi, A., Turroni, F., Mancabelli, L., Serafini, F., Secchi, A., van Sinderen, D., et al. (2015). Insights into teichoic acid biosynthesis by *Bifidobacterium bifidum* PRL2010. *FEMS Microbiol. Lett.* 362:fnv141. doi: 10.1093/femsle/fnv141

Collado, M. C., Gueimonde, M., Hernández, M., Sanz, Y., and Salminen, S. (2005). Adhesion of selected *Bifidobacterium* strains to human intestinal mucus and the role of adhesion in enteropathogen exclusion. *J. Food Prot.* 68, 2672–2678.

de O Ferreira, E., Araújo Lobo, L., Barreiros Petrópolis, D., dos S Avelar, K. E., Ferreira, M. C., e Silva Filho, F. C., et al. (2006). A *Bacteroides fragilis* surface glycoprotein mediates the interaction between the bacterium and the extracellular matrix component laminin-1. *Res. Microbiol.* 157, 960–966. doi: 10.1016/j.resmic.2006.09.005

Del Re, B., Sgorbati, B., Miglioli, M., and Palenzona, D. (2000). Adhesion, autoaggregation and hydrophobicity of 13 strains of *Bifidobacterium longum*. *Lett. Appl. Microbiol.* 31, 438–442. doi: 10.1046/j.1365-2672.2000.00845.x

Duranti, S., Milani, C., Lugli, G. A., Turroni, F., Mancabelli, L., Sanchez, B., et al. (2015). Insights from genomes of representatives of the human gut commensal *Bifidobacterium bifidum*. *Environ. Microbiol.* 17, 2515–2531. doi: 10.1111/1462-2920.12743

Duranti, S., Turroni, F., Lugli, G. A., Milani, C., Viappiani, A., Mangifesta, M., et al. (2014). Genomic characterization and transcriptional studies of the starch-utilizing *Bifidobacterium adolescentis* 22L. *Appl. Environ. Microbiol.* 80, 6080–6090. doi: 10.1128/AEM.01993-14

Fernández, L., Langa, S., Martín, V., Maldonado, A., Jiménez, E., Martín, R., et al. (2013). The human milk microbiota: origin and potential roles in health and disease. *Pharmacol. Res.* 69, 1–10. doi: 10.1016/j.phrs.2012.09.001

Ferreira, E. O., Falcão, L. S., Vallim, D. C., Santos, F. J., Andrade, J. R. C., Andrade, A. F. B., et al. (2002). *Bacteroides fragilis* adherence to Caco-2 cells. *Anaerobe* 8, 307–314. doi: 10.1016/S1075-9964(03)00008-8

Ferreira Ede, O., Teixeira, F. L., Cordeiro, F., Araujo Lobo, L., Rocha, E. R., Smith, J. C., et al. (2013). The Bfp60 surface adhesin is an extracellular matrix and plasminogen protein interacting in *Bacteroides fragilis*. *Int. J. Med. Microbiol.* 303, 492–497. doi: 10.1016/j.ijmm.2013.06.007

Foligné, B., Daniel, C., and Pot, B. (2013). Probiotics from research to market: the possibilities, risks and challenges. *Curr. Opin. Microbiol.* 16, 284–292. doi: 10.1016/j.mib.2013.06.008

Foroni, E., Serafini, F., Amidani, D., Turroni, F., He, F., Bottacini, F., et al. (2011). Genetic analysis and morphological identification of pilus-like structures in members of the genus *Bifidobacterium*. *Microb. Cell Fact.* 10(Suppl. 1), S16. doi: 10.1186/1475-2859-10-S1-S16

Gareau, M. G., Sherman, P. M., and Walker, W. A. (2010). Probiotics and the gut microbiota in intestinal health and disease. *Nat. Rev. Gastroenterol. Hepatol.* 7, 503–514. doi: 10.1038/nrgastro.2010.117

Gilad, O., Svensson, B., Viborg, A. H., Stuer-Lauridsen, B., and Jacobsen, S. (2011). The extracellular proteome of *Bifidobacterium animalis* subsp. lactis BB-12 reveals proteins with putative roles in probiotic effects. *Proteomics* 11, 2503–2514. doi: 10.1002/pmic.201000716

Gleinser, M., Grimm, V., Zhurina, D., Yuan, J., and Riedel, C. U. (2012). Improved adhesive properties of recombinant bifidobacteria expressing the *Bifidobacterium bifidum*-specific lipoprotein BopA. *Microb. Cell Fact.* 11, 80. doi: 10.1186/1475-2859-11-80

González-Rodríguez, I., Sánchez, B., Ruiz, L., Turroni, F., Ventura, M., Ruas-Madiedo, P., et al. (2012). Role of extracellular transaldolase from *Bifidobacterium bifidum* in mucin adhesion and aggregation. *Appl. Environ. Microbiol.* 78, 3992–3998. doi: 10.1128/AEM.08024-11

Grimm, V., Gleinser, M., Neu, C., Zhurina, D., and Riedel, C. U. (2014). Expression of fluorescent proteins in bifidobacteria for analysis of host-microbe interactions. *Appl. Environ. Microbiol.* 80, 2842–2850. doi: 10.1128/AEM.04261-13

Grönlund, M.-M., Grześkowiak, L., Isolauri, E., and Salminen, S. (2011). Influence of mother's intestinal microbiota on gut colonization in the infant. *Gut Microbes* 2, 227–233. doi: 10.4161/gmic.2.4.16799

Guglielmetti, S., Tamagnini, I., Minuzzo, M., Arioli, S., Parini, C., Comelli, E., et al. (2009). Study of the adhesion of *Bifidobacterium bifidum* MIMBb75 to human intestinal cell lines. *Curr. Microbiol.* 59, 167–172. doi: 10.1007/s00284-009-9415-x

Guglielmetti, S., Tamagnini, I., Mora, D., Minuzzo, M., Scarafoni, A., Arioli, S., et al. (2008). Implication of an outer surface lipoprotein in adhesion of *Bifidobacterium bifidum* to Caco-2 cells. *Appl. Environ. Microbiol.* 74, 4695–4702. doi: 10.1128/AEM.00124-08

Guglielmetti, S., Taverniti, V., Minuzzo, M., Arioli, S., Zanoni, I., Stuknyte, M., et al. (2010). A dairy bacterium displays in vitro probiotic properties for the pharyngeal mucosa by antagonizing group A streptococci and modulating the immune response. *Infect. Immun.* 78, 4734–4743. doi: 10.1128/IAI.00559-10

He, F., Ouwehan, A. C., Hashimoto, H., Isolauri, E., Benno, Y., and Salminen, S. (2001). Adhesion of *Bifidobacterium* spp. to human intestinal mucus. *Microbiol. Immunol.* 45, 259–262. doi: 10.1111/j.1348-0421.2001.tb02615.x

Henderson, B. (2014). An overview of protein moonlighting in bacterial infection. *Biochem. Soc. Trans.* 42, 1720–1727. doi: 10.1042/BST20140236

Henderson, B., and Martin, A. (2011). Bacterial virulence in the moonlight: multitasking bacterial moonlighting proteins are virulence determinants in infectious disease. *Infect. Immun.* 79, 3476–3491. doi: 10.1128/IAI.00179-11

Hidalgo-Cantabrana, C., Sánchez, B., Milani, C., Ventura, M., Margolles, A., and Ruas-Madiedo, P. (2014). Genomic overview and biological functions of exopolysaccharide biosynthesis in *Bifidobacterium* spp. *Appl. Environ. Microbiol.* 80, 9–18. doi: 10.1128/AEM.02977-13

Hirt, H., Erlandsen, S. L., and Dunny, G. M. (2000). Heterologous inducible expression of *Enterococcus faecalis* pCF10 aggregation substance asc10 in *Lactococcus lactis* and *Streptococcus gordonii* contributes to cell hydrophobicity and adhesion to fibrin. *J. Bacteriol.* 182, 2299–2306. doi: 10.1128/JB.182.8.2299-2306.2000

Holmes, E., Kinross, J., Gibson, G. R., Burcelin, R., Jia, W., Pettersson, S., et al. (2012). Therapeutic modulation of microbiota-host metabolic interactions. *Sci. Transl. Med.* 4:137rv6. doi: 10.1126/scitranslmed.3004244

Huang, J. Y., Lee, S. M., and Mazmanian, S. K. (2011). The human commensal *Bacteroides fragilis* binds intestinal mucin. *Anaerobe* 17, 137–141. doi: 10.1016/j.anaerobe.2011.05.017

Huberts, D. H. E. W., and van der Klei, I. J. (2010). Moonlighting proteins: an intriguing mode of multitasking. *Biochim. Biophys. Acta* 1803, 520–525. doi: 10.1016/j.bbamcr.2010.01.022

Izquierdo, E., Medina, M., Ennahar, S., Marchioni, E., and Sanz, Y. (2008). Resistance to simulated gastrointestinal conditions and adhesion to mucus as probiotic criteria for *Bifidobacterium longum* strains. *Curr. Microbiol.* 56, 613–618. doi: 10.1007/s00284-008-9135-7

Johansson, M. E. V. (2014). Mucus layers in inflammatory bowel disease. *Inflamm. Bowel Dis.* 20, 2124–2131. doi: 10.1097/MIB.0000000000000117

Jost, T., Lacroix, C., Braegger, C., and Chassard, C. (2015). Impact of human milk bacteria and oligosaccharides on neonatal gut microbiota establishment and gut health. *Nutr. Rev.* 73, 426–437. doi: 10.1093/nutrit/nuu016

Kainulainen, V., Reunanen, J., Hiippala, K., Guglielmetti, S., Vesterlund, S., Palva, A., et al. (2013). BopA has no major role in the adhesion of *Bifidobacterium bifidum* to intestinal epithelial cells, extracellular matrix proteins and mucus. *Appl. Environ. Microbiol.* 79, 6989–6997. doi: 10.1128/AEM.01993-13

Klaenhammer, T. R., and Kullen, M. J. (1999). Selection and design of probiotics. *Int. J. Food Microbiol.* 50, 45–57. doi: 10.1016/S0168-1605(99)00076-8

Kline, K. A., Fälker, S., Dahlberg, S., Normark, S., and Henriques-Normark, B. (2009). Bacterial adhesins in host-microbe interactions. *Cell Host Microbe* 5, 580–592. doi: 10.1016/j.chom.2009.05.011

Kouidhi, B., Zmantar, T., Hentati, H., and Bakhrouf, A. (2010). Cell surface hydrophobicity, biofilm formation, adhesives properties and molecular detection of adhesins genes in *Staphylococcus aureus* associated to dental caries. *Microb. Pathog.* 49, 14–22. doi: 10.1016/j.micpath.2010.03.007

Lawley, T. D., and Walker, A. W. (2013). Intestinal colonization resistance. *Immunology* 138, 1–11. doi: 10.1111/j.1365-2567.2012.03616.x

Liu, D., Wang, S., Xu, B., Guo, Y., Zhao, J., Liu, W., et al. (2011). Proteomics analysis of *Bifidobacterium longum* NCC2705 growing on glucose, fructose, mannose, xylose, ribose, and galactose. *Proteomics* 11, 2628–2638. doi: 10.1002/pmic.201100035

Macfarlane, S., Woodmansey, E. J., and Macfarlane, G. T. (2005). Colonization of mucin by human intestinal bacteria and establishment of biofilm communities in a two-stage continuous culture system. *Appl. Environ. Microbiol.* 71, 7483–7492. doi: 10.1128/AEM.71.11.7483-7492.2005

Maier, B., and Wong, G. C. L. (2015). How bacteria use type IV pili machinery on surfaces. *Trends Microbiol.* 23, 775–788. doi: 10.1016/j.tim.2015.09.002

Marchesi, J. R., Adams, D. H., Fava, F., Hermes, G. D. A., Hirschfield, G. M., Hold, G., et al. (2016). The gut microbiota and host health: a new clinical frontier. *Gut* 65, 330–339. doi: 10.1136/gutjnl-2015-309990

Matamoros, S., Gras-Leguen, C., Le Vacon, F., Potel, G., and de La Cochetiere, M.-F. (2013). Development of intestinal microbiota in infants and its impact on health. *Trends Microbiol.* 21, 167–173. doi: 10.1016/j.tim.2012.12.001

Niemann, H. H., Schubert, W. D., and Heinz, D. W. (2004). Adhesins and invasins of pathogenic bacteria: a structural view. *Microbes Infect.* 6, 101–112. doi: 10.1016/j.micinf.2003.11.001

O'Connell Motherway, M., Zomer, A., Leahy, S. C., Reunanen, J., Bottacini, F., Claesson, M. J., et al. (2011). Functional genome analysis of *Bifidobacterium breve* UCC2003 reveals type IVb tight adherence (Tad) pili as an essential and conserved host-colonization factor. *Proc. Natl. Acad. Sci. U.S.A.* 108, 11217–11222. doi: 10.1073/pnas.1105380108

Pan, W.-H., Li, P.-L., and Liu, Z. (2006). The correlation between surface hydrophobicity and adherence of *Bifidobacterium* strains from centenarians' faeces. *Anaerobe* 12, 148–152. doi: 10.1016/j.anaerobe.2006.03.001

Papadimitriou, K., Zoumpopoulou, G., Foligné, B., Alexandraki, V., Kazou, M., Pot, B., et al. (2015). Discovering probiotic microorganisms: in vitro, in vivo, genetic and omics approaches. *Front. Microbiol.* 6:58. doi: 10.3389/fmicb.2015.00058

Pérez, P. F., Minnaard, Y., Disalvo, E. A., and De Antoni, G. L. (1998). Surface properties of bifidobacterial strains of human origin. *Appl. Environ. Microbiol.* 64, 21–26.

Preising, J., Philippe, D., Gleinser, M., Wei, H., Blum, S., Eikmanns, B. J., et al. (2010). Selection of bifidobacteria based on adhesion and anti-inflammatory capacity in vitro for amelioration of murine colitis. *Appl. Environ. Microbiol.* 76, 3048–3051. doi: 10.1128/AEM.03127-09

Proft, T., and Baker, E. N. (2009). Pili in Gram-negative and Gram-positive bacteria - structure, assembly and their role in disease. *Cell. Mol. Life Sci.* 66, 613–635. doi: 10.1007/s00018-008-8477-4

Pumbwe, L., Skilbeck, C. A., and Wexler, H. M. (2006). The *Bacteroides fragilis* cell envelope: quarterback, linebacker, coach—or all three? *Anaerobe* 12, 211–220. doi: 10.1016/j.anaerobe.2006.09.004

Riedel, C. U., Foata, F., Goldstein, D. R., Blum, S., and Eikmanns, B. J. (2006). Interaction of bifidobacteria with Caco-2 cells-adhesion and impact on expression profiles. *Int. J. Food Microbiol.* 110, 62–68. doi: 10.1016/j.ijfoodmicro.2006.01.040

Ruas-Madiedo, P., Gueimonde, M., Margolles, A., de los Reyes-Gavilán, C. G., and Salminen, S. (2006). Exopolysaccharides produced by probiotic strains modify the adhesion of probiotics and enteropathogens to human intestinal mucus. *J. Food Prot.* 69, 2011–2015.

Ruiz, L., Couté, Y., Sánchez, B., de los Reyes-Gavilán, C. G., Sanchez, J.-C., and Margolles, A. (2009). The cell-envelope proteome of *Bifidobacterium longum* in an in vitro bile environment. *Microbiology* 155, 957–967. doi: 10.1099/mic.0.024273-0

Savijoki, K., Suokko, A., Palva, A., Valmu, L., Kalkkinen, N., and Varmanen, P. (2005). Effect of heat-shock and bile salts on protein synthesis of *Bifidobacterium longum* revealed by [35S]methionine labelling and two-dimensional gel electrophoresis. *FEMS Microbiol. Lett.* 248, 207–215. doi: 10.1016/j.femsle.2005.05.032

Sears, C. L., Geis, A. L., and Housseau, F. (2014). *Bacteroides fragilis* subverts mucosal biology: from symbiont to colon carcinogenesis. *J. Clin. Invest.* 124, 4166–4172. doi: 10.1172/JCI72334

Serafini, F., Strati, F., Ruas-Madiedo, P., Turroni, F., Foroni, E., Duranti, S., et al. (2013). Evaluation of adhesion properties and antibacterial activities of the infant gut commensal *Bifidobacterium bifidum* PRL2010. *Anaerobe* 21, 9–17. doi: 10.1016/j.anaerobe.2013.03.003

Sivan, A., Corrales, L., Hubert, N., Williams, J. B., Aquino-Michaels, K., Earley, Z. M., et al. (2015). Commensal *Bifidobacterium* promotes antitumor immunity and facilitates anti-PD-L1 efficacy. *Science* 350, 1084–1089. doi: 10.1126/science.aac4255

Stavru, F., Archambaud, C., and Cossart, P. (2011). Cell biology and immunology of *Listeria monocytogenes* infections: novel insights. *Immunol. Rev.* 240, 160–184. doi: 10.1111/j.1600-065X.2010.00993.x

Tailford, L. E., Crost, E. H., Kavanaugh, D., and Juge, N. (2015). Mucin glycan foraging in the human gut microbiome. *Front. Genet.* 6:81. doi: 10.3389/fgene.2015.00081

Tan, F. Y. Y., Tang, C. M., and Exley, R. M. (2015). Sugar coating: bacterial protein glycosylation and host–microbe interactions. *Trends Biochem. Sci.* 40, 342–350. doi: 10.1016/j.tibs.2015.03.016

Telford, J. L., Barocchi, M. A., Margarit, I., Rappuoli, R., and Grandi, G. (2006). Pili in gram-positive pathogens. *Nat. Rev. Microbiol.* 4, 509–519. doi: 10.1038/nrmicro1443

Tuomola, E., Crittenden, R., Playne, M., Isolauri, E., and Salminen, S. (2001). Quality assurance criteria for probiotic bacteria. *Am. J. Clin. Nutr.* 73, 393S–398S.

Turroni, F., Serafini, F., Foroni, E., Duranti, S., O'Connell Motherway, M., Taverniti, V., et al. (2013). Role of sortase-dependent pili of *Bifidobacterium bifidum* PRL2010 in modulating bacterium-host interactions. *Proc. Natl. Acad. Sci. U.S.A.* 110, 11151–11156. doi: 10.1073/pnas.1303897110

Turroni, F., Serafini, F., Mangifesta, M., Arioli, S., Mora, D., van Sinderen, D., et al. (2014). Expression of sortase-dependent pili of *Bifidobacterium bifidum* PRL2010 in response to environmental gut conditions. *FEMS Microbiol. Lett.* 357, 23–33. doi: 10.1111/1574-6968.12509

Tytgat, H. L. P., and Lebeer, S. (2014). The sweet tooth of bacteria: common themes in bacterial glycoconjugates. *Microbiol. Mol. Biol. Rev.* 78, 372–417. doi: 10.1128/MMBR.00007-14

van Tassell, M. L., and Miller, M. J. (2011). *Lactobacillus* adhesion to mucus. *Nutrients* 3, 613–636. doi: 10.3390/nu3050613

Vélez, M. P., De Keersmaecker, S. C. J., and Vanderleyden, J. (2007). Adherence factors of *Lactobacillus* in the human gastrointestinal tract. *FEMS Microbiol. Lett.* 276, 140–148. doi: 10.1111/j.1574-6968.2007.00908.x

Ventura, M., Turroni, F., Zomer, A., Foroni, E., Giubellini, V., Bottacini, F., et al. (2009). The *Bifidobacterium* dentium Bd1 genome sequence reflects its genetic adaptation to the human oral cavity. *PLoS Genet.* 5:e1000785. doi: 10.1371/journal.pgen.1000785

Walker, A. W., Martin, J. C., Scott, P., Parkhill, J., Flint, H. J., and Scott, K. P. (2015). 16S rRNA gene-based profiling of the human infant gut microbiota is strongly influenced by sample processing and PCR primer choice. *Microbiome* 3, 26. doi: 10.1186/s40168-015-0087-4

Wei, X., Wang, S., Zhao, X., Wang, X., Li, H., Lin, W., et al. (2016). Proteomic profiling of *Bifidobacterium bifidum* S17 cultivated under in vitro conditions. *Front. Microbiol.* 7:97. doi: 10.3389/fmicb.2016.00097

Wei, X., Yan, X., Chen, X., Yang, Z., Li, H., Zou, D., et al. (2014). Proteomic analysis of the interaction of *Bifidobacterium longum* NCC2705 with the intestine cells Caco-2 and identification of plasminogen receptors. *J. Proteomics* 108, 89–98. doi: 10.1016/j.jprot.2014.04.038

Weidenmaier, C., and Peschel, A. (2008). Teichoic acids and related cell-wall glycopolymers in Gram-positive physiology and host interactions. *Nat. Rev. Microbiol.* 6, 276–287. doi: 10.1038/nrmicro1861

Westermann, C., Zhurina, D. S., Baur, A., Shang, W., Yuan, J., and Riedel, C. U. (2012). Exploring the genome sequence of *Bifidobacterium bifidum* S17 for potential players in host-microbe interactions. *Symbiosis* 58, 191–200. doi: 10.1007/s13199-012-0205-z

Wexler, H. M. (2007). *Bacteroides*: the good, the bad, and the nitty-gritty. *Clin. Microbiol. Rev.* 20, 593–621. doi: 10.1128/CMR.00008-07

Whelan, K., and Quigley, E. M. M. (2013). Probiotics in the management of irritable bowel syndrome and inflammatory bowel disease. *Curr. Opin. Gastroenterol.* 29, 184–189. doi: 10.1097/MOG.0b013e32835d7bba

Yatsunenko, T., Rey, F. E., Manary, M. J., Trehan, I., Dominguez-Bello, M. G., Contreras, M., et al. (2012). Human gut microbiome viewed across age and geography. *Nature* 486, 222–227. doi: 10.1038/nature11053

Yuan, J., Wang, B., Sun, Z., Bo, X., Yuan, X., He, X., et al. (2008). Analysis of host-inducing proteome changes in *bifidobacterium longum* NCC2705 grown in Vivo. *J. Proteome Res.* 7, 375–385. doi: 10.1021/pr0704940

Yuan, J., Zhu, L., Liu, X., Li, T., Zhang, Y., Ying, T., et al. (2006). A proteome reference map and proteomic analysis of *Bifidobacterium longum* NCC2705. *Mol. Cell. Proteomics* 5, 1105–1118. doi: 10.1074/mcp.M500410-MCP200

Zhu, D., Sun, Y., Liu, F., Li, A., Yang, L., and Meng, X.-C. (2016). Identification of surface-associated proteins of *Bifidobacterium animalis* ssp. lactis KLDS 2.0603 by enzymatic shaving. *J. Dairy Sci.* 99, 5155–5172. doi: 10.3168/jds.2015-10581

Zhurina, D., Dudnik, A., Waidmann, M. S., Grimm, V., Westermann, C., Breitinger, K. J., et al. (2013). High-quality draft genome sequence of *Bifidobacterium longum* E18, isolated from a healthy adult. *Genome Announc.* 1, e1084-13. doi: 10.1128/genomeA.01084-13

**Conflict of Interest Statement:** The authors declare that the research was conducted in the absence of any commercial or financial relationships that could be construed as a potential conflict of interest.

# 15

# Anti-viral Effect of *Bifidobacterium adolescentis* against Noroviruses

*Dan Li[1]*, Adrien Breiman[2,3,4,5], Jacques le Pendu[2,3,4†] and Mieke Uyttendaele[1†]*

[1] Laboratory of Food Microbiology and Food Preservation, Faculty of Bioscience Engineering, Ghent University, Ghent, Belgium, [2] UMR 892, Institut National de la Santé et de la Recherche Médicale, Nantes, France, [3] UMR 6299, Centre National de la Recherche Scientifique, Nantes, France, [4] Faculty of Medicine, University of Nantes, Nantes, France, [5] Nantes University Hospital, Nantes, France

This study aims to investigate the effect of *Bifidobacterium adolescentis* against noroviruses (NoVs). Murine norovirus-1 (MNV-1) used as a surrogate was detected by plaque assay and RT-qPCR. Human NoV virus like particles (VLPs) were detected by cell-binding assay. It was shown that the presence of *B. adolescentis* could inhibit the multiplication of MNV-1 on RAW 264.7 cells within 48 h of co-incubation period at 37°C. This inhibition did not occur at the viral binding stage, as no difference was observed in MNV-1 genomic copies collected from washed RAW 264.7 cells without and with *B. adolescentis* after co-incubation for 1 h at room temperature. Meanwhile, the presence of *B. adolescentis* decreased the binding of human NoV GI.1 VLPs to both Caco-2 cells and HT-29 cells, while no reduction was induced for the binding of human NoV GII.4 VLPs to Caco-2 cells.

**Edited by:**
*David Berry, University of Vienna, Austria*

**Reviewed by:**
*Douglas Morrison, University of Glasgow, UK*
*Lorena Ruiz, Universidad Complutense de Madrid, Spain*

***Correspondence:**
*Dan Li*
*dan.li@ugent.be*

† These authors are co-senior authors.

**Keywords:** ***Norovirus*, *Bifidobacterium adolescentis*, probiotics, mechanisms**

## INTRODUCTION

Probiotics are defined as "living micro-organisms, which upon ingestion in certain numbers, exert health benefits beyond inherent basic nutrition." Most probiotic microorganisms belong to lactic acid bacteria (LAB), such as *Lactobacillus* sp., *Bifidobacterium* sp., and *Enterococcus* sp (Ljungh and Wadström, 2006). Probiotics may benefit the human and animal host directly, by preventing the infection and combating the causative agent of the intestinal disorder, or indirectly, by balancing the disrupted equilibrium of the enteric flora and augmenting the host's immune responses (Maragkoudakis et al., 2010; Sanders et al., 2013). Clinical evidence has reported that feeding of probiotics can prevent effectively for diarrhea and shedding of rotavirus (Saavedra et al., 1994; Van Niel et al., 2002; Sazawal et al., 2006; Grandy et al., 2010). Accordingly, with the use of animal models, multiple probiotic strains have shown anti-rotavirus effect (Muñoz et al., 2011; Kandasamy et al., 2014; Mao et al., 2016). In addition, *in vitro* studies have demonstrated that probiotics may have antiviral activity against rotavirus (Maragkoudakis et al., 2010; Muñoz et al., 2011), coxsackievirus (Kim et al., 2014), hepatitis C virus (El-Adawi et al., 2015), as well as noroviruses (NoVs; Aboubakr et al., 2014; Rubio-del-Campo et al., 2014).

Noroviruses, one genera of the *Caliciviridae* family, were reported as the cause of between 73% to greater than 95% of global epidemic nonbacterial gastroenteritis outbreaks and approximately half of all gastroenteritis outbreaks (Atmar and Estes, 2006). Although NoV infections are generally mild, it may require hospital care and can be associated with mortality in elderly, chronically ill or immune-compromised patients. Taking into consideration their widespreadness, NoVs are causing heavy disease burdens associated with large economic losses (Van Beek et al., 2013). Despite recent progress, several key challenges remain in assessing the efficacy of vaccines and antiviral drugs for human NoV infection, such as the lack of a robust cell culture system or animal model limits direct study of these viruses, and the extreme genetic heterogeneity among strains (Karst et al., 2014). Therefore, it is of high interest to investigate further if probiotics can be employed for NoV control and treatment.

In this study, due to the non-cultivability of human NoVs, murine norovirus-1 (MNV-1, a very commonly used human NoV surrogate, Kniel, 2014), and human NoV virus like particles (VLPs) were used to study the effect of *Bifidobacterium adolescentis* against NoVs.

## MATERIALS AND METHODS

### Bacteria

*Bifidobacterium adolescentis* (LMG10502, biological origin: adult intestines) was obtained from Belgian Coordinated Collection of Microorganisms (BCCM/LMG). *B. adolescentis* was cultured in tryptone soya broth (TSB, Oxoid, Thermo) at 37°C. The anaerobic atmosphere was generated with the use of ANAEROGEN$^{TM}$ COMPACT (Oxoid, Thermo).

Before each experiment, *B. adolescentis* were cultured for 48 h, normalized to an $OD_{570}$ of 0.4, and washed twice with phosphate buffered saline (PBS, pH 7.4).

### Cell Lines, Virus, and Virus-Like Particles (VLPs)

Cells of the murine macrophage cell line RAW 264.7 (ATCC TIB-71; kindly provided by Prof. H. W. Virgin, Washington University School of Medicine, St. Louis, MO, USA) were maintained in complete DMEM medium and grown at 37°C under a 5% $CO_2$ atmosphere. Complete DMEM consisted of Dulbecco's modified Eagle's medium (DMEM; Lonza, Walkersville, MD, USA) containing 10% low-endotoxin fetal bovine serum (HyClone, Logan, UT, USA), 100 U/ml penicillin, 100 μg/ml streptomycin (Lonza), 10 mM HEPES (Lonza), and 2 mM L-glutamine (Lonza).

RAW 264.7 cells were infected with MNV-1.CW1 and passaged seven times at a multiplicity of infection (MOI) of 0.05 (MNV-1:cells) for 2 days. After two freeze-thaw cycles, low speed centrifugation was used to remove cellular debris from the virus suspension, as described by Wobus et al. (2004). The lysate containing suspended MNV-1 was stored in aliquots at −75°C.

Cell line Caco-2 (ECACC 86010202) was cultured in Eagle's minimum essential medium with Earle's salts (EMEM; Lonza) supplemented with 10% low-endotoxin fetal bovine serum (HyClone), 100 U/ml penicillin, 100 μg/ml streptomycin (Lonza), and 2 mM L-glutamine (Lonza) and grown at 37°C under a 5% $CO_2$ atmosphere. HT-29 (ATCC HTB-38) cells were cultured in Dulbecco's modified Eagle's medium (DMEM; GIBCO®, Life Technologies.) supplemented with 10% low-endotoxin fetal bovine serum (GIBCO®, Life Technologies), 100 U/ml penicillin, 100 μg/ml streptomycin (GIBCO®, Life Technologies), and 2 mM L-glutamine (GIBCO®, Life Technologies) and grown at 37°C under a 5% $CO_2$ atmosphere.

Virus-like particles of human NoV GI.1 (Norwalk virus) and GII.4 (Dijon 1996) were generated using a previously described method (Jiang et al., 1992). Recombinant baculoviruses containing the VP1 protein from NoV GI.1 and GII.4 were generated, and VLPs were produced by infection of Hi5 insect cells. VLPs were released from the infected Hi5 cells by three rounds of freeze-thawing and then clarified by removal of cellular debris (6,000 × g for 30 min) and baculovirus (14,000 × g for 30 min). The VLPs were partially purified through a 30% (wt/vol) sucrose cushion in TNC buffer (50 mM Tris-HCl, pH 7.4, 150 mM NaCl, 10 mM $CaCl_2$) containing the protease inhibitor leupeptin at 150,000 × g for 2 h. The pelleted VLPs were resuspended in TNC and further purified by isopynic centrifugation in cesium chloride (150,000 × *g*; 18 h). The resultant VLP bands were collected by puncture, and the solution containing VLPs was dialyzed against PBS prior to quantification by bicinchoninic acid (BCA) protein assay (Thermo Scientific) and stored at −80°C.

### Cytotoxicity Test-Neutral Red Assay

The cytotoxicity of bacteria on RAW 264.7 cells was measured by a neutral red assay based on the description by Fotakis and Timbrell (2006) with a few modifications. The RAW 264.7 cells were seeded into 96-well plates at a density of $10^5$ viable cells per well. On the following day, the RAW 264.7 cells were washed with PBS to remove the antibiotics, the *B. adolescentis* re-suspended in DMEM (Lonza) were added onto the cells (50 μl per well) and incubated for 1 h at room temperature. Afterward the inoculum was aspirated and fresh complete DMEM without antibiotics was added (100 μl per well). Plates were incubated at 37°C and 5% $CO_2$.

After 2 days incubation, the medium of the cells was changed to neutral red dye (Sigma-Aldrich, St. Louis, MO, USA, 100 μg/ml) dissolved in DMEM (100 μl per well) and incubated for another 2 h at 37°C and 5% $CO_2$. Cells were then washed with PBS and the addition of elution medium (EtOH/AcCOOH, 50%/1%, 100 μl per well) followed by gentle shaking for 10 min so that complete dissolution was achieved. The optical density was read at 540 nm ($OD_{540}$).

### Viral Multiplication Inhibition Test on MNV-1 Detected by Plaque Assay

The RAW 264.7 cells were seeded into six-well plates at a density of 2 × $10^6$ viable cells per well. On the following day, the *B. adolescentis* re-suspended in DMEM (Lonza) was used to make ten-fold dilutions from MNV-1 lysate. The RAW 264.7 cells were washed with PBS to remove the antibiotics, the mixture of bacteria and MNV-1 was added onto the cells (500 μl per well, two wells per sample). Plates were incubated for 1 h at room temperature and manually rocked every 15 min before aspirating the inoculum and overlaying the cells with 1.5% SeaPlaque agarose (Cambrex, Rockland, ME, USA) in MEM (Lonza) supplemented with 10% low-endotoxin fetal bovine serum, 1% HEPES, 1% penicillin/streptomycin, and 2% glutamine (complete MEM) per well. Plates were incubated at 37°C and 5% $CO_2$ for 2 days. To visualize plaques, cells were stained with 1.5%

SeaKem agarose in complete MEM containing 1% neutral red (Sigma–Aldrich) per well for 6 h.

Plaque sizes were shown to be associated with the virulence in multiple studies (Takemoto, 1966; Lipton, 1980). In this study, from photos taken from each group with the same format and size, the diameters of plaques were measured and recorded by the use of ImageJ (National Institutes of Health, Bethesda, MD, USA).

## Viral Binding Inhibition Test on MNV-1 Detected by RT-qPCR

The RAW 264.7 cells were seeded into 96-well plates at a density of $10^5$ viable cells per well. On the following day, the *B. adolescentis* was re-suspended in MNV-1 dilutions with DMEM (50 μl per sample, with 5.8 ± 0.1 log MNV-1 genomic copies per sample). The RAW 264.7 cells were washed with PBS to remove the antibiotics, the mixture of bacteria and MNV-1 were added onto the cells (50 μl per well) and incubated for 1 h at room temperature. Afterward the inoculum was aspirated and the cells were washed by PBS for three times.

The RNAs of washed cells were extracted by using the RNeasy Mini kit (Qiagen, Hilden, Germany). For each sample the 100 μl PBS was firstly mixed with 350 μl lysis buffer, and the mixture was pipetted back onto the cells in the 96-well plates in order to lyse the cells. The following procedures were performed according to the RNA Cleanup protocol and the RNA were stored at -75°C.

The RT-qPCR of MNV-1 was performed by the use of RNA UltraSense™ One-Step Quantitative RT-PCR System (Life technologies). The primers and probe sequence of MNV-1 were previously shown by Baert et al. (2008). Twenty micro liter of reaction mixture [200 nM each primer, 200 nM probe, 50 nM ROX (Life technologies)] was added to 5 μl of RNA. The RT-qPCR assays were performed in an ABI 7300 system (Applied Biosystems). The amplification profile consisted of 50°C for 15 min, 95°C for 2 min and 45 cycles of 95°C for 15 s and 60°C for 30 s.

An absolute quantification of MNV-1 genomic copies was performed as described by Baert et al. (2008). To obtain representative positive control standards, the plasmid p20.3 containing a full-length cDNA clone of MNV-1.CW1 (Baert et al., 2008) was used for the quantifications. Ten-fold serial dilutions ranging from $10^6$ to 10 copies of plasmids per reaction were used to prepare the standard curves.

## Viral Binding Inhibition Test on Human NoV VLPs Detected by Fluorescence Measurement

The HT-29 cells were seeded into 96-well plates at a density of $10^5$ viable cells per well. On the following day, the *B. adolescentis* was re-suspended in NoV GI.1 VLP suspension [5 μg/ml, in PBS-0.1% bovine serum albumin (BSA)]. The mixture of bacteria and VLPs was added onto the cells (50 μl per well) and incubated for 1 h at room temperature. After washing with PBS for three times, the cells were fixed with 4% paraformaldehyde, and stained by anti-VLP rabbit polyclonal antibodies (lp130 for GI.1, diluted in PBS-0.1% BSA, 1-h incubation at 37°C) followed by Alexa Fluor® 488 Goat Anti-Rabbit IgG (H+L) antibody (Life Technologies, diluted in PBS-0.1% BSA, 1-h incubation at 37°C). Three times washing by PBS was always performed after each step. The fluorescence was read by a fluorimeter (Fluoroskan, Thermo Scientific; Ex/Em = 490/525 nm) in arbitrary unit (a.u.).

The binding test of human NoV GI.1 and GII.4 VLPs (anti-VLP rabbit polyclonal antibodies lp130 for GI.1 and lp132 for GII.4) on Caco-2 cell were similar with the procedures above except that the Caco-2 cells were incubated for 21 days post confluency to be used as differentiated cells.

The binding of NoV GI.1 VLPs to HT-29 cells was also observed with a fluorescence microscope. The HT-29 cells were seeded into eight-well Nunc® Lab-Tek® Chamber Slide System (Sigma–Aldrich) at a density of $10^5$ viable cells per well. On the following day, the binding and staining steps were the same as described above. After the last washing step (from the secondary antibody incubation), the upper structure was removed from the bottom glass slide. The stained cells were mounted on slides with Vectashield® (Vector laboratories, Burlingame, CA, USA). The sealed slides were observed under a Zeiss Axiovert 200 fluorescence microscope.

## Data Analysis

Each result was presented as the mean value of three independent replicates with the standard deviation. Statistical analyses were performed by Mann–Whitney *U* test with SPSS 22 for Windows (SPSS, Inc., Chicago, IL, USA). Significant differences were considered when $P$ was <0.05.

# RESULTS

## Effect of *B. adolescentis* on MNV-1 Infectivity

First of all, after 48 h incubation, the cell viability determined by neutral red assay showed no significant reduction after incubation of the RAW 264.7 cells with *B. adolescentis* ($OD_{540}$ 1.1 ± 0.1–1.1 ± 0.2, $P > 0.05$).

MNV-1 lysate was diluted to a concentration that can form countable plaques on the six-well plates, and was seeded onto the cells with or without *B. adolescentis*. The plaque assay showed that compared with the negative control (MNV-1 on RAW 264.7 cells without bacteria), the MNV-1 plaque forming units (PFU) from cell-culture wells in the presence of *B. adolescentis* were decreased significantly from 20 ± 3–7 ± 2 PFU/well (**Figure A**, $P < 0.05$).

It was also noticed that the sizes of the plaques from cells incubated with *B. adolescentis* were smaller than the negative control (examples shown in **Figure C**). The average sizes of the plaques measured by ImageJ showed that compared with the negative control (MNV-1 on RAW 264.7 cells without bacteria), the plaque diameters from cell-culture wells in the presence of *B. adolescentis* were decreased significantly from 9.2 ± 2.7–3.9 ± 0.6 pixel (**Figure B**, $P < 0.05$). These results indicate that the effect of *B. adolescentis* on MNV-1 may occur mainly in the viral replication phase instead of showing direct virucidal effect on the MNV-1 or preventing the viruses from binding to the cells.

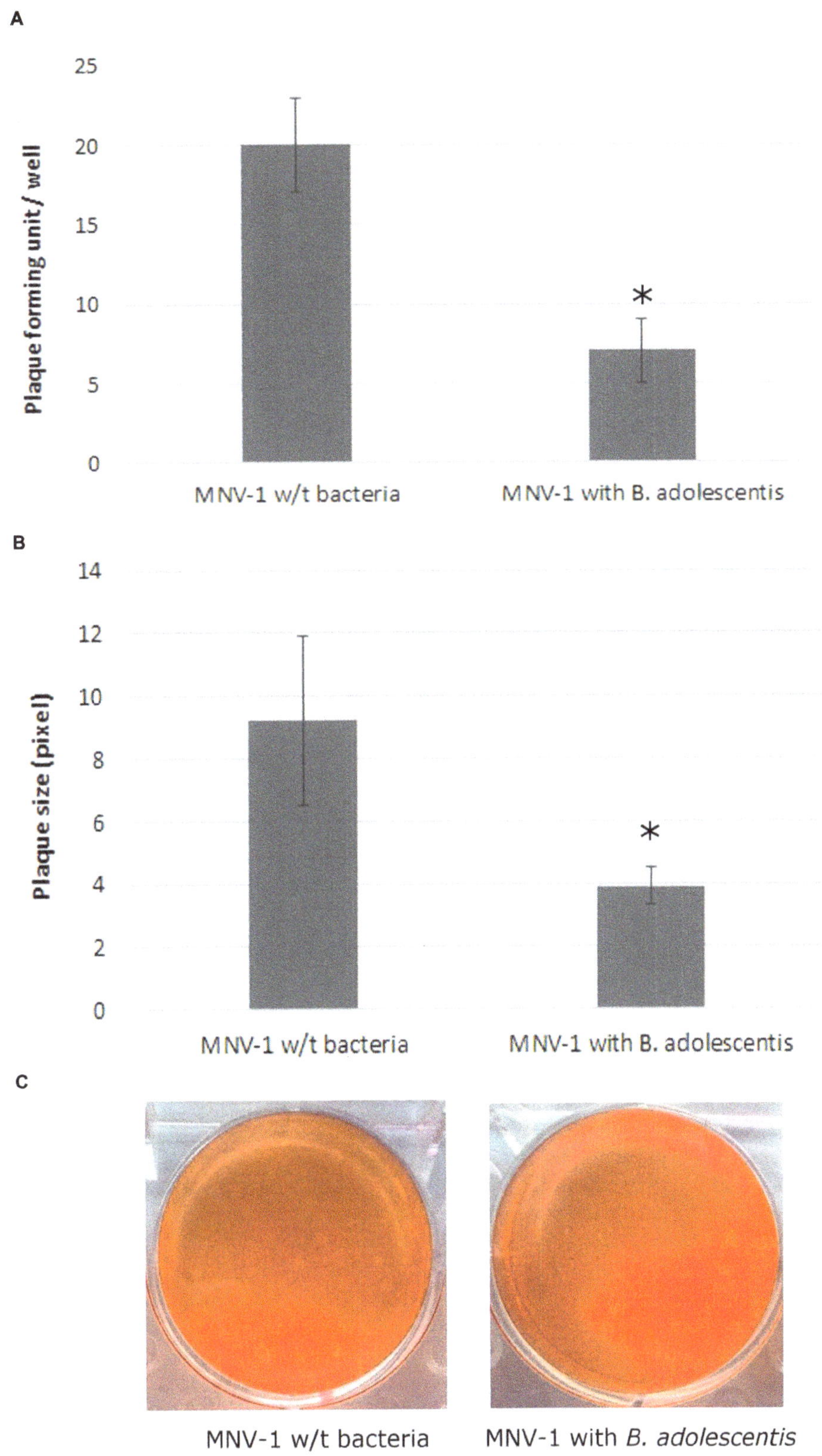


**FIGURE | MNV-1 multiplication inhibition by *B. adolescentis* determined by plaque assay. (A)** Plaque forming units (PFU) from cell-culture wells with and without *B. adolescentis*. Each data point is an average of three independent tests, and each error bar represents the data range. $^*P < 0.05$. **(B)** Plaque diameters from cell-culture wells with and without *B. adolescentis*. Each data point is an average of 15 measurements, and each error bar represents the data range. $^*P < 0.05$. **(C)** Plaque appearances from cell-culture wells with and without *B. adolescentis*.

Therefore lastly, a viral binding inhibition test was performed and indeed no significant difference of MNV-1 genomic copies collected from RAW 264.7 cells between the two groups with ($5.77 \pm 0.03$ log-genomic copies/ml) and without *B. adolescentis* ($5.74 \pm 0.01$ log-genomic copies/ml) was observed ($P > 0.05$).

## Effect of *B. adolescentis* on Human NoV VLPs Cell-Binding Ability

It has been previously reported that both VLPs of human NoV GI.1 and GII.4 could bind to Caco-2 cells (Tamura et al., 2004) and VLPs of human NoV GI.1 could bind to HT-29 cells (Rubio-del-Campo et al., 2014).

The presence of *B. adolescentis* decreased the binding of human NoV GI.1 VLPs to Caco-2 cells, measured by fluorescence intensity, significantly from $45.2 \pm 5.0$ (negative control without bacteria) to $33.0 \pm 4.6$ a.u. (incubation with *B. adolescentis*; **Figure A**, $P < 0.05$).

The fluorescence intensity of human NoV GI.1 VLPs on HT-29 cells also decreased, although not significantly, from $92.0 \pm 25.0$ to $52.0 \pm 5.0$ a.u. by the incubation of *B. adolescentis* (**Figure B**, $P = 0.05$). A visual example detected by fluorescence microscopy was shown in **Figure C**.

As for human NoV GII.4 VLPs, no significant reduction was observed for the binding of the VLPs to Caco-2 cells with the incubation of *B. adolescentis* ($19.0 \pm 1.7$ to $17.7 \pm 0.6$ a.u., $P > 0.05$).

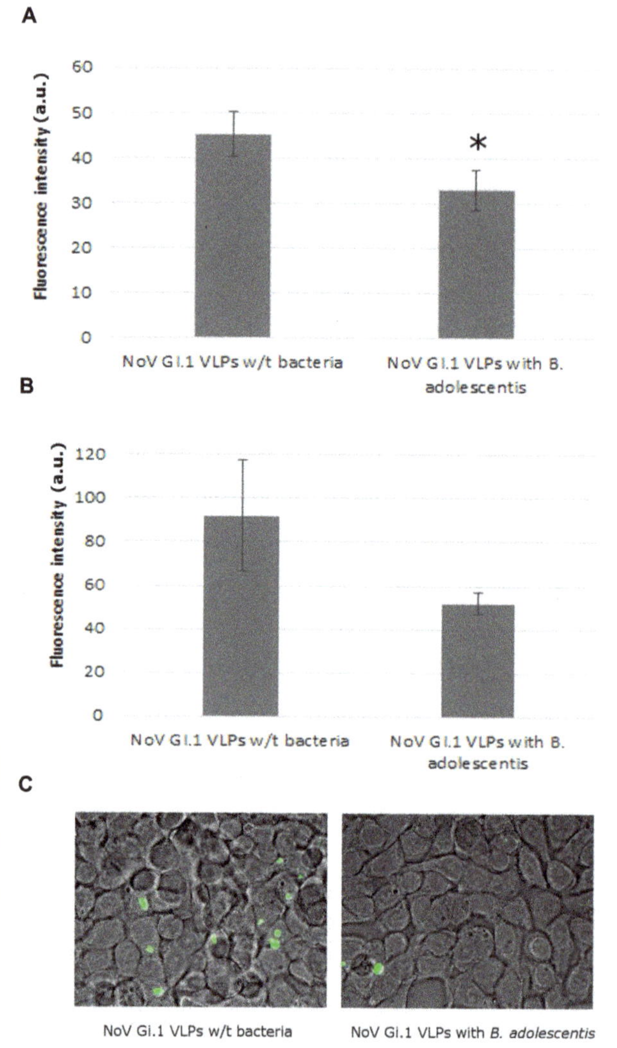


**FIGURE | Human NoV GI.1 VLPs binding inhibition by *B. adolescents*.** **(A)** Fluorescence measurement in arbitrary unit (a.u.) of human NoV GI.1 VLPs bound to Caco-2 cells. Each data point is an average of three independent tests, and each error bar represents the data range. *$P < 0.05$. **(B)** Fluorescence measurement in arbitrary unit (a.u.) of human NoV GI.1 VLPs bound to HT-29 cells. Each data point is an average of three independent tests, and each error bar represents the data range. **(C)** Fluorescence microscope observation of human NoV GI.1 VLPs bound to HT-29 cells.

## DISCUSSION

Currently, the research on the effect of probiotics on NoVs are still preliminary. On one hand, due to the non-cultivability of human NoVs, the *in vitro* studies were conducted with the use of surrogate viruses [e.g., feline calicivirus (FCV), Aboubakr et al., 2014; MNV-1 and Tulane virus (TV), Shearer et al., 2014] and artificially synthesized particles partially mimicking the viral structures (*P*-particles, Rubio-del-Campo et al., 2014), which may introduce gaps from the authentic NoV reactions. On the other hand, there are an insufficient number of in-depth studies on this topic before a general conclusion can be drawn. For instance, it was reported that the fermentation of *Donchimi* and oysters could effectively reduce the infectivity of FCV and MNV-1 (Lee et al., 2012; Seo et al., 2014). The population of LAB increased largely during the fermentation, however, it was neither clear on the role that the LAB played nor the associated mechanisms (Lee et al., 2012; Seo et al., 2014). For another instance, high infectivity reduction (6 -7-log reduction) was observed for FCV following co-incubation with *Lactococcus lactis* (both bacterial medium filtrate and cells suspension, Aboubakr et al., 2014). However, the cell-free supernatant of a commercial probiotic mixture of *Lactobacillus acidophilus, Lactobacillus rhamnosus, Bifidobacterium bifidum, Lactobacillus salivarius, and Streptococcus thermophiles* indicated no reduction for the infectivity of MNV and TV (Shearer et al., 2014). The reasons causing the inconsistence of the studies can be the differences between both of the studied viruses and bacteria strains, as well as the experimental settings such as time/sequence of addition, comparative ratio of viruses and bacteria, incubation conditions (time, atmosphere and temperature) and matrices, etc.

In this study, MNV-1 was employed as the surrogate as it was more persistent than FCV in the fermentation of both *Donchimi* and oysters (Lee et al., 2012; Seo et al., 2014). Human NoV VLPs from GI.1 and GII.4 were used as they could mimic the binding capacity of two representative NoV genotypes. *B. adolescentis* is a recognized probiotics (Lee et al., 1993; Kim et al., 2008) and has interested dairy manufacturers in producing "therapeutic fermented milk products" (Arunachalam, 1999; Puniya, 2015). Also it was reported recently to exhibit antiviral activity against Coxsackievirus (Kim et al., 2014). The bacterial cells were washed with PBS before being added to the viruses and cells in order to avoid any direct virucidal effect of the bacterial metabolites in the culture medium, such as organic acids, diacetyl, and bacteriocins. The bacteria and viruses (MNV-1 and human NoV VLPs) were co-incubated with the cells as this strategy was shown to be more effective than pre-treatment of cells, pre-treatment of viruses (Aboubakr et al., 2014; Rubio-del-Campo et al., 2014), or post-treatment of cells attached to viruses (Rubio-del-Campo et al., 2014).

It was demonstrated in this study that the presence of *B. adolescentis* could inhibit the multiplication of MNV-1 on RAW 264.7 cells. Meanwhile, it was shown that this inhibition effect did not occur on the binding stage of MNV-1 to RAW 264.7 cells. However, in contrast, the presence of *B. adolescentis* did decrease the binding of human NoV GI.1 VLPs to both Caco-2 cells and HT-29 cells. Although based on different models, these results indicate that the effects of *B. adolescentis* on different viruses might vary based on different mechanisms.

It was reported previously that NoV P-particles could bind to a series of probiotics with varied binding ability, which was postulated as one of the antiviral mechanisms of the probiotics (Rubio-del-Campo et al., 2014). However, based on a recent study of our group (Li et al., 2015), the *B. adolescentis* strain used in this study does not express histo-blood group antigen (HBGA) and could not bind to NoV VLPs of either GI.1 or GII.4. This result rules out the possibility of direct competition of *B. adolescentis* and NoV GI.1VLPs from binding to Caco-2 or HT-29 cells. Instead, as we and others have described a few bacterial lectins binding to HBGAs (Gilboa-Garber et al., 2011; Audfray et al., 2012) and Zhong et al. (2004) have reported that the adhesin of *B. adolescentis* 1027 could inhibit the adhesion of *Escherichia coli* to intestinal epithelial cell line Lovo, we assume that certain lectins secreted by *B. adolescentis* could bind to the HBGAs on the intestinal cell lines and compete with the binding of NoV GI.1VLPs.

As the binding of human NoV GI.1 VLPs to both Caco-2 cells and HT-29 cells was decreased in the presence of *B. adolescentis*, it is reasonable to assume that the viral multiplication of human NoV GI.1, if an *in vitro* model could be established, should also be decreased in the presence of *B. adolescentis*. On the other hand, although no reduction of the binding of human NoV GII.4 VLPs to Caco-2 cells was observed in the presence of *B. adolescentis*, the possibility cannot be ruled out that *B. adolescentis* may reduce the viral multiplication of human NoV GII.4 if an *in vitro* model could be established, or help combat NoVs *in vivo*. The differences observed between the two strains are not surprising as it is well known that the chemical structures of the receptor binding interfaces are different between NoV GI.1 and GII.4 (Tan and Jiang, 2011). However, the exact molecular mechanisms casing the influence of *B. adolescentis* on human NoV VLPs binding to intestinal cell lines still need further investigation.

## CONCLUSION

This study demonstrated the antiviral effect of *B. adolescentis* against NoVs. Due to the lack of models to study the infection of genuine human NoVs, the results were generated from studying surrogates detected by different methods (infection and binding of MNV-1 on murine macrophage cell line RAW 264.7and binding of human NoV VLPs on human intestinal epithelial cell lines Caco-2 and HT-29). Since the effects of *B. adolescentis* on MNV-1 and human NoV VLPs as well as their mechanisms were indicated to be different, this study shows the importance of establishing multiplication model for human NoVs in the future. In addition, more probiotic strains will be tested before a general conclusion can be drawn on the antiviral effect of probiotics on NoV infection.

## AUTHOR CONTRIBUTIONS

DL performed most of the experiments and composed the manuscript. AB supplied technical support for the experiments, especially for the fluorescence microscope as well as improvement of the manuscript. JP and MU gave strategic guidance for the work and suggestions to improve the manuscript.


## FUNDING

This study was supported by a postdoctoral grant (DL.) and a financial support for staying abroad (DL.) from the Research Foundation-Flanders (Fonds voor Wetenschappelijk Onderzoek-Vlaanderen) and by a grant from the Region des Pays de la Loire (ARMINA).

## ACKNOWLEDGMENT

The authors also acknowledge the help of imaging core facilities of the University of Nantes (Micropicell).


## REFERENCES


Aboubakr, H. A., El-Banna, A. A., Youssef, M. M., Al-Sohaimy, S. A., and Goyal, S. M. (2014). Antiviral effects of *Lactococcus lactis* on feline calicivirus, a human norovirus surrogate. *Food Environ. Virol.* 6, 282–289. doi: 10.1007/s12560-014-9164-2

Arunachalam, K. D. (1999). Role of bifidobacteria in nutrition, medicine and technology. *Nutrition Res.* 19, 1559–1597. doi: 10.1016/j.anaerobe.2015.05.012

Atmar, R. L., and Estes, M. K. (2006). The epidemiologic and clinical importance of norovirus infection. *Gastroenterol. Clin. North Am.* 35, 275–290. doi: 10.1016/j.gtc.2006.03.001

Audfray, A., Claudinon, J., Abounit, S., Ruvoën-Clouet, N., Larson, G., Smith, D. F., et al. (2012). Fucose-binding lectin from opportunistic pathogen *Burkholderia ambifaria* binds to both plant and human oligosaccharidic epitopes. *J. Biol. Chem.* 287, 4335–4347. doi: 10.1074/jbc.M111.314831

Baert, L., Wobus, C. E., Van Coillie, E., Thackray, L. B., Debevere, J., and Uyttendaele, M. (2008). Detection of murine norovirus 1 by using plaque assay, transfection assay, and real-time reverse transcription-PCR before and after heat treatment. *Appl. Environ. Microbiol.* 74, 543–546. doi: 10.1128/AEM.01039-07

El-Adawi, H., Nour, I., Fattouh, F., and El-Deeb, N. (2015). Investigation of the antiviral bioactivity of *Lactobacillus bulgaricus* 761N extracellular extract against hepatitis C virus (HCV). *Int. J. Pharm.* 11, 19–26. doi: 10.3923/ijp.2015.19.26

Fotakis, G., and Timbrell, J. A. (2006). In vitro cytotoxicity assays: comparison of LDH, neutral red, MTT and protein assay in hepatoma cell lines following exposure to cadmium chloride. *Toxicol. Lett.* 160, 171–177.

Gilboa-Garber, N., Zinger-Yosovich, K. D., Sudakevitz, D., Lerrer, B., Imberty, A., Wimmerova, M., et al. (2011). "The five bacterial lectins (PA-IL, PA-IIL, RSL, RS-IIL, and CV-IIL): interactions with diverse animal cells and glycoproteins," in *The Molecular Immunology of Complex Carbohydrates-3*, ed. A. M. Wu (New York, NY: Springer), 155–211.

Grandy, G., Medina, M., Soria, R., Terán, C. G., and Araya, M. (2010). Probiotics in the treatment of acute rotavirus diarrhoea. A randomized, double-blind, controlled trial using two different probiotic preparations in Bolivian children. *BMC Infect. Dis.* 10:253. doi: 10.1186/1471-2334-10-253

Jiang, X., Wang, M., Graham, D. Y., and Estes, M. K. (1992). Expression, self-assembly, and antigenicity of the Norwalk virus capsid protein. *J. Virol.* 66, 6527–6532.

Kandasamy, S., Chattha, K. S., Vlasova, A. N., Rajashekara, G., and Saif, L. J. (2014). Lactobacilli and Bifidobacteria enhance mucosal B cell responses and differentially modulate systemic antibody responses to an oral human rotavirus vaccine in a neonatal gnotobiotic pig disease model. *Gut Microbes* 5, 639–651. doi: 10.4161/19490976.2014.969972

Karst, S. M., Wobus, C. E., Goodfellow, I. G., Green, K. Y., and Virgin, H. W. (2014). Advances in norovirus biology. *Cell Host Microbe* 15, 668–680. doi: 10.1016/j.chom.2014.05.015

Kim, M. J., Lee, D. K., Park, J. E., Park, I. H., Seo, J. G., and Ha, N. J. (2014). Antiviral activity of *Bifidobacterium adolescentis* SPM1605 against Coxsackievirus B3. *Biotechnol. Biotechnol. Equip.* 28, 681–688. doi: 10.1080/13102818.2014.945237

Kim, Y., Lee, D., Kim, D., Cho, J., Yang, J., Chung, M., et al. (2008). Inhibition of proliferation in colon cancer cell lines and harmful enzyme activity of colon bacteria by *Bifidobacterium adolescentis* SPM0212. *Arch. Pharm. Res.* 31, 468–473. doi: 10.1007/s12272-001-1180-y

Kniel, K. E. (2014). The makings of a good human norovirus surrogate. *Curr. Opin. Virol.* 4, 85–90. doi: 10.1016/j.coviro.2014.01.002

Lee, J., Ametani, A., Enomoto, A., Sato, Y., Motoshima, H., Ike, F., et al. (1993). Screening for the immunopotentiating activity of food microorganisms and enhancement of the immune response by *Bifidobacterium adolescentis* M101-4. *Biosci. Biotechnol. Biochem.* 57, 2127–2132. doi: 10.1271/bbb.57.2127

Lee, M. H., Yoo, S.-H., Ha, S.-D., and Choi, C. (2012). Inactivation of feline calicivirus and murine norovirus during Dongchimi fermentation. *Food Microbial.* 31, 210–214. doi: 10.1016/j.fm.2012.04.002

Li, D., Breiman, A., Le Pendu, J., and Uyttendaele, M. (2015). Binding to histo-blood group antigen-expressing bacteria protects human norovirus from acute heat stress. *Front. Microbiol.* 6:659. doi: 10.3389/fmicb.2015.00659

Lipton, H. L. (1980). Persistent theiler's murine encephalomyelitis virus infection in mice depends on plaque Size. *J. Gen. Virol.* 46, 169–177. doi: 10.1099/0022-1317-46-1-169

Ljungh, A., and Wadström, T. (2006). Lactic acid bacteria as probiotics. *Curr. Issues Intest. Microbiol.* 7, 73–90.

Mao, X., Gu, C., Hu, H., Tang, J., Chen, D., Yu, B., et al. (2016). Dietary *Lactobacillus rhamnosus* GG supplementation improves the mucosal barrier function in the intestine of weaned piglets challenged by porcine rotavirus. *PLoS ONE* 11:e0146312. doi: 10.1371/journal.pone.0146312

Maragkoudakis, P. A., Chingwaru, W., Gradisnik, L., Tsakalidou, E., and Cencic, A. (2010). Lactic acid bacteria efficiently protect human and animal intestinal epithelial and immune cells from enteric virus infection. *Int. J. Food Microbiol.* 141, S91–S97. doi: 10.1016/j.ijfoodmicro.2009.12.024

Muñoz, J. A. M., Chenoll, E., Casinos, B., Bataller, E., Ramón, D., Genovés, S., et al. (2011). Novel probiotic *Bifidobacterium longum* subsp. infantis CECT 7210 strain active against rotavirus infections. *Appl. Environ. Microbiol.* 77, 8775–8783. doi: 10.1128/AEM.05548-11

Puniya, A. K. (2015). *Fermented Milk and Dairy Products*. Boca Raton, FL: CRC Press, 105.

Rubio-del-Campo, A., Coll-Marqués, J. M., Yebra, M. J., Buesa, J., Pérez-Martínez, G., Monedero, V., et al. (2014). Noroviral p-particles as an in vitro model to assess the interactions of noroviruses with probiotics. *PLoS ONE* 9:e89586. doi: 10.1371/journal.pone.0089586

Saavedra, J. M., Bauman, N., Perman, J., Yolken, R., and Oung, I. (1994). Feeding of *Bifidobacterium bifidum* and *Streptococcus thermophilus* to infants in hospital for prevention of diarrhoea and shedding of rotavirus. *Lancet* 344, 1046–1049. doi: 10.1016/S0140-6736(94)91708-6

Sanders, M. E., Guarner, F., Guerrant, R., Holt, P. R., Quigley, E. M., Sartor, R. B., et al. (2013). An update on the use and investigation of probiotics in health and disease. *Gut* 62, 787–796. doi: 10.1136/gutjnl-2012-302504

Sazawal, S., Hiremath, G., Dhingra, U., Malik, P., Deb, S., and Black, R. E. (2006). Efficacy of probiotics in prevention of acute diarrhoea: a meta-analysis of masked, randomised, placebo-controlled trials. *Lancet Infect. Dis.* 6, 374–382. doi: 10.1016/S1473-3099(06)70495-9

Seo, D. J., Lee, M. H., Seo, J., Ha, S.-D., and Choi, C. (2014). Inactivation of murine norovirus and feline calicivirus during oyster fermentation. *Food Microbiol.* 44, 81–86. doi: 10.1016/j.fm.2014.05.016

Shearer, A. E., Hoover, D. G., and Kniel, K. E. (2014). Effect of bacterial cell-free supernatants on infectivity of norovirus surrogates. *J. Food Protect.* 77, 145–149. doi: 10.4315/0362-028X.JFP-13-204

Takemoto, K. K. (1966). Plaque mutants of animal viruses. *Prog. Med. Virol.* 8, 314–348.

Tamura, M., Natori, K., Kobayashi, M., Miyamura, T., and Takeda, N. (2004). Genogroup II noroviruses efficiently bind to heparin sulfate proteoglycan associated with the cellular membrane. *J. Virol.* 78, 3817–3826. doi: 10.1128/JVI.78.8.3817-3826.2004

Tan, M., and Jiang, X. (2011). Norovirus-host interaction: multi-selections by human histo-blood group antigens. *Trends Microbiol.* 19, 382–388. doi: 10.1016/j.tim.2011.05.007

Van Beek, J., Ambert-Balay, K., Botteldoorn, N., Eden, J., Fonager, J., Hewitt, J., et al. (2013). Indications for worldwide increased norovirus activity associated with emergence of a new variant of genotype II. 4, late 2012. *Euro Surveill.* 18, 8–9.

Van Niel, C. W., Feudtner, C., Garrison, M. M., and Christakis, D. A. (2002). *Lactobacillus* therapy for acute infectious diarrhea in children: a meta-analysis. *Pediatrics* 109:e52. doi: 10.1542/peds.109.4.678

Wobus, C. E., Karst, S. M., Thackray, L. B., Chang, K., Sosnovtsev, S. V., Belliot, G., et al. (2004). Replication of norovirus in cell culture reveals a tropism for dendritic cells and macrophages. *PLoS Biol.* 2:e432. doi: 10.1371/journal.pbio.0020432

Zhong, S.-S., Zhang, Z.-S., Wang, J.-D., Lai, Z.-S., Wang, Q.-Y., Pan, L.-J., et al. (2004). Competitive inhibition of adherence of enterotoxigenic *Escherichia coli*, enteropathogenic *Escherichia coli* and *Clostridium difficile* to intestinal epithelial cell line Lovo by purified adhesin of *Bifidobacterium adolescentis* 1027. *World J. Gastroenterol.* 10, 1630–1633. doi: 10.3748/wjg.v10.i11.1630

**Conflict of Interest Statement:** The authors declare that the research was conducted in the absence of any commercial or financial relationships that could be construed as a potential conflict of interest.

# 16

# *Bifidobacterium animalis* ssp. *lactis* CNCM-I2494 Restores Gut Barrier Permeability in Chronically Low-Grade Inflamed Mice

**Edited by:**
Francesca Turroni,
University College Cork, Ireland

**Reviewed by:**
Miguel Gueimonde,
Consejo Superior de Investigaciones Científicas, Spain
Valentina Taverniti,
Universita degli Studi di Milano, Italy

***Correspondence:**
Rebeca Martín
rebeca.martinrosique@jouy.inra.fr

*Rebeca Martín[1]*, Laure Laval[1,2], Florian Chain[1], Sylvie Miquel[1†], Jane Natividad[3], Claire Cherbuy[1], Harry Sokol[1,4,5], Elena F. Verdu[3], Johan van Hylckama Vlieg[2†], Luis G. Bermudez-Humaran[1], Tamara Smokvina[2] and Philippe Langella[1]*

[1] Commensal and Probiotics-Host Interactions Laboratory, Micalis Institute, Institut National de la Recherche Agronomique, AgroParisTech, Université Paris-Saclay, Jouy-en-Josas, France, [2] Danone Nutricia Research, Palaiseau, France, [3] Farncombe Family Digestive Health Research Institute, McMaster University, Hamilton, ON, Canada, [4] ERL INSERM U 1057/UMR7203, Faculté de Médecine Saint-Antoine, Université Pierre et Marie Curie, Paris, France, [5] Service de Gastroentérologie, Hôpital Saint-Antoine, Assistance Publique – Hôpitaux de Paris, Paris, France

Growing evidence supports the efficacy of many probiotic strains in the management of gastrointestinal disorders associated with deregulated intestinal barrier function and/or structure. In particular, bifidobacteria have been studied for their efficacy to both prevent and treat a broad spectrum of animal and/or human gut disorders. The aim of the current work was thus to evaluate effects on intestinal barrier function of *Bifidobacterium animalis* ssp. *lactis* CNCM-I2494, a strain used in fermented dairy products. A chronic dinitrobenzene sulfonic acid (DNBS)-induced low-grade inflammation model causing gut dysfunction in mice was used in order to study markers of inflammation, intestinal permeability, and immune function in the presence of the bacterial strain. In this chronic low-grade inflammation mice model several parameters pointed out the absence of an over active inflammation process. However, gut permeability, lymphocyte populations, and colonic cytokines were found to be altered. *B. animalis* ssp. *lactis* CNCM-I2494 was able to protect barrier functions by restoring intestinal permeability, colonic goblet cell populations, and cytokine levels. Furthermore, tight junction (TJ) proteins levels were also measured by qRT-PCR showing the ability of this strain to specifically normalize the level of several TJ proteins, in particular for claudin-4. Finally, *B. lactis* strain counterbalanced $CD4^+$ lymphocyte alterations in both spleen and mesenteric lymphoid nodes. It restores the Th1/Th2 ratio altered by the DNBS challenge (which locally augments $CD4^+$ Th1 cells) by increasing the Th2 response as measured by the increase in the production of major representative Th2 cytokines (IL-4, IL-5, and IL-10). Altogether, these data suggest that *B. animalis* ssp. *lactis* CNCM-I2494 may efficiently prevent disorders associated with increased barrier permeability.

**Keywords: micro-inflammation, apical junction proteins, goblet cells**

## INTRODUCTION

The intestinal barrier is an effective defense mechanism that depends on the integrity of the cells and the junctional complexes between them. The gut barrier is a functional unit organized as a multilayer system composed by a physical barrier which prevents bacterial adhesion and regulates paracellular diffusion and a functional layer able to discriminate between pathogens and commensal microorganisms (Lopetuso et al., 2015). The physical barrier is formed by a mucus layer followed by a monolayer of epithelial cells (Denker and Nigam, 1998; Natividad and Verdu, 2013) performing the paracellular transport across the barrier controlled by apical junction proteins (Natividad and Verdu, 2013). The mucus protects the epithelium from harmful microorganisms and antigens being also a lubricant for intestinal motility (Lopetuso et al., 2015). Outer mucus is composed of the highly glycosylated mucin MUC2 protein produced by the goblet cells (Lopetuso et al., 2015). The regulation of its function is mediated by both endogenous and exogenous factors (Agostini et al., 2012; Distrutti et al., 2013) and is a key factor in the development of several diseases involving altered gut permeability and dysfunction such as irritable bowel syndrome (IBS), food allergies, type-1 diabetes, and obesity (Perrier and Corthésy, 2011; Camilleri et al., 2012; Vaarala, 2012). Diverse microorganisms have shown to protect barrier integrity and promote its restoration when damaged. Among them, increasing evidence points out that strains of lactic acid bacteria (Gaudier et al., 2004) and bifidobacteria regulate gut barrier function using different mechanisms (Agostini et al., 2012; Distrutti et al., 2013). For instance, *Lactobacillus rhamnosus* GG (LGG), *B. breve* NCC2950 and a mixture of lactobacilli and bifidobacteria (*L. casei, L. plantarum, L. acidophilus, L. delbrueckii ssp. bulgaricus, B. longum, B. breve, and B. infantis*) prevent the increase in intestinal permeability *in vivo* (Ukena et al., 2007; Mennigen et al., 2009; Donato et al., 2010; Natividad et al., 2013).

Bifidobacteria, naturally present in the colonic microbiota, correspond to up to 80% of the cultivable fecal microorganisms in full-term breastfed infants (Picard et al., 2005). They have been traditionally considered as safe microorganisms, due to their Generally Recognized As Safe (GRAS) status and are widely used as health-promoting bacteria in functional foods. Especially, *B. animalis* ssp. *lactis* (*B. lactis*) CNCM I-2494 has a long history of use in fermented dairy products and shows a high gastrointestinal survival (Picard et al., 2005; Rochet et al., 2008). A fermented milk product (FMP)-containing *B. lactis* CNCM I-2494 together with lactic acid bacterial starter cultures has shown positive effects on gut function in several randomized controlled studies (Picard et al., 2005) improving: (i) gastrointestinal well-being and digestive symptoms in women reporting minor digestive problems (Guyonnet et al., 2009a), (ii) abdominal girth and gastrointestinal transit (Agrawal et al., 2009), (iii) health related quality of life and symptoms in IBS in adults (Guyonnet et al., 2009b), and (iv) colonic transit time and minor digestive problems in healthy women (Marteau et al., 2002, 2013). The physiological effects of this strain have been also evaluated in animal studies where it has been capable to reduce the aberrant crypts incidence in chemically induced carcinogenesis models in rats (Tavan et al., 2002), improve colitis in mice (Veiga et al., 2010), hydrolyze bile salts in the gastrointestinal tract of pigs (Lepercq et al., 2004), and prevent the increase of intestinal permeability induced by partial restraint stress in rats (Agostini et al., 2012). The molecular mechanisms underlying the positive effects of strain CNCM I-2494 are far from being completely understood although its genome have been sequenced (Chervaux et al., 2011). Recent identification of several restriction and modification systems in this strain and development of specific molecular tools opened the way in studying specific bacterial mechanisms involved in the cross-talk of strain CNCM I-2494 with the host (O'Connell Motherway et al., 2014).

The clear relationship between *B. lactis* CNCM I-2494 and the protection of gut dysfunction in both animal models and clinical trials combined to the industrial importance of this strain has prompted us to deeper analyze its possible effects on an altered permeability and gut dysfunction model. Gut dysfunction was achieved thanks to a first inflammatory insult followed with a second subclinical chemical challenge as previously described (Laval et al., 2015; Martin et al., 2015). The aim of this work was to clarify the direct effect of the strain in the murine intestinal epithelium barrier and function.

## MATERIALS AND METHODS

### Bacterial Growth Conditions and Animals

*Bifidobacterium animalis* ssp. *lactis* CNCM-I2494 was grown in MRS medium (Difco, USA) supplemented with cysteine (0.5 mg/ml; Sigma–Aldrich) under anaerobic conditions at 37°C.

Male C57BL/six mice (6–8 weeks old; Janvier, Le Genest Saint Isle, France) were maintained at the animal care facilities of the National Institute of Agricultural Research (IERP, INRA, Jouy-en-Josas, France) under specific pathogen-free (SPF) conditions. Mice were housed under standard conditions for a minimum of 1 week before experimentation. All experiments were performed in accordance with European Community rules for animal care and were approved by the relevant local committee (Comethea). Protocol number 02550.01.

### Experimental Design

Inflammation was induced as previously described (Laval et al., 2015) (**Supplementary Figure S** ). Briefly, mice where challenged, under anesthesia, with a first intra-rectal dose of 100 mg/Kg of dinitrobenzene sulfonic acid (DNBS) solution (ICN, Biomedical Inc.) in 30% ethanol (EtOH). Control mice (without colitis) received only 30% EtOH. Thirteen days after the first DNBS injection, $5 \times 10^9$ CFU of viable bacteria in 200 μl of PBS or PBS alone were administered intra-gastrically, daily for 10 days (gavage period). Finally, 21 days after the first challenge, the mice were challenged again with a second administration of 50 mg/kg of DNBS or EtOH. Weight loss was monitored during 3 days following the second DNBS injection to assess possible clinical signs of distress.

To confirm the absence of over inflammation, colonic macroscopic and histological scores as well as colonic myeloperoxidase (MPO) activity (a marker of the degree of infiltration by polymorphonuclear neutrophils) and serum lipocalin-2 levels (an early inflammation marker) were determined as previously described (Shashidharamurthy et al., 2013; Martin et al., 2014; Laval et al., 2015).

## Histological Features Analysis

Flushed colons were fixed in 4% paraformaldehyde or Carnoy buffer, dehydrated and embedded in paraffin according to a standard protocol. Histological features were analyzed by hematoxylin–eosin–safran (Perrier and Corthésy, 2011) staining. Periodic acid-Schiff (PAS) and Alcian blue (AB) staining were performed as in Wrzosek et al. (2013).

## Intestinal Permeability *In Vivo*

Permeability *in vivo* was assessed using fluorescein isothiocyanate-conjugated dextran (FITC–dextran 3000–5000 Da, Sigma–Aldrich) tracer as previously described (Tambuwala et al., 2010). Briefly, at the endpoint 0.6 mg/g body weight of FITC–dextran dissolved in PBS was administered to mice by oral gavage. To measure the presence of FITC–dextran in blood, 3.5 h after the gavage blood samples were recovered from the retro-orbital venous plexus and kept in dark at 4°C until analysis. Mice were housed under standard conditions during this period with un-limited access to water and food. Serum has separated by centrifugation and plasma FITC levels were determined using a fluorescence microplate reader (excitation 485 nm and emission 530 nm; Tecan, Lyon, France).

## Apical Junctional Analysis by Quantitative Real-time PCR (qPCR)

Total RNA was isolated from 20 to 30 mg samples of colon with an RNeasy Mini Kit (Qiagen) as previously described (Laval et al., 2015). qPCR was performed with diluted cDNA (10×) in triplicate and with an iQ5 Real-Time Detection System (Bio-Rad). The reaction mix consisted of Ssofast Evagreen Supermix (Bio-Rad), primers at 0.5 μM (Martin et al., 2015), and 2 μL of diluted cDNA. Values are expressed as relative fold differences normalized to a housekeeping gene, *Gapdh,* by the $2^{-\Delta\Delta C_T}$ method. All procedures were performed according to the manufacturers' instructions.

## Analyses of Lymphoid Populations Present in the Spleen and in the Mesenteric Lymphoid Nodes (MLNs)

Mononuclear cells were isolated from spleens and MLN by gentle extrusion of the tissue through a 50 μm-mesh Nylon cell strainer (BD). Cells were suspended in Dulbecco's Modified Eagle Medium (DMEM) medium supplemented with 10% of fetal calf serum (FCS), 2 mM L-glutamine, 50 U/mg penicillin, and 50 U/mg streptomycin (Lonza, Levallois-Perret, France). Erythrocytes were lysed with red blood-cell lysing buffer (Sigma–Aldrich).

For flow cytometry analysis, aliquots of $10^6$–$10^7$ cells per sample were pre-incubated with purified anti-mouse CD16/CD32 (eBioscience, San Diego, CA, USA) and then labeled with anti-CD4-FITC, anti-CD3e-PE, and anti-CD8-PerCP (all from eBioscience) according to the manufacturer's instructions. The stained cells were analyzed by flow cytometry (Accuri, BDbioscience) with CFlow Sampler software (BD).

For stimulation experiments, $2 \times 10^5$ cells per well were cultured for 48 h (37°C, 10% $CO_2$) in DMEM medium in P24 plates pre-coated with anti-CD3/CD28 antibodies (4 μg/mL each; eBioscience) or phorbol 12-myristate 13-acetate (PMA)/ionomycin (cell stimulation cocktail, 1×, ebioscience). Culture supernatant was frozen at −80°C until processing.

## Cytokine Assays

Blood samples were obtained from the retro-orbital venous plexus before the mice were euthanized and centrifuged, and the sera stored at −80°C until analysis. One centimeter samples of distal colon were recovered and homogenized in an appropriate volume of PBS (final concentration of 50 mg/ml) in a Tissue Lyser (Qiagen). IL-6, IL-10, IFN-γ, TNF-α, IL-5, IL-2, IL-22, IL-1α, IL-13, IL-17, IL-4, IL-27, and IL-12p70 were assayed in blood and colon samples with a cytometric bead array system (Mouse Th1/Th2/Th17/Th22 13plex Flowcytomix; eBioscience, San Diego, CA, USA). For cytokine quantification in cell culture supernatants the following ELISA tests were performed according to manufacturer's instruction: IL-4, IL-5, IFNγ, IL-17, IL-12p70, and IL-10 (MabTech); TGFβ and IL-22 (ebioscience).

## Statistical Analysis

GraphPad software (GraphPad Sofware, La Jolla, CA, USA) was used for statistical analysis. Results are presented as bar graphs or dot plots with means ± SEM. Comparisons involved the non-parametric Kruskal-Wallis test followed by a Dunn's Multiple Comparison test. A $p$ value of less than 0.05 was considered significant.

# RESULTS

## Confirmation of Micro-inflammation in DNBS Challenged Mice

The induction of a low-grade inflammation status following a chronic low-dose DNBS in the mice was confirmed through the follow-up of health, histological and inflammatory parameters (**Supplementary Figure** ). In particular, weight loss (**Supplementary Figure** ), colonic macroscopic and histological scores (**Supplementary Figures** ) as well as the MPO activity in the colon ( ) and the Lipocalin-2 concentration in serum (**Supplementary Figure S2E**) were measured. The absence of differences for all these parameters among the groups, even in presence of *B. lactis* CNCM-I2494, added to the lack of detection of cytokine levels in serum samples (IL-6, IL-10, IFN-γ, TNF-α, IL-5, IL-2, IL-22, IL-1α, IL-13, IL-17, IL-4, IL-27, and IL-12p70, data not shown) discards the presence of an overt and active inflammation in this model.

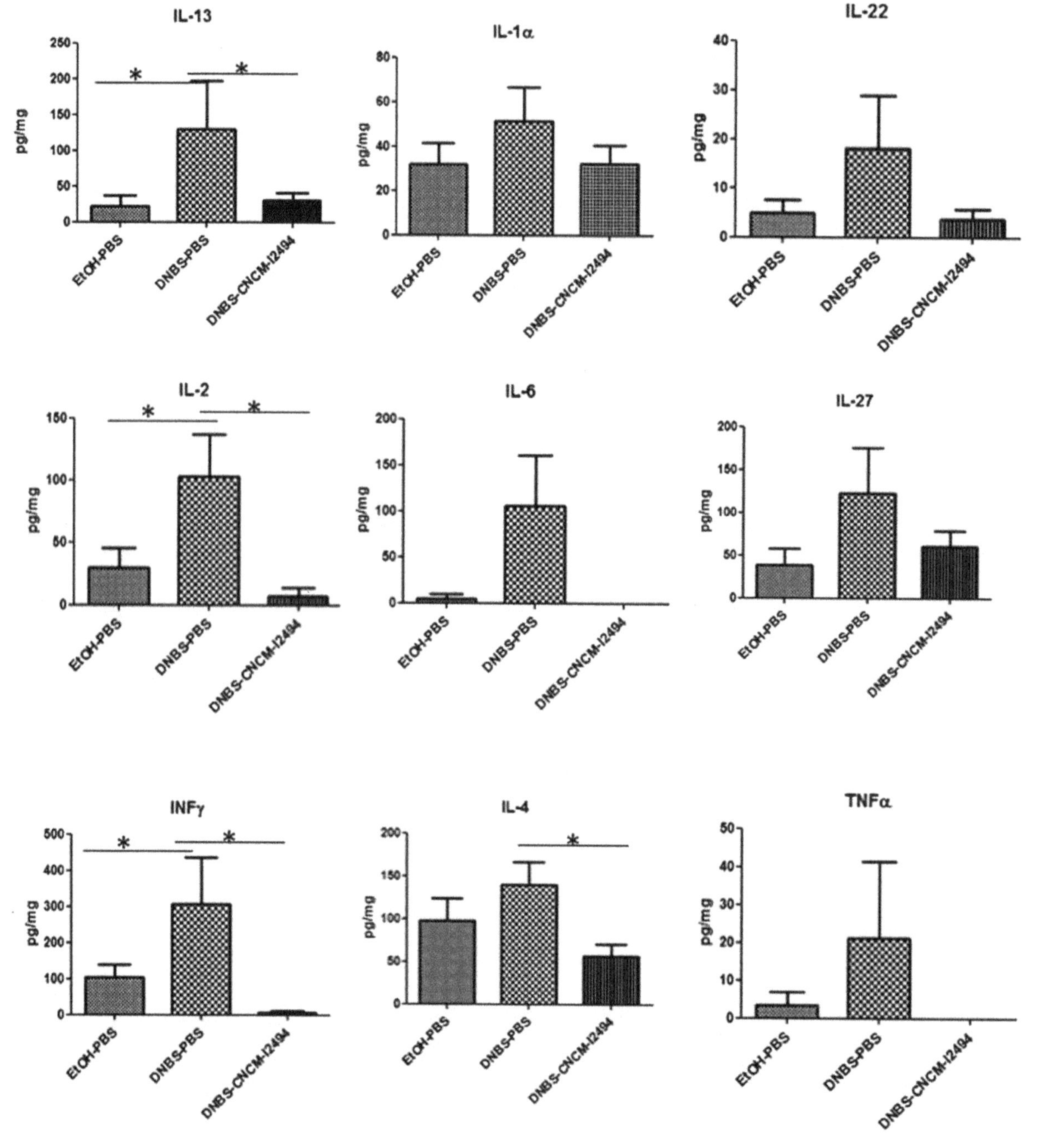


**FIGURE | Cytokine concentrations in colon in the dinitrobenzene sulfonic acid (DNBS) micro-inflammation model.** Control non-inflamed (EtOH–PBS), control inflamed (DNBS–PBS), and *B. lactis* CNCM I-2494 strain (DNBS–CNCM-I2494). $^{*}p < 0.05$ ($n = 8$).

However, the presence of slightly elevated, although no statistically significant, cytokines IL-13, IL-1α, IL-6, IL-22, IL-2, IL-27, IL-4, IFN-γ, and TNF-α levels in colonic tissues, compared to healthy controls, suggest a local low-grade inflammation (**Figure** ). Treatment with *B. lactis* CNCM-I2494 reduced these increases in cytokine production (**Figure** ): Notably, restoration was statistically significant for IL-2, IL-13, and IFNγ ($p < 0.05$).

## *Bifidobacterium lactis* CNCM-I2494 Restores Colonic Permeability by Modulating Apical Junction Protein Levels

The integrity of the gut barrier was assessed by the analysis of the permeability with the paracellular tracer FITC-dextran *in vivo* at the endpoint. Of note, all the animals were submitted to exactly the same protocol and waiting time to avoid differences due to a minimal possible clearance phenomenon due to renal function. The mice treated with DNBS showed high permeability to the tracer ($p < 0.05$) (**Figure** ) confirming an alteration in the barrier permeability as it has been previously observed (Laval et al., 2015; Martin et al., 2015). The oral administration of *B. lactis* CNCM-I2494 strain resulted in a decrease in permeability ($p < 0.05$). To further analyze the effect on the barrier function the expression of the relevant mRNAs of adherent junction (AJ) and tight junction (TJ) proteins were measured by qRT-PCR (**Figure** ). The mRNAs for Claudin-3, 4, E-cadherin, Occludin, and the zona occludens proteins (ZO-1) were all less abundant in DNBS-treated mice than in control mice ($p < 0.05$). CNCM-I2494 tends to partially re-establish the levels of all of them (**Figure** ). Notably, this effect was statistically significant for Claudin-4 ($p < 0.05$). Taken together, both the histological analysis and the transcriptional data demonstrate that strain *B. lactis* CNCM I-2494 protects against DNBS-induced chronic barrier dysfunction.

### *Bifidobacterium lactis* CNCM-I2494 Restores Goblet Cell Population Altered by DNBS Chronic Challenge

Histological features, analyzed by hematoxylin–eosin–safran (Perrier and Corthésy, 2011) staining, showed no significant differences in general morphology, crypt depth or total numbers of cells per crypt (data not shown). The numbers of goblet cells stained either by AB (**Figure** ), specific for acidic mucopolysaccharides, or PAS (**Figure** ), specific for neutral mucopolysaccharides, were significantly lower in DNBS challenged control group ($p < 0.05$). *B. lactis* CNCM-I2494 was able to enhance the percentage of AB or PAS positive cells per crypt ($p < 0.05$) reaching the values of the non-inflamed control group (**Figure** ).

### *Bifidobacterium. lactis* CNCM I-2494 Modulates CD3$^+$/CD4$^+$ T-Cell Populations in Spleen and MLNs by Increasing T Helper (Th) Profile 2

To study further the mechanism by which *B. lactis* CNCM-I2494 exerts protective function, T-cells from spleen and MLN were isolated and analyzed by flow cytometry. DNBS-treated mice showed lower CD3$^+$/CD4$^+$ T-cell percentages in spleen (**Figure** ) than the control group and higher CD3$^+$/CD4$^+$ cell percentages in MLN (**Figure** ; $p < 0.05$). *B. lactis* CNCM-I-2494 tends to reduce the CD3$^+$/CD4$^+$ decrease in spleen (**Figure** ) and significantly control CD3$^+$/CD4$^+$ increase in MLN ($p < 0.05$; **Figure** ). No variations were observed in CD3$^+$/CD8$^+$ T-cell percentages in spleen or MLN (data not shown).

As variations in CD4$^+$ T-cell populations were found, MLN and spleen cells were cultured after isolation in the presence of two different stimulators during 48 h: CD28$^+$/CD3$^+$ to specifically stimulate lymphocytes and PMA/IO to stimulate all the cells present in the organ disaggregate. Representative cytokines of the major Th profiles (IL-4, IL-5, IFNγ, IL-17, IL-12p70, IL-10, TGFβ, and IL-22) were determined in the culture supernatants (**Figures** and and data not shown). The IL-17, IL-22, IL-12p70, and TGFβ levels as well as IFNγ in spleen samples were under the ELISA detection limits (2.4, 5.5, 8.6, 10, and 6.5 pg/ml, respectively; data not shown). This fact, in addition to slight increases of Th1 levels (IFNγ) by the DNBS treatment in MLN (**Figure** ) confirms the low-grade inflammation status of the mice model. Differences were found in the levels of IL-4, IL-5, and IL-10 in both spleen (**Figure** ) and MLN (**Figure** ) ($p < 0.05$). Strain CNCM-I2494 increased Th2 levels as measured by IL-4 and IL-10 augmentation in both spleen (**Figure** ) and MLN (**Figure** ) and also IL-5 in spleen samples corresponding to an anti-inflammatory patter in this model. This anti-inflammatory patter has been confirmed locally by the INFγ/IL-4 ratio in MLN samples (**Figure** ). Nevertheless, CNCM-I2494 was not able to control the small increase in IFNγ caused by the DNBS challenge (**Figure** ). Finally, DNBS treatment caused also an increase in IL-5 in MLN samples. No significant differences were found between CD3$^+$/CD28$^+$ and PMA/IO stimulations, excepting IL-4 and IL-5 in spleen where a major level of stimulation was achieved with the first one (**Figures** and **5**). Taken together these data demonstrate that CNCM-I2494 strain is able to counterbalance the Th1/Th2 ratio altered by the DNBS challenge (which locally augments CD4$^+$ Th1 cells) by increasing the Th2 response as measured by the increase in the production of major representative Th2 cytokines.

## DISCUSSION

Epithelial barrier dysfunction is now considered as one of the major contributors to the development of several diseases and syndromes (Perrier and Corthésy, 2011; Camilleri et al., 2012; Vaarala, 2012). In several of them, such as IBS, studies suggest an interplay between luminal factors (e.g, foods and bacteria residing in the intestine), the epithelial barrier, and the mucosal immune system (Barbara et al., 2012). In a healthy state, the epithelial barrier allows a low translocation of luminal antigens by paracellular transport by receptor-mediated or non-selective endocytosis (Natividad and Verdu, 2013). Therefore, a higher local antigen exposure caused by an increase of intestinal permeability could activate intestinal immune system and inflammation may thus occur (Ohman and Simren, 2007; Natividad and Verdu, 2013). Preclinical studies have shown that selective probiotic strains exhibit the potential to improving mucosal barrier homeostasis (Barbara et al., 2012).

As related above, the administration of fermented milk containing *B. lactis* CNCM I-2494 has been found to prevent *in vivo* the increase of intestinal permeability in rats (Agostini et al., 2012). However, due to possible synergistic interplay of the different strains and/or metabolites contained in this product the specific effect of this *B. lactis* strain on gut

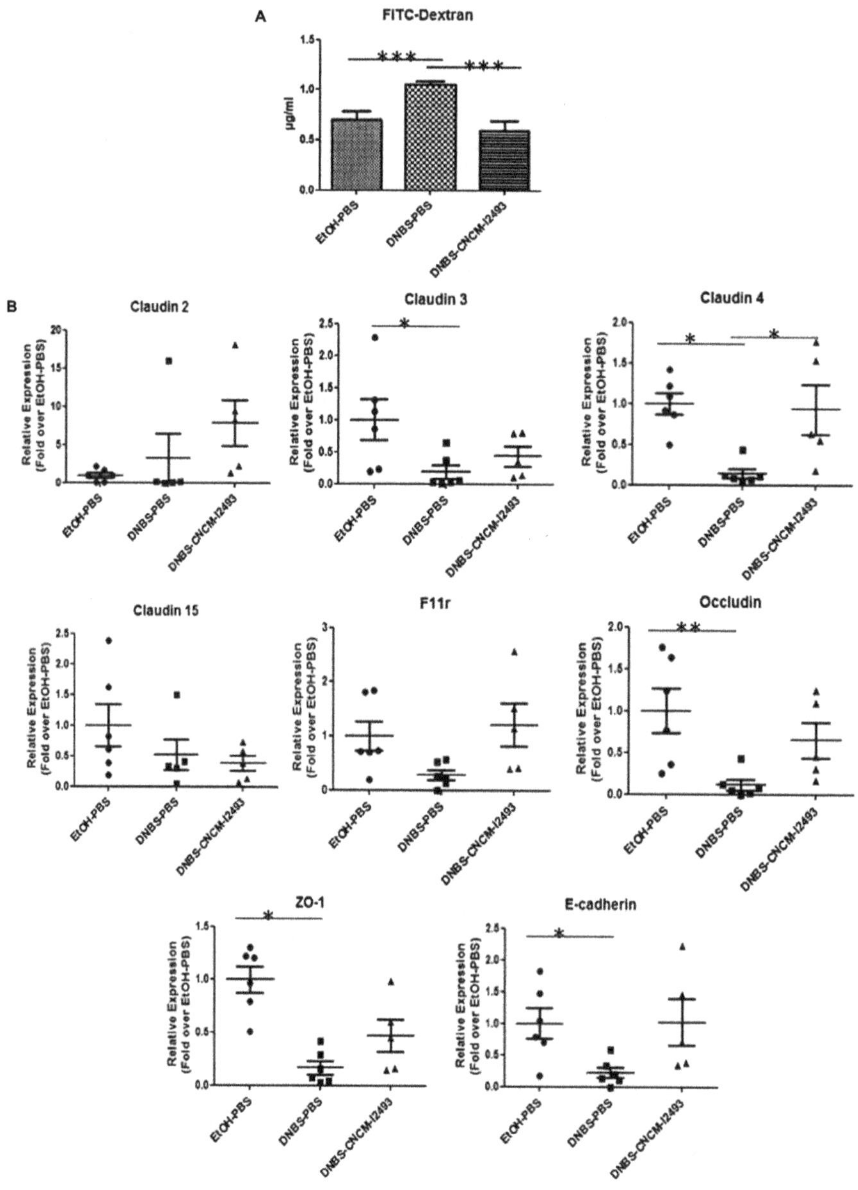


**FIGURE | *In vivo* permeability measurements and effect on apical junction protein mRNAs.** For *in vivo* measurements of gut permeability, animals were orally gavaged with fluorescein isothiocyanate (FITC)-dextran **(A)**. Apical junction protein expression levels were determined by real-time qPCR **(B)**. Control non-inflamed (EtOH-PBS, black circles) control inflamed (DNBS-PBS, black squares) *B. lactis* CNCM I-2494 strain (DNBS-CNCM-I2494, black triangles). $^*p < 0.05$, $^{**}p < 0.01$, $^{***}p < 0.001$ ($n = 8$).

barrier is still unknown. Here, we aimed to clarify the specific effect of *B. lactis* CNCM-I2494 strain on intestinal barrier function.

As previously observed, DNBS-treated mice showed alteration in gut permeability (Laval et al., 2015). *In vivo* values with the paracellular tracer FITC-dextran showed an increase in permeability in DNBS-challenged mice that was restored when mice were treated with *B. lactis* CNCM-I2494 suggesting an effect of the strain on gut barrier function. To better decipher the beneficial effect of *B. lactis* CNCM-I9434 strain on gut permeability, mucus producing cells were analyzed by two different specific staining: AB, which specifically stains acidic mucopolysaccharides and PAS staining, specific for neutral mucopolysaccharides. Both staining protocols reveal the decrease of goblet cell mucus producing cells in mice challenged with DNBS, confirming the functional abnormalities on the tissue despite the lack of macroscopic or microscopic damages. Mice treated with *B. lactis* CNCM-I2494 strain recover the same goblet cell accounts than control mice pointing out a positive effect of the strain in restoring epithelial normal cell composition

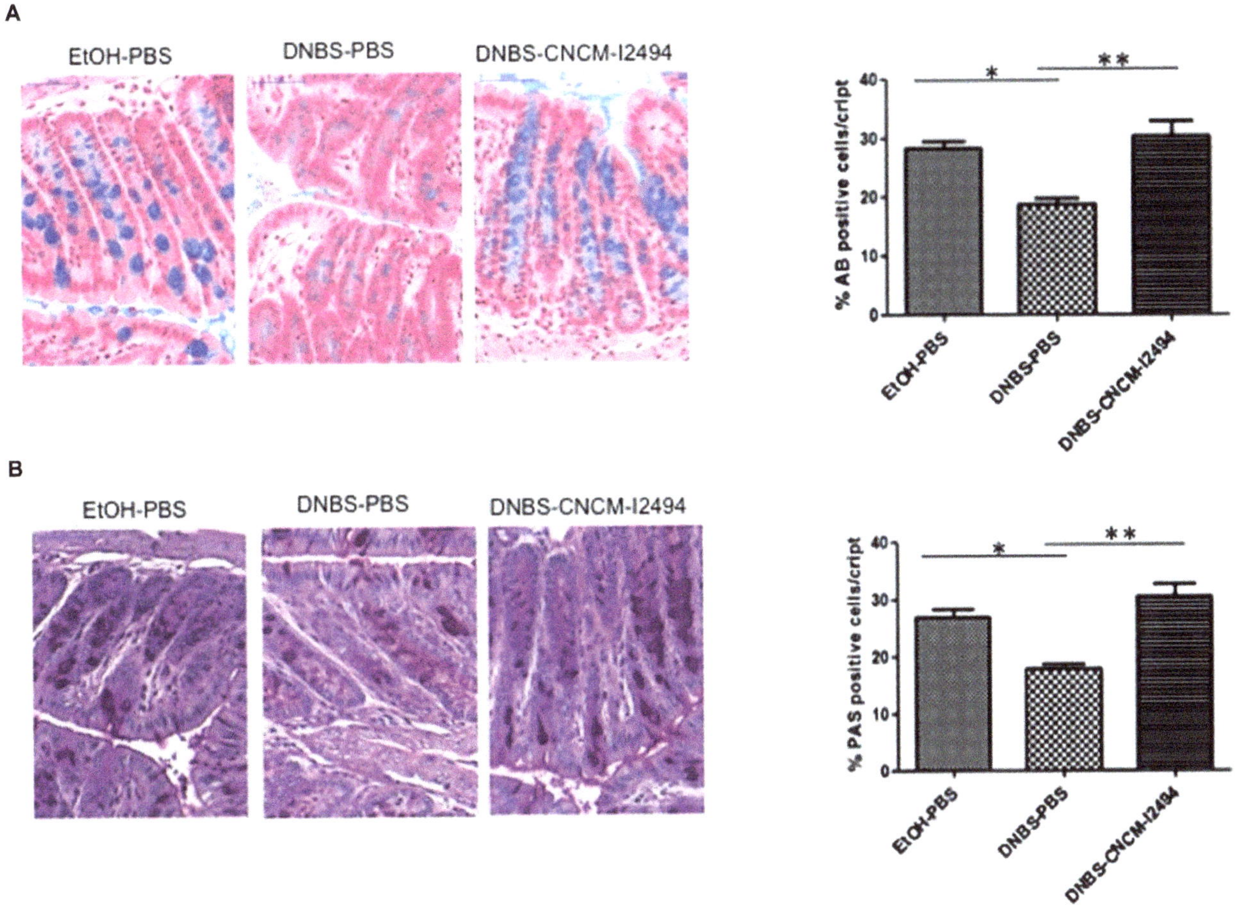


**FIGURE | Goblet cell detection.** Representative photos and % of positive cells stained with AB (Alcian Blue) **(A)** and PAS (Periodic Acid-Schiff) **(B)**. Control non-inflamed (EtOH–PBS), control inflamed (DNBS–PBS), *B. lactis* CNCM I-2494 strain (DNBS–CNCM-I2494). *$p < 0.05$ **$p < 0.01$ ($n = 4$).

and probably the mucus production. In fact, mucus production was previously shown to be affected during inflammation with intestinal dysbiosis (Fyderek et al., 2009). Several bifidobacteria strains, alone or in mixture, induce mucus production and/or are able to adhere to it (He et al., 2001; Gaudier et al., 2005). Even if the goblet cell depletion observed in DNBS challenged mice could explain the differences found in *in vivo* permeability, alterations in apical junction proteins have been also reported previously in this model (Laval et al., 2015). The apical junctions are formed by TJ and AJ proteins. Here, according to our previous results, the expression of TJ proteins measured by RT-qPCR is reduced by the DNBS intra-rectal administration in a protein-specific way (Laval et al., 2015). The treatment with *B. lactis* CNCM-I2494 strain tended to restore F11r, Occludin, E-cadherin and ZO-1 expression, showing this effect especially remarkable for claudin 4. These results are consistent with previous studies in which some lactic bacteria and bifidobacteria prevented changes in occludin, ZO-1, claudin-1, claudin-3, claudin-4, and claudin-5 proteins (Mennigen et al., 2009). Indeed, Agostini et al. (2012) showed that *B. lactis* CNCM-I2494 restored occludin and JAM-A concentrations to control levels after partial restrain stress in rat administration of fermented milk containing *Lactococcus lactis* CNCM-I1631 and two classical yogurt starters.

Changes on mucosal permeability as the ones observed in the DNBS low-dose model can be the cause or the consequence of a low immune activation. To assess the effect of *B. lactis* CNCM-I2494 strain on mucosal immunity and decipher its possible effect on host immune response, colonic cytokine levels, and spleen MLN lymphocyte populations were analyzed. In this specific context, *B. lactis* CNCN-I2494 treatment restored the mild increased IL-13, IL-2, IL-4, and INF-γ colonic values to normal. Several studies pointed out the cytokines as one of the causes of TJ protein modulation. For instance, *in vitro* test have shown a relationship between IL-13 and an increase in paracellular permeability (Prasad et al., 2005) and INF-γ or IL-4 increases have been linked to TJ protein expression alterations (Bruewer et al., 2005; Wisner et al., 2008; Suzuki et al., 2011). Therefore, the effect of *B. Lactis* CNCM-I2494 on cytokine down-regulation could be the factor which triggered permeability restoration.

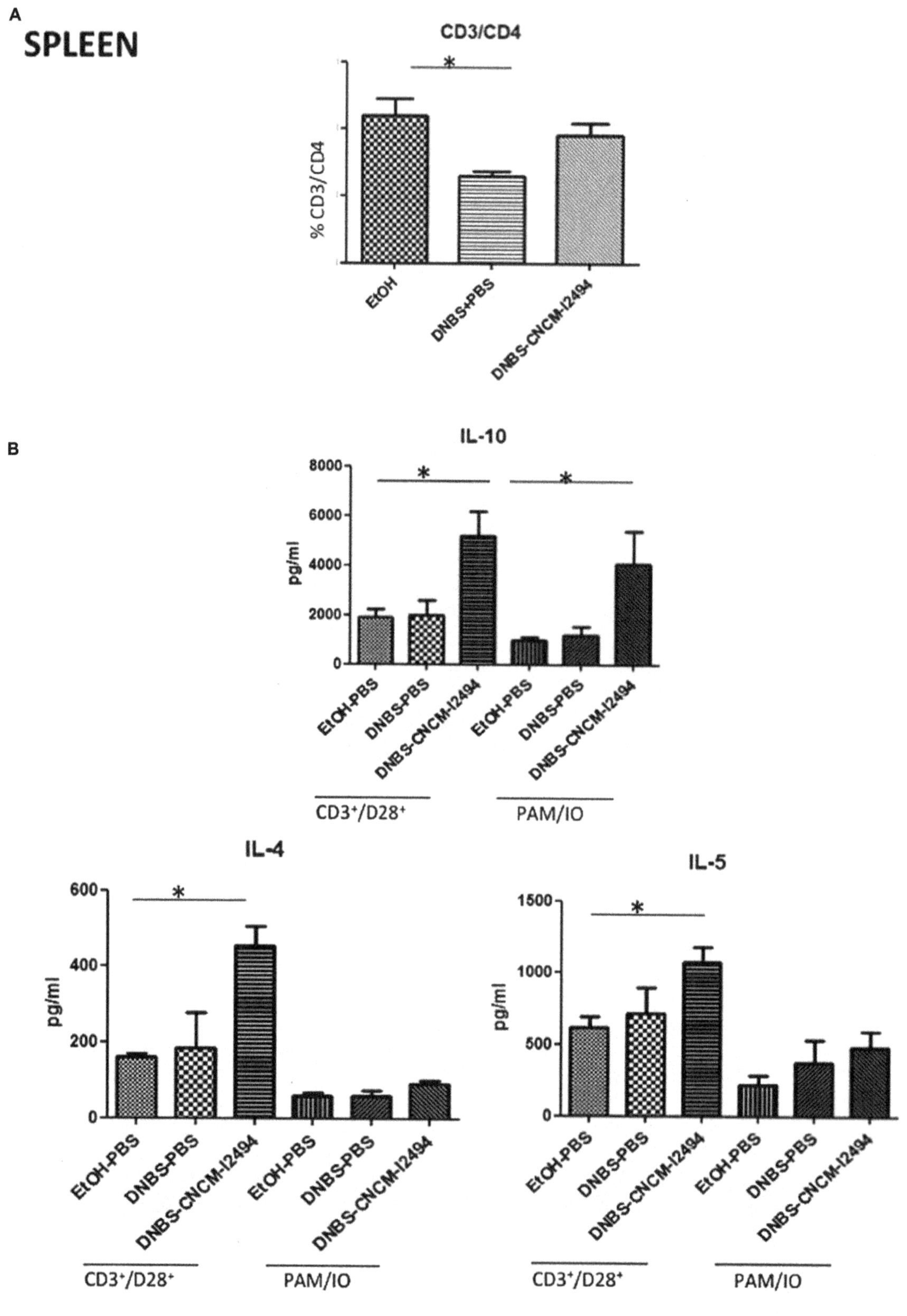


**FIGURE | Splenocyte population levels.** CD3/CD4 positive cells detected by flow cytometry **(A)** and cytokine production in spleen cell cultures stimulated with $CD3^+/CD28^+$ or PAM/IO **(B)**. Control non-inflamed (EtOH-PBS), control inflamed (DNBS-PBS), *B. lactis* CNCM I-2494 strain (DNBS-CNCM-I2494). $^*p < 0.05$ ($n = 8$).

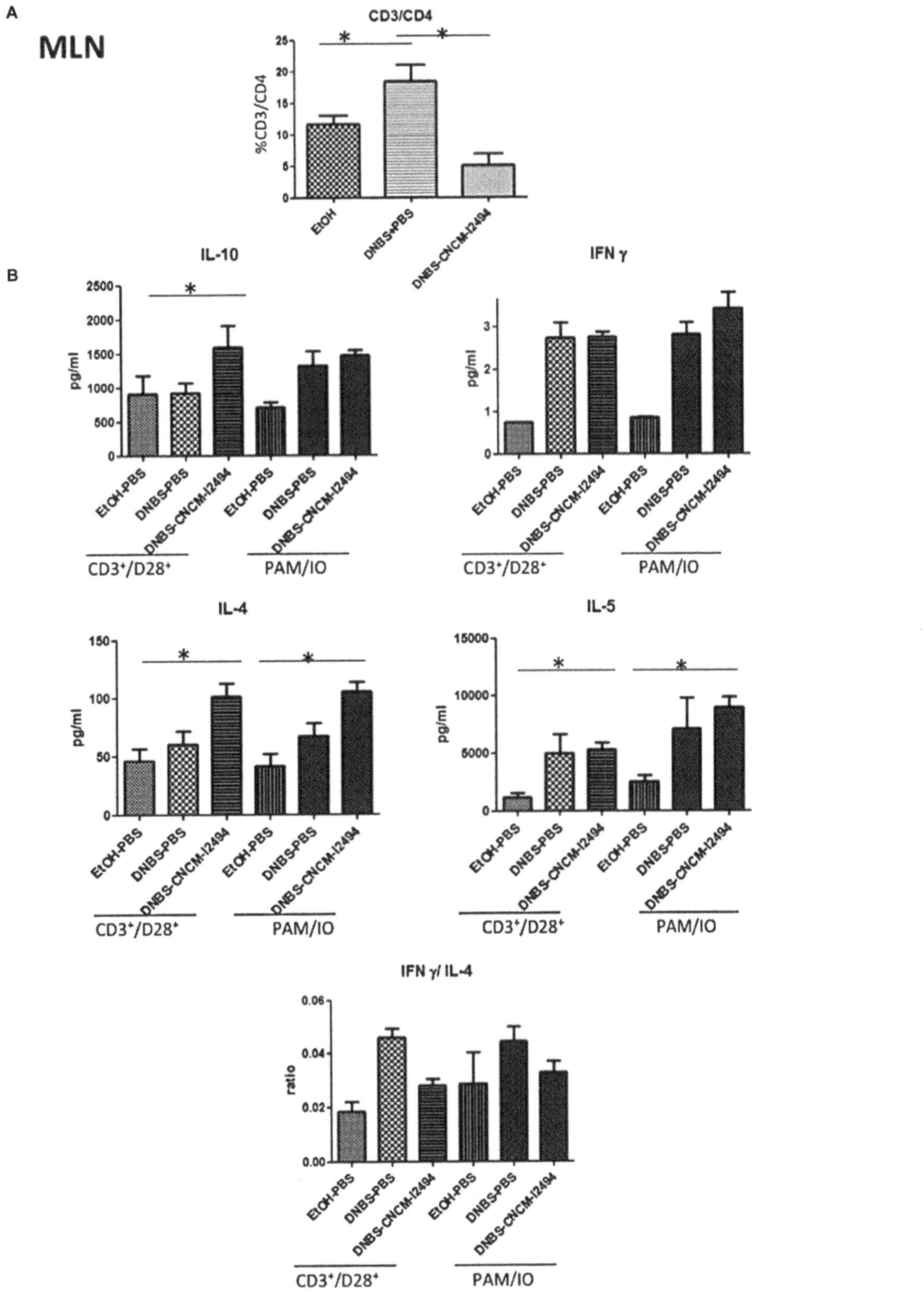


**FIGURE | MLN population levels.** CD3/CD4 positive cells detected by flow cytometry **(A)** and cytokine production in MLN cultures stimulated with $CD3^+/CD28^+$ or PMA/IO **(B)**. Control non-inflamed (EtOH–PBS), control inflamed (DNBS–PBS), *B. lactis* CNCM I-2494 strain (DNBS–CNCM-I2494). $^*p < 0.05$ ($n = 8$).

Mucosal dendritic cells present antigens to the adaptative immune system which directs the polarization of naïve CD4 T cells toward different T-helper cell subsets (Th1 and Th2 among others; Zhu and Paul, 2008). Classically, hapten-mediated colon inflammation protocols trinitrobenzene sulfonic acid (TNBS and DNBS) have been associated with Th1 response (Zuo et al., 2014). Our study confirms that, even in a gut dysfunction model provoked by a low-grade inflammation, DNBS challenge increase lightly Th1 response. Although *B. lactis* CNCM-I2494 was not able to decrease Th1 response, an upper-regulation of Th2 subset has been observed counterbalancing the Th1/Th2 ratio at local level. The increased Th2 cell subset may contribute to the decreased Th1 cell subset due to the mutual antagonizing effects of both Th substets (Donato et al., 2010). Several studies have been performed to assert the role of *Bifidobacterium* strains in modulating T-cell populations, being their results strain and model dependent (Lopez et al., 2011). Our results are consistent with those of Zheng et al. (2014) who showed that one strain of *B. breve* modulates T cell polarization toward Th2 and Treg cell-associated responses *in vitro* and *in vivo* in a murine model of DSS-induced colitis.

Our results support the hypothesis of Agostini et al. (2012) who pointed out the improvement of the intestinal barrier (epithelial cells and mucus layers) permeability as part of the beneficial effect of the fermented milk commercial product containing CNCM-I2494. In addition, here we firstly point to CNCM-I2494 strain as a possible responsible of this effect. Furthermore, the present study supports that the action mechanism of this protective effect may be mediated by improvement on apical junction proteins and goblet cell population. Finally, the modulation of the host T-cell composition by CNCM-I2494 strain may be the host pathway involved in this phenomenon.

## AUTHOR CONTRIBUTIONS

RM, TS, JH, EV, CC, LB-H, and PL designed all the experiments. RM, SM, FC, LL, JN, and HS performed the experiments. RM wrote the manuscript. SM, TS, JH, and PL corrected the manuscript. All authors read and approved the final manuscript.

## FUNDING

LL and RM receives a salary from Danone Nutricia Research in the framework of a CIFRE contract funded by ANRT and a postdoc contract, respectively. SM receives a salary from FPARIS collaborative project. EV holds a Canada Research Chair. The work was partially funded by CCFC grants to PL and EV.

## ACKNOWLEDGMENTS

We thank the histology facility of UMR 1313 GABI, 78350, Jouy-en-Josas, France and the MIMA2 platform for access to the virtual slide scanner (Pannoramic SCAN, 3DHISTECH). Authors would like to thank Sylvia Le Guin and all UEAR personal for their help.

## SUPPLEMENTARY MATERIAL

The Supplementary Material for this article can be found online at: http://journal.frontiersin.org/article/10.3389/fmicb.2016.00608

**FIGURE | Low-grade inflammation experimental protocol.** Colitis was induced by intra-rectal administration of 100 mg/kg of DNBS in solution in 30% ethanol. Control mice (without colitis) received only 30% EtOH. The effects of DNBS are highest during the first 3 days after its administration (DNBS period). Ten days after the end of the DNBS period bacterial culture or PBS were intra-gastrically administered daily for 10 days (gavage period). Colitis was reactivated 21 days after the first DNBS injection with a second injection of 50 mg/kg of DNBS solution. Three days after reactivation mice were sacrificed. Modified from Martin et al. (2015).

**FIGURE | Evaluation of inflammatory status of DNBS-challenged mice.** Inflammatory status assessed from **(A)** the recovery after reactivation (%weight change); **(B)** macroscopic score; **(C)** histological score; **(D)** MPO activity $n = 24$ mice per group, and **(E)** Lipocalin-2 levels ($n = 8$ mice per group) in the control non-inflamed group (EtOH–PBS), control inflamed group (DNBS–PBS), *B. lactis* CNCM I-2494 strain (DNBS–CNCM-I2494).

## REFERENCES

Agostini, S., Goubern, M., Tondereau, V., Salvador-Cartier, C., Bezirard, V., Leveque, M., et al. (2012). A marketed fermented dairy product containing *Bifidobacterium lactis* CNCM I-2494 suppresses gut hypersensitivity and colonic barrier disruption induced by acute stress in rats. *Neurogastroenterol. Motil.* 24:e172. doi: 10.1111/j.1365-2982.2011.01865.x

Agrawal, A., Houghton, L. A., Morris, J., Reilly, B., Guyonnet, D., Goupil Feuillerat, N., et al. (2009). Clinical trial: the effects of a fermented milk product containing *Bifidobacterium lactis* DN-173 010 on abdominal distension and gastrointestinal transit in irritable bowel syndrome with constipation. *Aliment. Pharmacol. Ther.* 29, 104–114. doi: 10.1111/j.1365-2036.2008.03853.x

Barbara, G., Zecchi, L., Barbaro, R., Cremon, C., Bellacosa, L., Marcellini, M., et al. (2012). Mucosal permeability and immune activation as potential therapeutic targets of probiotics in irritable bowel syndrome. *J. Clin. Gastroenterol.* 46(Suppl.), S52–S55. doi: 10.1097/MCG.0b013e318264e91800004836-201210001-00012

Bruewer, M., Utech, M., Ivanov, A. I., Hopkins, A. M., Parkos, C. A., and Nusrat, A. (2005). Interferon-gamma induces internalization of epithelial tight junction proteins via a macropinocytosis-like process. *FASEB J* 19, 923–933. doi: 10.1096/fj.04-3260com

Camilleri, M., Lasch, K., and Zhou, W. (2012). Irritable bowel syndrome: methods, mechanisms, and pathophysiology. The confluence of increased permeability, inflammation, and pain in irritable bowel syndrome. *Am. J. Physiol. Gastrointest. Liver Physiol.* 303, 775–785. doi: 10.1152/ajpgi.00155.2012

Chervaux, C., Grimaldi, C., Bolotin, A., Quinquis, B., Legrain-Raspaud, S., van Hylckama Vlieg, J. E., et al. (2011). Genome sequence of the probiotic strain *Bifidobacterium animalis* subsp. lactis CNCM I-2494. *J. Bacteriol.* 193, 5560–5561. doi: 10.1128/JB.05716-11

Denker, B. M., and Nigam, S. K. (1998). Molecular structure and assembly of the tight junction. *Am. J. Physiol. Renal Physiol.* 274, 1–9.

Distrutti, E., Cipriani, S., Mencarelli, A., Renga, B., and Fiorucci, S. (2013). Probiotics VSL#3 protect against development of visceral pain in murine model of irritable bowel syndrome. *PLoS ONE* 8:e63893. doi: 10.1371/journal.pone.0063893

Donato, K. A., Gareau, M., Wang, Y. J., and Sherman, P. M. (2010). *Lactobacillus rhamnosus* GG attenuates interferon-{gamma} and tumour necrosis factor-alpha-induced barrier dysfunction and pro-inflammatory signalling. *Microbiology* 156, 3288–3297. doi: 10.1099/mic.0.040139-0

Fyderek, K., Strus, M., Kowalska-Duplaga, K., Gosiewski, T., Wedrychowicz, A., Jedynak-Wasowicz, U., et al. (2009). Mucosal bacterial microflora and mucus layer thickness in adolescents with inflammatory bowel disease. *World J Gastroenterol* 15, 5287–5294. doi: 10.3748/wjg.15.5287

Gaudier, E., Jarry, A., Blottiere, H. M., de Coppet, P., Buisine, M. P., Aubert, J. P., et al. (2004). Butyrate specifically modulates MUC gene expression in intestinal epithelial goblet cells deprived of glucose. *Am. J. Physiol. Gastrointest. Liver Physiol.* 287, G1168–G1174. doi: 10.1152/ajpgi.00219.2004

Gaudier, E., Michel, C., Segain, J. P., Cherbut, C., and Hoebler, C. (2005). The VSL# 3 probiotic mixture modifies microflora but does not heal chronic dextran-sodium sulfate-induced colitis or reinforce the mucus barrier in mice. *J. Nutr.* 135, 2753–2761.

Guyonnet, D., Schlumberger, A., Mhamdi, L., Jakob, S., and Chassany, O. (2009a). Fermented milk containing *Bifidobacterium lactis* DN-173 010 improves gastrointestinal well-being and digestive symptoms in women reporting minor digestive symptoms: a randomised, double-blind, parallel, controlled study. *Br. J. Nutr.* 102, 1654–1662. doi: 10.1017/S0007114509990882

Guyonnet, D., Woodcock, A., Stefani, B., Trevisan, C., and Hall, C. (2009b). Fermented milk containing *Bifidobacterium lactis* DN-173 010 improved self-reported digestive comfort amongst a general population of adults. A randomized, open-label, controlled, pilot study. *J. Dig. Dis.* 10, 61–70. doi: 10.1111/j.1751-2980.2008.00366.x

He, F., Ouwehan, A. C., Hashimoto, H., Isolauri, E., Benno, Y., and Salminen, S. (2001). Adhesion of *Bifidobacterium* spp. to human intestinal mucus. *Microbiol. Immunol.* 45, 259–262. doi: 10.1111/j.1348-0421.2001.tb02615.x

Laval, L., Martin, R., Natividad, J., Chain, F., Miquel, S., de Maredsous, C. D., et al. (2015). Lactobacillus rhamnosus CNCM I-3690 and the commensal bacterium *Faecalibacterium prausnitzii* A2-165 exhibit similar protective effects to induced barrier hyper-permeability in mice. *Gut Microbes* 6, 1–9. doi: 10.4161/19490976.2014.990784

Lepercq, P., Relano, P., Cayuela, C., and Juste, C. (2004). Bifidobacterium animalis strain DN-173 010 hydrolyses bile salts in the gastrointestinal tract of pigs. *Scand. J. Gastroenterol.* 39, 1266–1271. doi: 10.1080/00365520410003515

Lopetuso, L. R., Scaldaferri, F., Bruno, G., Petito, V., Franceschi, F., and Gasbarrini, A. (2015). The therapeutic management of gut barrier leaking: the emerging role for mucosal barrier protectors. *Eur. Rev. Med. Pharmacol. Sci.* 19, 1068–1076.

Lopez, P., Gonzalez-Rodriguez, I., Gueimonde, M., Margolles, A., and Suarez, A. (2011). Immune response to *Bifidobacterium bifidum* strains support Treg/Th17 plasticity. *PLoS ONE* 6:e24776. doi: 10.1371/journal.pone.0024776

Marteau, P., Cuillerier, E., Meance, S., Gerhardt, M. F., Myara, A., Bouvier, M., et al. (2002). *Bifidobacterium animalis* strain DN-173 010 shortens the colonic transit time in healthy women: a double-blind, randomized, controlled study. *Aliment. Pharmacol. Ther.* 16, 587–593. doi: 10.1046/j.1365-2036.2002.01188.x

Marteau, P., Guyonnet, D., Lafaye, de Micheaux, P., and Gelu, S. (2013). A randomized, double-blind, controlled study and pooled analysis of two identical trials of fermented milk containing probiotic *Bifidobacterium lactis* CNCM I-2494 in healthy women reporting minor digestive symptoms. *Neurogastroenterol. Motil.* 25:e252. doi: 10.1111/nmo.12078

Martin, R., Chain, F., Miquel, S., Lu, J., Gratadoux, J. J., Sokol, H., et al. (2014). The Commensal Bacterium *Faecalibacterium prausnitzii* is protective in DNBS-induced chronic moderate and severe colitis models. *Inflamm. Bowel. Dis.* 20, 417–430. doi: 10.1097/01.MIB.0000440815.76627.64

Martin, R., Miquel, S., Chain, F., Natividad, J. M., Jury, J., Lu, J., et al. (2015). *Faecalibacterium prausnitzii* prevents physiological damages in a chronic low-grade inflammation murine model. *BMC Microbiol.* 15:67. doi: 10.1186/s12866-015-0400-1

Mennigen, R., Nolte, K., Rijcken, E., Utech, M., Loeffler, B., Senninger, N., et al. (2009). Probiotic mixture VSL#3 protects the epithelial barrier by maintaining tight junction protein expression and preventing apoptosis in a murine model of colitis. *Am. J. Physiol. Gastrointest. Liver Physiol.* 296, 1140–1149. doi: 10.1152/ajpgi.90534.2008

Natividad, J. M., Hayes, C. L., Motta, J. P., Jury, J., Galipeau, H. J., Philip, V., et al. (2013). Differential induction of antimicrobial REGIII by the intestinal microbiota and *Bifidobacterium breve* NCC2950. *Appl. Environ. Microbiol.* 79, 7745–7754. doi: 10.1128/AEM.02470-13

Natividad, J. M., and Verdu, E. F. (2013). Modulation of intestinal barrier by intestinal microbiota: pathological and therapeutic implications. *Pharmacol. Res.* 69, 42–51. doi: 10.1016/j.phrs.2012.10.007

O'Connell Motherway, M., Watson, D., Bottacini, F., Clark, T. A., Roberts, R. J., Korlach, J., et al. (2014). Identification of restriction-modification systems of *Bifidobacterium animalis* subsp. lactis CNCM I-2494 by SMRT sequencing and associated methylome analysis. *PLoS ONE* 9:e94875. doi: 10.1371/journal.pone.0094875

Ohman, L., and Simren, M. (2007). New insights into the pathogenesis and pathophysiology of irritable bowel syndrome. *Dig. Liver Dis.* 39, 201–215. doi: 10.1016/j.dld.2006.10.014

Perrier, C., and Corthésy, B. (2011). Gut permeability and food allergies. *Clin. Exp. Allergy* 41, 20–28. doi: 10.1111/j.1365-2222.2010.03639.x

Picard, C., Fioramonti, J., Francois, A., Robinson, T., Neant, F., and Matuchansky, C. (2005). Review article: bifidobacteria as probiotic agents – physiological effects and clinical benefits. *Aliment. Pharmacol. Ther.* 22, 495–512. doi: 10.1111/j.1365-2036.2005.02615.x

Prasad, S., Mingrino, R., Kaukinen, K., Hayes, K. L., Powell, R. M., MacDonald, T. T., et al. (2005). Inflammatory processes have differential effects on claudins 2, 3 and 4 in colonic epithelial cells. *Lab. Invest* 85, 1139–1162. doi: 10.1038/labinvest.3700316

Rochet, V., Rigottier-Gois, L., Ledaire, A., Andrieux, C., Sutren, M., Rabot, S., et al. (2008). Survival of Bifidobacterium animalis DN-173 010 in the faecal microbiota after administration in lyophilised form or in fermented product - a randomised study in healthy adults. *J. Mol. Microbiol. Biotechnol.* 14, 128–136. doi: 10.1159/000106092

Shashidharamurthy, R., Machiah, D., Aitken, J. D., Putty, K., Srinivasan, G., Chassaing, B., et al. (2013). Differential role of lipocalin 2 during immune complex-mediated acute and chronic inflammation in mice. *Arthritis Rheum.* 65, 1064–1073. doi: 10.1002/art.37840

Suzuki, T., Yoshinaga, N., and Tanabe, S. (2011). Interleukin-6 (IL-6) regulates claudin-2 expression and tight junction permeability in intestinal epithelium. *J. Biol. Chem.* 286, 31263–31271. doi: 10.1074/jbc.M111.238147

Tambuwala, M. M., Cummins, E. P., Lenihan, C. R., Kiss, J., Stauch, M., Scholz, C. C., et al. (2010). Loss of prolyl hydroxylase-1 protects against colitis through reduced epithelial cell apoptosis and increased barrier function. *Gastroenterology* 139, 2093–2101. doi: 10.1053/j.gastro.2010.06.068

Tavan, E., Cayuela, C., Antoine, J. M., Trugnan, G., Chaugier, C., and Cassand, P. (2002). Effects of dairy products on heterocyclic aromatic amine-induced rat colon carcinogenesis. *Carcinogenesis* 23, 477–483. doi: 10.1093/carcin/23.3.477

Ukena, S. N., Anurag, S., Dringenberg, U., Engelhardt, R., Seidler, U., Hansen, W., et al. (2007). Probiotic *Escherichia coli* Nissle 1917 inhibits leaky gut by enhancing mucosal integrity. *PLoS ONE* 2:e1308. doi: 10.1371/journal.pone.0001308

Vaarala, O. (2012). Is the origin of type 1 diabetes in the gut? *Immunol. Cell Biol.* 90, 271–276. doi: 10.1038/icb.2011.115

Veiga, P., Gallini, C. A., Beal, C., Michaud, M., Delaney, M. L., DuBois, A., et al. (2010). Bifidobacterium animalis subsp. lactis fermented milk product reduces inflammation by altering a niche for colitogenic microbes. *Proc. Natl. Acad. Sci. U.S.A.* 107, 18132–18137. doi: 10.1073/pnas.1011737107

Wisner, D. M., Harris, L. R., Green, C. L., and Poritz, L. S. (2008). Opposing regulation of the tight junction protein claudin-2 by interferon-gamma and interleukin-4. *J. Surg. Res.* 144, 1–7. doi: 10.1016/j.jss.2007.03.059

Wrzosek, L., Miquel, S., Noordine, M. L., Bouet, S., Joncquel Chevalier-Curt, M., Robert, V., et al. (2013). *Bacteroides thetaiotaomicron* and *Faecalibacterium prausnitzii* influence the production of mucus glycans and the development of goblet cells in the colonic epithelium of a gnotobiotic model rodent. *BMC Biol.* 11:61. doi: 10.1186/1741-7007-11-61

Zheng, B., van Bergenhenegouwen, J., Overbeek, S., van de Kant, H. J., Garssen, J., Folkerts, G., et al. (2014). *Bifidobacterium breve* attenuates murine dextran sodium sulfate-induced colitis and increases regulatory T cell responses. *PLoS ONE* 9:e95441. doi: 10.1371/journal.pone.0095441

Zhu, J., and Paul, W. E. (2008). CD4 T cells: fates, functions, and faults. *Blood* 112, 1557–1569. doi: 10.1182/blood-2008-05-078154

Zuo, L., Yuan, K. T., Yu, L., Meng, Q. H., Chung, P. C., and Yang, D. H. (2014). *Bifidobacterium infantis* attenuates colitis by regulating T cell subset responses. *World J. Gastroenterol.* 20, 18316–18329. doi: 10.3748/wjg.v20.i48.18316

**Conflict of Interest Statement:** The authors declare that the research was conducted in the absence of any commercial or financial relationships that could be construed as a potential conflict of interest.

# 17

# Effect of a Ropy Exopolysaccharide-producing *Bifidobacterium animalis* subsp. *lactis* Strain Orally Administered on DSS-Induced Colitis Mice Model

*Claudio Hidalgo-Cantabrana[1,2], Francesca Algieri[2], Alba Rodriguez-Nogales[2], Teresa Vezza[2], Pablo Martínez-Camblor[3,4], Abelardo Margolles[1], Patricia Ruas-Madiedo[1]* and Julio Gálvez[2]*

[1] Department of Microbiology and Biochemistry of Dairy Products, Instituto de Productos Lácteos de Asturias – Consejo Superior de Investigaciones Científicas, Villaviciosa, Spain, [2] CIBER-EHD, Department of Pharmacology, ibs.GRANADA, Center for Biomedical Research, University of Granada, Granada, Spain, [3] Geisel School of Medicine at Dartmouth, Hanover, NH, USA, [4] Universidad Autónoma de Chile, Santiago, Chile

**Edited by:**
*Marco Ventura, University of Parma, Italy*

**Reviewed by:**
*Douwe Van Sinderen, University College Cork, Ireland*
*Chiara Ferrario, University of Parma, Italy*

***Correspondence:**
*Patricia Ruas-Madiedo*
*ruas-madiedo@ipla.csic.es*

Exopolysaccharide (EPS)-producing bifidobacteria, particularly *Bifidobacterium animalis* subsp. *lactis* strains, are used in the functional food industry as promising probiotics with purported beneficial effects. We used three isogenic strains of *B. animalis* subsp. *lactis*, with different EPS producing phenotypes (mucoid-ropy and non-ropy), in order to determine their capability to survive the murine gastrointestinal tract transit, as well as to evaluate their role in improving clinical outcomes in a chemically-induced colitis model. The three strains were able to survive in the intestinal tract of C57BL/6J mice during the course of the intervention study. Furthermore, the disease activity index (DAI) of the animal group treated with the ropy strain was significantly lower than of the DAI of the placebo group at the end of the treatment. However, no significant differences were found among the three strains. The analysis of several immune parameters, such as TNFα and IL-10 quantified in blood plasma and lymphocyte populations enumerated in mesenteric nodes, showed some significant variations among the four experimental animal groups. Remarkably, a higher capability of the ropy strain to increase regulatory T-cells in mesenteric lymphoid nodes was demonstrated, suggesting a higher ability of this strain to regulate inflammatory responses at mucosal level. Our data indicate that strains of *B. animalis* subsp. *lactis* producing EPS that confer a mucoid-ropy phenotype could represent promising candidates to perform further studies targeting intestinal inflammatory processes.

**Keywords:** ***Bifidobacterium*, exopolysaccharide, mucoid, ropy, immune modulation, DSS-colitis, mouse model**

## INTRODUCTION

The human gut is one of the most densely populated ecosystems inhabited by a community of microorganisms that actively contributes to the health of the host. Microbiota establishes a symbiotic relationship with the host, but also their members must co-exist in a balanced state. Indeed, disturbances in microbiota composition (dysbiosis) and function, as well as in the microbiota-host cross-talk, are the basis of prevalent gastrointestinal diseases, and also of some extra-intestinal pathology (Tojo et al., 2014). Among others, microbiota shifts have been related to chronic inflammatory disorders, such as inflammatory bowel disease (IBD) (Huttenhower et al., 2014). Up to now, dietary interventions toward restoring the unbalanced microbiota, such as the oral administration of probiotics, have been a realistic approach for human applications (Reid et al., 2011; Collins, 2014). Hill et al. (2014) have recently reviewed the probiotic concept and they have adopted, with a minor grammatical correction, the definition "live microorganisms that, when administered in adequate amounts, confer a health benefit on the host," previously proposed by the FAO and WHO organizations (FAO/WHO. Food and Agricultural Organization of the United Nations and World Health Organization, 2001). Research on probiotics has generated a vast amount of literature, mainly related to the characterisation of the probiotic potential of *Lactobacillus* and *Bifidobacterium*, the genera most commonly used for human consumption. However, only a small number of strains have been evaluated for the prevention or treatment of gastrointestinal disorders, such as IBD, diverse diarrheas, irritable bowel syndrome (IBS) or necrotising enterocolitis (NEC) (WGO. World Gastroenterology Organisation, 2011; Sanders et al., 2013). In relation to bifidobacteria, specific strains belonging to the species *Bifidobacterium breve*, *Bifidobacterium longum*, *Bifidobacterium bifidum*, and *Bifidobacterium animalis* subsp. *lactis*, have been tested in human intervention studies resulting in positive effects on gastrointestinal dysfunctions (Tojo et al., 2014).

*B. animalis* subsp. *lactis* is the species that has been most successfully introduced in food formulations, since it is able to deal with technological challenges during food manufacture (Prasanna et al., 2014). However, analysis of the available genomes shows a very low genetic variability among members of this subspecies (Milani et al., 2013). One of the features shared in all of their genomes is the presence of a cluster involved in the synthesis of exopolysaccharides (EPS) which are carbohydrate polymers surrounding the cell wall (Hidalgo-Cantabrana et al., 2014b). These EPS play a relevant role for the producing strain because they are involved in protection and niche colonization, but EPS also act as intermediaries in the dialog established between bacteria and host (Hidalgo-Cantabrana et al., 2014b). It has been proposed that EPS are able to interact with receptors located in the gut epithelium (Lebeer et al., 2010b) and they could act as effector molecules eliciting different immune responses (Hidalgo-Cantabrana et al., 2012).

In recent years our group has been working on the study of biological properties of EPS synthesized by bifidobacteria. One of the *B. animalis* subsp. *lactis* strains from our collection displayed a mucoid-ropy phenotype, denoted by the formation of a long filament from the colony, and it was able to synthesize a high molecular weight (HMW)-EPS. Recently, working with three recombinant strains, we demonstrated the involvement of a mutation (C> 266 >T) in *wzz* gene in the occurrence of this phenotype and in the synthesis of the HMW-EPS fraction. This gene encodes a hypothetical membrane anchored protein, with a predicted large soluble domain, which is theoretically involved in the chain length determination of the polymer (Hidalgo-Cantabrana et al., 2015). The acquisition of the ropy phenotype favored the *in vitro* survival and adhesion of the recombinant strain to the gut environment. In turn, this strain also elicited an *in vitro* and *ex vivo* anti-inflammatory cytokine profile (Hidalgo-Cantabrana et al., 2015). Thus, in a step forward the current work aims to *in vivo* evaluate the probiotic potential of the ropy *B. animalis* subsp. *lactis* Balat_1410$^{S89L}$ checking its capability to counteract an inflamed state of colitis chemically induced by dextran sodium sulfate (DSS) administration.

## MATERIALS AND METHODS

### Bifidobacterial Strains and Preparation of Lyophilized Bacteria in Milk

Three isogenic EPS-producing *B. animalis* subsp. *lactis* strains from the IPLA-CSIC collection were used in this study to test their probiotic capabilities in an *in vivo* model. Based on the type strain DSM10140, recombinant strains were obtained from a previous study (Hidalgo-Cantabrana et al., 2015): strain ΔBalat_1410 (DSM10140 lacking the gene Balat_1410 complemented with pAM1, non ropy), strain Balat_1410 (strain ΔBalat_1410 complemented with pAM1 + Balat_1410, non ropy), and Balat_1410$^{S89L}$ (strain ΔBalat_1410 complemented with pAM1 + Balat_1410$^{S89L}$, ropy). Glycerol stocks kept at −80°C were plated on agar MRSC [MRS Difco (BD, Biosciences, San Diego, CA) containing 0.25% L-cysteine-HCl (Sigma-Chemical Co., St. Louis, MO)] supplemented with erythromycin (2.5 μg/ml) (MRSCE). Plates were incubated at 37°C under anaerobic conditions (80% $N_2$, 10% $CO_2$, 10% $H_2$) in an MG500 chamber (Don Whitley Scientific, West Yorkshire, UK). A single colony per strain was used to inoculate 50 ml MRSCE broth and after 20 h this culture was used to inoculate 2% (v/v) 1 L fresh medium. After 20 h incubation, cultures were centrifuged, washed once with PBS and resuspended in 100 ml tindalized 11% skimmed milk (Difco). Then, bacterial suspensions in milk were lyophilized for 48 h in the Virtis Freezemobile 12EL equipment (SP Scientific, NY). Each lyophilized bifidobacterial stock was tested for bacterial viability and enumeration by counting in agar-MRSCE and incubated under standard conditions. For oral administration, each lyophilized bifidobacteria was resuspended daily in water (at $5 \times 10^9$ cfu/ml) and 100 μl was administered per mouse.

### Animals, Experimental Designs and Sample Collection

All studies were carried out following the Directive 2010/63/EU of the European Parliament and the Council on the protection of animals used for scientific purposes (Directive 2010/63/EU, 2010). The experimental protocol was approved by the Ethical Committee of Laboratory Animals of the University of Granada (Spain) (Permit Number CEEA-2010-286). Male C57BL/6J mice (7–9 weeks old, approximately 20 g) were obtained from Janvier Labs (St Berthevin Cedex, France) and kept under conventional conditions with a standard pelleted diet and sterilized water for 1 week before beginning the experiments.

### Bifidobacteria Survival Study

Animals (32 mice) were randomly assigned to four experimental groups with 8 mice per group: three bifidobacterial groups, each orally receiving 100 μl/day ($5 \times 10^8$ cfu) of the corresponding milk-bifidobacterial suspension in water, and one placebo (control) group which received 100 μl/day of skimmed milk. Placebo and bifidobacteria suspensions were orally administered for 9 days by means of an intra-gastric cannula. Every day, the drinking water and food intake was measured for each group of mice, whereas the body weight was measured individually for each animal. Stool samples were taken, in duplicate, at group level every 2 days, including the first and last day. At the end of the intervention period (day 10), each mouse was anesthetized with isoflurane and then, the animals were sacrificed by cervical dislocation by expert and qualified persons, according to the Federation of European Laboratory Animal Science Associations (FELASA). The colon was excised, its content was collected and its length and weight was measured after washing with PBS. Stool and colon-content samples were immediately processed as described in the next section.

### Dextran Sodium Sulfate (DSS) Induced Colitis Study

Mice (40 animals) were randomly assigned to five groups, each with 8 animals, as described in **Figure** . The reference (non-colitic, non-placebo/bifidobacteria fed) group received daily, through an intra-gastric cannula, 100 μl water during the 15-day experimental period. The other four groups were fed with bifidobacteria, or a placebo, and treated with DSS. During the 15 days of the experimental period these four groups received a placebo (100 μl/day milk, group 1), or different bifidobacterial strains (100 μl/day $5 \times 10^8$ cfu milk-bifidobacterial suspensions in water, for groups 2, 3, and 4). After 9 days of bifidobacteria/placebo feeding, colitis was induced in the four groups by adding DSS (3%, w/v, 36–50 kDa, MP Biomedicals, CA) in the drinking water (Mähler et al., 1998). This DSS-treatment was kept for a period of 6 days unless the application of humane end-point, as defined below, was needed. At the end of the experimental period (day 15) animals were anesthetized and a blood sample was directly extracted from the heart using heparinized tubes. The blood plasma was obtained by centrifugation (9,300*xg*, 4°C, 20 min). The animals were then euthanatized in order to collect colon and colon-content, as described above, as well as mesenteric lymphoid nodes. Samples were immediately processed or stored at −20°C until their analysis.

Drinking water and food intake was measured at group level each day, whereas body weight, stool consistency and the presence of gross blood in feces were evaluated daily and scored (**Table** ) for each mouse during the experimental period. The disease activity index (DAI) was calculated as previously reported by Cooper et al. (1993): total score [body weight decrease + stool consistency + rectal bleeding] / 3. The ulcerative colitis (UC) disease is considered for DAI $\geq$ 1.5 and the humane endpoint was fixed at DAI = 3. Stool samples were taken at group level every 2 days to check viability of bifidobacteria.

## Microbial Quantification in Fecal Samples

### Bifidobacterial Quantification by Counting

Stool samples, collected in both studies as previously indicated, were homogenized in PBS (0.1 g/ml) using a Heidolph (Heidolph Instruments GmbH, Schwabach, Germany) stirrer for 1 min. Then, 100 μl was used to obtain serial dilutions in PBS which were plated onto the surface of TOS propionate agar medium (Merck, Darmstadt, Germany), supplemented with lithium-mupirocin as recommended by the manufacturer, to selectively enumerate total bifidobacteria. Additionally, agar lithium-mupirocin-TOS was supplemented with 2.5 μg/ml erythromycin (TOS+Ery) which is the selective marker of the plasmid present in the three *B. animalis* subsp. *lactis* strains under study. All plates were incubated at 37°C in an anaerobic jar using Oxoid-Anaerogen™ (Thermo Fisher Scientific Inc., Waltham, MA) sachets to generate anaerobic conditions for 72 h.

### Bifidobacterial Identification by 16S rDNA Amplification

Several colonies were picked up from TOS+Ery plates of each of the four experimental groups belonging to the bifidobacteria survival study. The DNA from a total of 67 colonies was isolated using the GenElute Bacterial Genomic DNA kit (Sigma) according to the manufacturer's instructions, but adding a prior step of bacterial lysis with mutanolysin and lysozyme. Primer uses were: pA/pH (Edwards et al., 1989). The PCR mixture (final volume 50 μl) was: 1 μl chromosomal DNA, 0.20 μM of each primer, 200 μM dNTPs (Amersham Bioscience, Upsala, Sweden) and 2.5 U Taq DNA-polymerase (Eppendorf, Hamburg, Germany). After an initial cycle at 95°C, the PCR amplification consisted of 30 cycles: 95°C for 1 min, 55°C for 50 s and 68°C for 2 min, and ended with a final extension step of 68°C for 10 min. Reactions were carried out in the thermal cycler UnoCycler (VWR International Eurolab S.L., Barcelona, Spain). Purification and sequencing of the amplicons were performed in Macrogen (Seoul, Korea). The free Chromas 1.45 software (Technelysium Pty Ltd., Australia) was used to process sequences which were compared to those held in the GenBank database (http://www.ncbi.nlm.nih.gov/genbank) using the BLASTn tool.

## Gene Expression Analyses in Colonic Samples by Reverse Transcriptase qPCR

After colon excision, PBS-washed colonic tissue was sectioned in small fragments which were placed in RNAlater stabilization reagent (Qiagen GmbH, Hilden, Germany) and then stored at −80°C until RNA extraction. Total RNA was isolated using

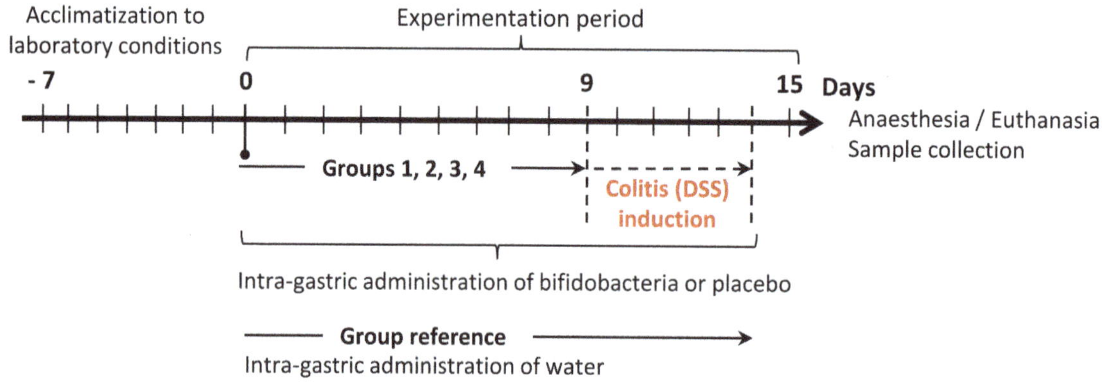


**FIGURE | Experimental design to check the capability of EPS-producing *B. animalis* subsp. *lactis* strains to counteract the dextran sodium sulfate (DSS)-induced colitis.** Group 1 (placebo) daily receiving 100 μl skimmed milk and Groups 2–4 daily receiving 100 μl milk with 5 × $10^8$ cfu bifidobacteria: group 2, strain ΔBalat_1410 (DSM10140 lacking the gene Balat_1410 complemented with pAM1, non ropy); group 3, strain Balat_1410 (strain ΔBalat_1410 complemented with pAM1 + Balat_1410, producing a non-ropy EPS); group 4, strain Balat_1410$^{S89L}$ (strain ΔBalat_1410 complemented with pAM1 + Balat_1410$^{S89L}$, producing a ropy EPS). Reference group (non DSS-induced colitis, non-placebo/bifidobacteria fed group) daily receiving 100 μl water.

**TABLE | Disease activity index (DAI) score used to evaluate the DSS-induced colitis. DAI index was calculated as total score (body weight decrease + stool consistency + rectal bleeding) divided by 3.**

| Score | Body weight decrease (%) | Stool consistency | Rectal bleeding |
|---|---|---|---|
| 0 | <1 | Normal | Normal |
| 1 | 1–5 | | |
| 2 | 5–10 | Loose stools | |
| 3 | 10–20 | | |
| 4 | >20 | Diarrhea | Gross bleeding |

*The humane end-point was established at DAI = 3.*

RNeasy Mini Kit (Qiagen) according to the manufacturer's recommendations. Purity and RNA concentration were determined with the NanoDropTM 2000 spectrophotometer. Then, 3 μg RNA was reverse transcribed to obtain cDNA using oligo(dT) primers and reagents from Promega (Promega, Southampton, UK) and the TProfessional basic thermocycler (Biometra GmbH, Goettingen Germany).

The amplification was performed in an Eco™ Real-Time PCR System (Illumina, San Diego, CA USA) using 20 ng of cDNA and specific primers (Table ). The $2^{-\Delta\Delta Ct}$ method was used to normalize expression results (Livak and Schmittgen, 2001). The values of the house-keeping glyceraldehyde-3-phosphate dehydrogenase (GAPDH) gene were first used to normalize the values obtained for each of the genes under study. These values were additionally normalized to those obtained in the reference group, thus this group had a normalized value of 0. Finally the relative expression of each gene was calculated as $2^{-\Delta\Delta Ct}$, thus the reference group had the relative expression value of 1.

## Cellular Population in Mesenteric Lymphoid Nodes

The mesenteric lymphoid nodes were carefully mashed with wet slides to decrease friction and the solutions were filtered through a 70 μM cell strainer. Cells were isolated, counted and plated on anti-CD3 (clone17A2, eBioscience, San Diego, CA, USA) and anti-CD28 (clone 37.51, eBioscience)—coated plates for FcγR blocking. Then, the cells were transferred to polystyrene tubes for surface staining with anti-CD4 (PerCP-Cy™5.5, clone RM4-5 BD Pharmigen™, Franklin Lake, NU, USA), anti-CD45 (APC-eFluor®780, clone 30-F11, eBioscience) and Viability Dye (eFluor®660, eBioscience) for 15 min at 4°C in the dark. The cells were then fixed, permeabilized with the Fixation/Permeabilization kit (eBioscience) and intracellular staining was done with anti-Foxp3 (PE, clone FJK-16s, eBioscience) for 30 min at 4°C in the dark. Data collection was performed using a flow cytometer CANTO II (BD Biosciences San Diego, CA, USA).

## Cytokine Levels of Blood Samples

The levels of tumor necrosis factor (TNF)-α and interleukin (IL)-10 cytokines in blood plasma were quantified using the mouse TNFα Quantikine and mouse IL-10 Quantikine ELISA kits (R&D Systems, Abingdon, UK) following the manufacturer's instructions.

## Statistical Analyses

All considered variables were distributed normally (Kolmogorov-Smirnov test was used to check normality); thus, they are described as mean ± standard deviation. Robust Student-Welch or ANOVA tests were used in order to check the equality among groups for all variables, and *post-hoc* T3 de Dunnet or LSD tests were used to compare each two groups. Standard linear regression models were employed to compute the size of the effect (95% confidence interval) of the different treatments on the DAI; to do this, crude data were adjusted by the initial conditions (DAI score the first day of DSS-induced colitis) and the placebo group was used as reference. Differences below 0.05 were considered statistically significant. All analyses were made with the free statistical software R (www.r-project.org).

# RESULTS AND DISCUSSION

## *B. animalis* subsp. *lactis* Strains Are Able to Colonize the Murine Gut

Orally administered probiotics should survive the upper gastrointestinal tract (GIT) transit and reach the colon alive, where they should persist for a certain time in order to be able to exert their beneficial effect. Therefore, an animal experimentation procedure was done to assess the viability of the three *B. animalis* subsp. *lactis* strains under study aiming to check whether the presence of the ropy HMW-EPS surrounding Balat_1410$^{S89L}$ strain could exert a protective role. Additionally, although *B. animalis* has QPS (qualified presumption of safety) status (EFSA BIOHAZ Panel, 2013), we evaluated the lack of toxicity of our strains, at this oral-dose administered to the animals, by measuring animal behavior and wellbeing parameters. In this regard, no differences ($p > 0.05$) were found in water and food intake, body weight, stool consistency or colon anatomy between the placebo (skimmed milk-fed) and each of the bifidobacteria-fed groups (**Table** ). These results support the fact that the three bifidobacterial strains used in this study had no side effects at the concentration level administered to this murine model.

The bifidobacterial population in the fecal samples was enumerated using agar-TOS (**Figure** ) which is a selective medium for bifidobacteria growing in milk products. After 2 days of administration, the bifidobacterial levels increased in all groups about 1 log unit, including the placebo receiving skimmed milk. However, during prolonged treatment the population of bifidobacteria in the placebo group tended to decline to reach the initial values, whereas counts in the three bifidobacteria-fed groups remained stable. Similarly, counts obtained in TOS+Ery, which is the medium containing the selective marker erythromycin of the plasmid present in the three strains under study, were significantly higher ($p < 0.05$) in the three bifidobacterial-fed groups with respect to the placebo (**Figure** ). Levels of erythromycin-resistant bacteria in the microbiota of placebo-fed mice were around 4 log units and remained unaltered for the duration of the experimentation time. After 2 days of administration, levels of erythromycin-resistant bifidobacteria increased (on average) 0.7 log cfu/g in the mice receiving bifidobacteria; the highest count increase (1.7 log cfu/g) was reached after 4 days, remaining at similar values until the end of the experimentation period. In general, no statistical differences among the three bifidobacteria-fed groups were found and, additionally, the colonies chosen to partially sequence the gene 16S rDNA were identified as members of *Bifidobacterium* genus; those isolated from Balat_1410$^{S89L}$ fed animals displayed a ropy phenotype as well (see Table ). Altogether, these results show that our bifidobacterial strains were able to survive as members of the intestinal microbiota of mice for at least 9 days, following a daily intake of $5 \times 10^8$ cfu. The increase in total bifidobacteria observed in this study could be related to the ability of EPS-producing *B. animalis* subsp. *lactis* to raise the number of other bifidobacterial species in the intestinal microbiota of rodents as previously described (Salazar et al., 2011). This effect could be related to the ability of members from this microbial community to use EPS as fermentable substrates (Salazar et al., 2009). However, the increase of erythromycin resistant members in feces is indicative of a higher survival rate in the gut of the specific strains used in this work. The *in vivo* performance of the three strains is rather similar, since there was no statistical difference in the number of bifidobacteria present in the fecal samples among the three groups of mice. Thus, it seems that the presence of HMW-EPS on the surface of the ropy Balat_1410$^{S89L}$ strain does not confer an additional protection of survival in mice gut. Fanning et al. (2012) found that fecal counts (made in Reinforced *Clostridium* agar with 50 mg/L mupirocin) of mice fed with EPS$^+$ *B. breve* UCC2003 were significantly higher than those obtained with two non-EPS producing mutant strains. Thus, these authors conclude that the presence of an EPS on the surface of the wild type *B. breve* was involved in its higher survival rate in the mice gut. On the contrary, *in vivo* studies carried out with different EPS-producing *Lactobacillus johnsonii* strains showed that mutants with deletion of genes from the *eps* cluster had an increased residence time in the animal's gut (Denou et al., 2008; Horn et al., 2013). It has also been shown that the loss of EPS production in *Lactobacillus rhamnosus* GG improves the *in vitro* adhesion of the strain to Caco2 cells (Lebeer et al., 2009); however, the presence of the EPS surrounding the wild type GG strain improves its *in vivo* performance allowing the strain to persist for longer periods because the EPS protects against the antimicrobial peptide LL-37 (Lebeer et al., 2010a). The intrinsic characteristics of EPS, such as composition and size, seem to promote different effects on bacterial survival under gut challenges. In some cases, polymers could prevent the exposure of molecules from the bacterial envelope involved in adhesion, but EPS can also act as a shield against host immune defense (Hidalgo-Cantabrana et al., 2012).

## The Ropy *B. animalis* subsp. *lactis* Balat_1410$^{S89L}$ Reduces DSS-Induced Damage

As mentioned above, *in vitro* and *ex vivo* tests (using human tissues) showed different immune modulation capabilities of the three isogenic bifidobacteria, the ropy strain Balat_1410$^{S89L}$ having an anti-inflammatory profile (Hidalgo-Cantabrana et al., 2015). To expand on these observations, we wanted to address whether this predicted anti-inflammatory ability is effective *in vivo,* modifying the immune response in a model of gut inflammation such as mice with DSS-induced colitis. This is a well-established animal model of acute mucosal inflammation that has been used for several decades (Wirtz et al., 2007) and it has been validated as a model for the translation of mice data to humans (Melgar et al., 2008).

In our experimental procedure, four groups of mice were pre-treated with an oral administration of bifidobacteria, or placebo, for 9 days before the administration of DSS in drinking water, which was continued for an additional 6 days (**Figure** ). As was expected, in accordance with results shown in the previous section, the pre-feeding for 9 days in the four groups did not modify the wellbeing and behavior parameters of the mice. This is also consistent with the absence of significant variations in the disease activity index (DAI) score of these groups (values

**TABLE | Mean and standard deviation of general parameters measured in four experimental groups of mice to evaluate the absence of toxicity of the bifidobacterial strains administered for 10 days at dose of 5 × 10$^8$ cfu per mouse and day.**

| Experimentation group | Food intake[a] (g/mouse) | Drink water[a] (ml/mouse) | Body weigh increase[b] (g) | Body weigh increase[b] (%) | Ratio weight/length of colon[c] |
|---|---|---|---|---|---|
| Placebo (skim milk) | 2.22 ± 0.39 | 2.52 ± 0.48 | 0.73 ± 0.17 | 3.24 ± 0.75 | 0.024 ± 0.005 |
| Strain ΔBalat_1410 | 2.49 ± 0.23 | 2.79 ± 0.34 | 0.73 ± 0.16 | 3.20 ± 0.70 | 0.024 ± 0.009 |
| Strain Balat_1410 | 2.24 ± 0.29 | 2.68 ± 0.45 | 0.85 ± 0.25 | 3.84 ± 1.13 | 0.025 ± 0.008 |
| Strain Balat_1410$^{S89L}$ | 2.18 ± 0.31 | 2.84 ± 0.51 | 0.66 ± 0.18 | 2.89 ± 0.78 | 0.026 ± 0.006 |

*No significant statistical differences ($p > 0.05$) were found among the four experimental groups after one-way ANOVA study.*
[a] *Values obtained for the whole mice group for each day of experimental procedure.*
[b] *body weight at day 10 subtracting the body weight measured at day 0 (27.71 ± 1.11) before the experimental procedure.*
[c] *colon weight and length determined after 10 days of experimental procedure.*

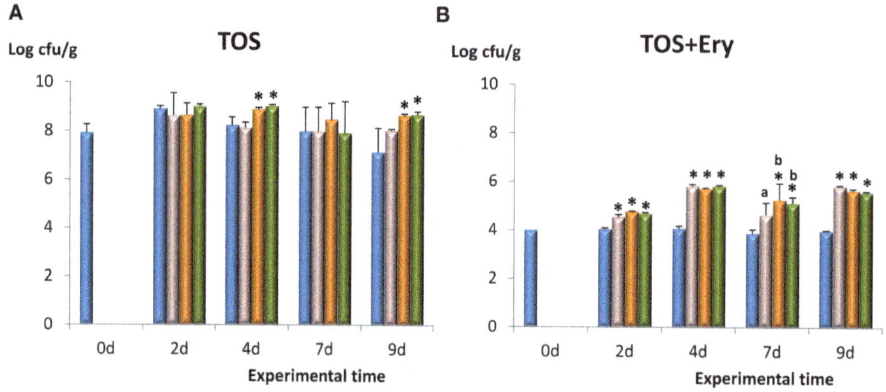


**FIGURE | Counts in agar-TOS to enumerate total bifidobacterial population (A) and agar-TOS+Ery (supplemented with 2.5 μg/ml erythromycin, the antibiotic marker of strains under study) to enumerate erythromycin-resistant bacteria (B) in fecal homogenates obtained from four experimental groups of mice along the experimental period.** Placebo group (■), feed *B. animalis* subsp. *lactis* ΔBalat_1410 (■), feed *B. animalis* subsp. *lactis* Balat_1410 (■), and feed *B. animalis* subsp. *lactis* Balat_1410$^{S89L}$ (■). Within each treatment day: statistical differences with respect to the control group are marked with an asterisk ($p < 0.05$) and differences among the three bifidobacterial groups are denoted with different letters ($p < 0.05$).

lower than 0.5, data not shown). The DAI score was kept close to 0 in the non-colitic, non-placebo/ bifidobacteria-fed (reference) group during all experimental procedures. In the four colitic groups, after the second day (day 11) of DSS treatment, the DAI values increased continuously (**Figure** ). The maximum score was obtained at the end of the experimental procedure, after 6 days of DSS colitic induction. In general, no statistical differences were detected on each day of treatment among the placebo and the three bifidobacteria-fed groups during DSS-treatment; the exception was the group fed with the strain Balat_1410$^{S89L}$, which showed a significantly ($p < 0.05$) lower DAI score than the placebo group at day 15. These results were confirmed when DAI data were adjusted by means of linear regression models taking into consideration the DAI of placebo group each day of DSS-induction (**Figure** ). This model allowed us to determine the effect of the treatments on the DAI with respect to the placebo (value = 0). Thus, negative values of the adjusted DAI indicate a reduction in the severity of the disease in comparison with the placebo group. The highest reduction in adjusted DAI obtained for *B. animalis* subsp. *lactis* Balat_1410$^{S89L}$ fed animals at the end of the treatment was mainly associated with a reduction in diarrhea and rectal bleeding. This led to a 44% reduction in the severity of the disease with respect to the placebo ($p < 0.007$), whereas, in the other two bifidobacteria-fed animals this reduction was not statistically significant, although the reduction of the illness was 23.2% for the strain ΔBalat_1410 and 16.7 % for the strain Balat_1410. In spite of the remarkable differences among the three strain-fed groups (disease severity in Balat_1410$^{S89L}$ was reduced by almost two and three-fold compared to ΔBalat_1410 and Balat_1410, respectively), these differences were not deemed statistically significant. Therefore, we cannot ascertain that the protective effect against DSS-induced colitis exerted by strain *B. animalis* subsp. *lactis* Balat_1410$^{S89L}$ is attributed to the HMW-EPS surrounding the bacteria. Some authors have reported the capability of probiotic strains to reduce the DAI index of animals under DSS-induced colitis. However, a mixture of several bacteria, or combinations of bacteria-bioactive compounds, are often tested, thus the positive effect cannot be assigned to a single bacterium (Dai et al., 2013). Only a few reports have shown the efficacy of single bifidobacterial strains (Hayes et al., 2014; Zheng et al., 2016).

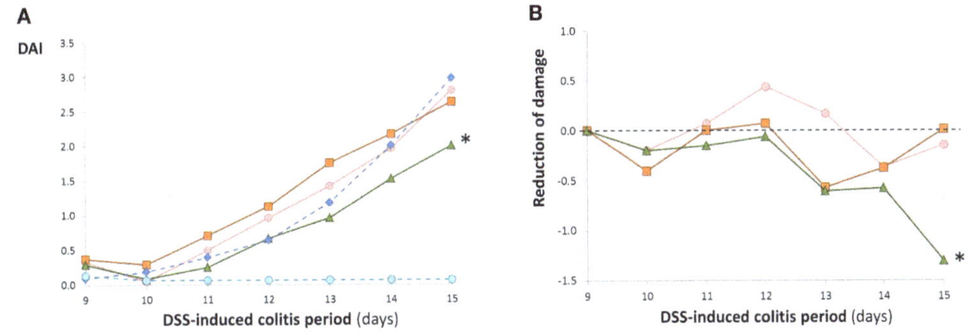


**FIGURE | Mean values of the disease activity index (DAI, Table ) calculated for each experimental group.** The coefficient of variation (SD/mean) percentage of these values varied between 9 and 70%. Each bifidobacteria-fed group was compared with the placebo throughout the DSS-treatment and statistical differences ($p < 0.05$) are marked with an asterisk **(A)**. The DAI score of each bifidobacterial group was adjusted, as indicated in material and methods, to values of placebo at the first day of DSS-treatment (value = 0). Negative values indicate a higher reduction in the disease severity with respect to the placebo. These adjusted data were compared with placebo during the 5 days of DSS-treatment and statistical differences ($p < 0.05$) are marked with an asterisk **(B)**. Reference group (-○-); DSS-induced colitis groups: placebo (-◆-), feed *B. animalis* subsp. *lactis* ΔBalat_1410 (-○-), feed *B. animalis* subsp. *lactis* Balat_1410 (-■-), feed *B. animalis* subsp. *lactis* Balat_1410$^{S89L}$ (-▲-).

## Different *B. animalis* subsp. *lactis* Strains Elicit Specific Host Responses

In order to understand the mechanism(s) involved in the attenuation of the DSS-colitis by the strain Balat_1410$^{S89L}$ several immune parameters, targeting adaptive as well as innate responses, were measured in different tissues. In our study, the expression of several genes related to the reinforcement of the intestinal epithelium barrier, such as those coding for chemokines, proteins of tight junctions and mucins, showed no statistical variations ($p > 0.05$) among the four groups of animals (**Table** ). Only the non-ropy EPS-producing strain Balat_1410 was able to slightly induce ($p < 0.05$) the expression of Mcp1, a chemokine able to attract monocytes, as well as the secretory protein TFF3, which is synthetized by goblet cells and acts to stabilize the mucus layer. Thus, it seems that, based on these results of gene expression, our strains were not able to modify the epithelial barrier in a remarkable way. However, it was reported that probiotics are able to reinforce the intestinal barrier function by multiple mechanisms (Lebeer et al., 2010a; Rao and Samak, 2013). The efficacy of *Bifidobacterium infantis* BB-02 to reduce intestinal injury in NEC was due to its ability to keep claudin-4 and occludin located in the tight-junction of bifidobacteria-fed mice (Bergmann et al., 2013). It was also demonstrated that bioactive factors released by *B. infantis* (one of the strains present in the VSL#3 mixture) were also able to reduce gut permeability by acting upon tight junction proteins (Ewaschuk et al., 2008). Similarly *B. bifidum* OLB6378 also favored the location of tight-junction and adherent-junction proteins, also being able to induce the synthesis of mucin-3, TFF3, and IL-6 (Khailova et al., 2009). These works, among others, show that bifidobacteria are able to improve the gut barrier function in a strain dependent manner. Regarding expression of cytokine genes in colonic tissue, statistical differences ($p < 0.05$) were detected only for pro-inflammatory IL-1β and IFNγ genes and the anti-inflammatory IL-10 (**Table** ); in general, the two first genes were expressed at a lower rate in the placebo group with respect to the three bifidobacteria-fed groups, whereas IL-10 gene had significantly lower expression rate only for the ropy strain Balat_1410$^{S89L}$. Thus, it seems that this strain induced the expression of genes that could lead to a mild pro-inflammatory response, since the ratio TNFα/IL-10 was significantly higher than that obtained for the other three groups (data not shown). A similar pro-Th1 profile was obtained with wild-type EPS-producing *B. animalis* subsp. *lactis* strains (one of them also producing a ropy HMW-EPS) co-cultivated *in vitro* with PBMC and GALT (gut associated lymphoid tissue) cells isolated from naïve rats (Hidalgo-Cantabrana et al., 2014a). Indeed, in the current study the concentration of these cytokines measured in blood plasma (**Figure** ) gives TNFα/ IL-10 ratio values of 1.69 ± 0.97 for placebo, 4.32 ± 2.15 for strain ΔBalat_1410, 2.77 ± 1.85 for strain Balat_1410, and 10.28 ± 9.15 for strain Balat_1410$^{S89L}$. The ratio obtained for the animals fed with the ropy strain was significantly ($p < 0.05$) higher than that obtained for the other three groups of mice. However, it is worth noting that none of the three strains, or the placebo, was able to induce a strong immune systemic response since the values of both cytokines were near (or lower) than value 1, which is the value of the reference animal group. Thus, this suggests that our strains, or the skimmed milk (placebo) used as a protectant for bifidobacteria delivery, did not contribute to increase the inflammatory process induced by DSS.

Immune cells isolated from the lymphoid nodes associated with the gastrointestinal tract were quantified by flow cytometry and significant ($p < 0.05$) variations among the four experimental groups were detected (**Figure** ). Mice orally receiving the ropy Balat_1410$^{S89L}$ or the non-ropy Balat_1410 strains presented a significant reduction in the population of $CD4^+/CD45^+$ cells, which are the markers for helper T-lymphocytes. The number of this population was even slightly lower than that of the reference group (value 1). In addition, the group receiving the strain Balat_1410$^{S89L}$ was the only one significantly different

**TABLE | Relative expression of genes related to immunity of intestinal mucosa in colonic tissue.**

| Parameter[a] | | Animal experimentation groups | | | | p-value |
|---|---|---|---|---|---|---|
| | | Placebo | Strain ΔBalat_1410 | Strain Balat_1410 | Strain Balat_1410$^{S89L}$ | |
| Cytokines | TNFα | 8.93 ± 5.42 | 8.64 ± 3.14 | 10.48 ± 3.99 | 7.57 ± 1.89 | 0.752 |
| | IL-10 | 3.85 ± 1.68[b] | 3.85 ± 1.63[b] | 2.19 ± 0.72[a,b] | 1.75 ± 0.26[a] | **0.041** |
| | IL-17 | 0.93 ± 0.29 | 0.86 ± 0.29 | 0.76 ± 0.23 | 0.83 ± 0.26 | 0.662 |
| | IL-1β | 10.98 ± 7.12[a] | 11.88 ± 2.89[a] | 28.94 ± 7.36[b] | 22.98 ± 3.94[b] | **0.001** |
| | IL-6 | 10.61 ± 8.42 | 11.88 ± 7.14 | 12.98 ± 6.62 | 15.27 ± 2.80 | 0.800 |
| | IL-12 | 0.91 ± 0.41 | 1.39 ± 1.00 | 1.52 ± 0.63 | 0.93 ± 0.28 | 0.168 |
| | TGFβ | 2.30 ± 0.98 | 3.19 ± 2.30 | 1.96 ± 1.12 | 1.03 ± 0.75 | 0.309 |
| | IFNγ | 0.77 ± 0.69[a] | 2.07 ± 1.72[b] | 2.44 ± 0.94[b] | 2.57 ± 0.95[b] | **0.029** |
| Chemokines | iCAM-1 | 1.91 ± 0.35 | 1.81 ± 0.93 | 1.48 ± 0.81 | 1.52 ± 0.38 | 0.552 |
| | Mcp1 | 5.12 ± 1.28[a] | 4.93 ± 2.74[a] | 11.41 ± 5.08[b] | 4.39 ± 1.87[a] | **0.001** |
| Tight junctions | Occludin | 0.66 ± 0.22 | 0.45 ± 0.22 | 0.62 ± 0.22 | 0.46 ± 0.09 | 0.106 |
| | ZO-1 | 0.69 ± 0.26 | 0.50 ± 0.24 | 0.60 ± 0.27 | 0.48 ± 0.21 | 0.314 |
| Secretory protein | TFF3 | 1.54 ± 0.56[a] | 2.41 ± 1.15[a] | 3.67 ± 1.29[b] | 1.76 ± 0.37[a] | **0.002** |
| Mucins | Muc1 | 2.92 ± 1.40 | 3.99 ± 2.00 | 6.11 ± 3.23 | 4.52 ± 2.42 | 0.148 |
| | Muc2 | 0.63 ± 0.38 | 0.42 ± 0.21 | 0.49 ± 0.29 | 0.52 ± 0.12 | 0.436 |
| | Muc3 | 0.88 ± 0.32 | 1.24 ± 0.66 | 1.69 ± 1.00 | 1.64 ± 1.09 | 0.158 |
| Enzymes | Mmp9 | 2.52 ± 0.74 | 3.31 ± 1.60 | 2.13 ± 0.90 | 2.78 ± 1.25 | 0.342 |
| | INOS | 3.37 ± 2.46 | 6.52 ± 3.94 | 10.08 ± 8.63 | 5.06 ± 1.97 | 0.096 |

*Values in each DSS-colitic group were referred to the average of those obtained in the control (no-colitic, value = 1) one. P-values from one-way ANOVA tests carried out among four groups; those values within the same row that do not share a common letter are statistically different according to the mean comparison LSD ($p < 0.05$).*

*[a]IL-, interleukin-; TNFα, tumor necrosis factor α; TGF-β, transforming growth factor β; IFNγ, interferon γ; MUC-, mucin-; TFF-3, trefoil factor 3; ZO-1, zonula occludens 1; MMP-9, matrix metallopeptidase 9; iNOS, inducible nitric oxide synthase; ICAM-1, intercellular adhesion molecule.*

*[b]The statistical differences among the four bifidobacterial groups are denoted with different letters ($p < 0.05$).*

*Bold values underlines statistical differences according to one-way ANOVA test.*

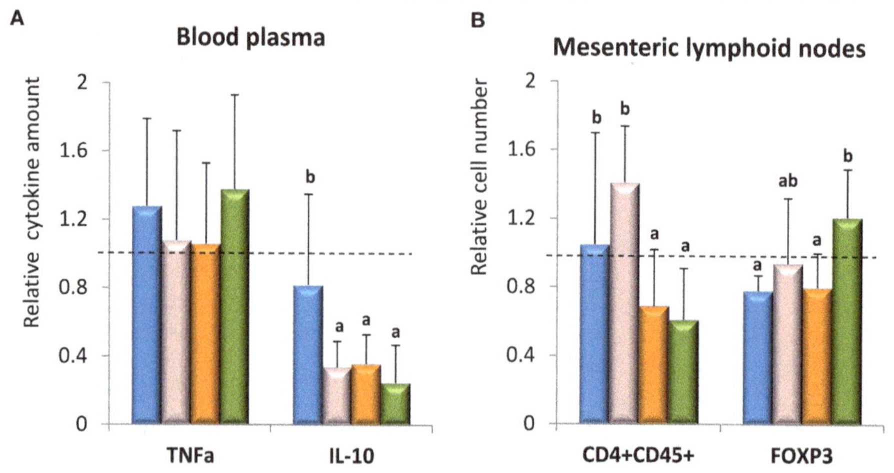


**FIGURE | Mean values of relative TNFα and IL-10 cytokines measured in blood samples (A) and relative number of $CD4^+CD25^+$ (T helper cells) and $FOXP_3$ (T regulatory cells) quantified in mesenteric lymphoid nodes (B).** Samples were collected after 6-days of DSS-treatment. Values of each DSS-colitic animal group were referred to as the average of those obtained in the reference group (value = 1). Placebo group (■), feed *B. animalis* subsp. *lactis* ΔBalat_1410 (■), feed *B. animalis* subsp. *lactis* Balat_1410 (■), feed *B. animalis* subsp. *lactis* Balat_1410$^{S89L}$ (■). Within each parameter: statistical differences among the four groups are denoted with different letters ($p < 0.05$).

from the placebo regarding the number of cells expressing Foxp3$^+$, which acts as a regulatory factor for differentiation and function of regulatory T-lymphocytes. Besides, values of Foxp3$^+$ cells in this group of animals, which also showed a reduced severity in DAI score, were slightly higher than the value of the animal reference group. Thus we suggest that the oral administration of the ropy strain could favor the recruitment of Treg cells to the intestinal mucosa, which could lead to a decrease in the severity of the damage caused by DSS. Jeon et al. (2012) showed *in vivo* that *B. breve* Yakult strain was able to ameliorate T cell-dependent colitis in mice through the induction of Treg cells toward producing IL-10. Similar results were obtained with a mixture of five probiotic strains and a mice model mimicking human Crohn's disease (Kwon et al., 2010). These authors also conclude that the efficacy of probiotics was due to an enrichment of CD4$^+$ Foxp3$^+$ Treg in the inflamed area. In recent years our group has been working with two models of isogenic, ropy and non-ropy EPS-producing *B. animalis* subsp. *lactis* strains: wild-types and the recombinant strains used in the current work. The results regarding the immune response that both types of strains could elicit are consistent when immune cells from the same origin are used. That is, all types of strains are able to induce a mild pro-inflammatory (Th1) response when a murine (rats or mice) model is used, either by *in vitro* (Hidalgo-Cantabrana et al., 2014a) or *in vivo* (the current work) approaches. However, the same EPS-producing strains tested *in vitro* or *ex vivo*, with immune cells and colonic tissue from human origin, had a reduced capability to elicit an immune response, or displayed an anti-inflammatory cytokine profile (López et al., 2012; Hidalgo-Cantabrana et al., 2015). These results underline the strong influence of the biological model used to test immune capability of strains with probiotic potential. Variations obtained with both models could be related to the differences in composition and function of the microbiota inhabiting the human and mouse intestines (Wos-Oxley et al., 2012).

Finally, due to the absence of statistical differences among the three strains we cannot conclude without doubt that the protective effect of DAI score promoted by strain Balat_1410$^{S89L}$ was due to the synthesis of the HWM-EPS. However, we can hypothesize that the ropy HMW-EPS (produced in higher amounts by this strain) could act as a physical barrier, like a biofilm, protecting from chemical insults and, consequently, reducing the severity of the mucosa damage. In this regard, we have previously shown that the EPS purified from the wild ropy strain *B. animalis* subsp. *lactis* A1dOxR was able to *in vitro* counteract the cytotoxic effect of bacterial toxins upon humans (Caco2 monolayers) or animals (rabbit erythrocyte) cells (Ruas-Madiedo et al., 2010). Synthesis of EPS by bifidobacteria in intestinal environmental conditions has not been proved to date; however, production of EPS by commensal *Bacteroides fragilis* in the gut of mice has recently been proved (Geva-Zatorsky et al., 2015). Thus, this fact would support our hypothesis that the HMW-EPS from strain *B. animalis* subsp. *lactis* Balat_1410$^{S89L}$ forms an EPS-biofilm protecting, to a certain degree, the intestinal epithelium from DSS-induced damage.

## CONCLUSION

In this work we have shown that the ropy strain *B. animalis* subsp. *lactis* Balat_1410$^{S89L}$ was able to protect mice from injury caused by DSS. To date, we have not found modifications in the expression of genes related to the reinforcement of the intestinal barrier that could explain the attenuation of chemically-induced damage; the most relevant immune parameter that could be related to this fact is the high capability of strain Balat_1410$^{S89L}$ to induce Treg cells in mesenteric lymphoid nodes which, in turn, could lead to a localized reduction of the inflammation at mucosal level. In spite of the only difference among the three isogenic recombinant strains tested in this study being the production of a ropy HMW-EPS, we cannot without doubt attribute the capability to reduce the DSS-induced damage to the polymer produced by Balat_1410$^{S89L}$.

There is a lack of consistency in the immune modulation capability of the ropy HMW-EPS-producing *B. animalis* subsp. *lactis* strains due to the high influence of the biological model used. However, the ability of the strain Balat_1410$^{S89L}$ to survive in the mice gut and the absence of adverse effects on the animals, together with previous *in vitro* and *ex vivo* evidence upon human cells, makes it feasible that the ropy (wild-type) *B. animalis* subsp. *lactis* could be a good candidate to check its anti-inflammatory ability in patients suffering from intestinal inflammation.

## AUTHOR CONTRIBUTIONS

JG, AM, and PR contributed with the conception, experimental design and results interpretation of this study. CH carried out all experiments, some of them performed with the collaboration of FA, AR, and TV. PM was in charge of the statistical analyses. PR was in charge of writing the drafted manuscript. All authors performed a critical revision of the manuscript and approved the final version.

## ACKNOWLEDGMENTS

This work was partially financed by FEDER European Union funds through the projects AGL2012-33278 and SAF2011-29648 of the Spanish Ministry of Economy and Competitiveness (MINECO), as well as to the projects AGR-6826 and CTS-164 of the "Junta de Andalucía." CH acknowledges his FPI fellowship to MINECO and FA acknowledges her fellowship to Junta de Andalucía. The CIBEREHD is funded by the "Instituto de Salud Carlos III."

## SUPPLEMENTARY MATERIAL

The Supplementary Material for this article can be found online at: http://journal.frontiersin.org/article/10.3389/fmicb.2016.00868

## REFERENCES

Bergmann, K. R., Liu, S. X., Tian, R., Kushnir, A., Turner, J. R., Li, H. L., et al. (2013). Bifidobacteria stabilize claudins at tight junctions and prevent intestinal barrier dysfunction in mouse necrotizing enterocolitis. *Am. J. Pathol.* 182, 1595–1606. doi: 10.1016/j.ajpath.2013.01.013

Collins, S. M. (2014). A role for the gut microbiota in IBS. *Nat. Rev. Gastroenterol. Hepatol.* 11, 497–505. doi: 10.1038/nrgastro.2014.40

Cooper, H. S., Murthy, S. N., Shah, R. S., and Sedergran, D. J. (1993). Clinicopathologic study of dextran sulfate sodium experimental murine colitis. *Lab. Invest.* 69, 238–249.

Dai, C., Zheng, C. Q., Meng, F. J., Zhou, Z., Sang, L. X., and Jiang, M. (2013). VSL#3 probiotics exerts the anti-inflammatory activity via PI3k/Akt and NF-kappa B pathway in rat model of DSS-induced colitis. *Mol. Cell. Biochem.* 374, 1–11. doi: 10.1007/s11010-012-1488-3

Denou, E., Pridmore, R. D., Berger, B., Panoff, J. M., Arigoni, F., and Brüssow, H. (2008). Identification of genes associated with the long-gut-persistence phenotype of the probiotic *Lactobacillus johnsonii* train NCC533 using a combination of genomics and transcriptome analysis. *J. Bacteriol.* 190, 3161–3168. doi: 10.1128/JB.01637-07

Directive. (2010/63/EU, 2010). Directive 2010/63/EU of the European Parliament and of the Council of 22 September 2010 on the protection of animals used for scientific purposes. *Official J. Eur. Union* 20.10.2010.

Edwards, U., Rogall, T., Blöcker, H., Emde, M., and Bottger, E. C. (1989). Isolation and direct complete nucleotide determination of entire genes. Characterization of a gene coding for 16S ribosomal RNA. *Nucleic Acids Res.* 17, 7843–7853. doi: 10.1093/nar/17.19.7843

EFSA BIOHAZ Panel (2013). Scientific Opinion on the maintenance of the list of QPS biological agents intentionally added to food and feed (2013 update). *EFSA J.* 11:3449. doi: 10.2903/j.efsa.2013.3449

Ewaschuk, J. B., Diaz, H., Meddings, L., Diederichs, B., Dmytrash, A., Backer, J., et al. (2008). Secreted bioactive factors from *Bifidobacterium infantis* enhance epithelial cell barrier function. *Am. J. Physiol. Gastrointest. Liver Physiol.* 295, G1025–G1034. doi: 10.1152/ajpgi.90227.2008

Fanning, S., Hall, L. J., Cronin, M., Zomer, A., MacSharry, J., Goulding, D., et al. (2012). Bifidobacterial surface-exopolysaccharide facilitates commensal-host interaction through immune modulation and pathogen protection. *Proc. Natl. Acad. Sci. U.S.A.* 109, 2108–2113. doi: 10.1073/pnas.1115621109

FAO/WHO. Food and Agricultural Organization of the United Nations and World Health Organization (2001). "Probiotics in food. Health and nutritional properties and guidelines for evaluation," in *FAO Food and Nutrition Paper 85*.

Geva-Zatorsky, N., Alvarez, D., Hudak, J. E., Reading, N. C., Erturk-Hasdemir, D., Dasgupta, S., et al. (2015). *In vivo* imaging and tracking of host-microbiota interactions via metabolic labeling of gut anaerobic bacteria. *Nat. Med.* 31, 1091–1100. doi: 10.1038/nm.3929

Hayes, C. L., Natividad, J. M. M., Jury, M. R., Langella, P., and Verdu, E. F. (2014). Efficacy of *Bifidobacterium breve* NCC2950 against DSS-induced colitis is dependent on bacterial preparation and timing of administration. *Benef. Microbes* 5, 79–88. doi: 10.3920/BM2013.0039

Hidalgo-Cantabrana, C., López, P., Gueimonde, M., de los Reyes-Gavilán, C. G., Suárez, A., Margolles, A., et al. (2012). Immune modulation capability of exopolysaccharides synthesised by lactic acid bacteria and bifidobacteria. *Probiotics Antimicrob. Proteins* 4, 227–237. doi: 10.1007/s12602-012-9110-2

Hidalgo-Cantabrana, C., Nikolic, M., López, P., Suárez, A., Miljkovic, M., Kojic, M., et al. (2014a). Exopolysaccharide-producing *Bifidobacterium animalis* subsp. *lactis* strains and their polymers elicit different responses on immune cells from blood and gut associated lymphoid tissue. *Anaerobe* 26, 24–30. doi: 10.1016/j.anaerobe.2014.01.003

Hidalgo-Cantabrana, C., Sánchez, B., Álvarez-Martín, P., López, P., Martínez-Álvarez, N., Delley, M., et al. (2015). A single mutation in the gene responsible for the mucoid phenotype of *Bifidobacterium animalis* subsp. *lactis* confers surface and functional characteristics. *Appl. Environ. Microbiol.* 81, 7960–7968. doi: 10.1128/AEM.02095-15

Hidalgo-Cantabrana, C., Sánchez, B., Milani, C., Ventura, M., Margolles, A., and Ruas-Madiedo, P. (2014b). Genomic overview and biological functions of exopolysaccharide biosynthesis in *Bifidobacterium* spp. *Appl. Environ. Microbiol.* 80, 9–18. doi: 10.1128/AEM.02977-13

Hill, C., Guarner, F., Reid, G., Gibson, G. R., Merenstein, D. J., Pot, B., et al. (2014). The international scientific association for probiotics and prebiotics consensus statement on the scope and appropriate use of the term probiotic. *Nat. Rev. Gastroenterol. Hepatol.* 11, 506–514. doi: 10.1038/nrgastro.2014.66

Horn, N., Wegmann, U., Dertli, E., Mulholland, F., Collins, S. R. A., Waldron, K. W., et al. (2013). Spontaneous mutation reveals influence of exopolysaccharide on *Lactobacillus johnsonii* surface characteristics. *PLoS ONE* 8:e59957. doi: 10.1371/journal.pone.0059957

Huttenhower, C., Kostic, A. D., and Xavier, R. J. (2014). Inflammatory bowel disease as a model for translating the microbiome. *Immunity* 40, 843–854. doi: 10.1016/j.immuni.2014.05.013

Jeon, S. G., Kayama, H., Ueda, Y., Takahashi, T., Asahara, T., Tsuji, H., et al. (2012). Probiotic *Bifidobacterium breve* induces IL-10-producing Tr1 cells in the colon. *PLoS Pathog.* 8:e1002714. doi: 10.1371/journal.ppat.1002714

Khailova, L., Dvorak, K., Arganbright, K. M., Halpern, M. D., Kinouchi, T., Yajima, M., et al. (2009). *Bifidobacterium bifidum* improves intestinal integrity in a rat model of necrotizing enterocolitis. *Am. J. Physiol. Gastrointest. Liver Physiol.* 297, G940–G949. doi: 10.1152/ajpgi.00141.2009

Kwon, H. K., Lee, C. G., So, J. S., Chae, C. S., Hwang, J. S., Sahoo, A., et al. (2010). Generation of regulatory dendritic cells and CD4+Foxp3+ T cells by probiotics administration suppresses immune disorders. *Proc. Natl. Acad. Sci. U.S.A.* 107, 2159–2164. doi: 10.1073/pnas.0904055107

Lebeer, S., Claes, I. J. J., Verhoeven, T. L. A., Vanderleyden, J., and De Keersmaecker, S. C. J. (2010a). Exopolysaccharide of *Lactobacillus rhamnosus* GG forms a protective shield against innate immune factors in the intestine. *Microb. Biotechnol.* 4, 368–374 doi: 10.1111/j.1751-7915.2010.00199.x

Lebeer, S., Vanderleyden, J., and De Keersmaecker, S. C. J. (2010b). Host interactions of probiotic bacterial surface molecules: comparison with commensals and pathogens. *Nat. Rev. Microbiol.* 8, 171–184. doi: 10.1038/nrmicro2297

Lebeer, S., Verhoeven, T. L. A., Francius, G., Schoofs, G., Lambrichts, I., Dufrêne, Y., et al. (2009). Identification of a gene cluster for the biosynthesis of a long, galactose-rich exopolysaccharide in *Lactobacillus rhamnosus* GG and functional analysis of the priming glycosyltransferase. *Appl. Environ. Microbiol.* 75, 3554–3563. doi: 10.1128/AEM.02919-08

Livak, K. J., and Schmittgen, T. D. (2001). Analysis of relative gene expression data using real-time quantitative PCR and the 2-ΔΔCT method. *Methods* 24, 402–408. doi: 10.1006/meth.2001.1262

López, P., Monteserín, D. C., Gueimonde, M., de los Reyes-Gavilán, C. G., Margolles, A., Suárez, A., et al. (2012). Exopolysaccharide-producing *Bifidobacterium* strains elicit different *in vitro* responses upon interaction with human cells. *Food Res. Int.* 46, 99–107. doi: 10.1016/j.foodres.2011.11.020

Mähler, M., Bristol, I. J., Leiter, E. H., Workman, A. E., Birkenmeier, E. H., Elson, C. O., et al. (1998). Differential susceptibility of inbred mouse strains to dextran sulfate sodium induced colitis. *Am. J. Physiol. Gastrointest. Liver Physiol.* 274, G544–G551.

Melgar, S., Karlsson, L., Rehnström, E., Karlsson, A., Utkovic, H., Jansson, L., et al. (2008). Validation of murine dextran sulfate sodium-induced colitis using four therapeutic agents for human inflammatory bowel disease. *Int. Immunopharmacol.* 8, 836–844. doi: 10.1016/j.intimp.2008.01.036

Milani, C., Duranti, S., Lugli, G. A., Bottacini, F., Strati, F., Arioli, S., et al. (2013). Comparative genomics of *Bifidobacterium animalis* subsp. *lactis* reveals a strict monophyletic bifidobacterial taxon. *Appl. Environ. Microbiol.* 79, 4304–4315. doi: 10.1128/AEM.00984-13

Prasanna, P. H. P., Grandison, A. S., and Charalampopoulos, D. (2014). Bifidobacteria in milk products: an overview of physiological and biochemical properties, exopolysaccharide production, selection criteria of milk products and health benefits. *Food Res. Int.* 55, 247–262. doi: 10.1016/j.foodres.2013.11.013

Rao, R. K., and Samak, G. (2013). Protection and restitution of gut barrier by probiotics: nutritional and clinical implications. *Cur. Nutr. Food Sci.* 9, 99–107. doi: 10.2174/1573401311309020004

Reid, G., Younes, J. A., Van der Mei, H. C., Gloor, G. B., Knight, R., and Busscher, H. J. (2011). Microbiota restoration: natural and supplemented recovery of human microbial communities. *Nat. Rev. Microbiol.* 9, 27–38. doi: 10.1038/nrmicro2473

Ruas-Madiedo, P., Medrano, M., Salazar, N., de los Reyes-Gavilán, C. G., Pérez, P., and Abraham, A. G. (2010). Exopolysaccharides produced by *Lactobacillus* and

*Bifidobacterium* strains abrogate *in vitro* the cytotoxic effect of bacterial toxins on eukaryotic cells. *J. Appl. Microbiol.* 109, 2079–2086. doi: 10.1111/j.1365-2672.2010.04839.x

Salazar, N., Binetti, A., Gueimonde, M., Alonso, A., Garrido, P., González del Rey, C., et al. (2011). Safety and intestinal microbiota modulation by the exopolysaccharide-producing strains *Bifidobacterium animalis* IPLA R1 and *Bifidobacterium longum* IPLA E44 orally administered to Wistar rats. *Int. J. Food Microbiol.* 144, 342–351. doi: 10.1016/j.ijfoodmicro.2010.10.016

Salazar, N., Ruas-Madiedo, P., Kolida, S., Collins, M., Rastall, R., Gibson, G., et al. (2009). Exopolysaccharides produced by Bifidobacterium longum IPLA E44 and *Bifidobacterium animalis* subsp. *lactis* IPLA R1 modify the composition and metabolic activity of human faecal microbiota in pH-controlled batch cultures. *Int. J. Food Microbiol.* 135, 260–267. doi: 10.1016/j.ijfoodmicro.2009.08.017

Sanders, M. E., Guarner, F., Guerrant, R., Holt, P. R., Quigley, E. M. M., Sartor, R. B., et al. (2013). An update on the use and investigation of probiotics in health and disease. *Gut* 62, 787–796. doi: 10.1136/gutjnl-2012-302504

Tojo, R., Suárez, A., Clemente, M. G., de los Reyes-Gavilán, C. G., Margolles, A., Gueimonde, M., et al. (2014). Intestinal microbiota in health and disease: role of bifidobacteria in gut homeostasis. *World J. Gastroenterol.* 20, 15163–15176. doi: 10.3748/wjg.v20.i,41.15163

WGO. World Gastroenterology Organisation (2011). *Global Guidelines: Probiotics and Prebiotics.* Available online at: http://www.worldgastroenterology.org/probiotics-prebiotics.html.

Wirtz, S., Neufert, C., Weigmann, B., and Neurath, M. F. (2007). Chemically induced mouse models of intestinal inflammation. *Nat. Protoc.* 2, 541–546. doi: 10.1038/nprot.2007.41

Wos-Oxley, M. L., Bleich, A., Oxley, A. P. A., Kahl, S., Janus, L. M., Smoczek, A., et al. (2012). Comparative evaluation of establishing a human gut microbial community within rodent models. *Gut Microb.* 3, 1–16. doi: 10.4161/gmic.19934

Zheng, B., van Bergenhenegouwen, J., van de Kant, H. J. G., Folkerts, G., Garssen, J., Vos, A. P., et al. (2016). Specific probiotic dietary supplementation leads to different effects during remission and relapse in murine chronic colitis. *Benef. Microbes* 7, 205–213. doi: 10.3920/bm2015.0037

**Conflict of Interest Statement:** The authors declare that the research was conducted in the absence of any commercial or financial relationships that could be construed as a potential conflict of interest.

# 18

# Effect of *Bifidobacterium* upon *Clostridium difficile* Growth and Toxicity when Co-cultured in Different Prebiotic Substrates

*L. Valdés-Varela, Ana M. Hernández-Barranco, Patricia Ruas-Madiedo and Miguel Gueimonde* *

*Microbiology and Biochemistry of Dairy Products, Probiotics and Prebiotics, Instituto de Productos Lácteos de Asturias–Consejo Superior de Investigaciones Científicas, Villaviciosa, Spain*

**Edited by:**
*David Berry,*
*University of Vienna, Austria*

**Reviewed by:**
*Amparo Latorre,*
*University of Valencia, Spain*
*Bärbel Stecher,*
*Ludwig Maximilians Universität München, Germany*

***Correspondence:**
*Miguel Gueimonde*
*mgueimonde@ipla.csic.es*

The intestinal overgrowth of *Clostridium difficile*, often after disturbance of the gut microbiota by antibiotic treatment, leads to *C. difficile* infection (CDI) which manifestation ranges from mild diarrhea to life-threatening conditions. The increasing CDI incidence, not only in compromised subjects but also in traditionally considered low-risk populations, together with the frequent relapses of the disease, has attracted the interest for prevention/therapeutic options. Among these, probiotics, prebiotics, or synbiotics constitute a promising approach. In this study we determined the potential of selected *Bifidobacterium* strains for the inhibition of *C. difficile* growth and toxicity in different carbon sources. We conducted co-cultures of the toxigenic strain *C. difficile* LMG21717 with four *Bifidobacterium* strains (*Bifidobacterium longum* IPLA20022, *Bifidobacterium breve* IPLA20006, *Bifidobacterium bifidum* IPLA20015, and *Bifidobacterium animalis* subsp. *lactis* Bb12) in the presence of various prebiotic substrates (Inulin, Synergy, and Actilight) or glucose, and compared the results with those obtained for the corresponding mono-cultures. *C. difficile* and bifidobacteria levels were quantified by qPCR; the pH and the production of short chain fatty acids was also determined. Moreover, supernatants of the cultures were collected to evaluate their toxicity using a recently developed model. Results showed that co-culture with *B. longum* IPLA20022 and *B. breve* IPLA20006 in the presence of short-chain fructooligosaccharides, but not of Inulin, as carbon source significantly reduced the growth of the pathogen. With the sole exception of *B. animalis* Bb12, whose growth was enhanced, the presence of *C. difficile* did not show major effects upon the growth of the bifidobacteria. In accordance with the growth data, *B. longum* and *B. breve* were the strains showing higher reduction in the toxicity of the co-culture supernatants.

**Keywords: probiotics, prebiotics, inhibition, *Clostridium difficile*, *Bifidobacterium*, toxin, HT29, RTCA**

## INTRODUCTION

*Clostridium difficile* is often present in the intestinal microbiota of both infants and adults, where it may be found in about 70 and 17% of the subjects, respectively (Ozaki et al., 2004; Jangi and Lamont, 2010). However, this microorganism is also the main causative agent of antibiotic associated diarrhea in nosocomial environments (Leffler and Lamont, 2015). The epidemiology of

*C. difficile* infection (CDI) is changing, with an increasing occurrence in populations traditionally considered of low-risk (Carter et al., 2012), likely due to the appearance of hipervirulent strains (Rupnik et al., 2009; Yakob et al., 2015). CDI is treated with antibiotics but a high rate of recurrence is present. In this context, new therapeutic alternatives for treating or preventing CDI are being continuously explored, among them the inhibition of *C. difficile* growth by the use of probiotics or prebiotics has been tested (Ambalam et al., 2015; Auclair et al., 2015; Forssten et al., 2015).

In general, probiotics and prebiotics have been proposed as biotherapeutic agents to prevent the dysbiosis caused by antibiotics or infections, and to help the microbiota restoration after it (Reid et al., 2011). The development of food products targeting at the inhibition of *C. difficile* constitutes an interesting approach in the context of the marketing of products bearing health claims. Reducing the intestinal levels of specific pathogens, such as *C. difficile*, has been considered by the European Food Safety Authority (EFSA) as a beneficial physiological effect [(EFSA Panel on Dietetic Products, Nutrition and Allergies (NDA), 2011)]. Therefore, such an effect would constitute an opportunity for the development of food products bearing a health claim in the area of gastrointestinal health.

To date, different probiotic strains and prebiotic substrates have been reported to increase colonization resistance against *C. difficile* (Hopkins and Macfarlane, 2003; Kondepudi et al., 2014; Auclair et al., 2015; Forssten et al., 2015). In addition to their microbiota-modulatory properties, probiotics have been found to protect against infections by other mechanisms, such as production of antimicrobial compounds or competition by adhesion sites or nutrients (Servin, 2004). The ability of certain probiotics, mainly bifidobacteria and lactobacilli, to inhibit *in vitro* the adhesion of *C. difficile* to intestinal epithelial cells or intestinal mucus is well established (Collado et al., 2005; Banerjee et al., 2009). Similarly, the ability to produce antimicrobials inhibiting the growth of *C. difficile in vitro* has been repeatedly reported (Lee et al., 2013; Schoster et al., 2013). However, other potential targets of probiotics and prebiotics on CDI, such as their impact on toxin production by the pathogen, and/or toxin activity, have been explored to a lesser extent and have not attracted attention until recently (Kondepudi et al., 2014; Yun et al., 2014; Andersen et al., 2015). Ambalam et al. (2015) recently reported the ability of cell-free supernatants from some *Lactobacillus* strains, and a probiotic mix, to inhibit the growth of *C. difficile* strains in variable way depending on the carbon source used. Moreover, the authors observed a reduction of toxin titers in those *C. difficile* cultures with inhibitory cell-free supernatants added. Moreover, we have demonstrated that incubation of toxigenic *C. difficile* cell-free culture supernatants with specific bifidobacterial strains reduces the cytotoxic effect upon human epithelial intestinal cells (Valdés et al., 2016). However, the influence of prebiotic substrates upon *C. difficile* growth and toxicity when co-cultured with bifidobacteria remains largely unknown.

In this context the aim of this study was to evaluate *in vitro* the potential of four bifidobacterial strains for inhibiting the growth of *C. difficile* when co-cultured with different prebiotics as carbon source. Moreover, the effect of the strains and prebiotics on the toxicity of the co-culture supernatants upon human intestinal epithelial cells (HT29) was also determined.

## MATERIALS AND METHODS

### Bacterial Strains and Culture Conditions

The widely used probiotic strain *Bifidobacterium animalis* subsp. *lactis* Bb12 and three strains of bifidobacteria from IPLA culture collection, two of them isolated from infant's feces (*Bifidobacterium longum* IPLA20022 and *Bifidobacterium bifidum* IPLA20015) (Solís et al., 2010) and the other one from breast-milk (*Bifidobacterium breve* IPLA20006) (Arboleya et al., 2011), were used. These last three strains were selected based on the good ability to reduce toxicity of *C. difficile* supernatants (Valdés et al., 2016). With regard to *C. difficile* we used the strain LMG21717, known to produce TcdA toxin and also, although at lower quantities, TcdB. This strain belongs to ribotype 001, which is one of the most common ones found in Europe (Martin et al., 2016). The *Bifidobacterium* strains were routinely grown in MRS (Biokar Diagnostics, Beauvois, France) supplemented with 0.25% L-cysteine (Sigma-Chemical Co., St. Louis, MO, USA) in an anaerobic chamber MG500 (Don Whitley Scientific, Yorkshire, UK) and *C. difficile* was grown in Reinforced Clostridial Medium (RCM, Oxoid, Thermo Fisher Scientific Inc., Waltham, MA) in Hungate tubes as previously described (Valdés et al., 2016). Overnight cultures (18 h) of the bifidobacterial strains and 13 h-old cultures of *C. difficile* were used to inoculate the batch culture fermentations.

For the batch mono- and co-culture fermentations a defined medium with the following composition was used: proteose peptone (10 g/L) (BD-Difco, New Jersey, EE.UU.), beef extract (10 g/L) (BD-Difco), yeast extract (5 g/L) (BD-Difco), polysorbate 80 (1 mL/L) (Sigma), ammonium citrate (2 g/L) (Sigma), sodium acetate (5 g/mL) (Sigma), magnesium sulfate (0.2 g/L) (Probus, Barcelona, Spain), manganese sulfate (0.056 g/L) (Panreac, Barcelona, Spain), and dipotassium phosphate (2 g/L) (Merck, New Jersey, EE.UU). Pairwise combinations of the *C. difficile* strain with the different *Bifidobacterium* strains, as well as the corresponding monocultures, were performed in the medium described above with a 2% (w/v) of different commercial prebiotic substrates added [Synergy 1 (Beneo-Orafti, Barcelona, Spain), Inulin (Sigma) and Actilight (Beghin Meiji and Tereos Syral, Marckolsheim, France)], glucose or without adding any carbon source (used as control). Each media was distributed into Hungate tubes which were inoculated with different *Bifidobacterium* strains at a final level of about $10^5$ CFU/ml in case of *B. longum*/*B. breve* and $10^4$ CFU/ml in case of *B. bifidum*/*B. animalis*, with *C. difficile* strain at final level of $10^6$ CFU/ml or with both of them, in the case of the co-culture. The bifidobacteria were inoculated at a different level depending on the strain with the aim of allowing a balanced growth of both microorganisms (bifidobacteria and clostridia). The appropriate inoculum size was determined in previous experiments (data not shown).

Co-cultures, and the corresponding mono-cultures, in different carbon sources were carried out in triplicate under anaerobic conditions at 37°C for 24 h. Samples were taken at 0 and 24 h for bacterial growth assessment by quantitative PCR (qPCR), quantification of SCFA by Gas Chromatography (GC), pH measurements (pH meter Basic 20+, Crison Instruments S.A., Barcelona, Spain), and toxigenicity determination. One milliliter of each mono-culture or co-culture was centrifuged (16,000 × g for 10 min), and pellets and supernatants were collected. For toxigenic activity upon HT29 cells, the pH of 0.7 ml cell-free supernatant from each batch culture was adjusted to 7.55 ± 0.05 with 1 and 0.1 N NaOH. All supernatants and pellets were immediately frozen at −80°C until use.

## Quantification of Bacterial Growth by qPCR

DNA was extracted from pellets of batch cultures using the GenElute Bacterial Genomic DNA Kit (Sigma) and kept at −80°C until analyzed. The levels of *C. difficile* and bifidobacteria in the cultures were determined as DNA copies per ml by qPCR using previously described primers and conditions (Arboleya et al., 2012). Reactions were performed on MicroAmp optical plates sealed with MicroAmp optical caps (Applied Biosystems, Foster City, CA, USA) with a 7500 Fast Real-Time PCR System (Applied Biosystems) using SYBR Green PCR Master Mix (Applied Biosystems). One microlitre of template DNA was used in the 25 mL PCR mixture. Standard curves were made with pure cultures of *B. longum* NCIMB8809 and *C. difficile* LMG 21717. In all cultures the levels of the microorganisms were above the corresponding detection limit of the technique ($1 \times 10^3$ and $3 \times 10^3$ for bifidobacteria and *C. difficile*, respectively). Samples were analyzed by duplicate in at least two independent PCR runs.

## Determination of the Production of Short Chain Fatty Acids by GC-MS

Cell-free supernatant (0.1 mL) from each batch culture was mixed with 1 ml methanol, 0.1 ml internal standard solution (2-ethylbutyric 1.05 mg/ml), and 0.1 ml 20% formic acid. This mixture was centrifuged and the supernatant obtained was used for quantification of SCFA by GC in a system composed of a 6890NGC injection module (Agilent Technologies Inc., Palo Alto, Ca, USA) connected to a flame injection detector (FID) and a mass spectrometry (MS) 5973N detector (Agilent) as described previously (Salazar et al., 2011).

## Monitoring the Cytotoxic Effect of the Culture Supernatants upon Intestinal Epithelial Cells

The intestinal cell line HT29 (ECACC 91072201) was purchased from the "European Collection of Cell Cultures" (Salisbury, UK) and stored under liquid $N_2$. McCoy's Medium (MM) supplemented with 10% fetal serum bovine, 3 mM L-glutamine and a mixture of antibiotics (50 μg/ml streptomycin-penicillin, 50 μg/ml gentamicin, and 1.25 μg/ml amphotericin B) was used for HT29 cultivation. All media and reagents were purchased from Sigma-Aldrich. Maintenance of the cell line, between passages 145 and 149, was performed under standard conditions at 37°C 5% $CO_2$ atmosphere, in a $CO_2$-Series Shel-Lab incubator (Sheldon Manufacturing Inc., OR, USA). The experimental procedures were carried out with the cell passage 149.

We used an RTCA (real time cell analyser) xCelligence (ACEA Bioscience Inc., San Diego, CA) system, introduced in a Heracell-240 Incubator (Thermo Electron LDD GmbH, Langenselbold, Germany) set at 37°C with 5% $CO_2$ atmosphere, to monitor HT29 cells behavior. A method previously described, allowing the assessment of the damage caused by *C. difficile* supernatants, was used (Valdés et al., 2015). This method is based in the real-time monitoring of the cell index (CI). This CI is an arbitrary unit that measures the impedance, in gold-microelectrodes coating the surface of E-plates, which changes as consequence of the HT29 cells attachment and growth.

In short, 16-well E-plates were seed with $2 \times 10^5$ HT29 cells (in 100 μl), hold in the RTCA equipment, incubated for 22 h to ensure the formation of a monolayer (confluent state) and the CI was monitored (recording signal every 15 min). After this incubation the medium was removed from the wells and the methodology followed was slightly different depending on the experiment. To determine the effect of the carbon source on the toxicity of *C. difficile,* 200 μL of MM containing different concentrations (from 0.63 to 40%, v/v) of cell-free neutralized-supernatants from *C. difficile* mono-cultures were added to the wells. EC50 values (concentration at which half of the maximum damage was detected) for the cultures, in the different carbon sources tested, were then calculated as previously described (Valdés et al., 2015). To determine the effect of bifidobacteria on the toxigenic capability of *C. difficile* in the different carbon sources, 200 μL of MM containing a 5% (v/v) of the neutralized supernatant from each mono- and co-culture were added to the wells. Additionally, wells filled with 200 μl of MM (non-cytotoxic control) were included in each experiment. Then, monitoring continued (recording signal every 10 min) up to 20 h under standard incubation conditions. The data analyses were carried out through RTCA software 1.2.1 (ACEA Bioscience). The CI values were normalized as previously described (Valdés et al., 2015) by dividing the CI at every point by the CI at time zero (the time of the supernatant addition, thus making the CI equal to 1 at this time) and then referred to the normalized CI of the control sample (MM) (the normalized-CI of the control sample is then the "0 line" shown in figures).

Toxin A concentration in the supernatant of *C. difficile* mono-cultures in different carbon sources was determined by ELISA test (tgcBIOMICS GmbH, Bingen, Germany).

## Statistical Analysis

To asses differences among carbon sources or between mono- and co-cultures, one-way ANOVAs followed by SNK (Student-Newman-Keuls, $p < 0.05$) mean comparison test were performed. The statistical package IBM SPSS Statistics for Window Version 22.0 (IBM Corp., Armonk NY) was used to carry out these analyses.

## RESULTS AND DISCUSSION

### Inhibition of *C. difficile* Growth When Co-cultured with *Bifidobacterium* Strains in Different Carbon Sources

There is a great scientific interest on the development of interventions for preventing or treating CDI, including vaccines (Senoh et al., 2015), antimicrobials (Gebhart et al., 2015; Vickers et al., 2015), anti-toxin antibodies (Yang et al., 2015), or genetically engineered bacteria producing them (Andersen et al., 2015), among others. Fecal transplants have demonstrated a high efficacy to treat recurrent CDI (Lee et al., 2016), underlining the importance of the gut microbiota in this disease. Probiotics and prebiotics constitute another interesting option although differences among strains and substrates seem to exist (Allen et al., 2013).

In our study the mono-culture of the *Bifidobacterium* strains (**Figure** ) in different substrates (dark colored bars) showed that all the strains grew well in glucose. In agreement with previous reports (Rossi et al., 2005; Kondepudi et al., 2012), the strains showed the ability to grow in short-chain fructooligosaccharides (Synergy and Actilight) (scFOS) but they were not able to grow, or did it poorly, in Inulin (**Figure** ). This observation was further supported by the production of bacterial metabolites (**Figure** , Supplementary File) and the pH (Supplementary Figure ), which in the case of Inulin remained similar to those of the negative control without carbon source added (WCS). Interestingly, *B. longum* IPLA20022 showed a significantly higher growth ($p < 0.05$) in the prebiotics Synergy and Actilight than in glucose (**Figure** ), whereas no statistically significant differences were observed for *B. breve* IPLA20006 or *B. bifidum* IPLA20015 between glucose and these two prebiotics. The mono-cultures of *B. animalis* Bb12 showed a significantly lower ($p < 0.05$) growth in all prebiotics than in glucose. This strain exhibited the lowest growth of all bifidobacteria in glucose, Synergy, and Actilight (**Figure** ), which correlates with the limited drop in pH observed for this strain after 24 h of incubation (Supplementary Figure 1). With regard to the pathogen, *C. difficile* grew well in Synergy, not differing significantly from glucose, and to a lower extent in Actilight (**Figure** ). Therefore, in spite of generally claimed high specific fermentation of prebiotic substrates, some intestinal pathogens may also be able to ferment and grow in some of them. This underlines the importance of a careful selection of the most appropriate strains, substrates, and combinations.

When co-cultured with *C. difficile* in the different carbon sources, the behavior of the bifidobacteria was, in general, similar to that observed in the mono-cultures. We observed increases in bifidobacterial counts in glucose, Synergy, and Actilight and poor grow in Inulin. Regarding *C. difficile*, it grew better in glucose, followed by Synergy, which is in agreement with the

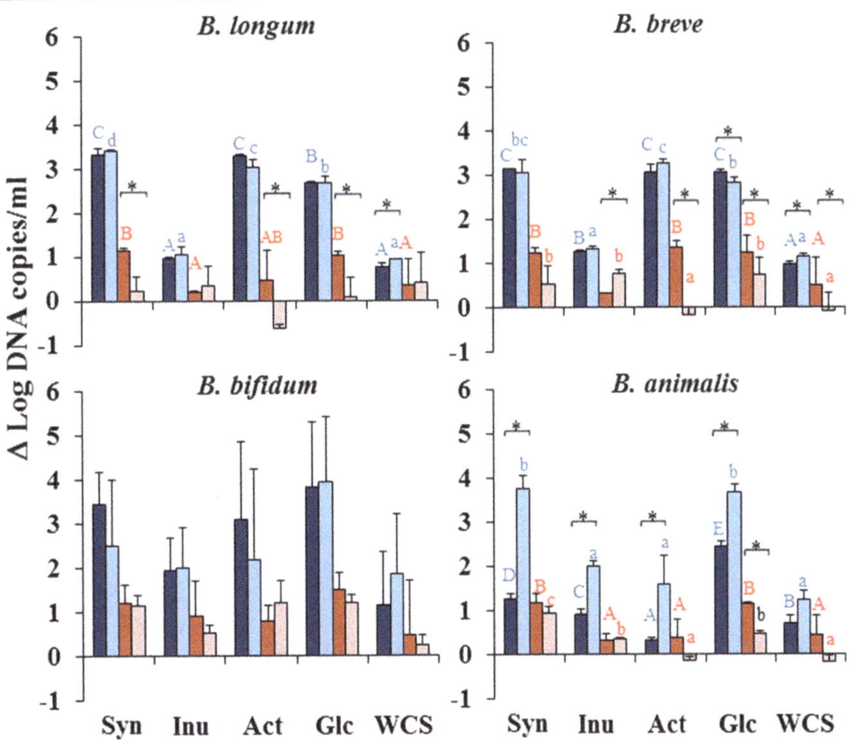


**FIGURE | Increments, with respect to time zero, on the levels (Log CFU/mL) of the strains when grown in mono-culture (*Bifidobacterium* dark-blue column and *C. difficile* dark-red) or co-culture (*Bifidobacterium* light-blue and *C. difficile* light-red column) in the prebiotics Synergy (Syn), Inulin (Inu), and Actilight (Act), in glucose (Glc) or without any carbon source added (WCS).** Different capital letters above columns denote statistically significant differences ($p < 0.05$) among carbon sources in the mono-cultures of the corresponding bacterial strain, whereas different lowercase letters indicate differences in the co-cultures (either Bifidobacterium in blue letters or *C. difficile* in red letters). *Indicates statistically significant differences ($p < 0.05$) for the corresponding bacterial strain between mono- and co-culture within the same substrate.

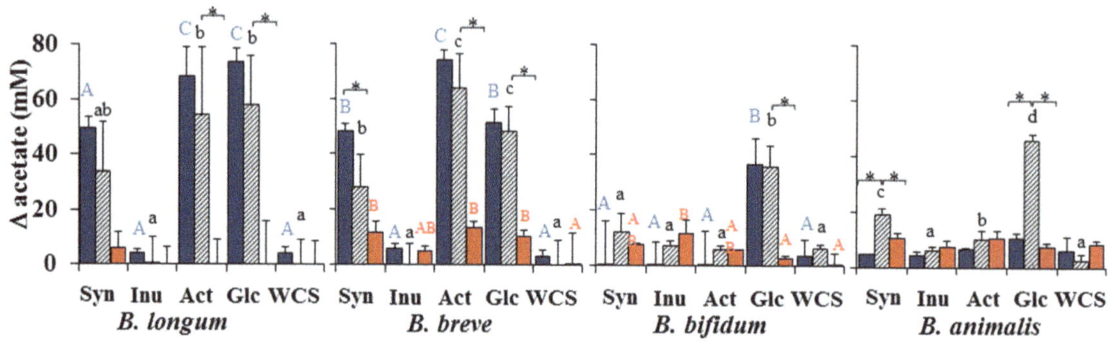


**FIGURE | Increments, with respect to time zero, in the concentration of acetate on the bacterial cultures when grown in mono-culture (*Bifidobacterium* blue-bars and *C. difficile* red-bars) or in co-culture (stripped bars) in the prebiotics Synergy (Syn), Inulin (Inu), and Actilight (Act), in glucose (Glc) or without any carbon source added (WCS).** Different letters above columns denote statistically significant differences ($p < 0.05$) among carbon sources in the corresponding bacterial cultures, either mono-cultures (capital letter; red color for bifidobacteria and blue color for *C. difficile*) or co-cultures (lowercase letters). *Indicates statistically significant differences ($p < 0.05$) for the corresponding bacterial strain between mono- and co-culture.

mono-culture data, but the growth in Actilight was, in general, significantly ($p < 0.05$) worse in co-culture, the contrary being true for Inulin (**Figure** ). This growth behavior of *C. difficile* in the different carbon sources was further confirmed by the metabolites production pattern (Supplementary File), showing in general a lower production of *C. difficile* metabolites, such as propionate or branched-SCFA, in co-culture with Actilight as carbon source than in the corresponding mono-culture, whilst the contrary was observed for Inulin.

When co- and mono-cultures were compared within the same carbon source, the growth of *C. difficile* was significantly reduced ($p < 0.05$) by *B. longum* IPLA20022, *B. breve* IPLA20006, or *B. animalis* Bb12 in glucose. The first two microorganisms also reduced *C. difficile* growth when co-cultured in Actilight and, in the case of *B. longum* also when Synergy was used as carbon source (**Figure** ). On the contrary, no statistically significant differences between mono- and co-cultures were observed for *B. bifidum* in any carbon source. These results showed a good correlation with the pattern of production of *C. difficile* metabolites and the drop in pH (Supplementary File). This suggests the production of organic acids, with the concomitant reduction of the pH, as an important mechanism of inhibition (Tejero-Sariñena et al., 2012).

These results point out at *B. longum* IPLA20022 and *B. breve* IPLA20006, and the prebiotics Synergy and Actilight, as the most promising alternatives for inhibiting the growth of *C. difficile*. Moreover, they suggest that the pathogen inhibition is strain and substrate specific, which is in agreement with previous reports (Kondepudi et al., 2012; Tejero-Sariñena et al., 2013; Ambalam et al., 2015). Interestingly, the growth of *C. difficile* was significantly increased ($p < 0.05$) by *B. breve* in the presence of Inulin, indicating a potential risk of such combination and underlining the importance of a careful strain and substrate specific assessment.

Interestingly, effects of the co-culture with *C. difficile* on the growth of the bifidobacterial strains were also observed. Whilst in glucose the co-culture with the clostridia did not affect the growth of *B. longum*, it significantly ($p < 0.05$) reduced that of *B. breve* but increased that of *B. animalis*. Moreover, the growth of the latter microorganism was also increased by the presence of *C. difficile* in the three prebiotics tested, mainly Synergy (**Figure** ) which was further confirmed by an enhanced production of acetate in the co-culture than in the corresponding monoculture (**Figure** ) and a higher drop in pH (Supplementary Figure ).

## The Carbon Source Determines the Toxicity of *C. difficile* Supernatants

In addition to bacterial growth, inhibiting the toxicity caused by *C. difficile*, for example by reducing toxin production or toxic activity, represents another target in CDI (Trejo et al., 2010, 2013). The toxicity of *C. difficile* culture supernatants has been found to be dependent on the culture media used (Valdés et al., 2015), suggesting a potential role of the carbon source available. Therefore, it is important to know whether the availability of different prebiotics as carbon source may have an impact on the toxicity of *C. difficile*. To clarify this point we determined the toxicity of neutralized cell-free supernatants, obtained from *C. difficile* monocultures after 24 h of incubation in the different carbon sources, upon the human epithelial cell line HT29 by using a real-time monitoring system (RTCA). To this end the EC50 values, defined as the concentration of supernatant causing 50% of the maximum cell damage, were calculated (Valdés et al., 2015). Supernatants obtained from the mono-cultures carried out without any carbon source or with Actilight added were significantly ($p < 0.05$) more toxic than the others (**Figure 3**). They showed EC50 values below 2%, which means that a concentration of monoculture supernatant lower than 2% already produced half of the maximum cell damage. On the contrary, the supernatant of the mono-culture in glucose resulted significantly ($p < 0.05$) less toxic than all the others (EC50 value over 6%), followed by that on Synergy and the one carried out with Inulin as carbon source (**Figure** ). The method used (Valdés et al., 2015) allowed us to determine that the *C. difficile* supernatants' toxicity was higher when no carbon source was added or when the available carbon source supported only a limited growth of

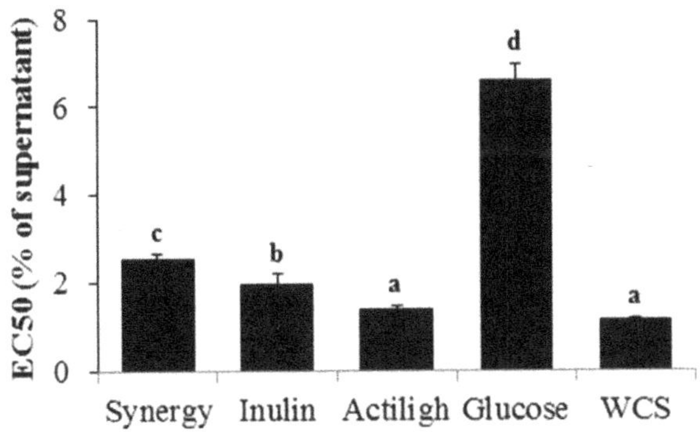


**FIGURE | Concentration (% v/v) of supernatants of *C. difficile* mono-cultures, in the different carbon sources tested, showing 50% of the maximum cell damage (EC50).** To calculate EC50s the cell indexes obtained after 12 h of incubation of the HT29 cells with supernatants were used. Different letters above the columns denote statistically significant differences ($p < 0.05$).

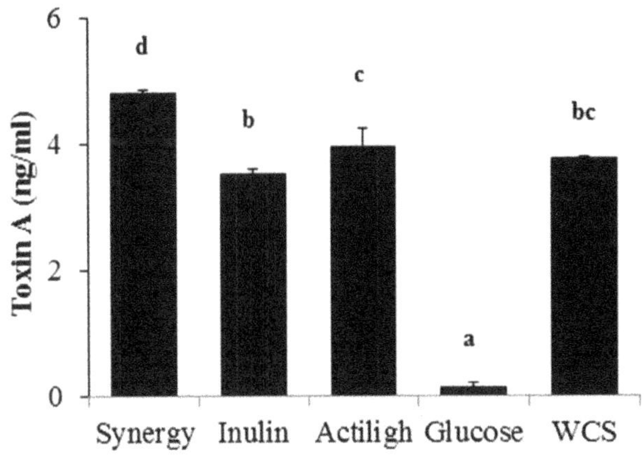


**FIGURE | Toxin A concentration in the different *C. difficile* supernatants obtained when the microorganism was growth in the different carbon sources.** Different letters above the columns denote statistically significant differences ($p < 0.05$).

the pathogen, such as in the case of Actilight. On the contrary, the supernatant obtained when the pathogen was grown in glucose, in spite of the good growth of *C. difficile*, resulted less toxic. The availability of rapidly metabolizable sugars has been reported to inhibit toxin synthesis in *C. difficile* (Bouillaut et al., 2015). This inhibition is mediated through repression of *treR* (also known as *tdcR*), an alternative sigma factor responsible for the positive regulation of *toxA* and *toxB* genes (Mani et al., 2002). Our results seem to confirm the higher production of toxins by *C. difficile* under nutrient limitation or stress conditions in which readily fermentable sugars are not available. Moreover, in *C. difficile* a co-induction of metabolic pathways, such as that of butyrate production, and toxin production has been reported (Karlsson et al., 2000). In our study the *C. difficile* monoculture grown in glucose showed, in general, lowest butyrate production than those carried out with Synergy, Actilight or WCS added, which is in good agreement with the lower toxin production in glucose. However, the culture of the strain in Inulin, in spite of a lower production of butyrate than that in glucose, showed higher toxin concentrations, comparable to those found WCS or in the other prebiotics tested. These results indicate that, at least in some circumstances, toxin production by *C. difficile* is uncoupled from the production of metabolites such as butyrate.

In accordance with the above mentioned toxicity data, the concentration of *C. difficile* toxin A showed the lowest value in the supernatant from the culture in glucose (**Figure** ). The supernatants obtained from cultures grown in Synergy, Actilight showed the highest toxin concentrations, whilst those from the growth on *C. difficile* in Inulin or WCS showed intermediate levels.

## Co-culture with Bifidobacteria in Different Carbon Sources Reduces *C. difficile* Toxicity

The ability of certain bifidobacterial strains, such as *B. longum* IPLA20022, to remove toxins from *C. difficile* cell-free supernatants, then diminishing their cytotoxicity, has been recently reported (Valdés et al., 2016). Now we compared the toxicity of the *Clostridium-Bifidobacterium* co-culture supernatants with that of the pathogen monoculture. In general we observed a significant reduction on the toxicity of the supernatants in co-culture. However, differences depending on the strain and the carbon source used were also observed, confirming the high specificity of these interactions (Trejo et al., 2010). The toxicities obtained for the co-cultures in the different carbon sources were compared by using the normalized cell index (CI) obtained after 12 h of incubation of HT29 cells with a 5% of the culture supernatants. As it was the case for the monocultures, supernatants from co-cultures carried out on the different carbon sources showed differences among them ($p < 0.05$) (**Table** ). Similarly to the mono-cultures, supernatants obtained in glucose showed the lowest toxicity whilst those in Inulin, or without any carbon source added, resulted the most toxic. When the supernatants of the co-cultures with the different bifidobacteria were compared with the *C. difficile* monoculture no statistically significant differences ($p > 0.05$) were obtained in media WCS added. However, in all the carbon sources tested, either glucose or prebiotics, statistically significant differences ($p < 0.05$) were observed depending on the bifidobacterial strain used (**Table** ). Co-culture in Synergy or Actilight of *C. difficile* with *B. longum* IPLA20022 or *B. breve* IPLA20006 significantly ($p < 0.05$) inhibited the toxicity of the supernatant (i.e., higher normalized CI) when compared with the mono-culture of *C. difficile*. However, *B. bifidum* IPLA20015 only was able to reduce ($p < 0.05$) the toxicity of the pathogen with Actiligh as carbon source whilst *B. animalis* Bb12 did not produce toxicity inhibition in any prebiotic. The four bifidobacteria tested were able to reduce ($p > 0.05$) the toxicity of the supernatant when co-cultured in glucose, in comparison to the *C. difficile* mono-culture, but none of them did it when the carbon source was Inulin. In the latter case, even, an increase in the toxicity was observed when the pathogen was co-incubated

**TABLE | Normalized cell index (mean ± sd) obtained after 12 h of incubation of HT29 cells with the supernatants (5%) of the *C. difficile* mono-culture or *C. difficile-Bifidobacterium* co-cultures grown in different prebiotics, glucose or without any carbon source added (WCS).**

| Culture | Normalized cell index | | | | |
|---|---|---|---|---|---|
| | Carbon source | | | | |
| | Synergy | Inulin | Actilight | Glucose | WCS |
| *C. difficile* | $-0.39 \pm 0.03^{a,1}$ | $-0.30 \pm 0.03^{a,2}$ | $-0.35 \pm 0.04^{a,1}$ | $-0.23 \pm 0.01^{b,1}$ | $-0.34 \pm 0.04^{a}$ |
| *C. difficile–B. longum* | $-0.06 \pm 0.04^{b,2}$ | $-0.32 \pm 0.07^{a,2}$ | $-0.13 \pm 0.05^{b,3}$ | $0.01 \pm 0.02^{b,3}$ | $-0.43 \pm 0.13^{a}$ |
| *C. difficile–B. breve* | $-0.02 \pm 0.02^{d,2}$ | $-0.32 \pm 0.03^{b,2}$ | $-0.07 \pm 0.01^{c,4}$ | $0.00 \pm 0.01^{d,3}$ | $-0.37 \pm 0.01^{a}$ |
| *C. difficile–B. bifidum* | $-0.40 \pm 0.08^{b,1}$ | $-0.56 \pm 0.02^{a,1}$ | $-0.24 \pm 0.01^{c,2}$ | $0.00 \pm 0.01^{d,3}$ | $-0.34 \pm 0.01^{b}$ |
| *C. difficile–B. animalis* | $-0.35 \pm 0.02^{a,1}$ | $-0.31 \pm 0.03^{a,2}$ | $-0.32 \pm 0.02^{a,1}$ | $-0.03 \pm 0.00^{b,2}$ | $-0.34 \pm 0.02^{a}$ |

*Different superscripts letters within the same row indicate statistically significant differences ($p < 0.05$) among carbon sources, whereas different superscript numbers within the same column denote differences ($p < 0.05$) among cultures.*

with *B. bifidum* (**Table** ), suggesting a potential risk for such combination.

Our results show that *B. longum* IPLA20022 and *B. breve* IPLA20006 reduced the toxicity of the co-cultures with sc-FOS as carbon source. Interestingly these two strains have previously shown the ability to remove *C. difficile* toxins from solution (Valdés et al., 2016). Although the putative mechanism behind toxin inactivation remains to be elucidated, it has been demonstrated that certain microorganisms produce compounds able to degrade *C. difficile* toxins or to reduce their toxicity (Castagliuolo et al., 1996; Banerjee et al., 2009; Carasi et al., 2012; Valdés et al., 2016). These mechanisms may be involved in the effect observed by us. However, given that in our case both microorganisms are co-incubated, the direct inhibition of the growth of the pathogen and/or an modulation of the expression of the toxin genes in *C. difficile* by the presence of bifidobacteria, similarly to that previously reported for *Lactobacillus acidophilus* (Yun et al., 2014), may also be involved. Previous studies pointed out a role of organic acids, such as lactic acid, in the inhibition of both growth and toxin production by *C. difficile* (Kolling et al., 2012; Yun et al., 2014). Therefore, the ability of bifidobacteria to produce acids, mainly acetic and lactic acids, and the pH drop caused by them may partially explain our observations. However, the role of other interactions cannot be overruled, especially since behaviors not explained by the acids, such as the increased toxicity of the co-culture *C. difficile-B. bifidum* in Inulin, were also observed.

## CONCLUSION

Co-culture with *B. longum* IPLA20022 or *B. breve* IPLA20006 in the presence of scFOS, but not of Inulin, reduces significantly the growth of *C. difficile*. Moreover, co-culture with these two strains in Synergy or Actilight reduced the toxicity of the *C. difficile* supernatants. Therefore, *B. longum* IPLA20022 and *B. breve* IPLA20006, in combination with Synergy or Actilight, are the most promising strains and compounds for the development of probiotic, prebiotic, or synbiotic products targeting at the reduction of CDI. However, future *in vitro* studies aiming at other clinically relevant *C. difficile* strains, as well as *in vivo* evaluation of the efficacy of the products, would be needed before drawing firm conclusions.

## AUTHOR CONTRIBUTIONS

MG and PR contributed with the conception, experimental design, and results interpretation of this study. LV carried out all experiments, AH performed chromatographic analyses. MG was in charge of writing the drafted manuscript. All authors performed a critical revision of the manuscript and approved the final version.

## ACKNOWLEDGMENTS

This work was fund by FEDER European Union funds through the projects AGL2012-33278 and AGL2013-43770R from the Spanish Ministry of Economy and Competitiveness (MINECO), and through the grants EQUIP11 and GRUPIN14-043 from the Program of Science, Technology and Innovation from the "Principado de Asturias." LV acknowledges her JAE-Pre fellowship to CSIC.

## SUPPLEMENTARY MATERIAL

The Supplementary Material for this article can be found online at: http://journal.frontiersin.org/article/10.3389/fmicb.2016.00738

## REFERENCES

Allen, S. J., Wareham, K., Wang, D., Bradley, C., Hutchings, H., Harris, W., et al. (2013). Lactobacilli and bifidobacteria in the prevention of antibiotic-associated diarrhoea and *Clostridium difficile* diarrhoea in older inpatients (PLACIDE): a randomised, double-blind, placebo-controlled, multicentre trial. *Lancet* 382, 1249–1257. doi: 10.1016/S0140-6736(13)61218-0

Ambalam, P., Kondepudi, K. K., Balusupati, P., Nilsson, I., Wadström, T., and Ljungh, A. (2015). Prebiotic preferences of human lactobacilli strains in co-culture with bifidobacteria and antimicrobial activity against

*Clostridium difficile. J. Appl. Microbiol.* 119, 1672–1682. doi: 10.1111/jam.12953

Andersen, K. K., Strokappe, N. M., Hultberg, A., Truusalu, K., Smidt, I., Mikelsaar, R. H., et al. (2015). Neutralization of *Clostridium difficile* Toxin B mediated by engineered Lactobacilli that produce single-domain antibodies. *Infect. Immun.* 84, 395–406. doi: 10.1128/IAI.00870-15

Arboleya, S., Binetti, A., Salazar, N., Fernández, N., Solís, G., Hernández-Barranco, A., et al. (2012). Establishment and development of intestinal microbiota in preterm neonates. *FEMS Microbiol. Ecol.* 79, 763–772. doi: 10.1111/j.1574-6941.2011.01261.x

Arboleya, S., Ruas-Madiedo, P., Margolles, A., Solis, G., Salminen, S., de los Reyes-Gavilan, C. G., et al. (2011). Characterization and *in vitro* properties of potentially probiotic *Bifidobacterium* strains isolated from breast-milk. *Int. J. Food Microbiol.* 149, 28–36. doi: 10.1016/j.ijfoodmicro.2010.10.036

Auclair, J., Frappier, M., and Millette, M. (2015). *Lactobacillus acidophilus* CL1285, *Lactobacillus casei* LBC80R, and *Lactobacillus rhamnosus* CLR2 (Bio-K+): characterization, manufacture, mechanisms of action, and quality control of a specific probiotic combination for primary prevention of *Clostridium difficile* infection. *Clin. Infect. Dis.* 60, S135–S143. doi: 10.1093/cid/civ179

Banerjee, P., Merkel, G. J., and Bhunia, A. K. (2009). *Lactobacillus delbrueckii* ssp. *bulgaricus* B-30892 can inhibit cytotoxic effects and adhesion of pathogenic *Clostridium difficile* to Caco-2 cells. *Gut Pathog.* 1:8. doi: 10.1186/1757-4749-1-8

Bouillaut, L., Dubois, T., Sononshein, A. L., and Dupuy, B. (2015). Integration of metabolism and virulence in *Clostridium difficile. Res. Microbiol.* 166, 375–383. doi: 10.1016/j.resmic.2014.10.002

Carasi, P., Trejo, F. M., Pérez, P. F., De Antoni, G. L., and Serradell, M. A. (2012). Surface proteins from *Lactobacillus kefir* antagonize *in vitro* cytotoxic effect of *Clostridium difficile* toxins. *Anaerobe* 18, 135–142. doi: 10.1016/j.anaerobe.2011.11.002

Carter, G. P., Rood, J. I., and Lyras, D. (2012). The role of toxin A and toxin B in the virulence of *Clostridium difficile. Trends Microbiol.* 20, 21–29. doi: 10.1016/j.tim.2011.11.003

Castagliuolo, I., Lamont, J. T., Nikulasson, S. T., and Pothoulakis, C. (1996). *Saccharomyces boulardii* protease inhibits *Clostridium difficile* Toxin A effects in the rat ileum. *Infect. Immun.* 64, 5225–5232.

Collado, M. C., Gueimonde, M., Hernández, M., Sanz, Y., and Salminen, S. (2005). Adhesion of selected *Bifidobacterium* strains to human intestinal mucus and its role in enteropathogen exclusion. *J. Food Protect.* 68, 2672–2678.

EFSA Panel on Dietetic Products, Nutrition and Allergies (NDA) (2011). Guidance on the scientific requirements for health claims related to gut and immune function. *EFSA J.* 9:1984. doi: 10.2903/j.efsa.2011.1984

Forssten, S. D., Röytiö, H., Hibberd, A. A., and Ouwehand, A. C. (2015). The effect of polydextrose and probiotic lactobacilli in a *Clostridium difficile*-infected human colonic model. *Microb. Ecol. Health Dis.* 26, 27988. doi: 10.3402/mehd.v26.27988

Gebhart, D., Lok, S., Clare, S., Tomas, M., Stares, M., Scholl, D., et al. (2015). A modified R-type bacteriocin specifically targeting *Clostridium difficile* prevents colonization of mice without affecting gut microbiota diversity. *MBio.* 6, e02368–e02314. doi: 10.1128/mBio.02368-14

Hopkins, M. J., and Macfarlane, G. T. (2003). Nondigestible oligosaccharides enhance bacterial colonization resistance against *Clostridium difficile in vitro. Appl. Environ. Microbiol.* 69, 1920–1927. doi: 10.1128/AEM.69.4.1920-1927.2003

Jangi, S., and Lamont, J. T. (2010). Asyntomatic colonization by *Clostridium difficile* in infants: implications for disease in later life. *J. Pediatr. Gastroenterol. Nutr.* 51, 2–7. doi: 10.1097/MPG.0b013e3181d29767

Karlsson, S. A., Lindberg, A., Norin, E., Burman, L. G., and Akerlund, T. (2000). Toxins, butyric acid, and other short-chain fatty acids are coordinatedly expressed and down-regulated by cysteine in *Clostridium difficile. Infect. Immun.* 68, 5881–5888. doi: 10.1128/IAI.68.10.5881-58 88.2000

Kolling, G. L., Wu, M., Warren, C. A., Durmaz, E., Klaenhammer, T. R., and Guerrant, R. L. (2012). Lactic acid production by *Streptococcus thermophilus* alters *Clostridium difficile* infection and *in vitro* Toxin A production. *Gut Microbes.* 3, 523–529. doi: 10.4161/gmic.21757

Kondepudi, K. K., Ambalam, P., Karagin, P. H., Nilsson, I., Wadström, T., and Ljungh, A. (2014). A novel multi-strain probiotic and synbiotic supplement for prevention of *Clostridium difficile* infection in a murine model. *Microbiol. Immunol.* 58, 552–558. doi: 10.1111/1348-0421.12184

Kondepudi, K. K., Ambalam, P., Nilsson, I., Wadstrom, T., and Ljungh, A. (2012). Prebiotic-non-digestible oligosaccharides preference of probiotics bifidobacteria and antimicrobial activity against *Clostridium difficile. Anaerobe* 18, 489–497. doi: 10.1016/j.anaerobe.2012.08.005

Lee, C. H., Steiner, T., Petrof, E. O., Smieja, M., Roscoe, D., Nematallah, A., et al. (2016). Frozen vs fresh fecal microbiota transplantation and clinical resolution of diarrhea in patients with recurrent *Clostridium difficile* infection: a randomized clinical trial. *J. Am. Med. Assoc.* 315, 142–149. doi: 10.1001/jama.2015.18098

Lee, J.-S., Chung, M.-J., and Seo, J.-G. (2013). *In vitro* evaluation of antimicrobial activity of lactic acid bacteria against *Clostridium difficile. Toxicol. Res.* 29, 99–106. doi: 10.5487/TR.2013.29.2.099

Leffler, D. A., and Lamont, J. T. (2015). *Clostridium difficile* infection. *N. Engl. J. Med.* 372, 1539–1548. doi: 10.1056/NEJMra1403772

Mani, N., Lyras, D., Barroso, L., Howarth, P., Wilkins, T., Rood, J. I., et al. (2002). Environmental response and autoregulation of *Clostridium difficile* TxeR, a sigma factor for toxin gene expression. *J. Bacteriol.* 184, 5971–5978. doi: 10.1128/JB.184.21.5971-5978.2002

Martin, J. S., Monaghan, T. M., and Wilcox, M. H. (2016). *Clostridium difficile* infection: epidemiology, diagnosis and understanding transmission. *Nat. Rev. Gastroenterol. Hepatol.* 13, 206–216. doi: 10.1038/nrgastro.2016.25

Ozaki, E., Kato, H., Kita, H., Karasawa, T., Maegawa, T., Koino, Y., et al. (2004). *Clostridium difficile* colonization in healthy adults: transient colonization and correlation with enterococcal colonization. *J. Med. Microbiol.* 53, 167–172. doi: 10.1099/jmm.0.05376-0

Reid, G., Younes, J. A., Van der Mei, H. C., Gloor, G. B., Knight, R., and Busscher, H. J. (2011). Microbiota restoration: natural and supplemented recovery of human microbial communities. *Nat. Rev. Microbiol.* 9, 27–38. doi: 10.1038/nrmicro2473

Rossi, M., Corradini, C., Amaretti, A., Nicolini, M., Pompei, A., Zanoni, S., et al. (2005). Fermentation of fructooligosaccharides and inulin by bifidobacteria: a comparative study of pure and fecal cultures. *Appl. Environ. Microbiol.* 71, 6150–6158. doi: 10.1128/AEM.71.10.6150-6158.2005

Rupnik, M., Wilcox, M. H., and Gerding, D. N. (2009). *Clostridium difficile* infection: new developments in epidemiology and pathogenesis. *Nat. Rev. Microbiol.* 7, 526–536. doi: 10.1038/nrmicro2164

Salazar, N., Binetti, A., Gueimonde, M., Alonso, A., Garrido, P., Gonzalez del Rey, C., et al. (2011). Safety and intestinal microbiota modulation by the exopolysaccharide-producing strains *Bifidobacterium animalis* IPLA R1 and *Bifidobacterium longum* IPLA E44 orally administered to Wistar rats. *Int. J. Food Microbiol.* 144, 342–351. doi: 10.1016/j.ijfoodmicro.2010.10.016

Schoster, A., Kokotovic, B., Permin, A., Pedersen, P. D., Dal Bello, F., and Guardabassi, L. (2013). *In vitro* inhibition of *Clostridium difficile* and *Clostridium perfringens* by commercial probiotic strains. *Anaerobe.* 20, 36–41. doi: 10.1016/j.anaerobe.2013.02.006

Senoh, M., Iwaki, M., Yamamoto, A., Kato, H., Fukuda, T., and Shibayama, K. (2015). Inhibition of adhesion of *Clostridium difficile* to human intestinal cells after treatment with serum and intestinal fluid isolated from mice immunized with nontoxigenic *C. difficile* membrane fraction. *Microb. Pathog.* 81, 1–5. doi: 10.1016/j.micpath.2015.03.001

Servin, A. L. (2004). Antagonistic activities of lactobacilli and bifidobacteria against microbial pathogens. *FEMS Microbiol. Rev.* 28, 405–440. doi: 10.1016/j.femsre.2004.01.003

Solís, G., de los Reyes-Gavilán, C. G., Fernández, N., Margolles, A., and Gueimonde, M. (2010). Establishment and development of lactic acid bacteria and bifidobacteria microbiota in breast-milk and the infant gut. *Anaerobe* 16, 307–310. doi: 10.1016/j.anaerobe.2010.02.004

Tejero-Sariñena, S., Barlow, J., Costabile, A., Gibson, G. R., and Rowland, I. (2012). *In vitro* evaluation of the antimicrobial activity of a range of probiotics against pathogens: evidence for the effects of organic acids. *Anaerobe* 18, 530–530. doi: 10.1016/j.anaerobe.2012.08.004

Tejero-Sariñena, S., Barlow, J., Costabile, A., Gibson, G. R., and Rowland, I. (2013). Antipathogenic activity of probiotics against *Salmonella Typhimurium* and *Clostridium difficile* in anaerobic batch culture systems: is it due to synergies in probiotic mixtures or the specificity of single strains? *Anaerobe* 24, 60–65. doi: 10.1016/j.anaerobe.2013.09.011

Trejo, F. M., De Antoni, G. L., and Pérez, P. F. (2013). Protective effect of bifidobacteria in an experimental model of *Clostridium difficile* associated colitis. *J. Dairy Res.* 80, 263–269. doi: 10.1017/S0022029913000216

Trejo, F. M., Pérez, P. F., and De Antoni, G. L. (2010). Co-culture with potentially probiotic microorganisms antagonises virulence factors of *Clostridium difficile in vitro*. *Antonie Van Leeuwenhoek.* 98, 19–29. doi: 10.1007/s10482-010-9424-6

Valdés, L., Alonso-Guervos, M., García-Suárez, O., Gueimonde, M., and Ruas-Madiedo, P. (2016). Selection of bifidobacteria and lactobacilli able to antagonise the cytotoxic effect of *Clostridium difficile* upon intestinal epithelial HT29 monolayer. *Front. Microbiol.* 7:577. doi: 10.3389/fmicb.2016.00577

Valdés, L., Gueimonde, M., and Ruas-Madiedo, P. (2015). Monitoring in real time the cytotoxic effect of *Clostridium difficile* upon the intestinal epithelial cell line HT29. *J. Microbiol. Methods* 119, 66–73. doi: 10.1016/j.mimet.2015.09.022

Vickers, R., Robinson, N., Best, E., Echols, R., Tillotson, G., and Wilcox, M. (2015). A randomised phase 1 study to investigate safety, pharmacokinetics and impact on gut microbiota following single and multiple oral doses in healthy male subjects of SMT19969, a novel agent for *Clostridium difficile* infections. *BMC Infect. Dis.* 15:91. doi: 10.1186/s12879-015-0759-5

Yakob, L., Riley, T. V., Paterson, D. L., Marquess, J., Soares-Magalhaes, R. J., Furuya-Kanamori, L., et al. (2015). Mechanisms of hypervirulent *Clostridium difficile* ribotype 027 displacement of endemic strains: an epidemiological model. *Sci. Rep.* 5:12666. doi: 10.1038/srep12666

Yang, Z., Ramsey, J., Hamza, T., Zhang, Y., Li, S., Yfantis, H. G., et al. (2015). Mechanisms of protection against *Clostridium difficile* infection by the monoclonal antitoxin antibodies actoxumab and bezlotoxumab. *Infect. Immun.* 83, 822–831. doi: 10.1128/IAI.02897-14

Yun, B., Oh, S., and Griffiths, M. W. (2014). *Lactobacillus acidophilus* modulates the virulence of *Clostridium difficile*. *J. Dairy Sci.* 97, 4745–4758 doi: 10.3168/jds.2014-7921

**Conflict of Interest Statement:** The authors declare that the research was conducted in the absence of any commercial or financial relationships that could be construed as a potential conflict of interest.

19

# Glycan Cross-Feeding Activities between Bifidobacteria under *In vitro* Conditions

*Francesca Turroni[1]*, Ezgi Özcan[2], Christian Milani[1], Leonardo Mancabelli[1], Alice Viappiani[3], Douwe van Sinderen[4], David A. Sela[2,5] and Marco Ventura[1]*

[1] Laboratory of Probiogenomics, Department of Life Sciences, University of Parma, Parma, Italy, [2] Department of Food Science, University of Massachusetts Amherst, Amherst, MA, USA, [3] GenProbio s.r.l., Parma, Italy, [4] Alimentary Pharmabiotic Centre and Department of Microbiology, Bioscience Institute, National University of Ireland, Cork, Ireland, [5] Center for Microbiome Research, University of Massachusetts Medical School, Worcester, MA, USA

**Edited by:**
Joerg Graf,
University of Connecticut, USA

**Reviewed by:**
Michael C. Nelson,
University of Connecticut, USA
Nicole Marie Koropatkin,
University of Michigan Medical School, USA

***Correspondence:**
Francesca Turroni,
Laboratory of Probiogenomics,
Department of Life Sciences,
University of Parma, Parma, Italy
francesca.turroni@unipr.it

Bifidobacteria colonize the gut of various mammals, including humans, where they may metabolize complex, diet-, and host-derived carbohydrates. The glycan-associated metabolic features encoded by bifidobacteria are believed to be strongly influenced by cross-feeding activities due to the co-existence of strains with different glycan-degrading properties. In this study, we observed an enhanced growth yield of *Bifidobacterium bifidum* PRL2010 when co-cultivated with *Bifidobacterium breve* 12L, *Bifidobacterium adolescentis* 22L, or *Bifidobacterium thermophilum* JCM1207. This enhanced growth phenomenon was confirmed by whole genome transcriptome analyses, which revealed co-cultivation-associated transcriptional induction of PRL2010 genes involved in carbohydrate metabolism, such as those encoding for carbohydrate transporters and associated energy production, and genes required for translation, ribosomal structure, and biogenesis, thus supporting the idea that co-cultivation of certain bifidobacterial strains with *B. bifidum* PRL2010 causes enhanced metabolic activity, and consequently increased lactate and/or acetate production. Overall, these data suggest that PRL2010 cells benefit from the presence of other bifidobacterial strains.

**Keywords: microbiota, microbe–microbe interactions, RNAseq, transcriptomics**

## Introduction

Bifidobacteria are key gut commensals of human beings, reaching a high-relative abundance when their host is an infant (Ventura et al., 2012; Turroni et al., 2014b). These bacteria have a saccharolytic lifestyle and their metabolism is consequently directed toward the utilization of carbohydrates that are naturally occurring in their ecological niches, which not only include dietary glycans, such as resistant-starch and xylan, but also host-derived glycans, i.e., mucin and human milk oligosaccharides (Turroni et al., 2010, 2011b; Duranti et al., 2014, 2015). In order to access these glycans, bifidobacteria have evolved an enzymatic repertoire of extracellular glycosyl hydrolases (GH) that catalyze the breakdown of such polysaccharides, with the production of mono, di/trisaccharides that are then imported into the cell through the action of specific carriers (Turroni et al., 2012). In addition, other species such as *Bifidobacterium longum* ssp. *infantis* possess apparently unique molecular mechanisms to capture intact oligosaccharides to be further processed intracellularly (Sela, 2011). It should, however, be considered that in their ecological niche bifidobacterial populations may interact with different strains/species which may lead to competition for or co-operative sharing of nutrients.

Biotic interactions between bacteria can either positively or negatively influence the fitness of the affected organisms (Pande et al., 2015). Several of these interactions rely on either the active or passive release of chemical molecules into the environment (Phelan et al., 2012; Morris et al., 2013). In this context, interspecies cross-feeding has been observed for *Bifidobacterium bifidum* PRL2010 and *Bifidobacterium breve* UCC2003 cells when cultivated on sialyllactose as the unique carbon source (Egan et al., 2014a,b). Moreover, previous studies have demonstrated metabolic cross-feeding between *Bifidobacterium adolescentis* and lactate-utilizing, butyrate-producing *Firmicutes* bacteria related to *Eubacterium hallii* and *Anaerostipes caccae* (Belenguer et al., 2006). This is significant and relevant to host health as butyrate is widely regarded as a beneficial short-chain fatty acid produced by elements of the microbiota. Furthermore, bacterial cross-feeding opportunities as facilitated by members of the colonic microbiota have been considered to be pivotal for carbohydrate turn-over in this ecological niche (De Vuyst and Leroy, 2011).

In recent years, extensive scientific efforts have been made to decode bifidobacterial genome sequences, which are part of a novel discipline called probiogenomics, directed to understand the genetics sustaining the adaptation of these bacteria to the intestine (Ventura et al., 2009; Milani et al., 2014). In this context, the genome sequence of *B. bifidum* PRL2010, an infant gut isolate, exhibits several genetic adaptations to the human gut and for this reason is employed as a model strain to investigate the biology of infant-associated bifidobacteria (Turroni et al., 2010, 2013; Serafini et al., 2014). The ability of PRL2010 to utilize host-derived glycans such as mucins and human milk oligosaccharides, and the capacity to produce pilus-like structures to facilitate gut colonization and immuno modulation, are clear examples of such genetic adaptations (Turroni et al., 2010, 2013; Serafini et al., 2014).

Here, we investigate possible cross-feeding activities of simple bifidobacterial communities under *in vitro* conditions, targeting specific complex, diet-associated carbohydrates. Such cross-feeding activities were investigated employing transcriptome analysis of bifidobacterial communities by means of RNAseq as well as by assessment of the metabolic profiles of these bifidobacterial consortia. The main findings of the current study are that *B. bifidum* PRL2010 does not utilize starch or xylan, unless co-cultured with a strain that produces extracellular glycoside hydrolases that can degrade these substrates. These findings provide evidence of mutalisitic cross-feeding between certain bifidobacterial strains when co-cultured in media containing starch or xylan.

## Materials and Methods

### Growth Conditions

*Bifidobacterium bifidum* PRL2010 (Turroni et al., 2010) on its own, or in combination with *B. breve* 12L. (Bottacini et al., 2014), *B. adolescentis* 22L (Duranti et al., 2014), or *Bifidobacterium thermophilum* JCM1207 were cultivated in an anaerobic atmosphere (2.99% $H_2$, 17.01% $CO_2$, and 80% $N_2$) in a chamber (Concept 400; Ruskin) at 37°C for 24 h in de Man–Rogosa–Sharpe (MRS; Scharlau Chemie, Barcelona, Spain) medium, supplemented with 0.05% (wt/vol) L-cysteine hydrochloride.

### Co-Cultivation

Viable cells of each of the following strains: *B. bifidum* PRL2010, *B. breve* 12L, *B. adolescentis* 22L, or *B. thermophilum* JCM1207, or these strains in co-cultivation with *B. bifidum* PRL2010 was inoculated in 6 ml of MRS (without any carbohydrate; Scharlau Chemie, Barcelona, Spain) supplemented with 1% of either RS2-resistant starch or xylan (Poly(β-D-xylopyranose[1→4]; Sigma–Aldrich) as the sole carbon source in triplicates. Cell suspensions were mixed and incubated at 37°C for 24 h under anaerobic conditions. Bacterial cell cultivations were performed in triplicate (biological replicates).

Bacterial strain enumerations at the beginning and at the end of a given growth experiment were determined by quantitative real-time PCR (qRT-PCR).

### Evaluation of PRL2010 Cell Numbers in Co-Cultivation Trials

Possible enhancement or reduction of PRL2010 growth as a consequence of co-cultivation with other bacteria was monitored by qRT-PCR at the begin as well as at the end of the growth experiments. The amounts of cells for each of the strain used at the begin of the growth evaluation trials determined by qRT-PCR assays are shown in **Supplementary Figure** . qRT-PCR experiments were based on strain-specific primers targeting genes present in single copy within the genomes of PRL2010 (BBPR_0282) 12L (B12L_0105), 22L (BADO_1546), and JCM1207 (BTHER_1915). The copy-number of a gene, and the deduced cell number (since the genes targeted were in single copy per genome) of a given strain used in the co-cultivation experiments was evaluated by comparing the cycle threshold (Ct) values obtained with those from a standard curve. Standard curves were calculated from serial dilutions of a culture with a known cell number (as determined by viable count assessment) for each bacterial strain vs. Ct produced for each target gene. In the case of PRL2010 we used the following primer couple: BBPR_0282-UNI (5′-GCGAACAATGATGGCACCTA-3′) and BBPR_282-REV (5′-GTCGAACACCACGACGATGT-3′). In the case of *B. breve* 12L, we used 12L-UNI (5′-CGAAGTTCCAGTTCACCAT-3′) and 12L-REV (5′-GTTCTTGGCGTTCCAGATGT-3′); for *B. adolescentis* 22L we employed the PCR primers 22L-UNI (5′-GACCAAGCCAACCAGTTCAT-3′) and 22L-REV (5′-TTGGTGGCCTTGTAGTAGCC-3′); and for *B. thermophilum* JCM1207 the following PCR primers BTHERfw (5′-TTACACGCATCCCAATACGC-3′) and BTHERrv (5′-CGTGAAGTATGGATGGTCGC-3′). These primers target genes of the sortase-dependent pilus loci identified in the genomes of these microorganisms (Turroni et al., 2014a).

## Metabolic Profiling

For quantitative determination of metabolites produced by bifidobacterial fermentation of starch and xylan, cell-free supernatants were analyzed using Agilent 1260 Infinity HPLC system equipped with Wyatt Optilab T-rEX Refractive Index detector. Separation was carried out using a Shodex Sugar SH1011 column (8.0 mm ID × 300 mm) at 60°C with the detector temperature maintained at 30°C. The mobile phase was prepared in 20 mM $H_2SO_4$ and run at a flow rate of 0.6 mL/min for 30 min. Injection volume was set at 10 μL and speed of draw and eject was set to 100 μL/min. External sugar standards (glucose, maltose, xylose, and maltotriose) and organic acid (acetic acid and lactic acid) standards were purchased from Sigma–Aldrich Co. (USA). Concentrations of individual sugars and organic acids in samples were calculated from calibration curves drawn from external standards for five different concentrations (0.5, 1, 5, 10, and 20 mg/L). Each metabolic profiling experiment was carried out in triplicate (three measurements were performed for each replicate). Glucose and maltose consumption was calculated by subtracting the values at 24 h from time zero. Since the concentration of metabolites produced is dependent upon the cell density at the end of fermentation (i.e., more cells would yield more endproducts), raw metabolite values were normalized by cell density to correct for differences in biomass.

## RNA Extraction and Purification

Total RNA was isolated using a previously described method (Turroni et al., 2011a). Briefly, cell pellets/tissue materials were resuspended in 1 mL of QUIAZOL (Qiagen, UK) and placed in a tube containing 0.8 g of glass beads (diameter, 106 μm; Sigma). Cells were lysed by shaking the mix on a BioSpec homogenizer at 4°C for 2 min (maximum setting). The mixture was then centrifuged at 12,000 rpm for 15 min, and the upper RNA-containing phase was recovered. The RNA sample was further purified by phenol extraction and ethanol precipitation according to an established method (Sambrook and Russel, 2001). RNA quality was checked by analyzing the integrity of rRNA molecules by a Tape Station (Agilent Technologies).

## RNAseq with Ion Torrent Personal Genome Machine (PGM)

One hundred nanogram of total RNA was used as the starting input for RNA-Seq library preparation. Briefly, 100 ng of total RNA was treated with MICROB*Express*™ Bacterial RNA Enrichment Kit (Ambion) to remove rRNA according to the supplier's instructions. The efficacy of rRNA depletion was checked by a Tape Station (Agilent Technologies), after which rRNA-depleted RNA samples were fragmented using RNaseIII (Life Technologies, USA) followed by size evaluation using a Tape Station (Agilent Technologies). Whole transcriptome libraries were constructed using the Ion Total-RNA Seq Kit v2 (Life Technologies, USA). Barcoded libraries were quantified by qRT-PCR and each library template was amplified on Ion Sphere Particles using Ion One Touch 200 Template Kit v2 (Life Technologies, USA). The samples were loaded on 316 Chips and sequenced by means of a PGM instrument (Life Technologies, USA). Sequencing reads were depleted of adapter sequences and quality filtered (with overall quality, quality window, and length filters) using a custom script implying fastq-mcf[1] (settings: –qual-mean=25, -w 5 –q 20 –l 100). The resulting processed sequences were aligned to the reference genomes through BWA (Li and Durbin, 2009) with mismatch and gap penalty increased to six and eight, respectively, in order to avoid cross-mapping during analysis of co-cultivation samples. Reference genomes used for RNASeq bioinformatics analyses are deposited under the following accession numbers: CP006711 (*B. breve* 12L), CP007443 (*B. adolescentis* 22L), CP001840 (*B. bifidum* PRL2010), and GCA_000741495.1 (*B. thermophilum* JCM1207). Counting of the number of reads that correspond to annotated open reading frames (ORFs) was performed using HTSeq[2] and analysis of the count data was performed using the R package DESeq (Anders and Huber, 2010).

## Statistical Analysis

Statistical significance between means was analyzed using the unpaired Student's $t$-test. Values are expressed as the means ± SE of the mean of a given experiment performed in triplicate.

## Sequence Accession Numbers

All RNAseq raw data from this study were deposited in the SRA database under the accession number SRP058697 (BioProject accession number PRJNA284883).

# Results and Discussion

## Evaluation of the Growth Performances on Dietary Polysaccharide

Growth capabilities of *B. bifidum* PRL2010, *B. adolescentis* 22L, *B. breve* 12L, and *B. thermophilum* JCM1207 cultivated on their own (mono-association) on MRS supplemented with starch or xylan as the sole carbon source were evaluated (**Figure** ; **Supplementary Figure** ) and compared to those achieved when these strains were co-cultivated in pairs (bi-associations) on identical substrates. In addition, cultivation experiments involving all the strains used in this study were also performed on MRS without carbon source, revealing, as expected, the absence of any sign of growth. These carbohydrates were selected in order to represent carbon sources that may occur in the human gut as they are derived from a plant-derived, glycan-based diet (Chassard and Lacroix, 2013). Notably, of all the possible strain pair combinations, only those bi-associations that included PRL2010 cells displayed significant differences in growth with respect to the respective mono-associations (**Figure** ; **Supplementary Figure** ). As expected, PRL2010 cells did not exhibit any significant growth on starch or xylan when grown in a mono-culture (Turroni et al., 2012). However, when this strain was co-cultivated with 22L or 12L cells on a MRS-supplemented with starch, the number of cells increased (about three- and fourfold, respectively, $p < 0.005$) compared to the situation of mono-association, respectively (**Figure** ).

[1]http://code.google.com/p/ea-utils

[2]http://www-huber.embl.de/users/anders/HTSeq/doc/overview.html

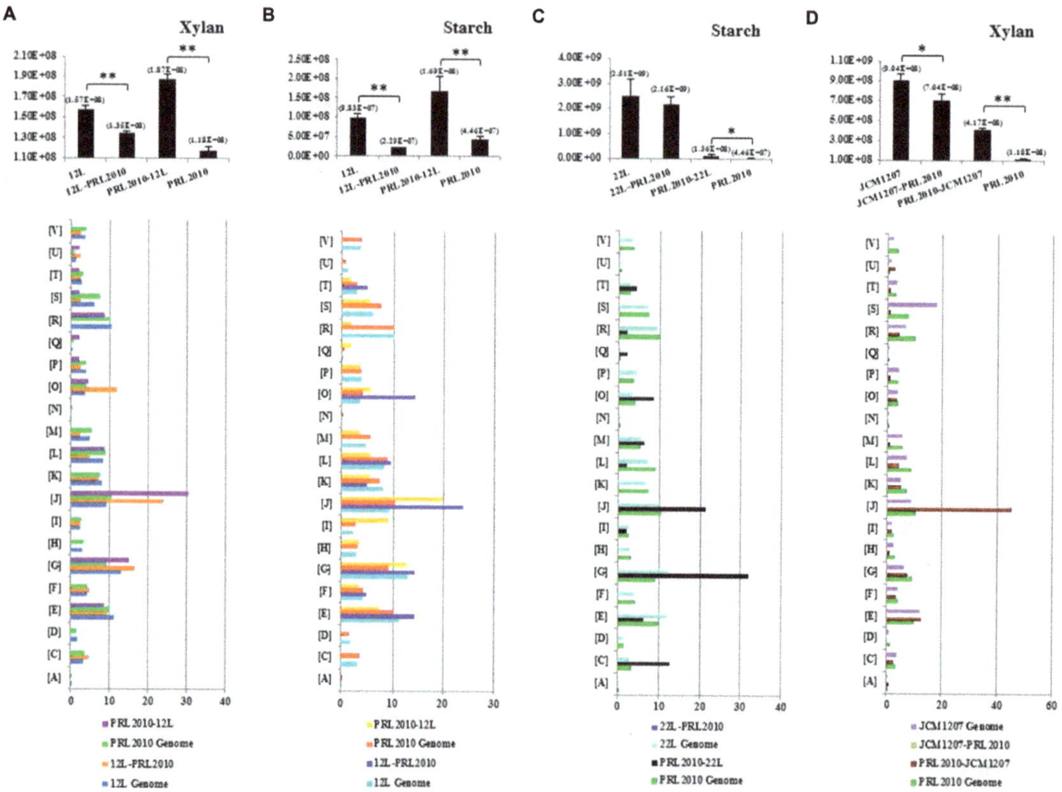

**FIGURE | Transcriptome analyses of co-cultivated bifidobacterial strains.** Only bi-associations showing a cross-feeding behavior between bifidobacterial strains are represented. **(A–D)** The cell number evaluation of *Bifidobacterium breve* 12L, *Bifidobacterium bifidum* PRL2010, *Bifidobacterium adolescentis* 22L, and *Bifidobacterium thermophilum* JCM1207 strains in mono- and co-cultivation on the glycans-based medium indicated above each panel by qRT-PCR. The bar plot placed below represented the functional annotation of expressed genes of bi-association according to their cluster orthologous gene (COG) categories. Results of qRT-PCR are represented in pillars in which the *y*-axis represents the genome copy number/ml of bacterial culture and *x*-axis showed the name of the strains involved in mono- and bi-associations. The value in parenthesis above each pillar represents the average cell numbers for that condition. The color of each COG family is indicated in the figure. Each COG family is identified by one-letter abbreviations: A, RNA processing and modification; B, chromatin structure and dynamics; C, energy production and conversion; D, cell cycle control and mitosis; E, amino acid metabolism and transport; F, nucleotide metabolism and transport; G, carbohydrate metabolism and transport; H, coenzyme metabolism; I, lipid metabolism; J, translation; K, transcription; L, replication and repair; M, cell wall/membrane/envelop biogenesis; N, cell motility; O, post-translational modification, protein turnover, chaperone functions; P, inorganic ion transport and metabolism; Q, secondary structure; T, signal transduction; U, intracellular trafficking and secretion; Y, nuclear structure; Z, cytoskeleton; R, general functional prediction only; S, function unknown. The percentage was calculated as the percentage of transcribed genes belonging to the indicated COG category with respect to all transcribed genes. Asterisks indicate that the presented data display a significant, either $*p < 0.05$ or $**p < 0.01$, deviation from the obtained values of the mono-association.

In contrast, 12L appears to significant decrease (compared to mono-association; fourfold, $p < 0.001$) during growth on starch when co-cultivated with the presence of PRL2010 cells (**Figure** ). Conversely, *B. bifidum* PRL2010 did not appear to influence the growth yields (changes less than twofold) of 22L cells on starch, or that of 12L when cultivated on xylan, as displayed by unchanged growth yields of bi-association compared to mono-association of these strains (**Figure** ). Another interesting increase in deduced PRL2010 cell numbers was observed for the bi-association of PRL2010 with *B. thermophilum* JCM1207 (threefold, $p < 0.001$), when grown on MRS supplemented with xylan (**Figure** ). In contrast, JCM1207 cells exhibit a modest reduction in growth ability (relative to mono-association), when co-cultivated with PRL2010 (**Figure** ; **Supplementary Figure** ). Overall, the concomitant presence of two different bifidobacterial strains was shown in some cases to increase the cell numbers of one partner when cultivated on complex carbohydrates, thus suggesting cross-feeding abilities of bifidobacterial strains.

## Assessing the Metabolic Profile of Bifidobacteria

In order to investigate if the co-occurrence of two strains influences bifidobacterial metabolism, we evaluated the production of the metabolic endproducts acetate and lactate, along with the depletion of various sugars from the culture supernatant (i.e., glucose and maltose). The results of this metabolic comparison between mono-associations and bi-associations (as collected from three independent experiments) are depicted in **Figure** . Bifidobacteria produce acetate and lactate as a result of their saccharoclastic fermentative metabolism through the so-called bifid shunt (Sela et al., 2008; Pokusaeva et al., 2011). Interestingly, lactate production from

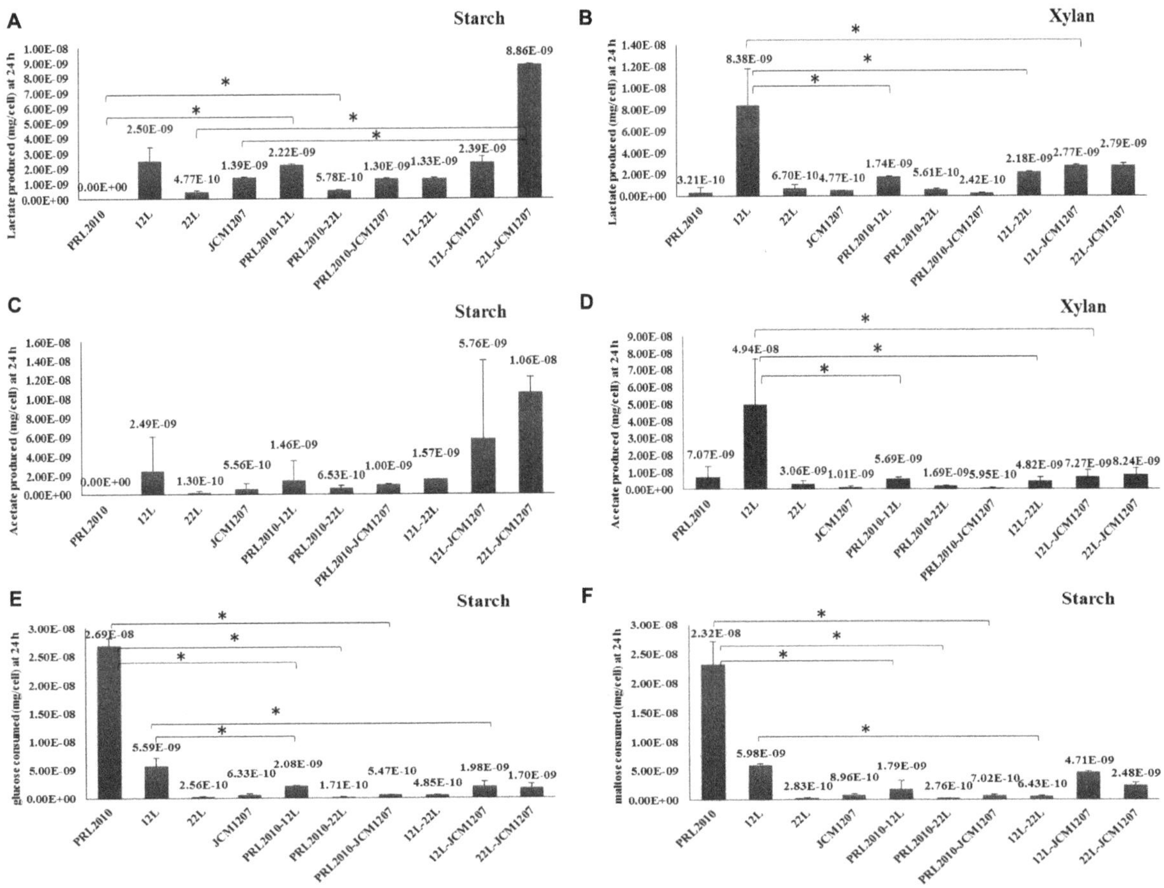


**FIGURE | Metabolic profiling of co-cultivated bifidobacteria. (A,B)** The evaluation of the lactate production of *B. bifidum* PRL2010, *B. breve* 12L, *B. adolescentis* 22L, and *B. thermophilum* JCM1207 strains in mono- and co-cultivation on starch-based and xylan-based medium at 24 h by HPLC, respectively. Values are expressed as mean ± SD mg per cell. **(C,D)** The evaluation of the acetate production of *B. bifidum* PRL2010, *B. breve* 12L, *B. adolescentis* 22L, and *B. thermophilum* JCM1207 strains in mono- and co-cultivation on starch-based and xylan-based medium at 24 h by HPLC, respectively. Values are expressed as mean ± SD mg per cell. **(E)** The evaluation of glucose consumption of *B. bifidum* PRL2010, *B. breve* 12L, *B. adolescentis* 22L, and *B. thermophilum* JCM1207 strains in mono- and co-cultivation on starch-based medium at 24 h by HPLC. Values are expressed as mean ± SD mg per cell. **(F)** The evaluation of maltose consumption of *B. bifidum* PRL2010, *B. breve* 12L, *B. adolescentis* 22L, and *B. thermophilum* JCM1207 strains in mono- and co-cultivation on starch-based medium at 24 h by HPLC, respectively. Values are expressed as mean ± SD mg per cell. Asterisks indicate that the presented data display a significant difference ($p < 0.05$) with respect to those obtained for the mono association. The value in parenthesis above each pillar represents the mean ± SD mg per cell for that condition.

starch and xylan fermentation showed significant differences between bifidobacterial species. Whereas acetate production from xylan fermentation varied significantly between certain species in mono-associations (PRL2010 vs. 12L, 12L vs. 22L, and 12L vs. JCM1207), and between 12L and all 12L co-cultivations, its production from starch did not show significant variation between species or bi-associations (**Figure** ). Since *B. bifidum* PRL2010 did not exhibit significant growth utilizing starch as a sole carbohydrate, acetate, and lactate production was not observed 24 h into the fermentation. Although growth of PRL2010 was enhanced when co-cultivated with 12L, 22L, or JCM1207 in a medium containing xylan as the sole carbon source, it did not appear to influence acetate and lactate production compared to the respective mono-associations (**Figure** ). Interestingly, lactate production during co-cultivation of 22L and JCM1207 in starch increased compared to the respective mono-associations (**Figure** ). This may be indicative of bacterial proto-cooperation when grown on starch, as the same relationship was not observed with xylan. *B. breve* 12L cells produced significantly more lactate and acetate while fermenting xylan in pure culture as compared to its bi-association with other species (**Figures** , ). This suggests that 12L energy metabolism may be inhibited by the presence of other bifidobacteria. Interestingly, 12L did not produce biomass when grown in axenic culture, suggesting metabolic flux in the absence of cellular growth. Glucose and maltose depletion during starch fermentation varied among species. *B. breve* 12L cells consumed more glucose alone than when co-cultivated with PRL2010, JCM1207, and 22L cells in starch fermentation (**Figures** , ). However, this did not coincide with a significant decrease in lactate production (**Figure** ). While fermenting starch individually, 22L and

JCM1207 consumed similar amounts of glucose (2.56E-10 and 6.33E-10 mg/cell, respectively) and maltose (2.83E-10 and 8.96E-10 mg/cell, respectively), whereas they consumed onefold more glucose and maltose than their consumption in pure cultures during co-culture (**Figures** ). This, in turn, resulted in a twofold increase in acetate and lactate production in co-culture compared to the situation in mono-associations (**Figures** ). Although PRL2010 did not produce significant biomass from xylan or starch fermentation, it appears that glucose and maltose from degradation of starch was depleted from the growth medium regardless. This may be explained by the fact that PRL2010 is utilizing free glucose and/or maltose that may be present in very low amounts in the starch-supplemented growth medium (carbohydrate contaminants), thereby allowing very limited growth (**Figures** ). In general, organic acid production, and sugar consumption did not exhibit a linear correlation among the tested strains. This is likely due to the hydrolysis of dietary oligosaccharides yielding increased concentrations of monomeric and dimeric sugars derived from xylan or starch before they enter the bifid shunt (Ze et al., 2012).

However, since we are measuring the initial and final metabolites produced, we may not have detected intermediate liberated/produced metabolites from the provided carbon sources. Thus, the consumption kinetics is unclear and we cannot exclude the possibility that metabolite concentrations deviate over time due to glycosyl hydrolase activity.

## Transcriptomics of the Cross-feeding Features

In order to evaluate the molecular aspects behind the cross-feeding activity as observed for some of the bi-associations involving the PRL2010–22L strain combination, or the PRL2010–12L strain combination when cultivated on starch, or the PRL2010–JCM1207 combination when grown on xylan, RNAseq experiments of these strain combinations cultivated on either of these substrates were performed. In order to increase the robustness of our RNAseq data, two technical replicates starting from the same library for each RNAseq trial were performed. When compared to the reference condition (mono-association) it was identified that the number of genes whose expression in either PRL2010 or the other strains, was significantly up-regulated (greater than or equal to twofold change, $p < 0.005$)

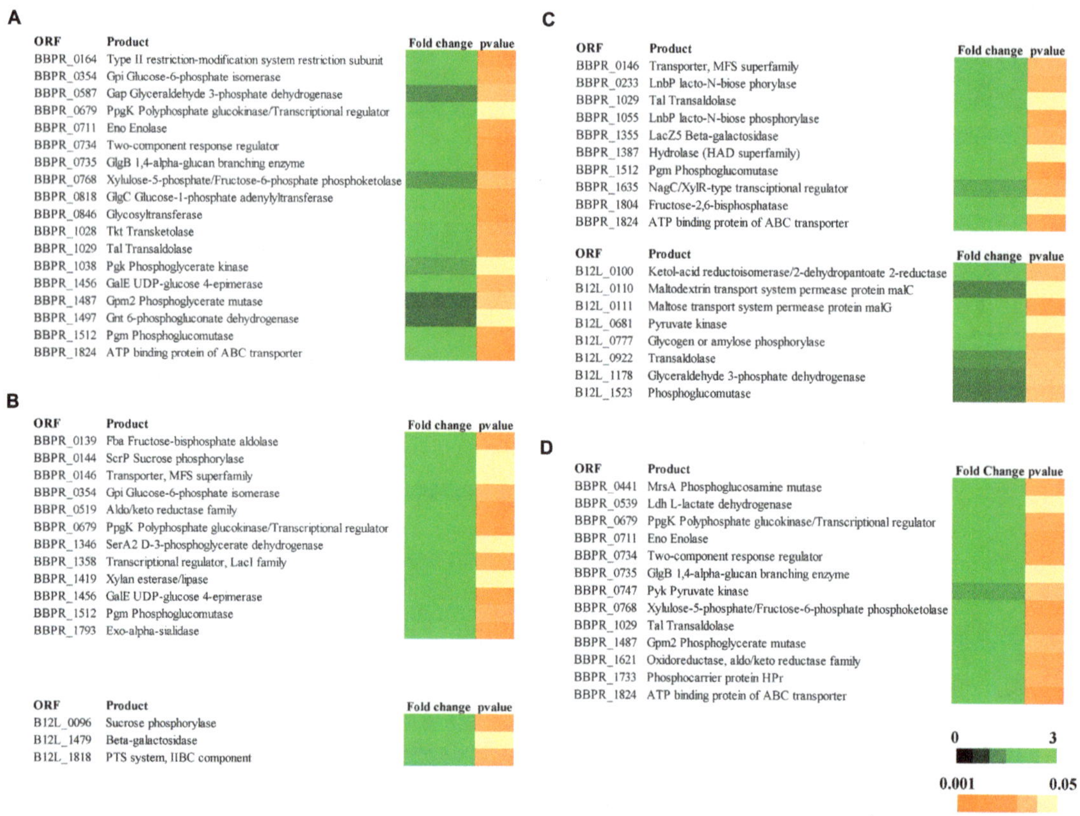


**FIGURE | Transcriptomic profiling of genes predicted to be involved in the metabolism of carbohydrates by bifidobacteria in response to the different bi- associations. (A)** The heat-map of transcriptional profiling of genes up-regulated in *B. bifidum* PRL2010 when this strain was co-cultivated with *B. adolescentis* 22L on starch-based medium. **(B)** The heat-map representing the transcriptional profiling of genes up-regulated in *B. bifidum* PRL2010 and in *B. breve* 12L when they were grown in bi-association on starch-based medium. **(C)** The heat-map representing the transcriptional profiling of genes up-regulated in *B. bifidum* PRL2010 and *B. breve* 12L when they were co-cultivated on xylan-based medium. **(D)** The heat-map representing the transcriptional profiling of the up-regulated genes in *B. bifidum* PRL2010 when was co-cultivated with *B. thermophilum* JCM1207 on xylan-based medium. Colors (black to green) represent the average signal intensity.

ranged from 0 to 121. We used cluster orthologous gene (COG) analysis in order to identify differentially transcribed genes that may contribute to specific biological functions within the gut. As illustrated in **Figure** , carbohydrate metabolism, corresponding to COG category [G], is one of the COG functions of PRL2010 most significantly affected by the interaction with another bifidobacterial strain. This is probably due to a response to the presence of specific breakdown capabilities exploited by 22L, 12L, or JCM1207 cells. In this context, we observed an up-regulation of an ABC-type transporter-encoding gene (BBPR_1824), as well as an major facilitator superfamily (MFS) transporter-encoding gene (BBPR_0146), thus possibly involved in the uptake of simple sugars when co-cultivated with 22L or 12L cells on MRS containing starch as a unique carbon source (**Figure** ). This finding can be explained by the fact that the extracellular amylases encoded by strains 22L (Duranti et al., 2014) and 12L (Bottacini et al., 2014) generate simple carbohydrates, which may then be imported by PRL2010 cells through its carbohydrate transporter arsenal (Turroni et al., 2012). Other transcriptionally induced genes of *B. bifidum* PRL2010 cells encompass enolase (BBPR_0711), glucose-6-phosphate isomerase (BBPR_354), phosphoglycerate kinase (BBPR_1038), glyceraldehyde 3-phosphate dehydrogenase (BBPR_0587), and phosphoglycerate mutase (BBPR_1487), whose functions are predicted to be related to (carbohydrate-dependent) energy production and conversion through the glycolytic pathway (**Supplementary Figure** ). These observations indicate that PRL2010 cells have enhanced flux through their central fermentative pathway in the presence of 22L or 12L cells. Notably, we also observed the enhanced transcription of genes encoding the various subunits of the ATPase system of PRL2010 strain, which may occur in response to medium acidification as a result of enhanced metabolic activity. Co-cultivation of PRL2010 cells with strain JCM1207 on a xylan-based medium also increased the transcription levels of genes that are predicted to be involved in the carbohydrate metabolism (**Figure** ). In particular, 121 genes of PRL2010 exhibited transcriptional up-regulation. The upregulated genes include several ORFs-encoding glycolytic enzymatic repertoire of PRL2010, such as enolase (BBPR_0711), phosphoglycerate mutase (BBPR_1487), and pyruvate kinase (BBPR_0747; **Figure** ).

Interestingly, strain 12L displayed the transcriptional up-regulation of 21 genes when it was co-cultivated with PRL2010 on a starch-based medium. Notably, among the up-regulated genes, we observed B12L_1818 encoding a putative IIBC component of a phosphotransferase system (PTS), which is involved in carbohydrate metabolism (**Figure** ). Co-cultivation of strain 12L with PRL2010 caused transcriptional up-regulation of 42 genes of *B. breve* 12L on xylan-based medium, some of which are known to be involved in energy generation through the bif shunt, such as pyruvate kinase (B12L_0681) and phosphoglucomutase (B12L_1523; **Figure** ). This is consistent with the observation that 12L produces significantly more organic acid endproducts in axenic culture than when co-cultured with another bifidobacterial species. In contrast, the transcriptomes of strain 22L when cultivated on starch, and that of JCM1207 strain when grown on xylan-based medium were shown to be unaltered when these strains were co-cultivated with PRL2010 cells compared to transcriptome data obtained when these strains were in mono-association. Such findings confirmed the results achieved by growth experiments (see above).

## Conclusion

In this study, we assessed possible glycan cross-feeding activities between simple bifidobacterial communities when metabolizing complex carbohydrates that, being present in the diet, are expected to be commonly found in the human gut. Our results highlight the existence of a gut commensal relationship between different bifidobacterial species (Supplementary Table ). We showed the *in vitro* ability of *B. bifidum* PRL2010 to cross-feed on sugars released by the starch- and/or xylan-degrading activities of *B. adolescentis* 22L, *B. breve* 12L, and *B. thermophilum* JCM1207. The generated information advances our knowledge on the metabolic adaptability and versatility of these strains, which no doubt will facilitate colonization of the human gut. Interestingly, when *B. bifidum* PRL2010 was co-cultivated with *B. breve* 12L we observed effects on the transcriptomes of both strains, apparently affecting the glycolytic pathway. The observed transcriptional changes may represent a molecular example of a mutualistic relationship between these two bifidobacterial strains, perhaps being a reflection of their common ecological origin, i.e., the infant gut.

The precise characterization of such complex interactions between gut microbiota members is pivotal in the process of modulation of the composition of the intestinal microbial population, especially during probiotic treatments.

## Acknowledgments

We thank GenProbio s.r.l. for financial support of the Laboratory of Probiogenomics. DS is a member of the APC Microbiome institute funded by Science Foundation Ireland (SFI) through the Irish Government's National Development Plan (Grant number SFI/12/RC/2273). We thank Dr. Albert Byung-Hun UM for his assistance and helpful discussions.

## Supplementary Material

The Supplementary Material for this article can be found online at: http://journal.frontiersin.org/article/10.3389/fmicb.2015.01030

**FIGURE | Evaluation of the cell density of bifidobacterial strains by qRT-PCR assays.** Evaluation of the cell density of mono- and co-cultivation of *Bifidobacterium breve* 12L, *Bifidobacterium adolescentis* 22L, *Bifidobacterium bifidum* PRL2010, and *Bifidobacterium thermophilum* JCM1207 strains on xylan- and starch-based medium by qRT-PCR experiments. *y*-axis represents the genome copy number/ml of bacterial culture, *x*-axis showed the name of the strains involved in mono- and bi-associations. Empty bars represent the amount of cells present at T0. Asterisks indicate that the presented data display a significant deviation, i.e., *$p < 0.05$ or **$p < 0.01$, with respect to the data obtained for the mono-association.

## References

Anders, S., and Huber, W. (2010). Differential expression analysis for sequence count data. *Genome Biol.* 11, R106. doi: 10.1186/gb-2010-11-10-r106

Belenguer, A., Duncan, S. H., Calder, A. G., Holtrop, G., Louis, P., Lobley, G. E., et al. (2006). Two routes of metabolic cross-feeding between *Bifidobacterium* adolescentis and butyrate-producing anaerobes from the human gut. *Appl. Environ. Microbiol.* 72, 3593–3599. doi: 10.1128/AEM.72.5.3593-3599.2006

Bottacini, F., O'Connell Motherway, M., Kuczynski, J., O'Connell, K. J., Serafini, F., Duranti, S., et al. (2014). Comparative genomics of the *Bifidobacterium breve* taxon. *BMC Genomics* 15:170. doi: 10.1186/1471-2164-15-170

Chassard, C., and Lacroix, C. (2013). Carbohydrates and the human gut microbiota. *Curr. Opin. Clin. Nutr. Metab. Care* 16, 453–460. doi: 10.1097/MCO.0b013e3283619e63

De Vuyst, L., and Leroy, F. (2011). Cross-feeding between bifidobacteria and butyrate-producing colon bacteria explains bifdobacterial competitiveness, butyrate production, and gas production. *Int. J. Food Microbiol.* 149, 73–80. doi: 10.1016/j.ijfoodmicro.2011.03.003

Duranti, S., Milani, C., Lugli, G. A., Turroni, F., Mancabelli, L., Sanchez, B., et al. (2015). Insights from genomes of representatives of the human gut commensal *Bifidobacterium bifidum*. *Environ. Microbiol.* 17, 2515–2531. doi: 10.1111/1462-2920.12743

Duranti, S., Turroni, F., Lugli, G. A., Milani, C., Viappiani, A., Mangifesta, M., et al. (2014). Genomic characterization and transcriptional studies of the starch-utilizing strain *Bifidobacterium adolescentis* 22L. *Appl. Environ. Microbiol.* 80, 6080–6090. doi: 10.1128/AEM.01993-14

Egan, M., Motherway, M. O., Kilcoyne, M., Kane, M., Joshi, L., Ventura, M., et al. (2014a). Cross-feeding by *Bifidobacterium breve* UCC2003 during co-cultivation with *Bifidobacterium bifidum* PRL2010 in a mucin-based medium. *BMC Microbiol.* 14:282. doi: 10.1186/s12866-014-0282-7

Egan, M., O'Connell Motherway, M., Ventura, M., and Van Sinderen, D. (2014b). Metabolism of sialic acid by *Bifidobacterium breve* UCC2003. *Appl. Environ. Microbiol.* 80, 4414–4426. doi: 10.1128/AEM.01114-14

Li, H., and Durbin, R. (2009). Fast and accurate short read alignment with Burrows-Wheeler transform. *Bioinformatics* 25, 1754–1760. doi: 10.1093/bioinformatics/btp324

Milani, C., Lugli, G. A., Duranti, S., Turroni, F., Bottacini, F., Mangifesta, M., et al. (2014). Genomic encyclopedia of type strains of the genus *Bifidobacterium*. *Appl. Environ. Microbiol.* 80, 6290–6302. doi: 10.1128/AEM.02308-2314

Morris, B. E., Henneberger, R., Huber, H., and Moissl-Eichinger, C. (2013). Microbial syntrophy: interaction for the common good. *FEMS Microbiol. Rev.* 37, 384–406. doi: 10.1111/1574-6976.12019

Pande, S., Shitut, S., Freund, L., Westermann, M., Bertels, F., Colesie, C., et al. (2015). Metabolic cross-feeding via intercellular nanotubes among bacteria. *Nat. Commun.* 6, 6238. doi: 10.1038/ncomms7238

Phelan, V. V., Liu, W. T., Pogliano, K., and Dorrestein, P. C. (2012). Microbial metabolic exchange–the chemotype-to-phenotype link. *Nat. Chem. Biol.* 8, 26–35. doi: 10.1038/nchembio.739

Pokusaeva, K., Fitzgerald, G. F., and Van Sinderen, D. (2011). Carbohydrate metabolism in *Bifidobacteria*. *Genes Nutr.* 6, 285–306. doi: 10.1007/s12263-010-0206-6

Sambrook, J., and Russel, D. W. (2001). *Molecular Cloning a Laboratory Manual.* New York, NY: Cold Spring Harbor Laboratory Press.

Sela, D. A. (2011). Bifidobacterial utilization of human milk oligosaccharides. *Int. J. Food Microbiol.* 149, 58–64. doi: 10.1016/j.ijfoodmicro.2011.01.025

Sela, D. A., Chapman, J., Adeuya, A., Kim, J. H., Chen, F., Whitehead, T. R., et al. (2008). The genome sequence of *Bifidobacterium longum* subsp. infantis reveals adaptations for milk utilization within the infant microbiome. *Proc. Natl. Acad. Sci. U.S.A.* 105, 18964–18969. doi: 10.1073/pnas.0809584105

Serafini, F., Turroni, F., Ruas-Madiedo, P., Lugli, G. A., Milani, C., Duranti, S., et al. (2014). Kefir fermented milk and kefiran promote growth of *Bifidobacterium bifidum* PRL2010 and modulate its gene expression. *Int. J. Food Microbiol.* 178, 50–59. doi: 10.1016/j.ijfoodmicro.2014.02.024

Turroni, F., Bottacini, F., Foroni, E., Mulder, I., Kim, J. H., Zomer, A., et al. (2010). Genome analysis of *Bifidobacterium bifidum* PRL2010 reveals metabolic pathways for host-derived glycan foraging. *Proc. Natl. Acad. Sci. U.S.A.* 107, 19514–19519. doi: 10.1073/pnas.1011100107

Turroni, F., Foroni, E., Montanini, B., Viappiani, A., Strati, F., Duranti, S., et al. (2011a). Global genome transcription profiling of bifidobacterium bifidum prl2010 under *in vitro* conditions and identification of reference genes for quantitative real-time PCR. *Appl. Environ. Microbiol.* 77, 8578–8587. doi: 10.1128/AEM.06352-11

Turroni, F., Milani, C., Van Sinderen, D., and Ventura, M. (2011b). Genetic strategies for mucin metabolism in *Bifidobacterium bifidum* PRL2010: an example of possible human-microbe co-evolution. *Gut. Microbes* 2, 183–189. doi: 10.4161/gmic.2.3.16105

Turroni, F., Serafini, F., Foroni, E., Duranti, S., O'Connell Motherway, M., Taverniti, V., et al. (2013). Role of sortase-dependent pili of *Bifidobacterium bifidum* PRL2010 in modulating bacterium-host interactions. *Proc. Natl. Acad. Sci. U.S.A.* 110, 11151–11156. doi: 10.1073/pnas.1303897110

Turroni, F., Serafini, F., Mangifesta, M., Arioli, S., Mora, D., Van Sinderen, D., et al. (2014a). Expression of sortase-dependent pili of *Bifidobacterium bifidum* PRL2010 in response to environmental gut conditions. *FEMS Microbiol. Lett.* 357, 23–33. doi: 10.1111/1574-6968.12509

Turroni, F., Ventura, M., Butto, L. F., Duranti, S., O'toole, P. W., Motherway, M. O., et al. (2014b). Molecular dialogue between the human gut microbiota and the host: a *Lactobacillus* and *Bifidobacterium* perspective. *Cell Mol. Life. Sci.* 71, 183–203. doi: 10.1007/s00018-013-1318-0

Turroni, F., Strati, F., Foroni, E., Serafini, F., Duranti, S., Van Sinderen, D., et al. (2012). Analysis of predicted carbohydrate transport systems encoded by *Bifidobacterium bifidum* PRL2010. *Appl. Environ. Microbiol.* 78, 5002–5012. doi: 10.1128/AEM.00629-12

Ventura, M., O'Flaherty, S., Claesson, M. J., Turroni, F., Klaenhammer, T. R., Van Sinderen, D., et al. (2009). Genome-scale analyses of health-promoting bacteria: probiogenomics. *Nat. Rev. Microbiol.* 7, 61–71. doi: 10.1038/nrmicro2047

Ventura, M., Turroni, F., Motherway, M. O., Macsharry, J., and Van Sinderen, D. (2012). Host-microbe interactions that facilitate gut colonization by commensal bifidobacteria. *Trends Microbiol.* 20, 467–476. doi: 10.1016/j.tim.2012.07.002

Ze, X., Duncan, S. H., Louis, P., and Flint, H. J. (2012). *Ruminococcus bromii* is a keystone species for the degradation of resistant starch in the human colon. *ISME J.* 6, 1535–1543. doi: 10.1038/ismej.2012.4

**Conflict of Interest Statement:** The authors declare that the research was conducted in the absence of any commercial or financial relationships that could be construed as a potential conflict of interest.

20

# High Iron-Sequestrating Bifidobacteria Inhibit Enteropathogen Growth and Adhesion to Intestinal Epithelial Cells *In vitro*

***Pamela Vazquez-Gutierrez, Tomas de Wouters, Julia Werder, Christophe Chassard and Christophe Lacroix****

*Laboratory of Food Biotechnology, Institute of Food, Nutrition and Health, ETH Zurich, Zürich, Switzerland*

**Edited by:**
*Marco Ventura, University of Parma, Italy*

**Reviewed by:**
*Silvia Arboleya Montes, Teagasc and APC Microbiome Institute, Ireland*
*Lorena Ruiz, Universidad Complutense de Madrid, Spain*

***Correspondence:**
*Christophe Lacroix*
*christophe.lacroix@hest.ethz.ch*

The gut microbiota plays an important role in host health, in particular by its barrier effect and competition with exogenous pathogenic bacteria. In the present study, the competition of *Bifidobacterium pseudolongum* PV8-2 (Bp PV8-2) and *Bifidobacterium kashiwanohense* PV20-2 (Bk PV20-2), isolated from anemic infant gut microbiota and selected for their high iron sequestration properties, was investigated against *Salmonella* Typhimurium (*S.* Typhi) and *Escherichia coli* O157:H45 (EHEC) by using co-culture tests and assays with intestinal cell lines. Single and co-cultures were carried out anaerobically in chemically semi-defined low iron (1.5 μM Fe) medium (CSDLIM) without and with added ferrous iron (30 μM Fe). Surface properties of the tested strains were measured by bacterial adhesion to solvent xylene, chloroform, ethyl acetate, and to extracellular matrix molecules, mucus II, collagen I, fibrinogen, fibronectin. HT29-MTX mucus-secreting intestinal cell cultures were used to study bifidobacteria competition, inhibition and displacement of the enteropathogens. During co-cultures in CSDLIM we observed strain-dependent inhibition of bifidobacterial strains on enteropathogens, independent of pH, organic acid production and supplemented iron. Bp PV8-2 significantly ($P < 0.05$) inhibited *S.* Typhi N15 and EHEC after 24 h compared to single culture growth. In contrast Bk PV20-2 showed less inhibition on *S.* Typhi N15 than Bp PV8-2, and no inhibition on EHEC. Affinity for intestinal cell surface glycoproteins was strain-specific, with high affinity of Bp PV8-2 for mucin and Bk PV20-2 for fibronectin. Bk PV20-2 showed high adhesion potential (15.6 ± 6.0%) to HT29-MTX cell layer compared to Bp PV8-2 (1.4 ± 0.4%). In competition, inhibition and displacement tests, Bp PV8-2 significantly ($P < 0.05$) reduced *S.* Typhi N15 and EHEC adhesion, while Bk PV20-2 was only active on *S.* Typhi N15 adhesion. To conclude, bifidobacterial strains selected for their high iron binding properties inhibited *S.* Typhi N15 and EHEC in co-culture experiments and efficiently competed with the enteropathogens on mucus-producing HT29-MTX cell lines. Further studies in complex gut ecosystems should explore host protection effects of Bp PV8-2 and Bk PV20-2 mediated by nutritional immunity mechanism associated with iron-binding.

**Keywords: iron sequestration, bifidobacteria, enteropathogens, inhibition, intestinal cell, adhesion**

## INTRODUCTION

Bifidobacteria are among the first commensal anaerobic bacteria that reach high levels in the infant gut within the first week of life, representing up to 50–80% of the gut bacteria (Jost et al., 2012; Turroni et al., 2012). The establishment of bifidobacteria in the gut has been associated with a broad range of beneficial effects on host health, such as modulation of intestinal microbiota composition, prevention of infection and immune-modulation (Broekaert and Walker, 2006; Yatsunenko et al., 2012). Inhibition of pathogens in the gut by bifidobacteria might be due to production of inhibitory substances, inhibition of epithelial and mucosal invasion of pathogens, competition for limited nutrients and/or the stimulation of mucosal immunity (Marco et al., 2006; Turroni et al., 2014). Potential inhibition mechanisms include the production of short-chain fatty acids and subsequent local pH decrease (Fukuda et al., 2011), or other antimicrobial compounds such as bacteriocins (Cheikhyoussef et al., 2008; Dobson et al., 2012; Martinez et al., 2013). Bifidobacteria can also compete with pathogens for adhesion to intestinal epithelial sites and nutrients, enhancing resistance to colonization of pathogenic bacteria (Collado et al., 2007; Aires et al., 2010).

The gut microbiota is constantly challenged by different stress factors, including enteropathogens, such as *Salmonella* and *Escherichia coli* O157:H45 (EHEC; Wardlaw et al., 2010). Pathogenesis of *Salmonella* requires its adhesion to host cell surfaces followed by invasion of intestinal epithelial cells, leading to systemic spreading (Sansonetti, 2004; Haraga et al., 2008; Santos et al., 2009). EHEC pathophysiology is attributed to the effects of shiga toxins encoded on the pO157 plasmid, survival to harsh conditions and the formation of attaching-and-effacing lesions on epithelial cells (Muller et al., 2009; Melton-Celsa et al., 2012; Thiennimitr et al., 2012). To inhibit pathogen infection in the gut, commensal intestinal microorganisms such as bifidobacteria, should be able to compete for corresponding niches. Bifidobacteria have been reported to occupy attachment sites, therefore preventing pathogen invasion and translocation (Bernet et al., 1994; Goto and Kiyono, 2012). The inhibitory activity and mechanisms of bifidobacteria against enteropathogens have been investigated by microbe-microbe and cell-microbe interaction models (Collado et al., 2007).

Different intestinal epithelial cell lines exhibiting specific characteristics and functions of the gut epithelium are used to study host-pathogen interactions. HT29-MTX cell line is a mucus-secreting clone of the HT-29 intestinal epithelial cell line suitable for mimicking the mucosal surface of the gut epithelium, which acts as the first line of interaction between the microbiota and its host (Lesuffleur et al., 1990; Gagnon et al., 2013). The intestinal mucus layer functions as a physical barrier, separating the epithelium from the bacterial load in the intestinal lumen. Mucus is also an important nutrient source for gut microbes and promotes selective adhesion of gut bacteria to the intestinal mucus layer. Interactions with the intestinal mucus layer is a property of commensal gut bacteria that can enhance the barrier function of the intestinal epithelium by limiting access of pathogens to this specific niche. The adherence to intestinal epithelial cells is therefore an important characteristic for beneficial gut bacteria, enhancing persistence in the gut, pathogen exclusion effects and specific bacterial and host-immune system interactions (Izquierdo et al., 2008; Bron et al., 2012).

The ability of bacteria to establish in the intestine is heavily dependent on competition for nutrients (Andrews et al., 2003). For example iron is an essential micronutrient for growth, proliferation, and persistence for most gut bacteria, including bifidobacteria and enteropathogens (Turroni et al., 2014). Pathogens such as S. Typhi and EHEC are known to possess efficient iron sequestration mechanisms that contribute to their pathogenicity and competitiveness in the gut (Berkley et al., 2005; Wardlaw et al., 2010; Cassat and Skaar, 2013; Monack and Hultgren, 2013; Winter et al., 2013). These systems have been directly linked to the ability of strains with high iron sequestration properties to establish efficiently in the gut (Weinberg, 2009; Kortman et al., 2012). In a previous study we reported isolation of 56 bifidobacterial strains from stools of breast fed, iron-deficient and anemic Kenyan infants (Vazquez-Gutierrez et al., 2015c). Isolated strains were characterized and compared to public culture collection strains. *Bifidobacterium kashiwanohense* PV20-2 (Bk PV20-2) and *Bifidobacterium pseudolongum* PV8-2 (Bp PV8-2) were selected for their high siderophore activity (iron-chelating molecules) and iron internalization. Analysis of the complete genome allowed to identify ferrous and specific ferric iron operons in both strains (Vazquez-Gutierrez et al., 2015a,b,c). Furthermore, a ferrous iron-binding protein and other proteins with adhesive properties were identified in the extracellular fraction of Bk PV20-2 together with. In the extracellular proteome of Bp PV8-2 a ferric iron-binding protein belonging to the ferric iron transport operon was shown. In the present study, the inhibitory activity of Bp PV8-2 and Bk PV20-2 was investigated during co-cultures with S. Typhi N15 and EHEC as a function of iron concentrations (1.5 and 30 $\mu$M ). Surface properties were tested by bacterial adhesion to solvent (BATS) and extracellular matrix molecules (ECMs) and the competition for epithelial binding sites was studied in HT29-MTX intestinal cellular model.

## MATERIALS AND METHODS

### Bacterial Strains and Growth Conditions

*Bifidobacterium pseudolongum* DSMZ20099 (Bp DSMZ20099) and *B. kashiwanohense* DSMZ21854 (Bk DSMZ21854) were obtained from the German collection of microorganisms (DSMZ; Leibniz, Germany). *B. pseudolongum* PV8-2 (Bp PV8-2) and *B. kashiwanohense* PV20-2 (Bk PV20-2), were obtained from the culture collection of the Laboratory of Food Biotechnology (ETH Zurich, Switzerland). *Salmonella enterica* ssp. *enterica* serovar Typhimurium N15 (S. Typhi N15) a clinical isolate obtained from the National Centre for Enteropathogenic Bacteria and *Listeria* (NENT, University of Zurich, Switzerland) and *E. coli* O157:H45 (EHEC) were kindly provided by Prof. Roger Stephan. Bifidobacteria were routinely cultured in de Man, Rogosa, and Sharpe (MRS) broth (Biolife, Italy) supplemented with 0.05% of L-cysteine hydrochloride monohydrate (cys; Sigma-Aldrich,

Switzerland). Enteropathogens were cultured in Luria-Bertani (LB) broth (Becton Dickinson, Switzerland) unless otherwise specified. Cells suspensions and serial dilutions were carried out in peptone water at pH 6.5, containing 1.5 g/L peptone water (CDH Bioscience, India) and 0.6 g/L cys (peptone-cys). Bifidobacterial viable cell counts were determined on MRS-cys agar (Becton Dickinson, Switzerland) plates, incubated for 72 h under anaerobiosis in anaerobic jars. *S.* Typhi and EHEC enumeration was done in Mac-Conkey agar (Oxoid, Switzerland) incubated 24 h at 37°C. A chemically semidefined low iron medium (CSDLIM) with a low iron concentration of 1.5 μM was used for co-culture interaction assays. The CSDLIM medium was previously used to test siderophore production with the CAS assay (Vazquez-Gutierrez et al., 2015c). Iron supplementation of the CSDLIM medium was achieved by adding 30 μM of ferrous iron (Sigma-Aldrich, Switzerland), corresponding to the iron concentration previously reported to increase *Salmonella* and EHEC pathogenicity (Cernat and Scott, 2012; Kortman et al., 2012). Iron concentration in CSDLIM was measured by graphic furnace atomic absorption spectrometry (Vazquez-Gutierrez et al., 2015c).

## Inhibitory Activity of *B. pseudolongum* PV8-2 and *B. kashiwanohense* PV20-2 during Co-cultures with Enteropathogens

Growth interactions of bifidobacteria and enteropathogens were investigated in CSDLIM with and without added ferrous iron as follow. The corresponding strains were first cultured twice at 37°C in MRS-cys for 24 h and LB broth for 12 h, respectively. Bacterial cells were harvested by centrifugation (Biofuge Primo, Heraeus, Switzerland) at 4°C, 16,000 × *g* for 10 min. The supernatant was discarded and the pellet was resuspended in peptone-cys water to an $OD_{600\,nm}$ of 1.0. Hungate tubes containing 10 mL of CSDLIM with headspace filled with $CO_2$ were inoculated with $\log_{10}$ 6.5 ± 0.05 CFU/mL Bp PV8-2, $\log_{10}$ 6.6 ± 0.13 CFU/mL Bk PV20-2, $\log_{10}$ 5.5 ± 0.06 CFU *S.* Typhi N15/mL and $\log_{10}$ 5.4 ± 0.15 CFU/mL EHEC for both mono- and co-cultures. Hungate tubes were incubated at 37°C for 24 h and samples were taken at 0, 12, and 24 h from the same tube through septum for absorbance determination at $OD_{600\,nm}$ (Biowave, CO8000, Biochrom, Ltd, England), pH and viable cell counts. Short chain fatty acid (SCFA) concentrations were measured by high performance liquid chromatography (HPLC; Thermo Fisher Scientific, Switzerland) as previously described (Cleusix et al., 2008). Briefly, 1 mL culture samples were centrifuged for 12 min at 10,000 × *g* and 4°C. Supernatant was filtered with a 0.45 μm nylon membrane (Infochroma AG, Switzerland) directly into HPLC vials. Analysis was performed at a flow rate of 0.4 mL/min with 10 mM sulphuric acid as eluent with an injection volume of 20 μL. Mean metabolite concentrations were expressed in millimolar (mM). Three independent repetitions of mono- and co-cultures in CSDLIM with and without ferrous iron supplementation (30 μM ferrous iron) were carried out.

To test the effects of pH decrease and SCFA on enteropathogens growth inhibition, *S.* Typhi N15 and EHEC were incubated at 37°C for 12 h, at pH 4.5 (pH measured at the end of co-cultures) and SCFA concentrations where enteropathogen counts began to decrease during co-cultures. After centrifugation at 4°C and 16,000 × *g* for 10 min, cell pellets were suspended in peptone-cys water and adjusted to $OD_{600\,nm}$ 1.0. Then hungate tubes containing 10 mL of CSDLIM pH 4.5, 7 mM lactate and 13 mM acetate, were inoculated with $\log_{10}$ 5.4 ± 0.06 CFU/mL *S.* Typhi N15 and $\log_{10}$ 5.3 ± 0.13 CFU/mL EHEC, which were the viable cell counts reached in co-cultures after 12 h incubation. Hungate tubes were incubated for 24 h at 37°C and 1 mL sample was taken every 4 h to determine pH, absorbance at 600 nm and viable cell counts. The experiment was performed in three independent replicates in CSDLIM with and without ferrous iron supplementation.

## Surface Properties of Bifidobacterial Strains

The BATS assay was used to investigate cell surface properties of bifidobacterial strains according to Xu et al. (2009), with slight modifications. Surface hydrophobicity, electron donor and acceptor properties were determined based on the affinity of bifidobacteria to xylene (apolar solvent), chloroform (polar acidic solvent) and ethyl-acetate (polar basic solvent). Bifidobacteria were cultured in MRS-cys and CDSLIM as described above. Briefly, cell pellets were re-suspended in phosphate-buffered saline (PBS), pH 7.3 to an $OD_{600\,nm}$ of 1.0. A volume of 3 mL of cell suspension was mixed with 1 mL of either xylene, chloroform (electron-acceptor), or ethyl-acetate (electron-donor; Sigma-Aldrich, Switzerland). The mixture was vortexed for 1 min and allowed to stand for 5 min to allow separation into two distinct phases. Then 1 mL of the aqueous phase was carefully collected with a pipette and $OD_{600\,nm}$ was measured using a UV-Visible spectrophotometer CARY 1Bio (Varian, Switzerland). The decrease in absorbance of the aqueous phase after contact with solvent was used as a measure of the cell surface hydrophobicity or electron-donor/electron-acceptor interaction. BATS was expressed by BATS (%) = $(1 - A_{5\,min}/A_{0\,min}) \times 100$, where $A_{0\,min}$ and $A_{5\,min}$ were the absorbance before and after extraction with the solvents, respectively (Xu et al., 2009). Three independent replicates of the experiment were carried out.

## Adhesion to Different Intestinal Cell Surface Molecules

The adhesion affinities of bacteria to the ECMs of intestinal epithelial cells were tested as described previously (Sillanpaa et al., 2008), with slight modifications. Briefly, a solution of type II mucus (Sigma-Aldrich, Switzerland) at 50 μg/mL was prepared in Tris-HCl (0.1 M, pH 8). Collagen I, fibrinogen and fibronectin (all from Sigma-Aldrich, Switzerland) were resuspended at 10 μg/mL in PBS (Gibco, Switzerland), pH 7.5. Bovine serum albumin (BSA; Sigma-Aldrich, Switzerland) was resuspended at 50 μg/mL in Tris-HCl and used as control for unspecific adhesion to protein surfaces. 100 μL of each suspension were applied to wells of a MaxiSorp™ 96-well microtiter plate (Nunc, Switzerland) and kept overnight at 4°C for adsorption. After removal of the liquid, adsorbed molecules were fixed for 10 min at 65°C and subsequently blocked with 100 μL PBS 1% tween 20 per well for 1 h at 37°C. Before application of bacteria, plates were washed three times with 100 μL PBS 0.005% tween 20 to remove unbound ECM, filled with 100 μL PBS and used within 24 h storage at 4°C.

Bifidobacteria and enteropathogen strains were first cultured in MRSc or LB media, respectively, then transferred and grown in CDLSIM, as presented above. Briefly, cell pellets (24°C, 16,000 × *g* during 10 min) were re-suspended in PBS at pH 5.5 and pH 7.5 to $OD_{600\ nm}$ of 1.0. 100 μl of bacterial suspension was applied in triplicates to coated plates and incubated for 1 h at 37°C to induce bacterial adhesion. Wells were washed three times with 100 μL PBS 0.005% Tween 20 to remove unattached bacteria and dried for 10 min at 65°C. Adhered bacteria were stained with 100 μL crystal violet (1 mg/mL, Sigma-Aldrich, Switzerland) per well for 45 min at room temperature. Crystal violet retained by fixed bacteria after three washing steps with 100 μL PBS was resolubilized in 100 μL citrate buffer (50 mM; pH 4) under continuous shaking at 37°C for 1 h. Absorbance of solubilized crystal violet was measured at $OD_{595\ nm}$ using a Powerwave XS spectrophotometer (Bio Tek, Switzerland). The experiment was carried out in three independent replicates.

### *In vitro* Antagonism of Bifidobacteria against Pathogen Adhesion to HT29-MTX Cells

The mucus-secreting intestinal epithelial cell line HT29-MTX was used to investigate adhesion properties of bifidobacteria and enteropathogens, as described previously with slight modifications (Gagnon et al., 2013). Briefly, HT29-MTX cells were seeded in 24-well tissue culture plates (Bioswisstec, Switzerland) at a concentration of 4 × $10^4$ cells/well, and grown in Dulbecco's Modified Eagle Medium (DMEM; Sigma-Aldrich, Switzerland) supplemented with 20% fetal bovine serum (Invitrogen, Switzerland), 1% penicillin/streptomycin (Life Technologies, Switzerland), and 1% non-essential amino acids (Life Technologies, Switzerland), at 37°C and 10% $CO_2$ in a humidified incubator (RB150, Revco, Switzerland). Culture medium was changed every other day and experiments were performed 21 days post-seeding on fully differentiated, confluent monolayers with mucus secretion verified using Alcian blue (stains acid mucopolysaccharides) and periodic acid Schiff (stains hexose and sialic acid-containing mucosubstances). After full differentiation the medium was exchanged to antibiotic free medium for 24 h. Tested bacterial cultures were grown in CDLSIM medium and prepared as described above. Bacterial cultures were washed with sterile 0.85% NaCl, and resuspended in DMEM for application to the cell monolayers. Cell monolayers were carefully washed with 500 μL of warm PBS. For all tests, bifidobacteria were added at $\log_{10}$ 7.7 ± 0.12 CFU/mL and *S.* Typhi N15 and EHEC were added at approximately $\log_{10}$ 6.3 ± 0.05 CFU/mL in DMEM to the HT29-MTX monolayer. After 2 h incubation at 37°C, HT29-MTX monolayers were washed twice with PBS to remove non-attached bacteria and detached using 0.25% trypsin-EDTA solution (Life Technologies, Switzerland). Bacterial cell counts were determined as described above. Adhesion was expressed as the percent ratio of adhered bacteria to number of bacteria added to the HT29-MTX cells monolayer. Experiments were performed in triplicates on three consecutive passages of the HT29-MTX cell line.

To determine the inhibition of pathogen adhesion by bifidobacteria the method of Gagnon et al. (2013) was used with slight modifications. Briefly, bifidobacteria were applied to the cell monolayer for 1 h. Then the well was washed once with PBS to remove non-adhering cells and the tested pathogen was added for a further incubation of 1 h. Enumeration of adhered bacteria was performed after serial dilution on respective media. To examine if adhered pathogenic bacteria could be displaced by the addition of bifidobacteria, enteropathogens were incubated 1 h and, after PBS washing, bifidobacteria were added and incubated 1 h. To investigate the ability of bifidobacteria to competitively exclude enteropathogens, bifidobacteria and pathogenic bacteria were added simultaneously to the HT29-MTX monolayer and incubated for 2 h. All incubations were done at 37°C and 10% $CO_2$. HT29-MTX monolayers were washed twice with PBS to remove non-attached bacteria, and treated with a 0.25% trypsin-EDTA solution for 15 min, for bifidobacteria and enteropathogens enumeration as stated earlier. Activity of bifidobacteria strains to compete with, displace and inhibit the adhesion of *S.* Typhi N15 and EHEC to the intestinal epithelial cell line HT29-MTX was expressed by the adhesion ratio. This corresponded to the ratio of the percentage of adhered bifidobacteria or pathogenic bacteria following simultaneous addition divided by the percentage of adhesion of the bacteria added alone to the cell culture (Serafini et al., 2013). All the above tests were carried out in triplicates on three consecutive passages of the HT29-MTX cell line.

### Statistical Analysis

To assess differences between treatments in inhibitory activity in mono- and co-culture experiments, and surface properties of bifidobacteria strains, the means of three independent repetitions were compared using un-paired Student's *t*-test. ANOVA with *post hoc* Tukey test was used to assess significant affinity of bifidobacteria and pathogens to ECM when compared to PBS control ($P$-value $< 0.05$). Significant differences in competition, inhibition and displacement experiments were tested by comparing the means of three independent repetition of the HT29-MTX cell line using un-paired Student's *t*-test. Statistical significance was established at $P$-value $< 0.05$ and SPSS software 17.0 (SPSS, Inc., Chicago, IL, USA) was used.

## RESULTS

### Inhibitory Activity of *B. pseudolongum* PV8-2 and *B. kashiwanohense* PV20-2 against Enteropathogens

The inhibitory activities of Bp PV8-2 and Bk PV20-2 against *S.* Typhi N15 and EHEC were tested in co-culture experiments and compared to mono-cultures of the same strains (**Figures** and ). Mono-cultures of Bp PV8-2 reached maximum viable

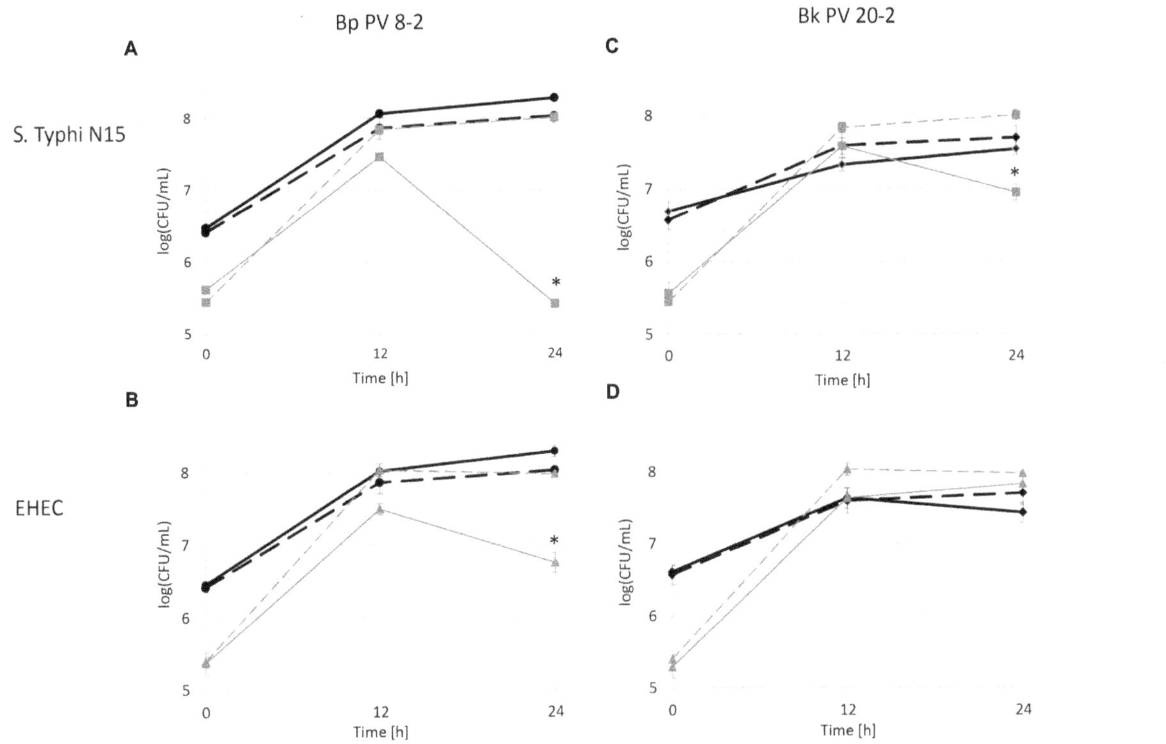


**FIGURE | Viable cell counts during mono- (dashed lines) and co-cultures (continuous lines) in low iron CSDLIM medium: (A) *Bifidobacterium pseudolongum* PV8-2 (Bp PV8-2) and *S.* Typhimurium N15 (*S.* Typhi N15); (B) Bp PV8-2 and *Escherichia coli* O157:H45 (EHEC); (C) *Bifidobacterium kashiwanohense* PV20-2 (Bk PV20-2) and *S.* Typhi N15; (D) Bk PV20-2 and EHEC.** Bp PV8-2, Bp PV8-2 in co-culture, *S.* Typhi N15, *S.* Typhi N15 in co-culture, Bk PV20-2, Bk PV20-2 in co-culture, EHEC, EHEC in co-culture. Stars (*) denote a significant ($P < 0.05$) difference compared with mono-cultures (mean ± SD, $n = 3$).

cell counts of $\log_{10}$ 8.1 ± 0.1 CFU/mL with and without iron supplementation, and pH of 4.5 ± 0.06 and 4.2 ± 0.01 after 24 h incubation in CDSLIM medium, respectively. In decreasing concentration the main organic acids produced were acetate, lactate and formate (**Table** ). Iron supplementation significantly ($P < 0.05$) increased production of all metabolites, by 12, 17, and 29% for acetate, lactate and formate, respectively, when compared to cultures in unsupplemented media. Similarly, Bk PV20-2 in mono-cultures reached viable cells counts of $\log_{10}$ 7.7 ± 0.2 CFU/mL with and without iron, and pH was 5.3 ± 0.04 and 4.7 ± 0.01, respectively. Organic acid productions were also significantly ($P < 0.05$) increased with iron, by 27% for acetate and 20% for lactate. No significant effect of iron on growth or metabolite production was observed during mono-cultures of enteropathogens. EHEC reached cell counts of $\log_{10}$ 7.9 ± 0.1 CFU/mL after 12 h and remained stable until 24 h of culture. pH after 24 h was 4.88 ± 0.01 and 5.05 ± 0.02, with and without iron supplementation, respectively. Main metabolites were lactate and acetate for S. Typhi N15 and EHEC (**Table** ).

Viable cell counts of *S.* Typhi N15 significantly decreased ($P < 0.05$) by $\log_{10}$ 2.5 ± 0.1 CFU/mL in co-culture with Bp PV8-2 and by $\log_{10}$ 1.1 ± 0.1 CFU/mL with Bk PV20-2 after 24 h when compared with mono-cultures of *S.* Typhi N15 (**Figures** , ). pH after 24 h of co-cultures of *S.* Typhi N15 with Bp PV8-2 was 4.19 ± 0.02 and 4.21 ± 0.02 and with Bk PV20-2 4.62 ± 0.01 and 4.59 ± 0.01, with or without iron supplementation, respectively. EHEC counts significantly decreased after 12 h when co-cultured with Bp PV8-2 compare to monocultures (**Figure** ). No significant differences were observed when co-culturing EHEC with Bk PV20-2. Metabolites produced during co-cultures of *S.* Typhi N15 and EHEC with Bp PV8-2 were in decreasing order acetate, lactate and formate (**Table** ), whereas with Bk PV20-2 only acetate and lactate were identified (**Tables** ). No significant differences in metabolites were observed following iron supplementation of the media. No significant effect of iron supplementation was detected on EHEC growth during co-cultures (**Figures** ).

*Salmonella* Typhi and EHEC were tested for the inhibitory conditions observed after 12 h of co-culture with bifidobacterial strains. CDSLIM medium was supplemented with 13 mM of acetate and 7 mM of lactate at pH 4.5 and growth tested for 24 h. *S.* Typhi N15 and EHEC viable cell counts remained constant ($\log_{10}$ 5.4 ± 0.12 CFU/mL) during 24 h. In contrast a significant viability decrease of *S.* Typhi N15 (Bp PV8-2 and Bk PV20-2) and EHEC (Bp PV8-2) was measured during co-cultures with both bifidobacteria between 12 and 24 h for both iron levels (**Figures** ).

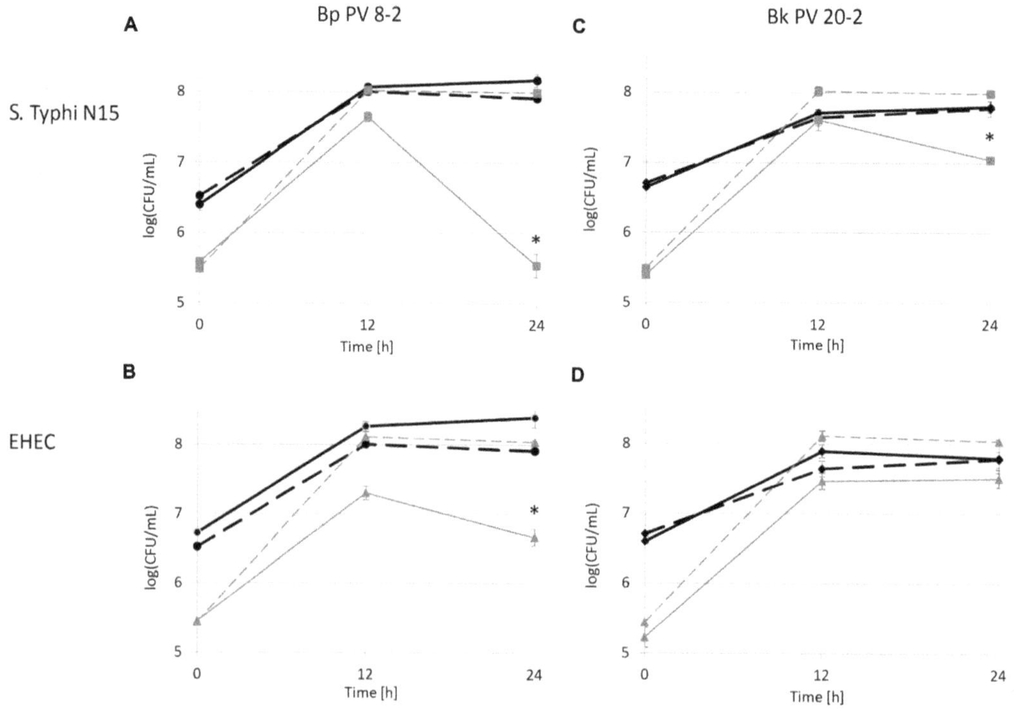


**FIGURE | Viable cell counts during co-cultures in iron supplemented (30 μM) CSDLIM medium: (A) *B. pseudolongum* PV8-2 (Bp PV8-2) and *S.* Typhimurium N15 (*S.* Typhi N15); (B) Bp PV8-2 and *E. coli* O157:H45 (EHEC); (C) *B. kashiwanohense* PV20-2 (Bk PV20-2) and *S.* Typhi N15; (D) Bk PV20-2 and EHEC.** –●– Bp PV8-2, —●— Bp PV8-2 in co-culture, –■– *S.* Typhi N15, —■— *S.* Typhi N15 in co-culture, –◆– Bk PV20-2, —◆— Bk PV20-2 in co-culture, –▲– EHEC, —▲— EHEC in co-culture. Stars (*) denote a significant ($P < 0.05$) difference compared with mono-cultures (mean ± SD, $n = 3$).

**TABLE | pH and concentrations of metabolites (mM) in culture supernatant measured with HPLC after 24 h incubation of mono- and co-cultures of *Bifidobacterium pseudolongum* PV8-2 in CSDLIM media (mean ± SD, $n = 3$).**

| ***B. pseudolongum* PV8-2** | **Acetate** | **Lactate** | **Formate** | **pH** |
|---|---|---|---|---|
| **with iron supplementation** | | | | |
| *B. pseudolongum* PV8-2 | 20.81 ± 0.35 | 8.04 ± 0.34 | 3.85 ± 0.23 | 4.50 ± 0.06 |
| *S.* Typhimurium N15 | 5.26 ± 0.09 | 14.82 ± 0.26 | – | 4.87 ± 0.02 |
| *B. pseudolongum* PV8-2/*S.* Typhimurium N15 | 21.81 ± 0.82 | 10.15 ± 0.57 | 3.20 ± 0.30 | 4.19 ± 0.02 |
| *E. coli* O157:H45 | 5.55 ± 0.07 | 12.72 ± 0.21 | – | 4.88 ± 0.01 |
| *B. pseudolongum* PV8-2/*E. coli* O157:H45 | 21.90 ± 0.75 | 9.87 ± 0.30 | 4.79 ± 0.21 | 4.15 ± 0.01 |
| **with iron supplementation (30 μM)** | | | | |
| *B. pseudolongum* PV8-2 | 23.28 ± 0.36* | 9.43 ± 0.20* | 4.96 ± 0.75* | 4.20 ± 0.01* |
| S. Typhimurium N15 | 5.47 ± 0.28 | 14.07 ± 0.56 | – | 5.16 ± 0.06 |
| *B. pseudolongum* PV8-2/*S.* Typhimurium N15 | 21.92 ± 0.72 | 10.64 ± 0.36 | 2.53 ± 0.25 | 4.21 ± 0.02 |
| *E. coli* O157:H45 | 5.93 ± 0.14 | 11.87 ± 0.09 | – | 5.05 ± 0.02 |
| *B. pseudolongum* PV8-2/*E. coli* O157:H45 | 22.81 ± 0.82 | 9.73 ± 0.43 | 4.82 ± 0.33 | 4.18 ± 0.01 |

*Stars (*) denote a significant ($P < 0.05$) difference in the respective metabolite production of cultures performed in iron supplemented compared with unsupplemented medium.*

## Surface Properties of Bifidobacteria Strains

Physico-chemical characteristics of bifidobacteria cell surfaces, such as hydrophobicity, electron-donor and electron-acceptor properties, are related to adhesion to intestinal epithelial cells. The affinity of bifidobacteria strains to different solvents, xylene, chloroform and ethyl-acetate, was determined using the BATS assay to quantify surface hydrophobicity. Bp PV8-2 showed similar surface properties as the type strain DSMZ20099 (**Figure** ). Under iron-limited conditions affinity to hydrophobic xylene was 84.4 ± 3.6% for Bp PV8-2 and 88.3 ± 11.6% for Bp DSMZ20099. Affinity to chloroform, an acidic solvent, was 99.2 ± 1% for Bp PV8-2 and 98.4 ± 3.5% for Bp DSMZ20099. No hydrophobic, electron donor/acceptor properties were observed for Bk PV20-2 (**Figure** ), whereas Bk DSMZ21854 showed hydrophobic and electron-donor properties only when grown

**TABLE | pH and concentrations of metabolites (mM) in culture supernatant measured with HPLC after 24 h incubation of mono- and co-cultures of *Bifidobacterium kashiwanohense* PV20-2 in CSDLIM media (mean ± SD, *n* = 3).**

| *B. kashiwanohense* PV20-2 | Acetate | Lactate | pH |
|---|---|---|---|
| **without iron supplementation** | | | |
| *B. kashiwanohense* PV20-2 | 8.69 ± 0.55 | 5.08 ± 0.20 | 5.30 ± 0.04 |
| *S.* Typhimurium N15 | 5.26 ± 0.09 | 14.82 ± 0.26 | 4.87 ± 0.02 |
| *B. kashiwanohense* PV20-2/*S.* Typhimurium N15 | 10.18 ± 0.40 | 10.04 ± 0.29 | 4.62 ± 0.01 |
| *E. coli* O157:H45 | 5.55 ± 0.07 | 12.72 ± 0.21 | 4.88 ± 0.01 |
| *B. kashiwanohense* PV20-2/*E. coli* O157:H45 | 9.55 ± 0.29 | 9.18 ± 0.28 | 4.64 ± 0.01 |
| **with iron supplementation (30 μM)** | | | |
| *B. kashiwanohense* PV20-2 | 11.00 ± 0.09* | 6.08 ± 0.06* | 4.70 ± 0.01* |
| *S.* Typhimurium N15 | 5.47 ± 0.28 | 14.07 ± 0.56 | 5.16 ± 0.06 |
| *B. kashiwanohense* PV20-2/*S.* Typhimurium N15 | 12.67 ± 0.34* | 8.63 ± 0.17* | 4.59 ± 0.01 |
| *E. coli* O157:H45 | 5.93 ± 0.14 | 11.87 ± 0.09 | 5.05 ± 0.02 |
| *B. kashiwanohense* PV20-2/*E. coli* O157:H45 | 12.29 ± 0.16* | 8.02 ± 0.27* | 4.61 ± 0.01 |

*Stars (*) denote a significant (P < 0.05) difference in the respective metabolite production of cultures performed in iron supplemented compared with unsupplemented medium.*

in CDSLIM (**Figure** ). Electron-acceptor properties were not observed for any of the strains tested. Bifidobacteria strains showed less affinity to all solvents when grown in MRS-cys compared to CDSLIM.

## Adhesion to Different Intestinal Cell Surface Molecules

The intestinal epithelial cell surface is covered by glycoproteins, such as type II mucus, collagen, fibrinogen and fibronectin, and can serve as attachment sites for microbes. The affinity of bacteria to glycoproteins can therefore influence strain capacity to compete for epithelial binding sites. No significant adhesion to the unspecific protein binding control BSA nor to collagen I was shown for the tested strains when compared with the uncoated control wells (**Figures** , ). Bp PV8-2, showed significant ($P < 0.05$) adhesion at pH 5.5 to type II mucin and Bk PV20-2 and Bk DSMZ21854 to fibronectin when compared with the uncoated control (**Figure** ). *S.* Typhi N15 showed significant ($P < 0.05$) adhesion to type II mucin, fibronectin, and fibrinogen. EHEC bound to type II mucin and fibronectin when compared with the uncoated control. At pH 7.5, all strains showed similar affinity to glycoproteins, BSA and the uncoated control (**Figure** ), except *S.* Typhi N15 that showed significantly ($P < 0.05$) higher adhesion to mucin II, fibrinogen, and fibronectin.

## *In vitro* Inhibition of Bifidobacteria Strains against *S.* Typhimurium N15 and EHEC

The ability to adhere to mucus and epithelial cells is an important feature for the barrier effect of bifidobacteria. The adhesion ratios of Bk PV20-2 and Bk DSMZ21854 to mucus-secreting HT29-MTX were significantly higher (15.6 ± 6.0% and 12.7 ± 2.4%, respectively) when compared to both Bp PV8-2 and Bp DSMZ20099 (1.4 ± 0.4% and 1.3 ± 0.3%, respectively). Very high adhesion ratios of *S.* Typhi N15 and EHEC were measured, with 87.8 ± 17.5% and 137.6 ± 51.7%, respectively, likely reflecting growth of the enteropathogens during the test.

The ability of bifidobacteria strains, to compete, displace and inhibit the adhesion of enteropathogens was tested on the HT29-MTX epithelial cell model. Both strains exhibited competitive abilities when added together with *S.* Typhi N15 in the competition assay, as shown by adhesion ratios significantly higher than 1 (1.88 ± 0.64 for Bp PV8-2 and 1.76 ± 0.51 for Bk PV20-2; **Figures** , ). In contrast, adhesion ratios lower than 1 ($P < 0.05$) were measured for *S.* Typhi N15 in competition with Bp PV8-2 (0.67 ± 0.08) and Bk PV20-2 (0.80 ± 0.22), indicating that enteropathogen adhesion was decreased in the presence of both bifidobacteria. In the displacement assay, Bp PV8-2 and Bk PV20-2 strains induced the release of *S.* Typhi N15 bound to HT29-MTX, indicated by adhesion ratios of 0.43 ± 0.15 and 0.44 ± 0.13, respectively. The inhibition assay showed that adhered bifidobacteria prevented the attachment of *S.* Typhi N15 and stably occupied a sufficient number of adhesion sites on the surface of HT29-MTX cells. Bp PV8-2 showed the highest degree of inhibition of *S.* Typhi N15 (0.08 ± 0.04) compared to Bk PV20-2 (0.21 ± 0.12).

Bk PV20-2 was not able to competitively exclude EHEC in the competition assay, as indicated in **Figure** . In the displacement assay with EHEC added first, Bp PV8-2 showed low adhesion ratio of 0.16 ± 0.03 compared with 0.51 ± 0.31 for EHEC which was not significantly different from 1 (**Figure** ). This data suggest that Bp PV8-2 could not displace previously adhered EHEC. In the inhibition assay (**Figure** ), EHEC adhesion could be significantly ($P < 0.05$) decreased by the presence of adhered Bp PV8-2. Bk PV20-2 did not reduce adhesion of EHEC when added simultaneously (competition assay) or after the addition of EHEC (displacement assay).

# DISCUSSION

## Inhibitory Activity of *B. pseudolongum* PV8-2 and *B. kashiwanohense* PV20-2 against Enteropathogens

Bifidobacteria play an essential role in the development and homeostasis of the host's immune system in infants where they represent one of the first commensal anaerobic bacteria colonizing the gut (Gupta and Garg, 2009). Efficient competition for iron is a key factor for bacterial growth, persistence and establishment in the intestine (Andrews et al., 2003). In our previous research, *B. pseudolongum* PV8-2 and *B. kashiwanohense* PV20-2 isolated from anemic infants in Kenya were therefore selected for this study based on their high iron sequestration capacity (Vazquez-Gutierrez et al., 2015c). Their inhibitory activities against two strains of enteropathogens,

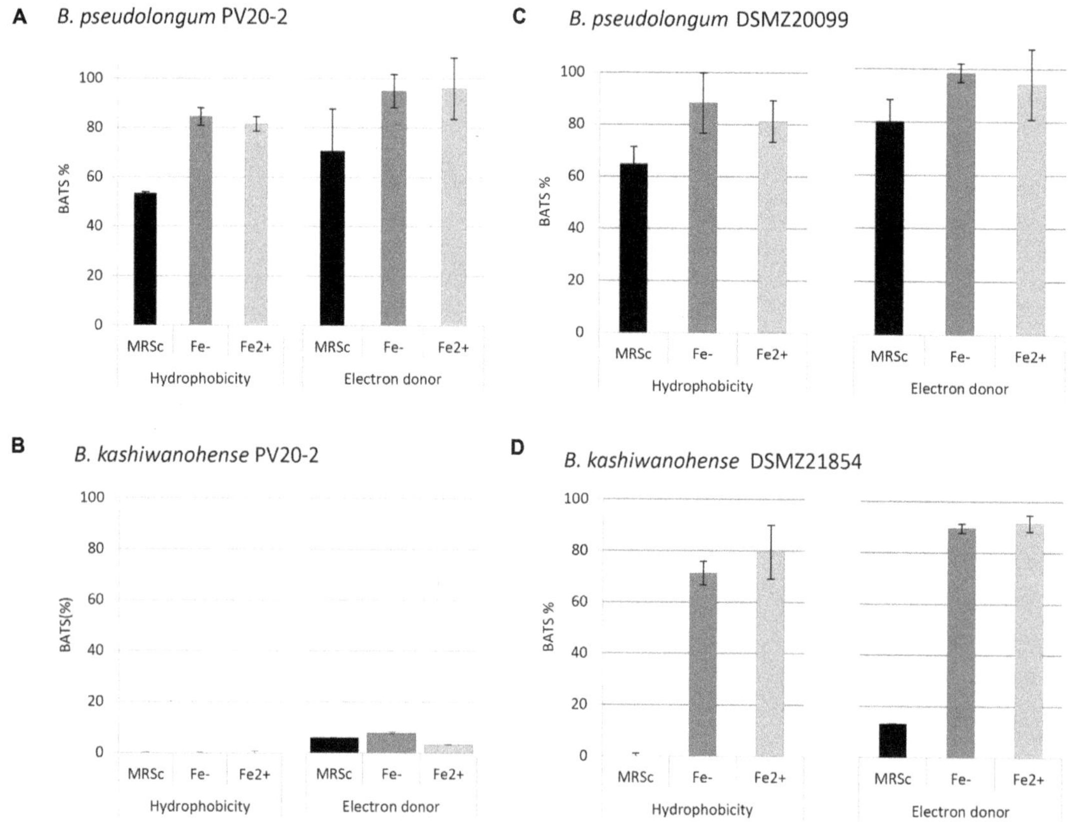


**FIGURE | Adhesion affinity of (A) *B. pseudolongum* PV8-2, (B) *B. pseudolongum* DSMZ20099, (C) *B. kashiwanohense* PV20-2, and (D) *B. kashiwanohense* DSMZ21854 to xylene (hydrophobicity) and chloroform (electron-donor properties).** Bifidobacterial strains were grown in MRS-cys and in CSDLIM without and with iron supplementation (mean ± SD, $n = 3$). [Fe-: without iron supplementation (1.5 μM), $Fe^{2+}$: with iron supplementation (30 μM).]

*S.* Typhi N15 and *E. coli* O157:H45 which are known to efficiently bind iron, were tested *in vitro*. Bifidobacteria may exert inhibitory activity against enteropathogens by production of organic acids, competition for essential growth nutrients, production of antibacterial peptides and co-aggregation with pathogens (Turroni et al., 2009; Butel, 2014; Ventura et al., 2014). Organic acids can prevent infections of pathogens by lowering the intestinal pH and hence restricting colonization of pathogenic bacteria that are sensitive to low pH (Bernet et al., 1993; Lievin et al., 2000; Shu et al., 2000; Gopal et al., 2001; Shu and Gill, 2001; Hammami et al., 2013).

Co-cultivation experiments revealed the inhibitory effects of Bp PV8-2 and Bk PV20-2 against *S.* Typhi N15 and EHEC. Both enteropathogens were not affected by incubation at low pH and organic acid concentrations produced by the bifidobacteria strains during co-cultures. While growth of EHEC was inhibited by Bp PV8-2 and slightly reduced by Bk PV 20-2 (not significantly), *S.* Typhi N15 was significantly reduced by both bifidobacterial strains during co-cultures compared with single cultures. The reduction of viability of enteropathogens in co-cultures could be the result of different factors combined, such as the fitness of the strain under test conditions, and high iron sequestration mechanisms and the production of inhibitory substances by bifidobacteria (Bailey et al., 2011). Additionally, lactate and acetate, may also function as a permeabilizer of the outer membrane of Gram-negative bacteria and may thus accentuate the effects of other inhibitory substances, such as bacteriocins (Oh et al., 2009).

For example, the extracellular proteome of Bp PV8-2 showed the expression of a lysozyme that might contribute to the inhibitory activity of the strain (Vazquez-Gutierrez, 2014), and the effect might be increased by the production of organic acids (Tejero-Sarinena et al., 2012). Both *Bifidobacterium* strains tested were less efficient against EHEC, possibly because EHEC has the ability to survive in many adverse conditions when it enters starvation, allowing EHEC to adapt to very harsh conditions with almost no available nutrients, including iron (Chekabab et al., 2013).

## Adhesion and Competition at the Intestinal Epithelium

The potential of Bp PV8-2 and Bk PV 20-2 to compete for adhesion sites was tested. Occupation of adhesion sites can reduce pathogen adhesion to intestinal epithelium and is mediated by bacterial surface properties like hydrophobicity and cell surface proteins (Botes et al., 2008; Xu et al., 2009). Bp PV8-2 had high affinity for the non-polar solvent xylene, indicating that this strain has hydrophobic cell surface properties. In contrast Bk PV20-2

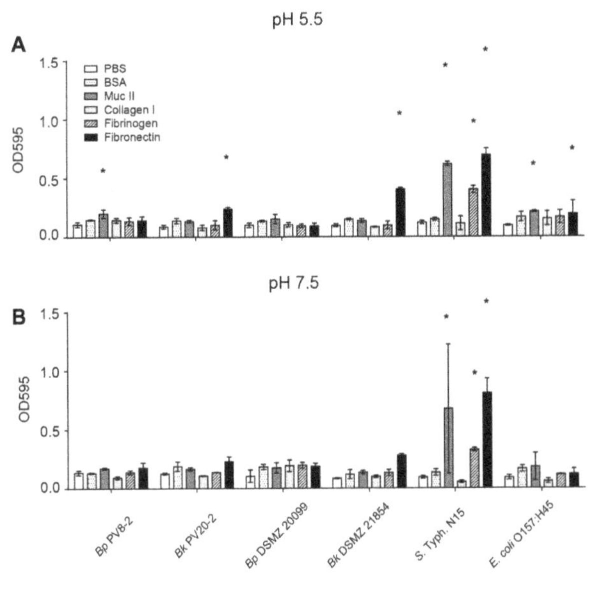


**FIGURE | Adhesion of bifidobacteria and enteropathogens strains: (A) to intestinal epithelial surface molecules (EMC) at pH 5.5 (mean ± SD, $n = 3$). (B)** To intestinal EMC at pH 7.5 (mean ± SD, $n = 3$). Stars denote significant ($P < 0.05$) adhesion to EMC compared with PBS.

was hydrophilic and showed no acid-base properties. The growth medium of bifidobacteria had a strong effect on strain surface property, in contrast with iron availability which did not affect the surface characteristics. Affinity to solvents data indicated that Bp PV8-2 has similar physico-chemical cell surface properties to type strain Bp DSMZ20099. In contrast, affinities to solvents of Bk PV20-2 and Bk DSMZ21854 were similar when both strains were grown in MRS-cys but very different when grown in CSDLIM, emphasizing the influence of growth conditions on surface properties of bacterial strains (Xu et al., 2009). Canzi et al. (2005) reported that even very close genetically related bifidobacteria strains can exhibit significantly different adhesion activities to hydrocarbons (xylene and hexadecane), supporting high strain specificity (Del Re et al., 2000).

Adhesion affinity to different binding sites of the intestinal epithelium was subsequently quantified by the adhesion affinity to a representative set of surface glycoproteins. Specific binding affinity to type II mucin was low for all strains, consistent with the findings of Collado et al. (2005) who observed weak adherence of bifidobacterial strains of human origin to human intestinal mucus glycoproteins. Bk PV20-2 and Bk DSMZ21854 showed affinity for fibronectin. Bk strains shared the binding affinity to fibronectin with both enteropathogens (Fujiwara et al., 2001), suggesting possible competition for intestinal binding sites by Bk PV20-2 that could prevent infections (Collado et al., 2005; Sperandio, 2012). Adhesion to extracellular glycoproteins of all strains was increased at pH 5.5 compared with pH 7.5. The acid environment resulting from the colonization of bifidobacteria could further support the competition for the epithelial binding sites, emphasizing the importance of the combined effect of physico-chemical affinity and surface properties (de Wouters et al., 2015; Jans et al., 2016).

Competition between bifidobacteria strains and entero pathogens by competition, displacement and inhibition was

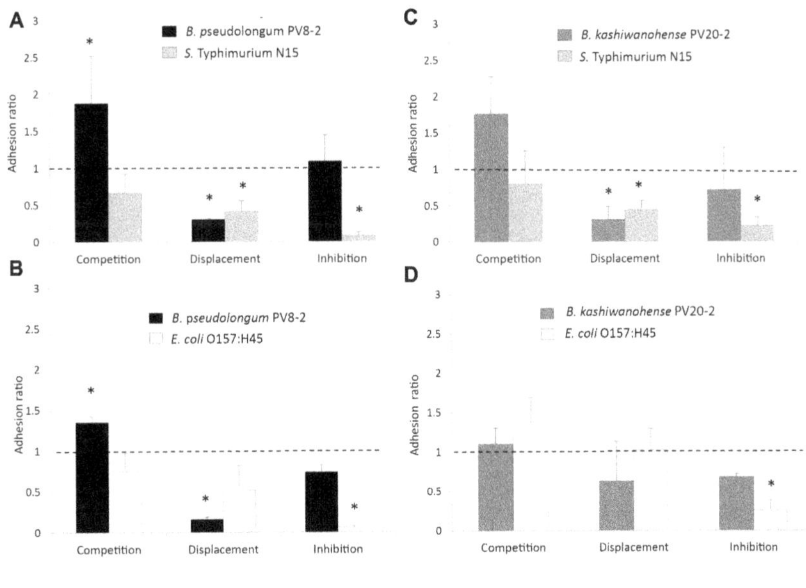


**FIGURE | Adhesion ratios of bifidobacteria and enteropathogens measured during competition, displacement and inhibition tests performed with mucus-producing HT29-MTX cell line: (A) *B. pseudolongum* PV8-2 (Bp PV8-2) and *S.* Typhimurium N15 (*S.* Typhi N15); (B) *B. kashiwanohense* PV20-2 (Bk PV20-2) and *E. coli* O157:H45 (EHEC); (C) Bk PV20-2 and *S.* Typhi N15; (D) Bk PV20-2 and EHEC.** The adhesion ratio corresponded to the ratio of the percentage of adhered bifidobacteria or pathogenic bacteria following simultaneous addition divided by the percentage of adhesion of the *Bifidobacterium* strain or pathogenic bacteria added alone to the cell culture. Dotted line (adhesion ratio = 1) indicates no effect of interactions of tested cultures. Columns with a star (*) indicate significantly different ($P < 0.05$) values when compared to 1 (mean ± SD, $n = 3$).

then studied on a differentiated, mucus-secreting HT29-MTX cells monolayer. In agreement with adhesion tests with single surface molecules, Bp strains showed only modest adhesion abilities to mucus-secreting HT29-MTX cells compared with Bk strains, *S.* Typhi N15 and EHEC. The high adhesion properties of both enteropathogens may reflect growth of the strain during the incubation test with cell layers as previously reported for *Salmonella* in a similar cell test (Dostal et al., 2014). Several studies suggest correlation between adhesion to intestinal cells and cell surface hydrophobicity measured with the BATS assay (Marin et al., 1997; Del Re et al., 2000), a result which was not confirmed in other studies (Savage, 1992; Ouwehand et al., 1999; Canzi et al., 2005; de Wouters et al., 2015). Even though hydrophobicity did not correlate with adhesion properties, BATS assay showed that cell surface properties of Bp PV8-2 and Bk PV20-2 are different, indicating strain-specificity. Bk PV20-2 strain showed no hydrophobic affinity suggesting that adhesion might be mediated by adhesion-like factors (Turroni et al., 2009; Ventura et al., 2014).

Adhesion properties of beneficial bifidobacteria to the mucosa have been shown to promote gut residence time, pathogen exclusion, protection of epithelial cells and immune modulation. Our data indicated that the degree of competition was dependent on bifidobacteria and enteropathogen strain. While both bifidobacteria strains were able to competitively exclude *S.* Typhi N15, only Bp PV8-2 was able to decrease the adhesion capacity of EHEC. In the presence of Bk PV20-2 adhesion of EHEC was increased, suggesting a sharing of metabolic activities leading to enhanced adhesion (Collado et al., 2005). Previous studies reported increased enteropathogen adhesion by bifidobacteria. For example under similar experimental conditions, Serafini et al. (2013) investigated antagonistic effects of *Bifidobacterium bifidum* PRL2010 against various enteropathogens, including *S.* Typhi and EHEC on HT-29 cells not secreting mucus (Gueimonde et al., 2006; Serafini et al., 2013). Our data suggests a direct competition for binding sites that protect the host against invasion of enteropathogens which might also be influenced by strain fitness related to iron sequestration mechanisms (Chauviere et al., 1992; Lee and Puong, 2002). Both Bp PV8-2 and Bk PV20-2 resulted in marked reductions in adhesion of *S.* Typhi N15 and EHEC, indicating that colonization with these potential probiotic candidates selected for high iron sequestration mechanisms might offer at least partial protection from infection with enteropathogenic bacteria (Collado et al., 2007). Further experiments have to be performed *in vivo* to support these effects.

## CONCLUSION

Ability of commensals such as bifidobacteria to restrain pathogen growth in the intestine is strongly affected by niche and nutrient competition. Our study showed that *B. pseudolongum* PV8-2 and *B. kashiwanohense* PV20-2, selected for their high iron sequestration mechanisms, exhibited strain-dependent inhibitory activity against *S.* Typhi N15 and EHEC. These strains may be potential probiotic candidates especially for inhibiting iron-dependent enteric pathogens such as enterobacteriaceae in the gut. The biological significance of such competitive probiotics and their potential as preventive or curative probiotics should be further investigated *in vitro* in presence of complex gut microbiota and *in vivo* with animal models.

## AUTHOR CONTRIBUTIONS

PV-G, TW, CC, and CL designed the experiment. PV-G, JW performed and analyzed the experiments. TW, CC, and CL supervised the experiments. PV-G, JW, TW, CC, and CL wrote the manuscript. All authors read and approved the final manuscript.

## ACKNOWLEDGMENT

This research was supported by own resources at ETH Zurich, Zurich Switzerland.

## REFERENCES

Aires, J., Anglade, P., Baraige, F., Zagorec, M., Champomier-Verges, M. C., and Butel, M. J. (2010). Proteomic comparison of the cytosolic proteins of three *Bifidobacterium longum* human isolates and B. longum NCC2705. *BMC Microbiol.* 10:29. doi: 10.1186/1471-2180-10-29

Andrews, S. C., Robinson, A. K., and Rodriguez-Quinones, F. (2003). Bacterial iron homeostasis. *FEMS Microbiol. Rev.* 27, 215–237. doi: 10.1016/S0168-6445(03)00055-X

Bailey, J. R., Probert, C. S., and Cogan, T. A. (2011). Identification and characterisation of an iron-responsive candidate probiotic. *PLoS ONE* 6:e26507. doi: 10.1371/journal.pone.0026507

Berkley, J. A., Lowe, B. S., Mwangi, I., Williams, T., Bauni, E., Mwarumba, S., et al. (2005). Bacteremia among children admitted to a rural hospital in Kenya. *N. Engl. J. Med.* 352, 39–47. doi: 10.1056/Nejmoa040275

Bernet, M. F., Brassart, D., Neeser, J. R., and Servin, A. L. (1993). Adhesion of human bifidobacterial strains to cultured human intestinal epithelial-cells and inhibition of enteropathogen-cell interactions. *Appl. Environ. Microbiol.* 59, 4121–4128.

Bernet, M. F., Brassart, D., Neeser, J. R., and Servin, A. L. (1994). *Lactobacillus acidophilus* LA 1 binds to cultured human intestinal cell lines and inhibits cell attachment and cell invasion by enterovirulent bacteria. *Gut* 35, 483–489. doi: 10.1136/gut.35.4.483

Botes, M., Loos, B., van, Reenen CA, Dicks, L. M., and Dicks, L. M. (2008). Adhesion of the probiotic strains *Enterococcus mundtii* ST4SA and *Lactobacillus plantarum* 423 to Caco-2 cells under conditions simulating the intestinal tract, and in the presence of antibiotics and anti-inflammatory medicaments. *Arch. Microbiol.* 190, 573–584. doi: 10.1007/s00203-008-0408-0

Broekaert, I. J., and Walker, W. A. (2006). Probiotics and chronic disease. *J. Clin. Gastroenterol.* 40, 270–274. doi: 10.1097/00004836-200603000-00021

Bron, P. A., van Baarlen, P., and Kleerebezem, M. (2012). Emerging molecular insights into the interaction between probiotics and the host intestinal mucosa. *Nat. Rev. Microbiol.* 10, 66–78. doi: 10.1038/nrmicro2690

Butel, M. J. (2014). Probiotics, gut microbiota and health. *Med. Mal. Infect.* 44, 1–8. doi: 10.1016/j.medmal.2013.10.002

Canzi, E., Guglielmetti, S., Mora, D., Tamagnini, I., and Parini, C. (2005). Conditions affecting cell surface properties of human intestinal bifidobacteria. *Antonie Van Leeuwenhoek.* 88(3–4), 207–219. doi: 10.1007/s10482-005-6501-3

Cassat, J. E., and Skaar, E. P. (2013). Iron in infection and immunity. *Cell Host Microb.* 13, 509–519. doi: 10.1016/j.chom.2013.04.010

Cernat, R. C., and Scott, K. P. (2012). Evaluation of novel assays to assess the influence of different iron sources on the growth of *Clostridium difficile*. *Anaerobe* 18, 298–304. doi: 10.1016/j.anaerobe.2012.04.007

Chauviere, G., Coconnier, M. H., Kerneis, S., Darfeuille-Michaud, A., Joly, B., and Servin, A. L. (1992). Competitive exclusion of diarrheagenic *Escherichia coli* (ETEC) from human enterocyte-like Caco-2 cells by heat-killed *Lactobacillus*. *FEMS Microbiol. Lett.* 70, 213–217. doi: 10.1016/0378-1097(92) 90700-X

Cheikhyoussef, A., Pogori, N., Chen, W., and Zhang, H. (2008). Antimicrobial proteinaceous compounds obtained from bifidobacteria: from production to their application. *In. J. Food Microbiol.* 125, 215–222. doi: 10.1016/j.ijfoodmicro.2008.03.012

Chekabab, S. M., Paquin-Veillette, J., Dozois, C. M., and Harel, J. (2013). The ecological habitat and transmission of *Escherichia coli* O157:H7. *FEMS Microbiol. Lett.* 341, 1–12. doi: 10.1111/1574-6968.12078

Cleusix, V., Lacroix, C., Vollenweider, S., and Le Blay, G. (2008). Glycerol induces reuterin production and decreases *Escherichia coli* population in an in vitro model of colonic fermentation with immobilized human feces. *FEMS Microbiol. Ecol.* 63, 56–64. doi: 10.1111/j.1574-6941.2007.00412.x

Collado, M. C., Gueimonde, M., Hernandez, M., Sanz, Y., and Salminen, S. (2005). Adhesion of selected *Bifidobacterium* strains to human intestinal mucus and the role of adhesion in enteropathogen exclusion. *J. Food Prot.* 68, 2672–2678.

Collado, M. C., Meriluoto, J., and Salminen, S. (2007). Role of commercial probiotic strains against human pathogen adhesion to intestinal mucus. *Lett. Appl. Microbiol.* 45, 454–460. doi: 10.1111/j.1472-765X.2007.02212.x

de Wouters, T., Jans, C., Niederberger, T., Fischer, P., and Ruhs, P. A. (2015). Adhesion potential of intestinal microbes predicted by physico-chemical characterization methods. *PLoS ONE* 10:e0136437. doi: 10.1371/journal.pone.0136437

Del Re, B., Sgorbati, B., Miglioli, M., and Palenzona, D. (2000). Adhesion, autoaggregation and hydrophobicity of 13 strains of *Bifidobacterium longum*. *Lett. Appl. Microbiol.* 31, 438–442. doi: 10.1046/j.1365-2672.2000.00845.x

Dobson, A., Cotter, P. D., Ross, R. P., and Hill, C. (2012). Bacteriocin production: a probiotic trait? *Appl. Environ. Microbiol.* 78, 1–6. doi: 10.1128/AEM.05576-11

Dostal, A., Gagnon, M., Chassard, C., Zimmermann, M. B., O'Mahony, L., Lacroix, C., et al. (2014). *Salmonella* adhesion, invasion and cellular immune responses are differentially affected by iron concentrations in a combined in vitro gut fermentation-cell model. *PLoS ONE* 9:e93549. doi: 10.1371/journal.pone.0093549

Fujiwara, S., Hashiba, H., Hirota, T., and Forstner, J. F. (2001). Inhibition of the binding of enterotoxigenic *Escherichia coli* Pb176 to human intestinal epithelial cell line HCT-8 by an extracellular protein fraction containing BIF of *Bifidobacterium longum* SBT2928: suggestive evidence of blocking of the binding receptor gangliotetraosylceramide on the cell surface. *Int. . J. Food Microbiol.* 67, 97–106.

Fukuda, S., Toh, H., Hase, K., Oshima, K., Nakanishi, Y., Yoshimura, K., et al. (2011). Bifidobacteria can protect from enteropathogenic infection through production of acetate. *Nature* 469, 543–547. doi: 10.1038/Nature09646

Gagnon, M., Zihler Berner, A., Chervet, N., Chassard, C., and Lacroix, C. (2013). Comparison of the Caco-2, HT-29 and the mucus-secreting HT29-MTX intestinal cell models to investigate *Salmonella* adhesion and invasion. *J. Microbiol. Methods* 94, 274–279. doi: 10.1016/j.mimet.2013.06.027

Gopal, P. K., Prasad, J., Smart, J., and Gill, H. S. (2001). In vitro adherence properties of *Lactobacillus rhamnosus* DR20 and *Bifidobacterium lactis* DR10 strains and their antagonistic activity against an enterotoxigenic *Escherichia coli*. *Int. J. Food Microbiol.* 67, 207–216. doi: 10.1016/S0168-1605(01)00440-8

Goto, Y., and Kiyono, H. (2012). Epithelial barrier: an interface for the cross-communication between gut flora and immune system. *Immunol. Rev.* 245, 147–163. doi: 10.1111/j.1600-065X.2011.01078.x

Gueimonde, M., Jalonen, L., He, F., Hiramatsu, M., and Salminen, S. (2006). Adhesion and competitive inhibition and displacement of human enteropathogens by selected lactobacilli. *Food Res. Int.* 39, 467–471. doi: 10.1016/j.foodres.2005.10.003

Gupta, V., and Garg, R. (2009). Probiotics. *Indian J. Med. Microbiol.* 27, 202–209. doi: 10.4103/0255-0857.53201

Hammami, R., Fernandez, B., Lacroix, C., and Fliss, I. (2013). Anti-infective properties of bacteriocins: an update. *Cell. Mol. Life Sci.* 70, 2947–2967. doi: 10.1007/s00018-012-1202-3

Haraga, A., Ohlson, M. B., and Miller, S. I. (2008). *Salmonellae* interplay with host cells. *Nat. Rev. Microbiol.* 6, 53–66. doi: 10.1038/nrmicro1788

Izquierdo, E., Medina, M., Ennahar, S., Marchioni, E., and Sanz, Y. (2008). Resistance to simulated gastrointestinal conditions and adhesion to mucus as probiotic criteria for *Bifidobacterium longum* strains. *Curr. Microbiol.* 56, 613–618. doi: 10.1007/s00284-008-9135-7

Jans, C., de Wouters, T., Bonfoh, B., Lacroix, C., Kaindi, D. W. M., Anderegg, J., et al. (2016). Phylogenetic, epidemiological and functional analyses of the *Streptococcus bovis/Streptococcus equinus* complex through an overarching MLST scheme. *BMC Microbiol.* 16:117. doi: 10.1186/s12866-016-0735-2

Jost, T., Lacroix, C., Braegger, C. P., and Chassard, C. (2012). New insights in gut microbiota establishment in healthy breast fed neonates. *PLoS ONE* 7:e44595. doi: 10.1371/journal.pone.0044595

Kortman, G. A., Boleij, A., Swinkels, D. W., and Tjalsma, H. (2012). Iron availability increases the pathogenic potential of *Salmonella typhimurium* and other enteric pathogens at the intestinal epithelial interface. *PLoS ONE* 7:e29968. doi: 10.1371/journal.pone.0029968

Lee, Y. K., and Puong, K. Y. (2002). Competition for adhesion between probiotics and human gastrointestinal pathogens in the presence of carbohydrate. *Br. J. Nutr.* 88(Suppl. 1), S101–S108. doi: 10.1079/BJN2002635

Lesuffleur, T., Barbat, A., Dussaulx, E., and Zweibaum, A. (1990). Growth adaptation to methotrexate of HT-29 human colon carcinoma cells is associated with their ability to differentiate into columnar absorptive and mucus-secreting cells. *Cancer Res.* 50, 6334–6343.

Lievin, V., Peiffer, I., Hudault, S., Rochat, F., Brassart, D., Neeser, J. R., et al. (2000). *Bifidobacterium* strains from resident infant human gastrointestinal microflora exert antimicrobial activity. *Gut* 47, 646–652. doi: 10.1136/gut.47.5.646

Marco, M. L., Pavan, S., and Kleerebezem, M. (2006). Towards understanding molecular modes of probiotic action. *Curr. Opin. Biotechnol.* 17, 204–210. doi: 10.1016/j.copbio.2006.02.005

Marin, M. L., Benito, Y., Pin, C., Fernandez, M. F., Garcia, M. L., Selgas, M. D., et al. (1997). Lactic acid bacteria: hydrophobicity and strength of attachment to meat surfaces. *Lett. Appl. Microbiol.* 24, 14–18. doi: 10.1046/j.1472-765X.1997.00334.x

Martinez, F. A., Balciunas, E. M., Converti, A., Cotter, P. D., and de Souza Oliveira, R. P. (2013). Bacteriocin production by *Bifidobacterium* spp. A review. *Biotechnol. Adv.* 31, 482–488. doi: 10.1016/j.biotechadv.2013.01.010

Melton-Celsa, A., Mohawk, K., Teel, L., and O'Brien, A. (2012). Pathogenesis of Shiga-toxin producing *Escherichia coli*. *Curr. Top. Microbiol. Immunol.* 357, 67–103. doi: 10.1007/82_2011_176

Monack, D. M., and Hultgren, S. J. (2013). The complex interactions of bacterial pathogens and host defenses. *Curr. Opin. Microbiol.* 16, 1–3. doi: 10.1016/j.mib.2013.03.001

Muller, D., Benz, I., Liebchen, A., Gallitz, I., Karch, H., and Schmidt, M. A. (2009). Comparative analysis of the locus of enterocyte effacement and its flanking regions. *Infect. Immun.* 77, 3501–3513. doi: 10.1128/IAI.00090-09

Oh, D. H., Pan, Y. W., Berry, E., Cooley, M., Mandrell, R., and Breidt, F. (2009). *Escherichia coli* O157:H7 Strains isolated from environmental sources differ significantly in acetic acid resistance compared with human outbreak strains. *J. Food Prot.* 72, 503–509.

Ouwehand, A. C., Kirjavainen, P. V., Gronlund, M. M., Isolauri, E., and Salminen, S. J. (1999). Adhesion of probiotic micro-organisms to intestinal mucus. *Int. Dairy J.* 9, 623–630. doi: 10.1016/S0958-6946(99)00132-6

Sansonetti, P. J. (2004). War and peace at mucosal surfaces. *Nat. Rev. Immunol.* 4, 953–964. doi: 10.1038/Nri1499

Santos, R. L., Raffatellu, M., Bevins, C. L., Adams, L. G., Tukel, C., Tsolis, R. M., et al. (2009). Life in the inflamed intestine, *Salmonella* style. *Trends Microbiol.* 17, 498–506. doi: 10.1016/j.tim.2009.08.008

Savage, D. C. (1992). Growth phase, cellular hydrophobicity, and adhesion in vitro of lactobacilli colonizing the keratinizing gastric epithelium in the mouse. *Appl. Environ. Microbiol.* 58, 1992–1995.

Serafini, F., Strati, F., Ruas-Madiedo, P., Turroni, F., Foroni, E., Duranti, S., et al. (2013). Evaluation of adhesion properties and antibacterial activities of the infant gut commensal *Bifidobacterium bifidum* PRL2010. *Anaerobe* 21, 9–17. doi: 10.1016/j.anaerobe.2013.03.003

Shu, Q., and Gill, H. S. (2001). A dietary probiotic (*Bifidobacterium lactis* HN019) reduces the severity of *Escherichia coli* O157:H7 infection in mice. *Med. Microbiol. Immunol.* 189, 147–152. doi: 10.1007/s430-001-8021-9

Shu, Q., Lin, H., Rutherfurd, K. J., Fenwick, S. G., Prasad, J., Gopal, P. K., et al. (2000). Dietary *Bifidobacterium lactis* (HN019) enhances resistance to oral *Salmonella* Typhimurium infection in mice. *Microbiol. Immunol.* 44, 213–222. doi: 10.1111/j.1348-0421.2000.tb02486.x

Sillanpaa, J., Nallapareddy, S. R., Singh, K. V., Ferraro, M. J., and Murray, B. E. (2008). Adherence characteristics of endocarditis-derived *Streptococcus gallolyticus* ssp. *gallolyticus (Streptococcus bovis* biotype I) isolates to host extracellular matrix proteins. *FEMS Microbiol. Lett.* 289, 104–109. doi: 10.1111/j.1574-6968.2008.01378.x

Sperandio, V. (2012). Microbiology. Virulence or competition? . *Science* 336, 1238–1239. doi: 10.1126/science.1223303

Tejero-Sarinena, S., Barlow, J., Costabile, A., Gibson, G. R., and Rowland, I. (2012). In vitro evaluation of the antimicrobial activity of a range of probiotics against pathogens: evidence for the effects of organic acids. *Anaerobe* 18, 530–538. doi: 10.1016/j.anaerobe.2012.08.004

Thiennimitr, P., Winter, S. E., and Baumler, A. J. (2012). *Salmonella*, the host and its microbiota. *Curr. Opin. Microbiol.* 15, 108–114. doi: 10.1016/j.mib.2011.10.002

Turroni, F., Marchesi, J. R., Foroni, E., Gueimonde, M., Shanahan, F., Margolles, A., et al. (2009). Microbiomic analysis of the bifidobacterial population in the human distal gut. *ISME J.* 3, 745–751. doi: 10.1038/ismej.2009.19

Turroni, F., Peano, C., Pass, D. A., Foroni, E., Severgnini, M., Claesson, M. J., et al. (2012). Diversity of bifidobacteria within the infant gut microbiota. *PLoS ONE* 7:e36957. doi: 10.1371/journal.pone.0036957

Turroni, F., Ventura, M., Butto, L. F., Duranti, S., O'Toole, P. W., Motherway, M. O., et al. (2014). Molecular dialogue between the human gut microbiota and the host: a *Lactobacillus* and *Bifidobacterium* perspective. *Cell. Mol. Life Sci.* 71, 183–203. doi: 10.1007/s00018-013-1318-0

Vazquez-Gutierrez, P. (2014). Screening and Characterization of Bifidobacteria with High Iron Binding Properties. Zürich: ETH-Zürich.

Vazquez-Gutierrez, P., Lacroix, C., Chassard, C., Klumpp, J., Jans, C., and Stevens, M. J. (2015a). Complete and assembled genome sequence of *Bifidobacterium kashiwanohense* PV20-2 isolated from the feces of an anemic Kenyan infant. *Genome Announc.* 3:e1467-14. doi: 10.1128/genomeA.01467-14

Vazquez-Gutierrez, P., Lacroix, C., Chassard, C., Klumpp, J., Stevens, M. J., and Jans, C. (2015b). *Bifidobacterium pseudolongum* strain PV8-2 isolated from a stool sample of an anemic Kenyan infant. *Genome Announc.* 3:e1469-14. doi: 10.1128/genomeA.01469-14

Vazquez-Gutierrez, P., Lacroix, C., Jaeggi, T., Zeder, C., Zimmerman, M. B., and Chassard, C. (2015c). Bifidobacteria strains isolated from stools of iron deficient infants can efficiently sequester iron. *BMC Microbiol.* 15:3. doi: 10.1186/s12866-014-0334-z

Ventura, M., Turroni, F., Lugli, G. A., and van Sinderen, D. (2014). Bifidobacteria and humans: our special friends, from ecological to genomics perspectives. *J. Sci. Food Agric.* 94, 163–168. doi: 10.1002/jsfa.6356

Wardlaw, T., Salama, P., Brocklehurst, C., Chopra, M., and Mason, E. (2010). Diarrhoea: why children are still dying and what can be done. *Lancet* 375, 870–872. doi: 10.1016/S0140-6736(09)61798-0

Weinberg, E. D. (2009). Iron availability and infection. *Biochem. Biophys. Acta* 1790, 600–605. doi: 10.1016/j.bbagen.2008.07.002

Winter, S. E., Lopez, C. A., and Baumler, A. J. (2013). The dynamics of gut-associated microbial communities during inflammation. *EMBO Rep.* 14, 319–327. doi: 10.1038/embor.2013.27

Xu, H., Jeong, H. S., Lee, H. Y., and Ahn, J. (2009). Assessment of cell surface properties and adhesion potential of selected probiotic strains. *Lett. Appl. Microbiol.* 49, 434–442. doi: 10.1111/j.1472-765X.2009.02684.x

Yatsunenko, T., Rey, F. E., Manary, M. J., Trehan, I., Dominguez-Bello, M. G., Contreras, M., et al. (2012). Human gut microbiome viewed across age and geography. *Nature* 486, 222–227. doi: 10.1038/nature11053

**Conflict of Interest Statement:** The authors declare that the research was conducted in the absence of any commercial or financial relationships that could be construed as a potential conflict of interest.

# 21

# Proteinaceous Molecules Mediating *Bifidobacterium*-Host Interactions

*Lorena Ruiz[1]*, Susana Delgado[2], Patricia Ruas-Madiedo[2], Abelardo Margolles[2] and Borja Sánchez[2]*

[1] Department of Nutrition, Food Science and Food Technology, Universidad Complutense de Madrid, Spain, [2] Department of Microbiology and Biochemistry of Dairy Products, Instituto de Productos Lácteos de Asturias-Consejo Superior de Investigaciones Científicas, Villaviciosa, Spain

Bifidobacteria are commensal microoganisms found in the gastrointestinal tract. Several strains have been attributed beneficial traits at local and systemic levels, through pathogen exclusion or immune modulation, among other benefits. This has promoted a growing industrial and scientific interest in bifidobacteria as probiotic supplements. However, the molecular mechanisms mediating this cross-talk with the human host remain unknown. High-throughput technologies, from functional genomics to transcriptomics, proteomics, and interactomics coupled to the development of both *in vitro* and *in vivo* models to study the dynamics of the intestinal microbiota and their effects on host cells, have eased the identification of key molecules in these interactions. Numerous secreted or surface-associated proteins or peptides have been identified as potential mediators of bifidobacteria-host interactions and molecular cross-talk, directly participating in sensing environmental factors, promoting intestinal colonization, or mediating a dialogue with mucosa-associated immune cells. On the other hand, bifidobacteria induce the production of proteins in the intestine, by epithelial or immune cells, and other gut bacteria, which are key elements in orchestrating interactions among bifidobacteria, gut microbiota, and host cells. This review aims to give a comprehensive overview on proteinaceous molecules described and characterized to date, as mediators of the dynamic interplay between bifidobacteria and the human host, providing a framework to identify knowledge gaps and future research needs.

**Edited by:**
*Marco Ventura, University of Parma, Italy*

**Reviewed by:**
*Ines Martinez, University of Nebraska—Lincoln, USA*
*Christophe Lacroix, Eidgenössische Technische Hochschule Zürich, Switzerland*

***Correspondence:**
*Lorena Ruiz*
*loruiz@ucm.es*

**Keywords: *Bifidobacterium*, host interaction, proteome, immunomodulation, adhesin**

## INTRODUCTION

The human gut is inhabited by a trillion of microorganisms which constitute the gut microbiota. These microorganisms are in close contact with the intestinal mucosa, which represents the largest extension of the human body exposed to external stimuli. A complex molecular interplay is established among microbiota, dietary components and host cells, which regulates immune and metabolic functions in the host (Furusawa et al., 2015). Dysbiosis, defined as changes in the gut microbiota structure associated to healthy individuals, disrupts the microbiome-host cross-talk homeostasis and correlates with metabolic and inflammatory disorders (Evans et al., 2013; Patel et al., 2013; Levy et al., 2015).

Pro- and pre-biotics can improve host health through microbiota modulation and immune system boosting (Picard et al., 2005). Some strains of bifidobacteria, which are among the first colonizers of the human intestine and one of the dominant groups in the breast-fed infant microbiota (Garrido et al., 2012), have been attributed several health benefits, encouraging interest in their use as probiotics. Pathogen inhibition and diarrhea amelioration are their best established outcomes and have been related to the production of organic acids (Fukuda et al., 2011), antibacterial peptides (Moroni et al., 2006), quorum-sensing inhibitors (Cotar et al., 2010), pathogen displacement (Ruas-Madiedo et al., 2006), and virulence attenuation (Tanner et al., 2016). Bifidobacteria also regulate host functions and ferment complex polysaccharides from our diet (Ménard et al., 2005; Heuvelin et al., 2009, 2010; Bermudez-Brito et al., 2013; Furusawa et al., 2015), although there is still limited knowledge on the molecular mechanisms triggering these effects.

Delineating the specific molecules mediating *Bifidobacterium* cross-talk with the host, will help to understand their beneficial effects and establish microbiome-targeted therapies for human diseases. This review gives an overview on molecules behind the bifidobacterial-host cross-talk, providing a framework to design safe and efficacious probiotic-derived supplements (Licciardi et al., 2010; Shenderov, 2013).

## INTESTINAL MUCOSA ADHESION

Bacterial adhesion to the intestinal surface is mediated by non-specific, hydrophobic or electrostatic interactions, and specific mechanisms involving macromolecular interactions between bacterial and host receptors. Functionally characterized *Bifidobacterium* adhesins, surface-associated proteins that facilitate bifidobacteria attachment to intestinal cells and/or the extracellular matrixes surrounding them, are reviewed below and summarized in **Figure** .

Involvement of pili in bifidobacterial adhesion has been studied in *Bifidobacterium bifidum* and *Bifidobacterium breve* (O'Connell Motherway et al., 2011b; Turroni et al., 2013). The *B. bifidum* PRL2010 genome harbors three pilus clusters. Of these, *pil2* and *pil3,* encode putative sortase-dependent pili that are expressed under both *in vitro* and *in vivo* conditions. Heterologous expression of *pil3* in *Lactococcus lactis* significantly increased this bacterium adhesion to the human epithelial cell line Caco-2 (Turroni et al., 2013). The type IVb tight adherence (Tad) pilus-encoding gene cluster from *B. breve* UCC2003, was found to be essential for the colonization of, and persistence in, the murine gut. Tad inactivation impaired the strain ability to stably colonize the murine intestine, as reflected by reduced shedding level and bifidobacterial numbers in the gut (O'Connell Motherway et al., 2011b). Another surface-anchored protein potentially involved in intestinal adhesion of bifidobacteria is BopA. This purified lipoprotein competes with *B. bifidum* MIMBb75 adhesion to Caco-2 cells, and probably facilitates *B. bifidum* adhesion (Guglielmetti et al., 2008). Indeed, *Bifidobacterium* strains overexpressing *bopA* adhere better (Gleinser et al., 2012). However, *B. bifidum* treatment with anti-BopA antibodies does not reduce the attachment to intestinal cells (Kainulainen et al., 2013), thus the specific adhesion mechanism mediated through BopA must be further elucidated.

Remarkably, bifidobacteria can use some moonlighting proteins, those with multiple functions (Jeffery, 2003), as adhesin-like factors. Surface-exposed glycolytic enzymes, including transaldolase from *B. bifidum* and enolase from *Bifidobacterium animalis*, are adhere to mucin and plasminogen, respectively (Candela et al., 2009; González-Rodríguez et al., 2012). Other surface-exposed moonlighting proteins, including the chaperone DnaK from *B. animalis* and the elongation factor Tu from *Bifidobacterium longum*, showed high affinity for human plasminogen *in vitro* and have been proposed as mediators of intestinal attachment (Candela et al., 2010; Wei et al., 2014).

It is worth highlighting that most of the adhesins herein described have been identified on *in vitro* assays and their relevance for intestinal colonization has not been tested *in vivo*. Since laboratory models do not accurately mimic all the factors that can affect bacterial attachment to the intestinal mucosa (e.g., due to the absence of a mucus layer and resident microbiota), functional confirmation of the adhesion capacity *in vivo* is still required (Ouwehand and Salminen, 2003).

## TIGHT-JUNCTIONS

A single layer of epithelial cells provides a selective barrier separating intestinal lumen from subjacent tissues. Tight-junctions (TJ) are multi-protein complexes that control molecule translocation across this barrier (Lee, 2015), and their disruption leads to uncontrolled trafficking of noxious molecules triggering inflammation (Bergmann et al., 2013).

Specific *Bifidobacterium* strains promote TJ enhancing epithelial barrier integrity (Ohland and Macnaughton, 2010; Mokkala et al., 2016). In animal models, *B. bifidum* and *B. longum* strains preserved TJ localization, attenuating intestinal permeability, and decreasing necrotizing enterocolitis incidence (Khailova et al., 2009; Bergmann et al., 2013; Srutkova et al., 2015). Preliminary work suggested that *B. bifidum* metabolites, like acetate, induced TJ expression in intestinal cells (Hsieh et al., 2015). Soluble factors present in *B. longum* lysates or secreted by *B. infantis* or *B. breve* strains, also mediate epithelial barrier maintenance (Ménard et al., 2005; Ewaschuk et al., 2008; Sultana et al., 2013). Further research to identify the specific molecules mediating this TJ promotion is needed.

## MUCUS, HMO, AND NON-DIGESTIBLE CARBOHYDRATES DEGRADATION

The spatial distribution of bacteria throughout the gastrointestinal tract, is partly controlled by nutrients availability for resident microbiota (Donaldson et al., 2016). Indeed, our wellbeing relates with the nutrient harvesting capability of our gut microbes. These microorganisms, overall, can utilize dietary and host carbohydrates, and glycans produced by other gut bacteria. Indeed, 89 "carbohydrate active enzymes" (CAZyme) have recently been identified in 85% of the microbiomes obtained from 488 individuals (Bhattacharya et al., 2015), suggesting that gut bacteria are highly specialized in using available glycans as their main sustenance. Bifidobacterial genomes are abundant in saccharolytic features whose expression is tightly regulated by available carbohydrates (Khoroshkin et al., 2016), supporting that host glycans were a potent evolutionary force driving their successful gut colonization (Sánchez et al., 2013; Milani et al., 2016).

Numerous studies demonstrated bifidobacteria's capability to use dietary non-digestible oligosaccharides, which is on the basis of the prebiotic concept (Rastall and Gibson, 2015). Glycosyl hydrolases (GH, or glycosidades), many of which are extracellular, have high specificity for the oligosaccharides constituents and cleave the glycosidic bonds (**Table** ). Special attention has been paid to the GH acting on human milk oligosaccharides (HMO) since these serve as substrates for bifidobacteria, which are the initial gut colonizers in breast-fed infants. HMO are structurally diverse and composed of several monosaccharides (glucose, galactose, N-acetylglucosamine, fucose, or sialic acid). They mainly consist of a lactose core linked to units ($n$ = 0–15) of lacto-N-biose (type I) or to N-acetyl-lactosamine (type II; Smilowitz et al., 2014). Bifidobacteria secrete GH that cleave specific linkages within the HMO molecules and the best characterized are those synthesized by *B. bifidum* which, together with *B. longum* subsp. *infantis*, are two abundant species in breast-fed neonates (**Table** ). These species employ different strategies for HMO utilization. Whereas *B. bifidum* has an array of membrane-associated GH, *B. longum* subsp. *infantis* is specialized in the import and intracellular breakdown of HMO (Garrido et al., 2013; Jae-Han et al., 2013). Moreover *B. longum* strains have similar HMO-utilization patterns, whilst *B. bifidum* strains are more diverse with some unable to use fucosylated or sialylated HMO (Garrido et al., 2015). Similarly, the *B. breve* HMO utilization profile is strain dependent and, contrary to *B. bifidum*, some strains consume fucosylated or sialilated HMOs. *B. breve*'s capability to use these HMOs explains its abundance in breast-fed babies (Ruiz-Moyano et al., 2013).

Some bifidobacteria can also utilize mucins from the mucus layer coating the intestine. Mucin composition and structure resemble that of HMO; consisting of a core of different O-glycans, built on α- and β-linked N-acetyl-galactosamine, galactose, and N-acetyl-glucosamine residues, which can incorporate fucose and sialic acid residues (Tailford et al., 2015). *B. longum* and *B. breve* strains' capability to effectively use mucin carbohydrates, has been confirmed *in vitro* (Ruas-Madiedo et al., 2008). However, GH able to degrade mucins have only been described in *B. bifidum* (**Table** ). Indeed, in a comparative genomic study 60% of the GH-encoding genes from *B. bifidum* were predicted to breakdown mucin-like glycans and most of them were exclusively present in this bifidobacterial species (Turroni et al., 2014). Remarkably, other species could use the mono- and oligosaccharides released by *B. bifidum* GH thus evidencing the existence of cross-feeding mechanisms, as it has been demonstrated in *B. breve* and *B. bifidum* co-cultures (Egan et al., 2014).

*Bifidobacterium* capacity to metabolize specific dietary and host-derived carbohydrates is also dependent on the presence of specific sugar transport systems. These are crucial for their competitive establishment in the gut, thus representing one of the molecular mechanisms by which bifidobacteria interact with the host. Import sugar mechanisms in bifidobacteria are herein described (Bottacini et al., 2014).

First, ATP-binding cassette (ABC) systems are active transporters which couple ATP hydrolysis to translocation uptake across the cell membrane. They are the most frequent sugar transporters in bifidobacteria and have been described for mono- and oligosaccharides in different species (Nishimoto and Kitaoka, 2007; Wada et al., 2008; Wei et al., 2012), although only a few of them have been functionally characterized at protein level (Suzuki et al., 2008; Ejby et al., 2013).

Secondly, some secondary transporters, predicted to consist of single integral membrane-associated proteins, have been characterized at protein level in bifidobacteria. These include permease systems for the uptake of lactose, glucose, and sucrose (Parche et al., 2006). Secondary transporters encoding genes have been identified in different bifidobacterial species, although most of them have not been characterized at protein level (Turroni et al., 2012).

Proton symporters of the glycoside-pentoside-hexuronide (GPH) cation symporter family for melibiose and pentosides were also described in *Bifidobacterium* (Lee and O'Sullivan, 2010; Turroni et al., 2012), although they remain to be characterized.

Finally, phosphoenolpyruvate-phosphotransferase (PEP-PTS) systems were first characterized in the 90's in *B. breve* and *B. bifidum* at protein level (Lee and O'Sullivan, 2010). Later, genome sequence availability revealed their wide spread distribution in bifidobacteria. In particular, *B. breve* UCC2003 genome contains four PEP-PTS systems, one of which has been characterized as a fructose-specific transporter (Mazé et al., 2007). Also, *in silico* analysis revealed a putative glucose-specific PEP-PTS uptake system in *B. longum* (Lorca et al., 2007). However, genome analysis of different *B. longum* strains showed that glucose-specific PTS transporters are minor in comparison with ABC transporters (Pokusaeva et al., 2011), thus glucose may be transported preferentially by secondary permeases (Parche et al., 2006).

Comparative genomic analysis revealed that sugar PEP-PTS systems are present in all bifidobacterial genomes, except for *B. animalis* subsp. *lactis* (Lee and O'Sullivan, 2010) which is hypothesized to have lost most of their carbohydrate transporters due to extended cultivation under industrial conditions.. In fact, the capability to utilize variable carbon sources is considered an adaptation to the gut environment. For instance, the dominant *Bifidobacterium* species in infant fecal samples (*B. longum* and *B. bifidum*) is consistent with their inherent ability to use host-derived oligosaccharides such as mucin and HMO (Bottacini et al., 2014), and their possession of a wide range of host-derived carbohydrate transporters, such as those involved in N-biose import (Suzuki et al., 2008).

## REGULATION T-EFFECTOR CELLS AND $T_{reg}$

In the absence of disease, the ensemble of molecular interactions taking place in the human gut results in

**TABLE | Gycosyl hydrolases and sugar transporters characterized and/or described in *Bifidobacterium* genus.**

**Glycosyl hydrolases**

| Substrate[a] | Glycosyl hydrolase (family) | Bifidobacteria species | References |
|---|---|---|---|
| **NON-DIGESTIBLE DIETARY CARBOHYDRATES** | | | |
| α-glycans: palatinose (1→6); turanose (1→3); maltotriose and maltose (1→4) linkages, etc. | α-1,6-glucosidase (GH13) | *B. breve* UCC2003 | Pokusaeva et al., 2009; Kelly et al., 2016 |
| Starch and starch-like carbohydrates (pullulan, maltodextrin, etc.) | α-amylases, amylopullanases, etc. | *B. adolescentis* 22L | Duranti et al., 2014 |
| Starch hydrolysates (maltodextrins, malto-OS, isomalto-OS, maltose, etc.) | α-glucosidases, α-amylases, etc. | *B. longum* subsp. *longum* BBMN68 | Liu et al., 2015 |
| Plant ginsenoside and cellobiose | β-glucosidase (GH1, GH3) | *B. animalis* subsp. *lactis* AD011 | Kim et al., 2012 |
| Isoflavone glycosides (daidzin) | β-glucosidases (GH3) | *B. pseudocatenulatum* IPLA36007 | Alegría et al., 2014 |
| β-glucosides (mycotoxins from cereal-based foods) | β-glucosidases | *B. adolescentis* DSM20083 | Michlmayr et al., 2015 |
| β-galactans, β-galacto-OS: (1→4) linkages | Endogalactanase (GH53) | *B. longum* NCC2705 | Hinz et al., 2005 |
| β-galactans (potato) | β-1,4-endogalactanase | *B. breve* UCC2003 | O'Connell Motherway et al., 2011a |
| β-1,3-galactooligossacharides and arabinogalactan | exo-β-1,3-galactanase | *B. longum* JCM1217 | Fujita et al., 2014 |
| Arabinoxylan [β-(1,4)-linked xylosyl backbone with arabinosyl side chains] | Arabinofuranohydrolase | *B. adolescentis* DSM20083 | van den Broek et al., 2005 |
| β-L-arabinofuranosides | β-L-arabinobiosidase (GH121) | *B. longum* JCM 1217 | Fujita et al., 2011 |
| α-1,5-linked arabino-OS | α-L-arabinofuranosidase (GH1) | *B. adolescentis* ATCC 15703 | Suzuki et al., 2013 |
| Plant ginsenoside | β-D-xylosidase | *B. breve* K-110 | Hyun et al., 2012 |
| Xylo-OS | β-D-xylosidase (GH43) | *B. animalis* subsp. *lactis* BB-12 | Viborg et al., 2013 |
| β-(2,1) in short-chain inulin-type fructans, Raffinose | β -fructofuranosidase (GH32) | *B. longum* KN29.1 | Bujacz et al., 2011 |
| Flavonoid rhamnoglycosides: (1→6) linkage | α-L-rhamnosidase | *B. dentium* | Bang et al., 2015 |
| β-Mannans (plants) | Mannanase (GH5_8) | *B. animalis* subsp. *lactis* Bl-04 | Morrill et al., 2015 |
| **HUMAN CARBOHYDRATES: MUCIN AND HMO** | | | |
| α-L-Fucosyl termini residues from glycoconjugates | 1,2-α-L-fucosidase (GH95) | *B. bifidum* JCM1254 | Katayama et al., 2004 |
| Mucin-OS (Core 1 type O-glycans) | Endo-α-N-acetylgalactosaminidase (GH101) | *B. longum* JCM 1217 | Fujita et al., 2005 |
| Mucin 2 (Core 3 type O-glycans) | α-N-acetylgalactosaminidase (GH129) | *B. bifidum* JCM1254 | Kiyohara et al., 2012 |
| Gastroduodenal mucin (terminal GlcNAcα1-4Gal) | α-N-acetylglucosaminidase (GH89) | *B. bifidum* JCM 1254 | Shimada et al., 2015 |
| HMO and lacto-N-tetraose (type I chain) | Lacto-N-biosidase (GH20) | *B. bifidum* JCM1254 | Wada et al., 2008 |
| HMO α1,3/4-fucosylated OS | 1,3–1,4-α-L-fucosidase | *B. bifidum* JCM1254 | Ashida et al., 2009; Ito et al., 2013 |
| HMO and lacto-N-neotetraose (type II chain) | β-galactosidase + β-N-acetylhexosaminidases | *B. bifidum* JCM1254 | Miwa et al., 2010 |
| HMO sialylOS | Exo-α-sialidase (GH33) | *B. bifidum* JCM1254 | Kiyohara et al., 2011 |
| Fucosylated HMO | α- L-fucosidases (GH29, GH95) | *B. longum* subsp. *infantis* ATCC15697 | Sela et al., 2011 |
| HMO (type I chain) + (type II chain) | β-1,3-galactosidase + β-galactosidase | *B. longum* subsp. *infantis* ATCC15697 | Yoshida et al., 2012 |

*(Continued)*

**TABLE | Continued**

| Substrate[a] | Transporter Family | Bifidobacteria species | References |
|---|---|---|---|
| **SUGAR TRANSPORTERS** | | | |
| Arabinoxylo-OS | ABC transporter | *Bifidobacterium animalis* subsp. *lactis* Bl-04 | Ejby et al., 2013 |
| Xylo-OSs | ABC transporter | *B. animalis* subsp. *lactis* BB-12 | Gilad et al., 2010 |
| Galacto-OS | ABC transporter | *B. breve* | O'Connell Motherway et al., 2011a |
| β-glucans | ABC transporter | *B. longum* subsp. *infantis* | Zhao and Cheung, 2013 |
| Galacto-OS, HMO, fructo-OS | ABC transporter | *B. longum* subsp. *infantis* ATCC15697 | Kim et al., 2012 |
| HMOs, inulin, Galacto-OS | ABC transporter | *B. longum* subsp. *infantis* ATCC15697 | Garrido et al., 2011 |
| galacto-N-biose/lacto-N-biose | ABC transporter | *B. longum* JCM1217 | Wada et al., 2007 |
| 4′-galactosyllactose | ABC transporter | *B. breve* Yakult | Shigehisa et al., 2015 |
| Cellobiose, galacto-OS, isomaltose, maltotriose, melibiose, panose, raffinose, stachyose, xylobiose β-xylo-OS | ABC transporter(s) | *B. lactis* Bl-04 | Andersen et al., 2013 |
| lacto-N-biose, galacto-N-biose | ABC-transporter | *B. longum* JCM1217 | Suzuki et al., 2008 |
| Fructose | ABC-transporter | *B. longum* NCC2705 | Liu et al., 2011; Wei et al., 2012 |
| Ribose | ABC transporter | *B. breve* UCC2003 | Pokusaeva et al., 2010 |
| Glucose | Secondary transporter | *B. animalis* DSMZ10140 | Briczinski et al., 2008 |
| Fructose | PTS | *B. breve* UCC2003 | Mazé et al., 2007 |
| Glucose | PTS | *B. longum* NCC2705 | Parche et al., 2007 |
| Glucose | PTS | *B. longum* NCC2705 | Parche et al., 2006 |
| Glucose | PTS | *B. animalis* subsp. *lactis* | Briczinski et al., 2008 |

*[a]OS, oligosaccharide(s).*
*Human carbohydrates: mucin and HMO.*
*PTS, phosphotransferase system.*

the intestinal homeostasis. Specialized epithelial cells denominated M-cells and antigen presenting cells (APCs) from the gut-associated lymphoid tissue (GALT) continuously sample the intestinal content. Interaction of APCs with the rest of GALT effectors, mainly T and B cells, leads to immunotolerance against commensal microbes and dietary components, whilst the capacity of mounting an acute, quick, and powerful response against enteropathogens is developed.

Differentiation of commensal and pathogenic bacteria is based on the presence of pattern recognition receptors (PRR) on the APC and epithelial cell surfaces. Among them, Toll-like receptors, NOD-like receptors, C-type lectin receptors, and RIG-I-like receptors are in charge of recognizing specific microbial-associated molecular patterns (MAMPs), such as flagellin, teichoic acids, or lipopolysaccharide among others. The type and intensity of the downstream and intracellular signaling cascades deployed after MAMPs-PRR interaction is essential for the APCs interaction with T-cells, which will finally determine the nature of the T-cell response. Roughly, T-cell responses are divided into effector ($T_h$) and regulatory ($T_{reg}$), its balance being key in the intestinal homeostasis maintenance (Maloy and Powrie, 2011). It is generally accepted that commensal microbiota, by inducing $T_{reg}$ response, modulates the Th1/Th2 balance favoring immune tolerance against the gut microbiota (Ventura et al., 2012). Indeed, the classical MAMP triggering $T_{reg}$ response is the exopolysaccharide A of the commensal bacterium *Bacteroides fragilis*, molecule also involved in the GALT maturation (Mazmanian et al., 2008).

Bifidobacteria may drive species-specific T-cell responses, as it was revealed by a series of experiments in which the cytokine secretion profiles of monocyte-derived dendritic cells (MoDCs) and full fractions of peripheral blood mononuclear cells (PBMCs) were determined (López et al., 2010). Relative levels of key cytokines (IL-10, IL-17, TNFα among others) suggested a specific immunomodulation mechanism for each species, as reported recently for probiotics (Hill et al., 2014). Challenging immature MoDCs with different strains, followed by co-culture with allogeneic naïve $CD4^+$ cells and cytokine determination, further confirmed this effect (López et al., 2011).

Remarkably, *B. bifidum* LMG13195 appeared to induce a $T_{reg}$ response *in vitro* (**Figure** ; López et al., 2011). Dendritic cells challenged with membrane vesicles from this strain induced naïve $CD4^+$ cells polarization into $T_{reg}$, as deduced from the increases in the expression of *foxP3* regulation factor and the CD25 marker (López et al., 2012). Most likely, surface-associated proteins play a role in this process. Several proteins have been identified in the bifidobacterial membrane, among which moonlighting proteins such as fructose-6-phosphate phosphoketolase or enolase, might be behind the immunomodulatory effects of the membrane vesicles (Sánchez et al., 2004). However, the particular proteins involved in this T-cell polarization have not been identified.

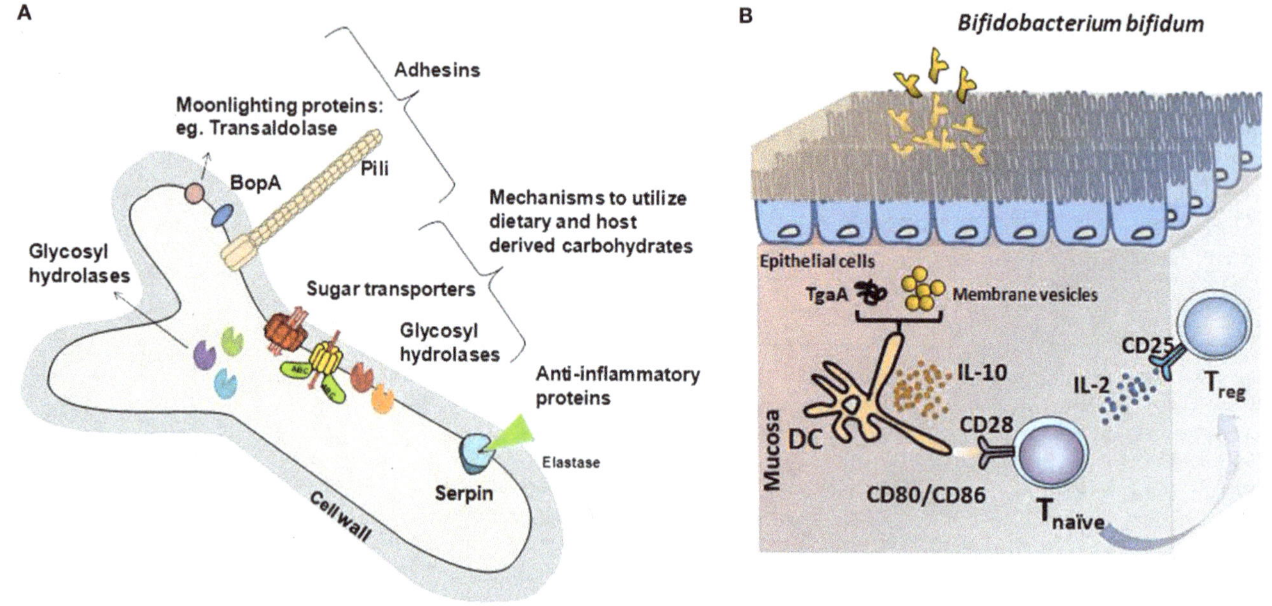


**FIGURE | (A)** Schematic representation of the bifidobacterial proteins identified as key mediators of the cross-talk mechanisms with the intestinal environment. Adhesin-like factors, proteins with immunomodulatory capabilities and glycosyl hydrolases specific for carbon sources encountered in the gastrointestinal tract are represented. **(B)** Graphical illustration of the immunomodulatory mechanisms driven by model *Bifidobacterium bifidum* strains. Different *B. bifidum* fractions and molecules induces $T_{reg}$ response, key in maintaining the balance of effector T-cell responses. Membrane vesicles or the extracellular protein TgaA affects dendritic cells, which induces $T_{reg}$ differentiation after interaction with naïve T-cells. In this process, increased IL-10 secretion, recognition of CD80, and CD86 by CD28 in naïve T-cells and release of IL-2 are key for $T_{reg}$ response development.

Other immunogenic extracellular proteins are pili, proteinaceous structures that self-assemble into filaments on the bacterial surface (Ventura et al., 2012). Specifically, one sortase-dependent pili from *B. bifidum* PRL2010 induced TNFα production during transient colonization of the murine mucosa, which acted as a macrophage-activating factor during Th1 (Turroni et al., 2013). Another surface-protein able to influence T-cell responses is TgaA from *B. bifidum*, a peptidoglycan-derived enzyme able to induce DC activation and IL-2 production (Guglielmetti et al., 2014). IL-2 is one of the main cytokines supporting $T_{reg}$ proliferation, which are characterized by the presence of CD25, the T-cell receptor for that interleukin (Zelante et al., 2012). Despite the evidence supporting an immunomodulation role of bifidobacteria, only a few surface-associated proteins have been identified as possible mediators of this effect on a limited number of *in vitro* experiments. Identifying the molecules behind this effect and confirming their efficacy in clinical trials, might provide keys to ameliorate diseases characterized by exacerbated immune responses.

## INDUCTION OF IgA PRODUCTION

Immunoglobulin A (IgA) is the most abundant antibody in human mucosa and modulates immune responses against commensal bacteria, preventing direct contact with immune cells (Peterson et al., 2007; Brandtzaeg, 2013). Globally, 40% of gut bacteria are IgA-coated although these values are species- and strain-dependent (Talja et al., 2014). In healthy individuals, IgA coating of bifidobacteria is higher than that of other commensals (van der Waaij et al., 2004; De Palma et al., 2010), explaining the immune tolerance to high densities of bifidobacteria. In fact, 44 proteins from *B. longum* and 24 from *B. adolescentis* were recognized by IgA (Talja et al., 2014). IgA-coated bifidobacteria also enhanced probiotic attachment to Caco-2 cells and increased production of mucosal defense molecules (Mathias et al., 2010).

Levels of IgA-coated gut commensals are altered in dysbiosis states such as those described in coeliac disease (De Palma et al., 2010), inflammatory bowel disease (van der Waaij et al., 2004), or autoimmunity disorders (Talja et al., 2014). Coeliac children showed reduced levels of bifidobacteria and IgA-coated bacteria (De Palma et al., 2010). Conversely, IgA from children developing islet autoimmunity, bound to more *B. adolescentis* antigens than those from healthy controls (Talja et al., 2014). Morevoer, bifidobacterial supplements modulate IgA production (Holscher et al., 2012; Kandasamy et al., 2014). A probiotic mixture containing bifidobacteria increased IgA and reduced diarrhea following rotavirus vaccination in a gnotobiotic pig model (Kandasamy et al., 2014). *B. animalis* Bb12 supplementation to formula-fed infants increased IgA in feces and, in those delivered by C-section, enhanced immune responses as reflected by higher anti-rotavirus and anti-poliovirus IgA production following vaccination (Holscher et al., 2012). The bifidogenic effect of galactooligosaccharides also correlated to increased IgA production (Vulevic et al., 2013; Paineau et al., 2014). Further, research to specifically delineate the bifidobacterial molecules mediating IgA induction and interaction is necessary.

## OTHER *BIFIDOBACTERIUM* EFFECTORS OF THE HOST-MICROBE DIALOGUE

A few extracellular proteins, with important physiological roles not discussed in the previous sections deserve further attention. Some *Bifidobacterium* strains produce surface-exposed Serine Protease Inhibitors of proteinaceous nature (serpins), which participate in a variety of physiological processes. The serpin produced by *B. longum* NCC2705 inhibits elastase-like proteases, including neutrophil or pancreatic elastases, thus suggesting a role in protecting bifidobacteria against exogenous proteases and potential anti-inflammatory activity (Ivanov et al., 2006). Remarkably, serpins are widely distributed in bifidobacteria and several species harbor serpin-encoding genes in their genomes (Turroni et al., 2010).

## CONCLUSIONS

Some *Bifidobacterium* proteins have been identified as mediators of the cross-talk bifidobacteria-host, providing bases to understand their beneficial traits and opening new avenues to conceive bifidobacterial-based therapeutic strategies. However, in most cases, the molecular mechanisms triggered remain unknown what has limited their translation into improved functional supplements. Identifying targets for intervention at intestinal level and developing appropriate models to search for bifidobacterial mediators is required to delineate strategies that fine-tune disease-associated alterations in the microbiota-host interplay.

## AUTHOR CONTRIBUTIONS

LR, SD, PM, AM, and BS contributed to the design and organization of the manuscript, drafted, reviewed, and accepted the final version of the manuscript.

## ACKNOWLEDGMENTS

This research was funded by Grant AGL2013-44039-R from the Spanish "Plan Estatal de I+D+I." LR has received funding from the People Programme (Marie Curie Actions) of the European Union's Seventh Framework Programme FP7/2007-2013/ under REA grant agreement n°624773. BS was recipient of a Ramón y Cajal postdoctoral contract from the Spanish Ministry of Economy and Competitiveness. SD is hired on a contract supported by BIO2014-55019-JIN project from the Spanish "Plan Estatal de I+D+I."

## REFERENCES

Alegría, A., Delgado, S., Guadamuro, L., Flórez, A. B., Felis, G. E., Torriani, S., et al. (2014). The genome of *Bifidobacterium pseudocatenulatum* IPLA 36007, a human intestinal strain with isoflavone-activation activity. *Gut Pathog.* 6:31. doi: 10.1186/1757-4749-6-31

Andersen, J. M., Barrangou, R., Abou Hachem, M., Lahtinen, S. J., Goh, Y. J., Svensson, B., et al. (2013). Transcriptional analysis of oligosaccharide utilization by *Bifidobacterium lactis* Bl-04. *BMC Genomics* 14:12. doi: 10.1186/1471-2164-14-312

Ashida, H., Miyake, A., Kiyohara, M., Wada, J., Yoshida, E., Kumagai, H., et al. (2009). Two distinct α-L-fucosidases from *Bifidobacterium bifidum* are essential for the utilization of fucosylated milk oligosaccharides and glycoconjugates. *Glycobiology* 9, 1010–1017. doi: 10.1093/glycob/cwp082

Bang, S.-H., Hyun, Y. J., Shim, J., Hong, S.-W., and Kim, D. H. (2015). Metabolism of rutin and poncirin by human intestinal microbiota and cloning of their metabolizing α-L-rhamnosidase from *Bifidobacterium dentium. J. Microbiol. Biotechnol.* 25, 18–25. doi: 10.4014/jmb.1404.04060

Bergmann, K. R., Liu, S. X., Tian, R., Kushnir, A., Turner, J. R., Li, H. L., et al. (2013). Bifidobacteria stabilize claudins at tight junctions and prevent intestinal barrier dysfunction in mouse necrotizing enterocolitis. *Am. J. Pathol.* 182, 1595–1606. doi: 10.1016/j.ajpath.2013.01.013

Bermudez-Brito, M., Muñoz-Quezada, S., Gomez-Llorente, C., Matencio, E., Bernal, M. J., Romero, F., et al. (2013). Cell-free culture supernatant of *Bifidobacterium breve* CNCM I-4035 decreases pro-inflammatory cytokines in human dendritic cells challenged with *Salmonella typhi* through TLR activation. *PLoS ONE* 8:e59370. doi: 10.1371/journal.pone.0059370

Bhattacharya, T., Ghosh, T. S., and Mande, S. S. (2015). Global profiling of carbohydrate active enzymes in human gut microbiome. *PLoS ONE* 10:e0142038. doi:10.1371/journal.pone.0142038

Bottacini, F., Ventura, M., van Sinderen, D., and O'Connell Motherway, M. (2014). Diversity, ecology and intestinal function of bifidobacteria. *Microb. Cell Fact.* 13(Suppl. 1):S4. doi: 10.1186/1475-2859-13-S1-S4

Brandtzaeg, P. (2013). Secretory IgA: designed for anti-microbial defense. *Front. Immunol.* 4:222. doi: 10.3389/fimmu.2013.00222

Briczinski, E. P., Phillips, A. T., and Roberts, R. F. (2008). Transport of glucose by *Bifidobacterium animalis* subsp. *lactis* occurs via facilitated diffusion. *Appl. Environ. Microbiol.* 74, 6941–6948. doi: 10.1128/AEM.01280-08

Bujacz, A., Jedrzejczak-Krzepkowska, M., Bielecki, S., Redzynia, I., and Bujacz, G. (2011). Crystal structures of the apo form of β-fructofuranosidase from *Bifidobacterium longum* and its complex with fructose. *FEBS J.* 278, 1728–1744. doi: 10.1111/j.1742-4658.2011.08098

Candela, M., Biagi, E., Centanni, M., Turroni, S., Vici, M., Musiani, F., et al. (2009). Bifidobacterial enolase, a cell surface receptor for human plasminogen involved in the interaction with the host. *Microbiology* 155, 3294–3303. doi: 10.1099/mic.0.028795-0

Candela, M., Centanni, M., Fiori, J., Biagi, E., Turroni, S., Orrico, C., et al. (2010). DnaK from *Bifidobacterium animalis* subsp. *lactis* is a surface-exposed human plasminogen receptor upregulated in response to bile salts. *Microbiology* 156, 1609–1618. doi: 10.1099/mic.0.038307-0

Cotar, A. I., Chifiriuc, M. C., Dinu, S., Pelinescu, D., Banu, O., and Lazăr, V. (2010). Quantitative real-time PCR study of the influence of probiotic culture soluble fraction on the expression of *Pseudomonas aeruginosa* quorum sensing genes. *Roum. Arch. Microbiol. Immunol.* 69, 213–223.

De Palma, G., Nadal, I., Medina, M., Donat, E., Ribes-Koninckx, C., Calabuig, M., et al. (2010). Intestinal dysbiosis and reduced immunoglobulin-coated bacteria associated with coeliac disease in children. *BMC Microbiol.* 10:63. doi: 10.1186/1471-2180-10-63

Donaldson, G. P., Lee, S. M., and Mazmanian, S. K. (2016). Gut biogeography of the bacterial microbiota. *Nature Rev. Microbiol.* 14, 20–32. doi: 10.1038/nrmicro3552

Duranti, S., Turroni, F., Lugli, G. A., Milani, C., Viappiani, A., Mangifesta, M., et al. (2014). Genomic characterization and transcriptional studies of the starch-utilizing strain *Bifidobacterium adolescentis* 22L. *Appl. Environ. Microbiol.* 80, 6080–6090. doi: 10.1128/AEM.01993-14

Egan, M., O'Connell-Motherway, M., Kilcoyne, M., Kane, M., Joshi, L., Ventura, M., et al. (2014). Cross-feeding by *Bifidobacterium breve* UCC2003 during co-cultivation with *Bifidobacterium bifidum* PRL2010 in a mucin-based medium. *BMC Microbiol.* 14:282. doi: 10.1186/s12866-014-0282-7

Ejby, M., Fredslund, F., Vujicic-Zagar, A., Svensson, B., Slotboom, D. J., and Abou Hachem, M. (2013). Structural basis for arabinoxylo-oligosaccharide capture by the probiotic *Bifidobacterium animalis* subsp. *lactis* Bl-04. *Mol. Microbiol.* 90, 1100–1112. doi: 10.1111/mmi.12419

Evans, J. M., Morris, L. S., and Marchesi, J. R. (2013). The gut microbiome: the role of a virtual organ in the endocrinology of the host. *J. Endocrinol.* 218, R37–R47. doi: 10.1530/JOE-13-0131

Ewaschuk, J. B., Diaz, H., Meddings, L., Diederichs, B., Dmytrash, A., Backer, J., et al. (2008). Secreted bioactive factors from *Bifidobacterium infantis* enhance epithelial cell barrier function. *Am. J. Physiol. Gastrointest. Liver Physiol.* 295, G1025–G1034. doi: 10.1152/ajpgi.90227.2008

Fujita, K., Oura, F., Nagamine, N., Katayama, T., Hiratake, J., Sakata, K., et al. (2005). Identification and molecular cloning of a novel glycoside hydrolase family of core 1 type O-glycan-specific endo-alpha-N-acetylgalactosaminidase from *Bifidobacterium longum*. *J. Biol. Chem.* 280, 37415–37422. doi: 10.1074/jbc.M506874200

Fujita, K., Sakaguchi, T., Sakamoto, A., Shimokawa, M., and Kithara, K. (2014). *Bifidobacterium longum* subsp. *longum* exo-β-1,3-Galactanase, an enzyme for the degradation of type II arabinogalactan. *Appl. Environ. Microbiol.* 80, 4577–4584. doi: 10.1128/AEM.00802-14

Fujita, K., Sakamoto, S., Ono, Y., Wakao, M., Suda, Y., Kitahara, K., et al. (2011). Molecular cloning and characterization of a β-L-arabinobiosidase in *Bifidobacterium longum* that belongs to a novel glycoside hydrolase family. *J. Biol. Chem.* 286, 5143–5150. doi: 10.1074/jbc.M110.190512

Fukuda, S., Toh, H., Hase, K., Oshima, K., Nakanishi, Y., Yoshimura, K., et al. (2011). Bifidobacteria can protect from enteropathogenic infection through production of acetate. *Nature* 469, 543–547. doi: 10.1038/nature09646

Furusawa, Y., Obata, Y., and Hase, K. (2015). Commensal microbiota regulates T cell fate decision in the gut. *Semin. Immunopathol.* 37, 17–25. doi: 10.1007/s00281-014-0455-3

Garrido, D., Barile, D., and Mills, D. A. (2012). A molecular basis for bifidobacterial enrichment in the infant gastrointestinal tract. *Adv. Nutr.* 3, 415S–421S. doi: 10.3945/an.111.001586

Garrido, D., Dallas, D. C., and Mills, D. A. (2013). Consumption of human milk glycoconjugates by infant-associated bifidobacteria: mechanisms and implications. *Microbiology* 159, 649–664. doi: 10.1099/mic.0.064113-0

Garrido, D., Kim, J. H., German, J. B., Raybould, H. E., and Mills, D. A. (2011). Oligosaccharide binding proteins from *Bifidobacterium longum* subsp. *infantis* reveal a preference for host glycans. *PLoS ONE* 6:e17315. doi: 10.1371/journal.pone.0017315

Garrido, D., Ruiz-Moyano, S., Lemay, D. G., Sela, D. A., German, J. B., and Mills, D. A. (2015). Comparative transcriptomics reveals key differences in the response to milk oligosaccharides of infant gut-associated bifidobacteria. *Sci. Rep.* 5:13517. doi: 10.1038/srep13517

Gilad, O., Jacobsen, S., Stuer-Lauridsen, B., Pedersen, M. B., and Svensson, B. (2010). Combined transcriptome and proteome analysis of *Bifidobacterium animalis* subsp. *lactis* BB-12 grown on xylo-oligosaccharides and a model of their utilization. *Appl. Environ. Microbiol.* 76, 7285–7291. doi: 10.1128/AEM.00738-10

Gleinser, M., Grimm, V., Zhurina, D., Yuan, J., and Riedel, C. U. (2012). Improved adhesive properties of recombinant bifidobacteria expressing the *Bifidobacterium bifidum*-specific lipoprotein BopA. *Microb. Cell Fact.* 11:80. doi: 10.1186/1475-2859-11-80

González-Rodríguez, I., Sánchez, B., Ruiz, L., Turroni, F., Ventura, M., Ruas-Madiedo, P., et al. (2012). Role of extracellular transaldolase from *Bifidobacterium bifidum* in mucin adhesion and aggregation. *Appl. Environ. Microbiol.* 78, 3992–3998. doi: 10.1128/AEM.08024-11

Guglielmetti, S., Tamagnini, I., Mora, D., Minuzzo, M., Scarafoni, A., Arioli, S., et al. (2008). Implication of an outer surface lipoprotein in adhesion of *Bifidobacterium bifidum* to Caco-2 cells. *Appl. Environ. Microbiol.* 74, 4695–4702. doi: 10.1128/AEM.00124-08

Guglielmetti, S., Zanoni, I., Balzaretti, S., Miriani, M., Taverniti, V., De Noni, I., et al. (2014). Murein lytic enzyme TgaA of *Bifidobacterium bifidum* MIMBb75 modulates dendritic cell maturation through its cysteine- and histidine-dependent amidohydrolase/peptidase (CHAP) amidase domain. *Appl. Environ. Microbiol.* 80, 5170–5177. doi: 10.1128/AEM.00761-14

Heuvelin, E., Lebreton, C., Bichara, M., Cerf-Bensussan, N., and Heyman, M. A. (2010). *Bifidobacterium* probiotic strain and its soluble factors alleviate chloride secretion by human intestinal epithelial cells. *J. Nutr.* 140, 7–11. doi: 10.3945/jn.109.114553

Heuvelin, E., Lebreton, C., Grangette, C., Pot, B., Cerf-Bensussan, N., and Heyman, M. (2009). Mechanisms involved in alleviation of intestinal inflammation by *Bifidobacterium breve* soluble factors. *PLoS ONE* 4:e5184. doi: 10.1371/journal.pone.0005184

Hill, C., Guarner, F., Reid, G., Gibson, G. R., Merenstein, D. J., Pot, B., et al. (2014). Expert consensus document: the international scientific association for probiotics and prebiotics consensus statement on the scope and appropriate use of the term probiotic. *Nat. Rev. Gastroenterol. Hepatol.* 11, 9. doi: 10.1038/nrgastro.2014.66

Hinz, S. W. A., Pastink, M. I., van den Broek, L. A. M., Vincken, J.-P., and Voragen, A. G. J. (2005). *Bifidobacterium longum* endogalactanase liberates galactotriose from type I galactans. *Appl. Environ. Microbiol.* 71, 5501–5510. doi: 10.1128/AEM.71.9.5501-5510.2005

Holscher, H. D., Czerkies, L. A., Cekola, P., Litov, R., Benbow, M., Santema, S., et al. (2012). *Bifidobacterium lactis* Bb12 enhances intestinal antibody response in formula-fed infants: a randomized, double-blind, controlled trial. *J. Parenter. Enteral Nutr.* 36(1 Suppl.), 106S–117S. doi: 10.1177/014860711 1430817

Hsieh, C. Y., Osaka, T., Moriyama, E., Date, Y., Kikuchi, J., and Tsuneda, S. (2015). Strengthening of the intestinal epithelial tight junction by *Bifidobacterium bifidum*. *Physiol. Rep.* 3:e12327. doi: 10.14814/ phy2.12327

Hyun, Y. J., Kim, B., and Kim, D. H. (2012). Cloning and characterization of ginsenoside Ra1-hydrolyzing beta-D-xylosidase from *Bifidobacterium breve* K-110. *J. Microbiol. Biotechnol.* 22, 535–540.

Ito, T., Katayama, T., Hattie, M., Sakurama, H., Wada, J., Suzuki, R., et al. (2013). Crystal structures of a glycoside hydrolase family 20 lacto-N-biosidase from *Bifidobacterium bifidum*. *J. Biol. Chem.* 288, 11795–11706. doi: 10.1074/jbc.M112.420109

Ivanov, D., Emonet, C., Foata, F., Affolter, M., Delley, M., Fisseha, M., et al. (2006). A serpin from the gut bacterium *Bifidobacterium longum* inhibits eukaryotic elastase-like serine proteases. *J. Biol. Chem.* 281, 17246–17252. doi: 10.1074/jbc.M601678200

Jae-Han, K., An, H. J., Garrido, D., German, J. B., Lebrilla, C. B., and Mills, D. A. (2013). Proteomic analysis of *Bifidobacterium longum* subsp. *infantis* reveals the metabolic insight on consumption of prebiotics and host glycans. *PLoS ONE* 8:e57535. doi: 10.1371/journal.pone.0057535

Jeffery, C. J. (2003). Moonlighting proteins: old proteins learning new tricks. *Trends Genet.* 19, 415–417. doi: 10.1016/S0168-9525(03)00167-7

Kainulainen, V., Reunanen, J., Hiippala, K., Guglielmetti, S., Vesterlund, S., Palva, A., et al. (2013). BopA does not have a major role in the adhesion of *Bifidobacterium bifidum* to intestinal epithelial cells, extracellular matrix proteins, and mucus. *Appl. Environ. Microbiol.* 79, 6989–6997. doi: 10.1128/AEM.01993-13

Kandasamy, S., Chattha, K. S., Vlasova, A. N., Rajashekara, G., and Saif, L. J. (2014). Lactobacilli and Bifidobacteria enhance mucosal B cell responses and differentially modulate systemic antibody responses to an oral human rotavirus vaccine in a neonatal gnotobiotic pig disease model. *Gut Microbes* 5, 639–651. doi: 10.4161/19490976.2014.969972.

Katayama, T., Sakuma, A., Kimura, T., Makimura, Y., Hiratake, J., Sakata, K., et al. (2004). Molecular cloning and characterization of *Bifidobacterium bifidum* 1,2-α-L-Fucosidase (AfcA), a novel inverting glycosidase (Glycoside Hydrolase family 95). *J. Bacteriol.* 186, 4885–4893. doi: 10.1128/JB.186.15.4885-4893.2004

Kelly, E. D., Bottacini, F., O'Callaghan, J., O'Connell Motherway, M., O'Connell, K. J., Stanton, K., et al. (2016). Glycoside hydrolase family 13 α-glucosidases encoded by *Bifidobacterium breve* UCC2003. A comparative analysis of function, structure and phylogeny. *Int. J. Food Microbiol.* 224, 55–65. doi: 10.1016/j.ijfoodmicro.2016.02.014

Khailova, L., Dvorak, K., Arganbright, K. M., Halpern, M. D., Kinouchi, T., Yajima, M., et al. (2009). *Bifidobacterium bifidum* improves intestinal integrity in a rat model of necrotizing enterocolitis. *Am. J. Physiol. Gastrointest. Liver Physiol.* 297, G940–G949. doi: 10.1152/ajpgi.00141.2009

Khoroshkin, M. S., Leyn, S. A., Van Sinderen, D., and Rodionov, D. A. (2016). Transcriptional regulation of carbohydrate utilization pathways in the *Bifidobacterium* genus. *Front. Microbiol.* 7:120. doi: 10.3389/fmicb.2016.00120

Kim, J. Y., Wang, Y., Park, S. J., Ji, G. E., and Park, M. S. (2012). Cloning and expression of β-glucosidases from *Bifidobacterium lactis* AD011. *Food Sci. Biotechnol.* 21, 731–738. doi: 10.1007/s10068-012-0095-0

Kiyohara, M., Nakatomi, T., Kurihara, S., Fushinobu, S., Suzuki, H., Tanaka, T., et al. (2012). α-N-acetylgalactosaminidase from infant-associated bifidobacteria belonging to novel glycoside hydrolase family 129 is implicated in alternative mucin degradation pathway. *J. Biol. Chem.* 287, 693–700. doi: 10.1074/jbc.M111.277384

Kiyohara, M., Tanigawa, K., Chaiwangsri, T., Katayama, T., Ashida, H., and Yamamoto, K. (2011). An exo-alpha-sialidase from bifidobacteria involved in the degradation of sialyloligosaccharides in human milk and intestinal glycoconjugates. *Glycobiology.* 21, 437–447. doi: 10.1093/glycob/cwq175

Lee, J. H., and O'Sullivan, D. J. (2010). Genomic insights into bifidobacteria. *Microbiol. Mol. Biol. Rev.* 74, 378–416. doi: 10.1128/MMBR.00004-10

Lee, S. H. (2015). Intestinal permeability regulation by tight junction: implication on inflammatory bowel diseases. *Intest. Res.* 13, 1–8. doi: 10.5217/ir.2015.13.1.11

Levy, M., Thaiss, C. A., and Elinav, E. (2015). Metagenomic cross-talk: the regulatory interplay between immunogenomics and the microbiome. *Genome Med.* 7, 120. doi: 10.1186/s13073-015-0249-9

Licciardi, P. V., Wong, S. S., Tang, M. L., and Karagiannis, T. C. (2010). Epigenome targeting by probiotic metabolites. *Gut Pathog.* 2:24. doi: 10.1186/1757-4749-2-24

Liu, D., Wang, S., Xu, B., Guo, Y., Zhao, J., Liu, W., et al. (2011). Proteomics analysis of *Bifidobacterium longum* NCC2705 growing in glucose, fructose, mannose, xylose, ribose, and galactose. *Proteomics* 11, 2628–2638. doi: 10.1002/pmic.201100035

Liu, S., Ren, F., Zhao, L., Jiang, L., Hao, Y., Jin, J., et al. (2015). Starch and starch hydrolysates are favorable carbon sources for bifidobacteria in the human gut. *BMC Microbiol.* 15:54. doi: 10.1186/s12866-015-0362-3

López, P., González-Rodríguez, I., Gueimonde, M., Margolles, A., and Suárez, A. (2011). Immune response to *Bifidobacterium bifidum* strains support Treg/Th17 plasticity. *PLoS ONE* 6:e24776. doi:10.1371/journal.pone.0024776

López, P., González-Rodríguez, I., Sánchez, B., Gueimonde, M., Margolles, A., and Suárez, A. (2012). Treg-inducing membrane vesicles from *Bifidobacterium bifidum* LMG13195 as potential adjuvants in immunotherapy. *Vaccine* 30, 825–829. doi:10.1016/j.vaccine.2011.11.115

López, P., Gueimonde, M., Margolles, A., and Suárez, A. (2010). Distinct *Bifidobacterium* strains drive different immune responses *in vitro*. *Int. J. Food Microbiol.* 138, 157–165. doi:10.1016/j.ijfoodmicro.2009.12.023

Lorca, G. L., Barabote, R. D., Zlotopolski, V., Tran, C., Winnen, B., Hvorup, R. N., et al. (2007). Transport capabilities of eleven gram-positive bacteria: comparative genomic analyses. *Biochim. Biophys. Acta* 1768, 1342–1366. doi: 10.1016/j.bbamem.2007.02.007

Maloy, K. J., and Powrie, F. (2011). Intestinal homeostasis and its breakdown in inflammatory bowel disease. *Nature* 474, 298–306. doi: 10.1038/nature10208

Mathias, A., Duc, M., Favre, L., Benyacoub, J., Blum, S., and Corthésy, B. (2010). Potentiation of polarized intestinal Caco-2 cell responsiveness to probiotics complexed with secretory IgA. *J. Biol. Chem.* 285, 33906–33913. doi: 10.1074/jbc.M110.135111

Mazé A., O'Connell-Motherway, M., Fitzgerald, G. F., Deutscher, J., and van Sinderen, D. (2007). Identification and characterization of a fructose phosphotransferase system in *Bifidobacterium breve* UCC2003. *Appl. Environ. Microbiol.* 73, 545–553. doi: 10.1128/AEM.01496-06

Mazmanian, S. K., Round, J. L., and Kasper, D. L. (2008). A microbial symbiosis factor prevents intestinal inflammatory disease. *Nature* 453, 620–625. doi: 10.1038/nature07008

Ménard, S., Laharie, D., Asensio, C., Vidal-Martinez, T., Candalh, C., Rullier, A., et al. (2005). *Bifidobacterium breve* and *Streptococcus thermophilus* secretion products enhance T helper 1 immune response and intestinal barrier in mice. *Exp. Biol. Med.* 230, 749–756.

Michlmayr, H., Varga, E., Malachova, A., Nguyen, N. T., Lorenz, C., Haltrich, D., et al. (2015). A versatile family 3 Glycoside Hydrolase from *Bifidobacterium adolescentis* hydrolyzes β-glucosides of the *Fusarium* mycotoxins deoxynivalenol, nivalenol, and HT-2 toxin in cereal matrices. *Appl. Environ. Microbiol.* 81, 4885–4893. doi: 10.1128/AEM.01061-15

Milani, C., Turroni, F., Duranti, S., Lugli, G. A., Mancabelli, L., Ferrario, C., et al. (2016). Genomics of the genus *Bifidobacterium* reveals species-specific adaptation to the glycan-rich gut environment. *Appl. Environ. Microbiol.* 82, 980–991. doi: 10.1128/AEM.03500-15

Miwa, M., Horimoto, T., Kiyohara, M., Katayama, T., Kitaoka, M., Ashida, H., et al. (2010). Cooperation of β-galactosidase and β-N-acetylhexosaminidase from bifidobacteria in assimilation of human milk oligosaccharides with type 2 structure. *Glycobiology* 20, 1402–1409. doi: 10.1093/glycob/cwq101

Mokkala, K., Laitinen, K., and Röytiö H. (2016). *Bifidobacterium lactis* 420 and fish oil enhance intestinal epithelial integrity in Caco-2- cells. *Nutr. Res.* 36, 246–252. doi: 10.1016/j.nutres.2015.11.014

Moroni, O., Kheadr, E., Boutin, Y., Lacroix, C., and Fliss, I. (2006). Inactivation of adhesion and invasion of food-borne *Listeria monocytogenes* by bacteriocin-producing *Bifidobacterium* strains of human origin. *Appl. Environ. Microbiol.* 72, 6894–6901. doi: 10.1128/AEM.00928-06

Morrill, J., Kulcinskaja, E., Sulewska, A. M., Lahtinen, S., Stalbrand, H., Svensson, B., et al. (2015). The GH5 1,4-β-mannanase from *Bifidobacterium animalis* subsp. *lactis* Bl-04 possesses a low-affinity mannan-binding module and highlights the diversity of mannanolytic enzymes. *BMC Biochem.* 16:26. doi: 10.1186/s12858-015-0055-4

Nishimoto, M., and Kitaoka, M. (2007). Identification of N-acetylhexosamine 1-kinase in the complete lacto-N-biose I/galacto-N-biose metabolic pathway in *Bifidobacterium*. *Appl. Environ. Microbiol.* 73, 6444–6449. doi: 10.1128/AEM.01425-07

O'Connell Motherway, M., Fitzgerald, G. F., and van Sinderen, D. (2011a). Metabolism of a plant derived galactose-containing polysaccharide by *Bifidobacterium breve* UCC2003. *Microb. Biotechnol.* 4, 403–416. doi: 10.1111/j.1751-7915.2010.00218

O'Connell Motherway, M., Zomer, A., Leahy, S. C., Reunanen, J., Bottacini, F., Claesson, M. J., et al. (2011b). Functional genome analysis of *Bifidobacterium breve* UCC2003 reveals type IVb tight adherence (Tad) pili as an essential and conserved host-colonization factor. *Proc. Natl. Acad. Sci. U.S.A.* 108, 11217–11222. doi: 10.1073/pnas.1105380108

Ohland, C. L., and Macnaughton, W. K. (2010). Probiotic bacteria and intestinal epithelial barrier function. *Am. J. Physiol. Gastrointest. Liver Physiol.* 298, G807–G819. doi: 10.1152/ajpgi.00243.2009

Ouwehand, A. C., and Salminen, S. (2003). *In vitro* adhesion assay for probiotics and their *in vivo* relevance: a review. *Microb. Ecol. Health Dis.* 15, 175–184. doi: 10.1080/08910600310019886

Paineau, D., Respondek, F., Menet, V., Sauvage, R., Bornet, F., and Wagner, A. (2014). Effects of short-chain fructooligosaccharides on faecal bifidobacteria and specific immune response in formula-fed term infants: a randomized, double-blind, placebo-controlled trial. *J. Nutr. Sci. Vitaminol.* 60, 167–175. doi: 10.3177/jnsv.60.167

Parche, S., Amon, J., Jankovic, I., Rezzonico, E., Beleut, M., Schendel, I., et al. (2007). Sugar transport systems of *Bifidobacterium longum* NCC2705. *J. Mol. Microbiol. Biotechnol.* 12, 9–19. doi: 10.1159/000096455

Parche, S., Beleut, M., Rezzonico, E., Jacobs, D., Arigoni, F., Titgemeyer, F., et al. (2006). Lactose-over-glucose preference in *Bifidobacterium longum* NCC2705: glcP, encoding a glucose transporter, is subject to lactose repression. *J. Bacteriol.* 188, 1260–1265. doi: 10.1128/JB.188.4.1260-1265.2006

Patel, P. H., Maldera, J. A., and Edgar, B. A. (2013). Stimulating cROSstalk between commensal bacteria and intestinal stem cells. *EMBO J.* 32, 3009–3010. doi: 10.1038/emboj.2013.244

Peterson, D. A., McNulty, N. P., Guruge, J. L., and Gordon, J. I. (2007). IgA response to symbiotic bacteria as a mediator of gut homeostasis. *Cell Host Microbe* 2, 328–339. doi: 10.1016/j.chom.2007.09.013

Picard, C., Fioramonti, J., Francois, A., Robinson, T., Neant, F., and Matuchansky, C. (2005). Review article: bifidobacteria as probiotic agents — physiological effects and clinical benefits. *Aliment. Pharmacol. Ther.* 22, 495–512. doi: 10.1111/j.1365-2036.2005.02615.x

Pokusaeva, K., Fitzgerald, G. F., and van Sinderen, D. (2011). Carbohydrate metabolism in Bifidobacteria. *Genes Nutr.* 6, 285–306. doi: 10.1007/s12263-010-0206-6

Pokusaeva, K., Neves, A. R., Zomer, A., O'Connell-Motherway, M., MacSharry, J., Curley, P., et al. (2010). Ribose utilization by the human commensal *Bifidobacterium breve* UCC2003. *Microb. Biotechnol.* 3, 311–323. doi: 10.1111/j.1751-7915.2009.00152

Pokusaeva, K., O'Connell-Motherway, M., Zomer, A., Fitzgerald, G. F., and van Sinderen, D. (2009). Characterization of two Novel α-glucosidases from *Bifidobacterium breve* UCC2003. *Appl. Environ. Microbiol.* 75, 1135–1143. doi: 10.1128/AEM.02391-08

Rastall, R. A., and Gibson, G. R. (2015). Recent developments in prebiotics to selectively impact beneficial microbes and promote intestinal health. *Curr. Opin. Biotechnol.* 32, 42–49. doi: 10.1016/j.copbio.2014.11.002

Ruas-Madiedo, P., Gueimonde, M., Fernández-García, M., de los Reyes-Gavilán, C. G., and Margolles, A. (2008). Mucin degradation by *Bifidobacterium* strains isolated from the human intestinal microbiota. *Appl. Environ. Microbiol.* 74, 1936–1940. doi: 10.1128/AEM.02509-07

Ruas-Madiedo, P., Gueimonde, M., Margolles, A., de los Reyes-Gavilán, C. G., and Salminen, S. (2006). Exopolysaccharides produced by probiotic strains modify the adhesion of probiotics and enteropathogens to human intestinal mucus. *J. Food Prot.* 69, 2011–2015.

Ruiz-Moyano, S., Totten, S. M., Garrido, D. A., Smilowitz, J. T., German, J. B., Lebrilla, C. B., et al. (2013). Variation in consumption of human milk oligosaccharides by infant gut-associated strains of *Bifidobacterium breve. Appl. Environ. Microbiol.* 79, 6040–6049. doi: 10.1128/AEM.01843-13

Sánchez, B., Noriega, L., Ruas-Madiedo, P., de los Reyes-Gavilán, C. G., and Margolles, A. (2004). Acquired resistance to bile increases fructose-6-phosphate phosphoketolase activity in *Bifidobacterium. FEMS Microbiol. Lett.* 235, 35–41. doi: 10.1016/j.femsle.2004.04.009

Sánchez, B., Ruiz, L., Gueimonde, M., Ruas-Madiedo, P., and Margolles, A. (2013). Adaptation of bifidobacteria to the gastrointestinal tract and functional consequences. *Pharmacol. Res.* 69, 127–136. doi: 10.1016/j.phrs.2012.11.004

Sela, D. A., Garrido, D., Lerno, L., Wu, S., Tan, K., Eom, H.-J., et al. (2011). *Bifidobacterium longum* subsp. *infantis* ATCC 15697 α-fucosidases are active on fucosylated human milk oligosaccharides. *Appl. Environ. Microbiol.* 78, 795–803. doi: 10.1128/AEM.06762-11

Shenderov, B. A. (2013). Metabiotics: novel idea or natural development of probiotic conception. *Microb. Ecol. Health Dis.* 24:8. doi: 10.3402/mehd.v24i0.20399

Shigehisa, A., Sotoya, H., Sato, T., Hara, T., Matsumoto, H., and Matsuki, T. (2015). Characterization of a bifidobacterial system that utilizes galacto-oligosaccharides. *Microbiology* 161(Pt 7), 1463–1470. doi: 10.1099/mic.0.000100

Shimada, Y., Watanabe, Y., Wakinaka, T., Funeno, Y., Kubota, M., Chaiwangsri, T., et al. (2015). α-N-Acetylglucosaminidase from *Bifidobacterium bifidum* specifically hydrolyzes α-linked N-acetylglucosamine at nonreducing terminus of O-glycan on gastric mucin. *Appl. Microbiol. Biotechnol.* 99, 3941–3948. doi: 10.1007/s00253-014-6201

Smilowitz, J. T., Lebrilla, C. B., Mills, D. A., German, J. B., and Freeman, S. L. (2014). Breast milk oligosaccharides: structure-function relationships in the neonate. *Annu. Rev. Nutr.* 34, 143–169. doi: 10.1146/annurev-nutr-071813-105721

Srutkova, D., Schwarzer, M., Hudcovic, T., Zakostelska, Z., Drab, V., Spanova, A., et al. (2015). *Bifidobacterium longum* CCM 7952 promotes epithelial barrier function and prevents acute DSS-induced colitis in strictly strain-specific manner. *PLoS ONE* 10:e0134050. doi: 10.1371/journal.pone.0134050

Sultana, R., McBain, A. J., and O'Neill, C. A. (2013). Strain-dependent augmentation of tight-junction barrier function in human primary epidermal keratinocytes by *Lactobacillus* and *Bifidobacterium* lysates. *Appl. Environ. Microbiol.* 79, 4887–4894. doi: 10.1128/AEM.00982-13

Suzuki, H., Murakami, A., and Yoshida, K. (2013). Motif-guided identification of a glycoside hydrolase family 1 α-L-arabinofuranosidase in *Bifidobacterium adolescentis. Biosci. Biotechnol. Biochem.* 77, 1709–1714. doi: 10.1271/bbb.130279

Suzuki, R., Wada, J., Katayama, T., Fushinobu, S., Wakagi, T., Shoun, H., et al. (2008). Structural and thermodynamic analyses of solute-binding Protein from *Bifidobacterium longum* specific for core 1 disaccharide and lacto-N-biose I. *J. Biol. Chem.* 283, 13165–13173. doi: 10.1074/jbc.M709777200

Tailford, L. E., Crost, E. H., Kavanaugh, D., and Juge, N. (2015). Mucin glycan foraging in the human gut microbiome. *Front. Genet.* 6:81. doi: 10.3389/fgene.2015.00081

Talja, I., Kubo, A. L., Veijola, R., Knip, M., Simell, O., Ilonen, J., et al. (2014). Antibodies to Lactobacilli and Bifidobacteria in young children with different propensity to develop islet autoimmunity. *J. Immunol. Res.* 2014:325938. doi: 10.1155/2014/325938

Tanner, S. A., Chassard, C., Rigozzi, E., Lacroix, C., and Stevens, M. J. (2016). *Bifidobacterium thermophilum* RBL67 impacts on growth and virulence gene expression of *Salmonella enterica* subsp. *enterica* serovar Typhimurium. BMC Microbiol. 16:46. doi: 10.1186/s12866-016-0659-x

Turroni, F., Duranti, S., Bottacini, F., Guglielmetti, S., van Sinderen, D., and Ventura, M. (2014). *Bifidobacterium bifidum* as an example of a specialized human gut commensal. *Front. Microbiol.* 5:437. doi: 10.3389/fmicb.2014.00437

Turroni, F., Foroni, E., O'Connell-Motherway, M., Bottacini, F., Giubellini, V., and Zomer, A. (2010). Characterization of the serpin-encoding gene of *Bifidobacterium breve* 210B. *Appl. Environ. Microbiol.* 76, 3206–3219. doi: 10.1128/AEM.02938-09

Turroni, F., Serafini, F., Foroni, E., Duranti, S., O'Connell Motherway, M., and Taverniti, V. (2013). Role of sortase-dependent pili of *Bifidobacterium bifidum* PRL2010 in modulating bacterium-host interactions. *Proc. Natl. Acad. Sci. U.S.A.* 110, 11151–11156. doi: 10.1073/pnas.1303897110

Turroni, F., Strati, F., Foroni, E., Serafini, F., Duranti, S., van Sinderen, D., et al. (2012). Analysis of predicted carbohydrate transport systems encoded by *Bifidobacterium bifidum* PRL2010. *Appl. Environ. Microbiol.* 78, 5002–5012. doi: 10.1128/AEM.00629-12

van den Broek, L. A. M., Lloyd, R. M., Beldman, G., Verdoes, J. C., McCleary, B. V., and Voragen, A. G. J. (2005). Cloning and characterization of arabinoxylan arabinofuranohydrolase-D3 (AXHd3) from *Bifidobacterium adolescentis* DSM20083. *Appl. Microbiol. Biotechnol.* 67, 641–647. doi: 10.1007/s00253-004-1850-9

van der Waaij, L. A., Kroese, F. G., Visser, A., Nelis, G. F., Westerveld, B. D., Jansen, P. L., et al. (2004). Immunoglobulin coating of faecal bacteria in inflammatory bowel disease. *Eur. J. Gastroenterol. Hepatol.* 16, 669–674. doi: 10.1097/01.meg.0000108346.41221.19

Ventura, M., Turroni, F., Motherway, M. O. C., MacSharry, J., and van Sinderen, D. (2012). Host-microbe interactions that facilitate gut colonization by commensal bifidobacteria. *Trends Microbiol.* 20, 467–476. doi: 10.1016/j.tim.2012.07.002

Viborg, A. H., Sørensen, K. I., Gilad, O., Steen-Jensen, D. B., Dilokpimol, A., Jacobsen, S., et al. (2013). Biochemical and kinetic characterisation of a novel xylooligosaccharide-upregulated GH43 β-Dxylosidase/ α-L-arabinofuranosidase (BXA43) from the probiotic *Bifidobacterium animalis* subsp. *lactis* BB-12. *AMB Express* 3:56. doi: 10.1186/2191-0855-3-56

Vulevic, J., Juric, A., Tzortzis, G., and Gibson, G. R. (2013). A mixture of trans-galactooligosaccharides reduces markers of metabolic syndrome and modulates the fecal microbiota and immune function of overweight adults. *J. Nutr.* 143, 324–331. doi: 10.3945/jn.112.166132

Wada, J., Ando, T., Kiyohara, M., Ashida, H., Kitaoka, M., Yamaguchi, M., et al. (2008). *Bifidobacterium bifidum* lacto-N-biosidase, a critical enzyme for the degradation of human milk oligosaccharides with a Type 1 structure. *Appl. Environ. Microbiol.* 74, 3996–3904. doi: 10.1128/AEM.00149-08

Wada, J., Suzuki, R., Fushinobu, S., Kitaoka, M., Wakagi, T., Shoun, H., et al. (2007). Purification, crystallization and preliminary X-ray analysis of the galacto-N-biose/lacto-N-biose I- binding protein (GL-BP) of the ABC transporter from *Bifidobacterium longum* JCM1217. *Acta Crystallogr. Sect. F. Struct. Biol. Cryst. Commun.* 63(Pt 9), 751–753. doi: 10.1107/S1744309107036263

Wei, X., Guo, Y., Shao, C., Sun, Z., Zhurina, D., Liu, D., et al. (2012). Fructose uptake in *Bifidobacterium longum* NCC2705 is mediated by an ATP-binding cassette transporter. *J. Biol. Chem.* 287, 357–367. doi: 10.1074/jbc.M111.266213

Wei, X., Yan, X., Chen, X., Yang, Z., Li, H., Zou, D., et al. (2014). Proteomic analysis of the interaction of *Bifidobacterium longum* NCC2705 with the intestine cells Caco-2 and identification of plasminogen receptors. *J. Proteome Res.* 7, 375–385. doi: 10.1016/j.jprot.2014.04.038

Yoshida, E., Sakurama, H., Kiyohara, M., Nakajima, M., Kitaoka, M., Ashida, H., et al. (2012). *Bifidobacterium longum* subsp. *infantis* uses two different β-galactosidases for selectively degrading type-1 and type-2 human milk oligosaccharides. *Glycobiol.* 22, 361–368. doi: 10.1093/glycob/cwr116

Zelante, T., Fric, J., Wong, A. Y. W., and Ricciardi-Castagnoli, P. (2012). Interleukin-2 production by dendritic cells and its immuno-regulatory functions. *Front. Immunol.* 3:161. doi: 10.3389/fimmu.2012.00161

Zhao, J., and Cheung, P. C. (2013). Comparative proteome analysis of Bifidobacterium longum subsp. *infantis* grown on β-glucans from different sources and a model for their utilization. *J. Agric. Food Chem.* 61, 4360–4370. doi: 10.1021/jf400792j

**Conflict of Interest Statement:** The authors declare that the research was conducted in the absence of any commercial or financial relationships that could be construed as a potential conflict of interest.

# Permissions

All chapters in this book were first published in BTRTHGM, by Frontiers Media SA; hereby published with permission under the Creative Commons Attribution License or equivalent. Every chapter published in this book has been scrutinized by our experts. Their significance has been extensively debated. The topics covered herein carry significant findings which will fuel the growth of the discipline. They may even be implemented as practical applications or may be referred to as a beginning point for another development.

The contributors of this book come from diverse backgrounds, making this book a truly international effort. This book will bring forth new frontiers with its revolutionizing research information and detailed analysis of the nascent developments around the world.

We would like to thank all the contributing authors for lending their expertise to make the book truly unique. They have played a crucial role in the development of this book. Without their invaluable contributions this book wouldn't have been possible. They have made vital efforts to compile up to date information on the varied aspects of this subject to make this book a valuable addition to the collection of many professionals and students.

This book was conceptualized with the vision of imparting up-to-date information and advanced data in this field. To ensure the same, a matchless editorial board was set up. Every individual on the board went through rigorous rounds of assessment to prove their worth. After which they invested a large part of their time researching and compiling the most relevant data for our readers.

The editorial board has been involved in producing this book since its inception. They have spent rigorous hours researching and exploring the diverse topics which have resulted in the successful publishing of this book. They have passed on their knowledge of decades through this book. To expedite this challenging task, the publisher supported the team at every step. A small team of assistant editors was also appointed to further simplify the editing procedure and attain best results for the readers.

Apart from the editorial board, the designing team has also invested a significant amount of their time in understanding the subject and creating the most relevant covers. They scrutinized every image to scout for the most suitable representation of the subject and create an appropriate cover for the book.

The publishing team has been an ardent support to the editorial, designing and production team. Their endless efforts to recruit the best for this project, has resulted in the accomplishment of this book. They are a veteran in the field of academics and their pool of knowledge is as vast as their experience in printing. Their expertise and guidance has proved useful at every step. Their uncompromising quality standards have made this book an exceptional effort. Their encouragement from time to time has been an inspiration for everyone.

The publisher and the editorial board hope that this book will prove to be a valuable piece of knowledge for researchers, students, practitioners and scholars across the globe.

# List of Contributors


**Amy O'Callaghan and Douwe van Sinderen**
Alimentary Pharmabiotic Centre and School of Microbiology, University College Cork, Cork, Ireland

**Silvia Arboleya and Catherine Stanton**
APC Microbiome Institute, University College Cork, Cork, Ireland
Teagasc Food Research Centre, Moorepark, Fermoy, Cork, Ireland

**Claire Watkins**
APC Microbiome Institute, University College Cork, Cork, Ireland
Teagasc Food Research Centre, Moorepark, Fermoy, Cork, Ireland
School of Microbiology, University College Cork, Cork, Ireland

**R. Paul Ross**
APC Microbiome Institute, University College Cork, Cork, Ireland
Teagasc Food Research Centre, Moorepark, Fermoy, Cork, Ireland
School of Science, Engineering and Food Science, University College Cork, Cork, Ireland

**Matteo Soverini, Simone Rampelli, Silvia Turroni, Sara Quercia, Andrea Castagnetti, Elena Biagi, Patrizia Brigidi and Marco Candela**
Department of Pharmacy and Biotechnology, University of Bologna, Bologna, Italy

**Stephanie L. Schnorr**
Department of Anthropology, University of Oklahoma, Norman, OK, USA

**Gerald W. Tannock**
Department of Microbiology and Immunology, University of Otago, Dunedin, New Zealand
Riddet Centre for Research Excellence, Massey University, Palmerston North, New Zealand

**Pheng Soon Lee**
Mead Johnson Nutrition, Singapore, Singapore
Department of Human Nutrition, University of Otago, Dunedin, New Zealand

**Khai Hong Wong**
Mead Johnson Nutrition, Singapore, Singapore

**Blair Lawley**
Department of Microbiology and Immunology, University of Otago, Dunedin, New Zealand

**Audrey Rivière, Marija Selak, David Lantin, Frédéric Leroy and Luc De Vuyst**
Research Group of Industrial Microbiology and Food Biotechnology, Faculty of Sciences and Bioengineering Sciences, Vrije Universiteit Brussel, Brussels, Belgium

**Pauline Bondue, Sébastien Crèvecoeur, François Brose, Georges Daube and Véronique Delcenserie**
Department of Food Science, Fundamental and Applied Research for Animal and Health, Faculty of Veterinary Medicine, University of Liège, Liège, Belgium

**Marie-Christine Seghaye**
Department of Pediatrics, University Hospital Liège-Notre-Dame des Bruyères, Belgium

**Mansel W. Griffiths and Gisèle LaPointe**
Canadian Research Institute for Food Safety, University of Guelph, Guelph, Canada

**Chiara Ferrario, Sabrina Duranti, Christian Milani, Leonardo Mancabelli, Gabriele A. Lugli, Francesca Turroni and Marco Ventura**
Laboratory of Probiogenomics, Department of Life Sciences, University of Parma, Parma, Italy

**Marta Mangifesta and Alice Viappiani**
GenProbio Ltd., Parma, Italy

**Maria C. Ossiprandi**
Department of Medical-Veterinary Science, University of Parma, Parma, Italy

**Douwe van Sinderen**
APC Microbiome Institute and School of Microbiology, University College Cork, National University of Ireland, Cork, Ireland

**Michael L. Connolly, Xenofon Tzounis and Julie A. Lovegrove**
Hugh Sinclair Unit of Human Nutrition, University of Reading, Reading, UK
Institute for Cardiovascular and Metabolic Research, University of Reading, Reading, UK

**Kieran M. Tuohy**
Hugh Sinclair Unit of Human Nutrition, University of Reading, Reading, UK
Department of Food Quality and Nutrition, Research and Innovation Centre - Fondazione Edmund Mach, Trento, Italy

**Katelyn Brandt and Rodolphe Barrangou**
Functional Genomics Graduate Program, North Carolina State University, Raleigh, NC, USA
Department of Food, Bioprocessing and Nutrition Sciences, North Carolina State University, Raleigh, NC, USA

**Grace Zhang, Mobolaji Adeolu, Bijendra Khadka and Radhey S. Gupta**
Department of Biochemistry and Biomedical Sciences, McMaster University, Hamilton, ON, Canada

**Beile Gao**
CAS Key Laboratory of Tropical Marine Bio-resources and Ecology, Guangdong Key Laboratory of Marine Materia Medica, South China Sea Institute of Oceanology, Chinese Academy of Sciences, Guangzhou, China

**Timothy W. Yeung and Elif F. Üçok**
Department of Food Science, University of Massachusetts, Amherst, MA, USA

**Kendra A. Tiani**
Department of Food Science, University of Massachusetts, Amherst, MA, USA
Commonwealth Honors College, University of Massachusetts, Amherst, MA, USA

**David J. McClements**
Department of Food Science, University of Massachusetts, Amherst, MA, USA
Center for Bioactive Delivery, Institute of Applied Life Science, University of Massachusetts, Amherst, MA, USA

**David A. Sela**
Department of Food Science, University of Massachusetts, Amherst, MA, USA
Center for Bioactive Delivery, Institute of Applied Life Science, University of Massachusetts, Amherst, MA, USA
Center for Microbiome Research, University of Massachusetts Medical School, Worcester, MA, USA

**Alberto Sabbioni, Enzo Riccardi, Francesco Di Ianni, Valentino Beretti and Paola Superchi**
Department of Veterinary Medical Science, University of Parma, Parma, Italy

**Xiao Wei, Simiao Wang, Xiangna Zhao, Xuesong Wang, Huan Li, Weishi Lin, Jing Lu, Boxing Li, Yansong Sun and Jing Yuan**
Institute of Disease Control and Prevention, Academy of Military Medical Sciences, Beijing, China

**Daria Zhurina and Christian U. Riedel**
Institute of Microbiology and Biotechnology, University of Ulm, Ulm, Germany

**Christina Westermann and Marita Gleinser**
Institute of Microbiology and Biotechnology, University of Ulm, Ulm, Germany

**Sinéad C. Corr**
Department of Microbiology, Moyne Institute of Preventative Medicine, School of Genetics and Microbiology, Trinity College Dublin, Dublin, Ireland

**Dan Li and Mieke Uyttendaele**
Laboratory of Food Microbiology and Food Preservation, Faculty of Bioscience Engineering, Ghent University, Ghent, Belgium

**Adrien Breiman**
UMR 892, Institut National de la Santé et de la Recherche Médicale, Nantes, France
UMR 6299, Centre National de la Recherche Scientifique, Nantes, France
Faculty of Medicine, University of Nantes, Nantes, France
Nantes University Hospital, Nantes, France

**Jacques le Pendu**
UMR 892, Institut National de la Santé et de la Recherche Médicale, Nantes, France
UMR 6299, Centre National de la Recherche Scientifique, Nantes, France
Faculty of Medicine, University of Nantes, Nantes, France

**Rebeca Martín, Florian Chain, Sylvie Miquel, Claire Cherbuy, Luis G. Bermudez-Humaran and Philippe Langella**
Commensal and Probiotics-Host Interactions Laboratory, Micalis Institute, Institut National de la Recherche Agronomique, AgroParisTech, Université Paris-Saclay, Jouy-en-Josas, France

**Laure Laval**
Commensal and Probiotics-Host Interactions Laboratory, Micalis Institute, Institut National de la Recherche Agronomique, AgroParisTech, Université Paris-Saclay, Jouy-en-Josas, France
Danone Nutricia Research, Palaiseau, France

**Jane Natividad and Elena F. Verdu**
Farncombe Family Digestive Health Research Institute, McMaster University, Hamilton, ON, Canada

**Harry Sokol**
Commensal and Probiotics-Host Interactions Laboratory, Micalis Institute, Institut National de la Recherche Agronomique, AgroParisTech, Université Paris-Saclay, Jouy-en-Josas, France
ERL INSERM U 1057/UMR7203, Faculté de Médecine Saint-Antoine, Université Pierre et Marie Curie, Paris, France
Service de Gastroentérologie, Hôpital Saint-Antoine, Assistance Publique - Hôpitaux de Paris, Paris, France

**Johan van Hylckama Vlieg and Tamara Smokvina**
Danone Nutricia Research, Palaiseau, France

**Claudio Hidalgo-Cantabrana**
Department of Microbiology and Biochemistry of Dairy Products, Instituto de Productos Lácteos de Asturias - Consejo Superior de Investigaciones Científicas, Villaviciosa, Spain
CIBER-EHD, Department of Pharmacology, ibs. GRANADA, Center for Biomedical Research, University of Granada, Granada, Spain

**Francesca Algieri, Alba Rodriguez-Nogales, Teresa Vezza and Julio Gálvez**
CIBER-EHD, Department of Pharmacology, ibs. GRANADA, Center for Biomedical Research, University of Granada, Granada, Spain

**Pablo Martínez-Camblor**
Geisel School of Medicine at Dartmouth, Hanover, NH, USA
Universidad Autónoma de Chile, Santiago, Chile

**Abelardo Margolles and Patricia Ruas-Madiedo**
Department of Microbiology and Biochemistry of Dairy Products, Instituto de Productos Lácteos de Asturias - Consejo Superior de Investigaciones Científicas, Villaviciosa, Spain

**L. Valdés-Varela, Ana M. Hernández-Barranco and Miguel Gueimonde**
Microbiology and Biochemistry of Dairy Products, Probiotics and Prebiotics, Instituto de Productos Lácteos de Asturias–Consejo Superior de Investigaciones Científicas, Villaviciosa, Spain

**Ezgi Özcan**
Department of Food Science, University of Massachusetts Amherst, Amherst, MA, USA

**Douwe van Sinderen**
Alimentary Pharmabiotic Centre and Department of Microbiology, Bioscience Institute, National University of Ireland, Cork, Ireland

**Pamela Vazquez-Gutierrez, Tomas de Wouters, Julia Werder, Christophe Chassard and Christophe Lacroix**
Laboratory of Food Biotechnology, Institute of Food, Nutrition and Health, ETH Zurich, Zürich, Switzerland

**Lorena Ruiz**
Department of Nutrition, Food Science and Food Technology, Universidad Complutense de Madrid, Spain

**Susana Delgado and Borja Sánchez**
Department of Microbiology and Biochemistry of Dairy Products, Instituto de Productos Lácteos de Asturias-Consejo Superior de Investigaciones Científicas, Villaviciosa, Spain

# Index

www.ingramcontent.com/pod-product-compliance
Lightning Source LLC
LaVergne TN
LVHW061511220926
848485LV00017B/874
* 9 7 8 1 6 3 9 2 7 0 1 0 1 *